INTRODUCTION TO HYDROGEN TECHNOLOGY

INTRODUCTION TO HYDROGEN TECHNOLOGY

Second Edition

K.S.V. Santhanam
School of Chemistry and Materials Science,
Rochester Institute of Technology, USA

Roman J. Press
School of Chemistry and Materials Science,
Rochester Institute of Technology, USA

Massoud J. Miri
School of Chemistry and Materials Science,
Rochester Institute of Technology, USA

Alla V. Bailey
School of Chemistry and Materials Science,
Rochester Institute of Technology, USA

Gerald A. Takacs
School of Chemistry and Materials Science,
Rochester Institute of Technology, USA

WILEY

This edition first published 2018

First Edition published: 2009

Registered Office
John Wiley & Sons, Inc., 111 River Street, Hoboken, NJ 07030, USA

Editorial Office
111 River Street, Hoboken, NJ 07030, USA

For details of our global editorial offices, customer services, and more information about Wiley products visit us at www.wiley.com.

Wiley also publishes its books in a variety of electronic formats and by print-on-demand. Some content that appears in standard print versions of this book may not be available in other formats.

Library of Congress Cataloging-in-Publication Data

Names: Santhanam, K. S. V. (Kalathur S. V.), author. | Press, Roman J., author. | Miri, Massoud J., 1954- author. | Bailey, Alla V., 1949- author. | Takacs, Gerald A., 1943- author.
Title: Introduction to hydrogen technology / K.S.V. Santhanam, Roman J. Press, Massoud J. Miri, Alla V. Bailey, Gerald A. Takacs.
Description: Second edition. | Hoboken, NJ : John Wiley & Sons, 2018. | Includes bibliographical references and index. |
Identifiers: LCCN 2017022078 (print) | LCCN 2017036357 (ebook) | ISBN 9781119265580 (pdf) | ISBN 9781119265573 (epub) | ISBN 9781119265542 (cloth)
Subjects: LCSH: Hydrogen. | Renewable energy sources. | Hydrogen as fuel.
Classification: LCC TP245.H9 (ebook) | LCC TP245.H9 S27 2017 (print) | DDC 665.8/1–dc23
LC record available at https://lccn.loc.gov/2017022078

Cover Design: Wiley
Cover Images: (Background) © annasir/Gettyimages;
(Image) © Artwork studio BKK/Shutterstock

Set in size of 10pt and TimesLTStd by SPi Global, Chennai.

Printed in the United States of America

10 9 8 7 6 5 4 3 2 1

CONTENTS

PREFACE

Hydrogen gas continues to occupy a unique place in the world as it possesses properties that other elements do not. Consequently, scientists and engineers have been working on using it for improving the existing technology. The scientific literature on hydrogen technology is astronomically growing with the result that a large number of hydrogen-powered devices have entered the market and are being used in mobile phones, laptops, automobiles, utility vehicles, and so on. In addition, fuel cells have been developed to produce stationary power in a number of countries; it is predicted that the utilization will grow to 1.25 million fuel cells in the next 5 years.

The second edition of this book is to keep pace with the above developments and contains updated information on renewable energies, world petroleum production, and greenhouse gases. The new generation of solar cells is included in this book. Each chapter of the first edition is updated by including new developments. Of particular importance is fuel value of biodiesels, solid carbon fuel cells that is being considered for new developments, new Nafion membranes produced by grafting for polymer electrolyte membrane fuel cell, improved electrode materials for molten carbonate, and solid oxide fuel cells. The chapter on hydrogen technology is modified to address infrastructure technology.

Two new chapters—Hydrogen Production and Batteries are added in this edition. The Hydrogen Production chapter reviews the various developments in improving the water decomposition efficiency and the resulting environmental impact factor. In this chapter, nine different methods of hydrogen production are considered, which, hopefully, will lower the cost of hydrogen. The Batteries chapter is included to provide a deeper examination of how hydrogen is used as a fuel in a fuel cell to generate electricity. With battery, electricity is generated from the stored energy. In essence, both are generating electricity for practical applications. Hence, for comparison of the chemistries involved in the two ways of generating electricity, this chapter hopefully will stimulate the reader for appreciating the suitability of the two methods toward fulfilling the greenhouse gas effect in the atmosphere.

In the first edition, it was mentioned that UN's Intergovernmental Panel on Climate Change had been enforcing the need to reduce the greenhouse gases in the atmosphere. The second phase (2013–2020) of Kyoto agreement has begun, and the United Nations Framework Convention on Climatic Change (UNFCC) passed a resolution in Paris in 2015 (http://unfccc.int/paris_agreement/items/9485.php) to stabilize the concentrations of greenhouse gases in the atmosphere such that there would be minimum interference with the climatic system. The hydrogen technology can go a long way to fulfilling this goal.

Finally, this book, if it is used as a textbook for a course, contains problems at the end of each chapter.

The authors thank Profs. Paul Craig and Sophia Maggelakis for their support in bringing out this edition.

K.S.V. Santhanam, Roman J. Press, Massoud J. Miri, Alla V. Bailey, and Jerry A. Takacs
June 3, 2017,
Rochester, NY

ABOUT THE COMPANION WEBSITE

This book is accompanied by Instructor website:

www.wiley.com/go/santhanam/hydrogentech_2e

On this website you will find

- Multiple choice questions as a preparation for class room examinations.
- Problems and their solutions.
- Appendix

CHAPTER 1

Available Energy Resources

1.1 CIVILIZATION AND THE SEARCH FOR SUSTAINABLE ENERGY

Many thousands of years ago, our ancestors knew how to produce fire and they used it for several different purposes, including warming themselves and preparing food. They discovered that energy could be liberated from burning wood. The energy-liberating material was defined as fuel, and this led to the recognition that wood is a fuel. Early civilizations depended on this fuel for a long time. To improve their living conditions, humans searched for new forms of sustainable energy. This exploration resulted in the invention of wind-driven wheels that could be used to pump water from wells. Before this discovery, water was pulled from wells by human energy. This led to a correlation that wind is a source of energy. The wheel was also found useful for transportation forming a part of a chariot that could be rotated when drawn by horses.

During the 18th century, the most commonly used forms of energy were derived from wood, water, horses, and mills. The composition and structure of these materials were mysteries, and more so how the energy was liberated from them. These mysteries led to detailed investigations into the structure of matter by numerous scientists, including J.J. Thompson, J. Dalton, M. Faraday, M. Curie, N. Bohr, A. Einstein, and J. Gibbs. This search for understanding the composition and structure of matter resulted in astounding discoveries in science, including the discovery and understanding of molecules and atoms.

The energy liberates upon combustion and products of combustion were established during this period. During the 18th century, as mentioned in the beginning of the paragraph, it was demonstrated that alcohol could be produced by the destructive distillation of wood, and that alcohol could be used as a source of energy. A realization that wood could be replaced by alcohol and that it could do the job much more effectively resulted in the use of alcohol as a source of energy. Coal was used as a source of energy for running steam engines.

In the 19th century, organic chemists synthesized hydrocarbons and determined the energies available from them. The 20th century led to the search for naturally available sources of hydrocarbons, and the discovery, that oil and natural gas contain them, paved the way for their utilization as energy sources in transportation. The rapid utilization and resulting depletion of these naturally occurring sources by mankind is leading to the search for viable alternatives. In addition, hydrocarbon-based energy sources are responsible for pollution of

Introduction to Hydrogen Technology, Second Edition.
K.S.V. Santhanam, Roman J. Press, Massoud J. Miri, Alla V. Bailey, and Gerald A. Takacs

Book Companion site : www.wiley.com/go/santhanam/hydrogentech_2e

the atmosphere. These energy sources release carbon dioxide and carbon monoxide gases. Such gases are causing global warming (Section 1.3).

The 21st century is facing challenging problems, with faster depletion of fossil fuels and pollution arising from their use. Energy sources that are sustainable and producing negligible pollution are needed. In this context, hydrogen and fuel cells are being considered, but their exploration and use require policy decisions. Historically, the United States depends heavily on imported oils, and the infrastructure has been built on the imported oils and natural gases. In order to switch over to other fuels free from the restrictions discussed earlier, a smooth transitional infrastructure needs to be evolved.

A symbol of early human ingenuity is the first step pyramid, built for King Zoser in 2750 BC in Saqqará/Egypt. Similarly, the "energy pyramid" represents another advancement in human ingenuity. As the "food pyramid" represents a balanced approach to a healthy lifestyle, the energy pyramid (Figure 1.1) represents a balanced approach to consuming renewable and nonrenewable energy sources. With the gradual depletion of most nonrenewable sources of hydrocarbon-based fuels, the energy pyramid contains a diverse proportion of renewable fuels—hydro, solar, and wind power, along with various biomass-produced fuels.

During the 19th century, hydrogen was experimented as an energy source, and Sir William Grove demonstrated in 1839 that hydrogen and oxygen would combine to produce electricity. The product of the reaction was water. He called the device a fuel cell. In this method of producing electricity, there is no pollution generated and it is environmentally friendly for transportation. These two factors are very important in the 21st century. President George W. Bush spoke of the potential of hydrogen as a future energy source in his address to the National Building Museum on February 6, 2003. He stated, "Hydrogen fuel cells represent one of the most encouraging, innovative technologies of our era," and predicted that any obstacles in building hydrogen-based technology could be overcome by thoughtful research by scientists and engineers. This trend is continued by President Obama's administration by speeding up the hydrogen-powered transportation and energy production.

The United States is in a unique situation in its energy consumption. Growth was exponential in the second half of the 20th century. The United States consumes 25% of the

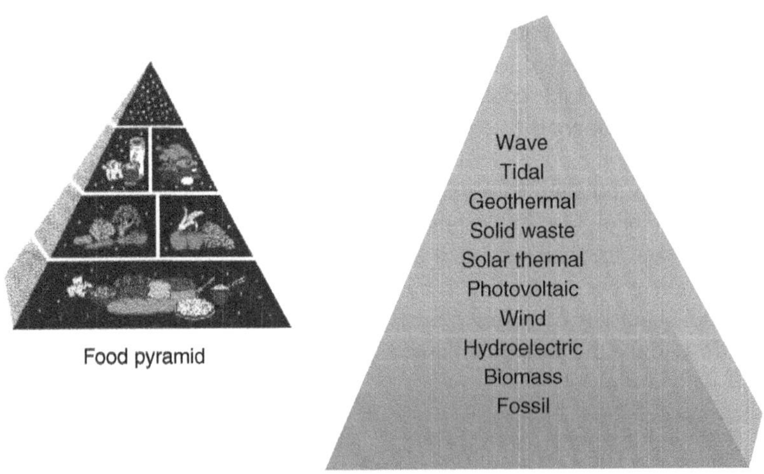

Figure 1.1 Energy pyramid.

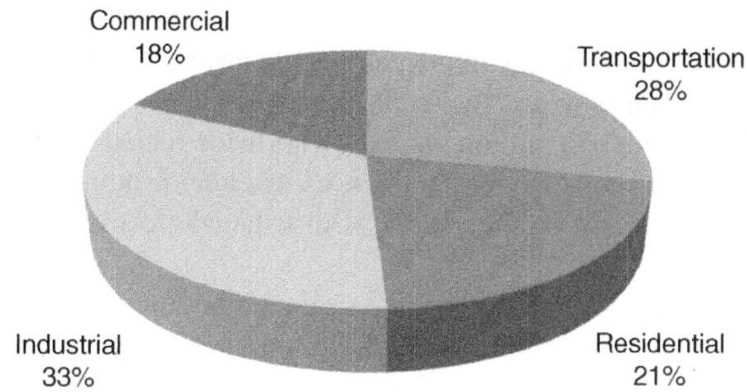

Figure 1.2 US energy use by sector.

world's energy supplies, which are distributed over the following four sectors: industrial commercial, transportation, and residential use (Figure 1.2). A deeper analysis shows that these four sectors showed a 300% increase in the annual usage since 1950. This trend has resulted in faster depletion of fossil fuels and greater environmental effects. Petroleum and gas reserves (fuels) are being rapidly depleted at a rate of a thousand times faster than the fuels are formed and stored. With the economic viability of the United States closely linked to fuel supplies from unstable regions around the globe, additional problems are likely to arise in the future. If domestic supplies of fuels decline, the need for importing fuels will increase. With current evidence for the imported fuel prices increasing year by year, the fuel needs and cost are likely to severely escalate in the near future.

Increased use of fossil fuels has had negative environmental effects: oil spills endanger aquatic and plant life, contaminate beaches and soil, and cause erosion of large masses of land. It also results in global warming effects. If we wish to solve all these problems, then we have to find alternative sources of energy. Hydrogen is one of the alternative energy sources that the world could rely on safely.

Industrialized society is built on the existing infrastructure and is primarily fueled by petroleum. If fuel prices are stable, then the infrastructure requires very little change and the status quo can be maintained. Unfortunately, the status quo does not address the problems of the future. Future needs can be met only by recognizing the problems generated by petroleum-based technology and making efforts to find energy sources free from these problems. Hydrogen-based technology appears to be an ideal solution in this context.

Hydrogen-based technology offers attractive options for use in an economically and socially viable world with negligible environmental effects. Hydrogen is everywhere on earth in the form of water and hydrocarbons. In other words, hydrogen as fuel produces water as the by-product, and water is the source for hydrogen. It is an ideal energy carrier and hence could play a major role in a new decentralized infrastructure that would provide power to vehicles, homes, and industries. Hydrogen is nontoxic, renewable, clean, and provides more energy per dollar. Hydrogen is also the fuel for energy-efficient fuel cells.

Fossil fuels such as oil and gas are being currently used to harvest hydrogen. This is not ideal as it does not solve environmental issues that arise with the usage of fossil fuels. In the future, it will be necessary to use renewable energy sources such as wind, hydro, solar, biomass, and geothermal instead.

The stationary power generation based on fuel cell technology is a viable energy source and has been implemented in several places in the world. This technology provides a drastic reduction in carbon dioxide output in comparison to the existing technology.

Leading automotive companies, such as GM, Ford, Opel, Daimler-Chrysler, and Toyota, have even made significant progress in developing advanced fuel cell propulsion systems using hydrogen. Hydrogen-powered fuel cells are approximately two times more efficient than gasoline engines. With 650 million vehicles worldwide fueled by gasoline, the market potential is immense. Fuel cells power modules, using either proton exchange membranes or solid oxide, can potentially be the source of distributed electric power generation for business and home use.

The purpose of this book is to introduce the reader to the fundamental, chemistry-based aspects of hydrogen technology. It also provides information on renewable energy, hydrogen production, and fuel cells. The latest developments and current research on alternative fuels are discussed. The core topics include acid–base chemistry, reaction topics, chemical equilibrium, thermodynamics, electrochemistry, organic chemistry, polymers, photochemistry, and environmental chemistry. The topics covered in this text are highly relevant to current international and national concerns about overconsumption of our planet's natural resources and the political implications of the United States' dependence on foreign oil to meet the majority of its energy needs. There are many reasons to search for renewable sources of energy—including, but not limited to, energy conservation, pollution avoidance, and prevention. Hydrogen, being one of the cleanest and most abundant alternative energy, will most likely play a critical role in a new energy infrastructure by providing a cleaner source of power to vehicles, homes, and industries.

The authors are members of the Rochester Institute of Technology Renewable Energy Enterprise (RITree). They sincerely hope that this book will give a very good background on chemical aspects of hydrogen technology, including its potential in fuel cells and impact on environment. It is also the hope of the authors that this publication will contribute to the preparation of a workforce ready for future challenges in the areas of energy consumption, generation, and the rapid commercialization of both hydrogen-powered transportation and nonautomotive applications.

1.2 THE PLANET'S ENERGY RESOURCES AND ENERGY CONSUMPTION

On this planet, sources of energy are fossil fuels, the sun, the wind, water, and the earth (the latter includes geothermal and nuclear energy). Fossil fuels-oil, natural gas, and coal-and nuclear energy are abundantly used at present. Since these energy sources are expected to be depleted within a couple of centuries, they have been called "nonrenewable" energy sources. Only about 20% of our energy needs come from renewable sources. Examples in this category are solar energy, wind energy, hydro energy, biomass, geothermal energy, tidal energy, and *hydrogen*. These sources are not very efficient and research needs to be done to improve their efficiencies.

1.2.1 Energy Consumption

The total world consumption of energy amounted to 400 Quad (=quadrillion) Btu in 2000.[1] A human being consumes about 0.9 GJ/day of energy, equivalent to burning 32 kg of coal per day, or as average energy supply, 10.4 kW. Any human being needs as nutrition only 0.14 kW or about 1% of the energy consumed per capita. Essentially, all human activities

1 1 Quad = 1.055 × 1018 J.

involve consumption of energy, for example, construction of buildings, production of consumer goods, medicine, food, packaging of products, transportation, heating and cooling, administrative work, and even activities in our leisure time. Between 1850 and 1970, the world population tripled with the result that energy consumption increased by a factor of 12.

CRITICAL THINKING QUESTIONS

A What are the sources of energy considered in the global energy consumption?
B Which continent is consuming the highest Quadrillion Btu?
Answer: A

(a) Crude oil (b) oil-derived sources (c) natural gas (d) coal and lignite (e) power production and (f) renewables.

Answer: B

Based on US Energy Information Administration (EIA),
North America: 116.191 Quadrillion Btu
Central and South America: 28.674 Quadrillion Btu
Europe: 81.45 Quadrillion Btu
Eurasia: 46.09 Quadrillion Btu
Middle East: 32.213 Quadrillion Btu
Africa: 17.335 Quadrillion Btu
Asia and Oceania: 202.19 Quadrillion Btu
World: 524.076 Quadrillion Btu

Based on the 2012 data provided by EIA, Asia and Oceania consume the highest, and Africa is the lowest ranked in total primary energy consumption.

1.2.2 Regional Differences

Energy consumption is not evenly distributed over the countries of the world. The developed rich countries, for example, the United States, Europe, and Japan, consume about 80% of the worldwide energy and represent 20% of the world's population. Consumer habits differ from region to region. In the United States, people drive larger, lesser-fuel-efficient automobiles and buy larger homes than in many other countries, such as China or India. Currently, an American on the average uses 10 times more energy than the average Chinese and 20 times more than the average Indian. However, energy consumption is rising faster in the developing countries.

1.2.3 Distribution by Economic Sector

Transportation accounts for 30% of the world's energy consumption, mainly due to the use of passenger cars. About one billion cars are used worldwide and about a quarter of all of the world's automobiles are driven in the United States. In Europe and Japan, more mass transportation is used, often encouraged by government policies such as high taxes for car registration and subsidies for mass transport. Since most automobiles run on petroleum-based gasoline, the higher use of mass transportation significantly reduces production of greenhouse gases and global warming (see Section 1.3) and causes less pollution.

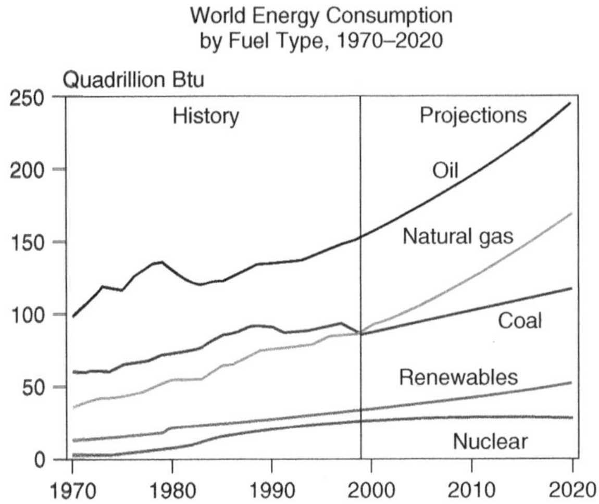

Figure 1.3 Consumption by type of energy source. [*Source*: EIA (Energy Information Administration), International Energy Outlook 2000, PPT by J. E. Hakes; http://tonto.eia.doe.gov/FTPROOT/presentations/ieo2000/sld002.htm.]

Approximately a third of the world's energy is used in residential and commercial buildings, for heating, cooling, cooking, and other appliances. Americans use about 2.4 times the energy of Europeans, due to larger homes and more appliances. The average size of the living space for a citizen in the United States is about 25 times that of a person on the African continent. Another third of the world's energy is used in industry for the production of various goods, such as consumer products, cars, buildings, and food.

1.2.4 Differentiation by Type of Energy Resource

Figure 1.3 shows the consumption by energy source for the last three decades along with projections up to 2020. In 2000, close to 150 Quad Btu of the energy we used was from petroleum, followed by another 70 Quad Btu from natural gas, about 70 Quad Btu from coal, and 15 Quad Btu nuclear fuel. The remainder was from renewable energy resources.

1.2.5 Meeting the Energy Demands of the Future

Improved technology has helped to increase energy efficiency, particularly of the renewable energy resources. However, since the world's economy is also steadily increasing, the consumption of energy grows by about 2% every year, and the demand for energy will only increase.

1.3 THE GREENHOUSE EFFECT AND ITS INFLUENCE ON QUALITY OF LIFE AND THE ECOSPHERE

We live on a planet that derives energy for all our activities from the sun. We wash our clothes in water and dry them in the sun. We get hydroelectric power from the evaporation of water by the heat of the sun. The green plants (trees, algae, etc.) on our planet perform photosynthesis using solar energy. There are many other applications of solar energy, such as in solar heaters, photovoltaic cells producing electricity, photogalvanic, and photobiological processes.

CRITICAL THINKING QUESTIONS

1 What is global warming?
It is a term used to describe the rapid change of the earth's climate.
2 Explain the different factors associated with the global warming.
a) Increased global air temperatures
b) Increased annual precipitation
c) Shorter winters
d) Shrinking ice covers
e) Presence of mosquito-borne diseases at higher altitudes
f) Rising sea levels

The sun produces solar radiation by a nuclear process, and the solar spectrum spans a wavelength of about 0.03–14,000 nm. Of these different wavelengths of radiation emitted by the sun, the highly energetic ones (γ-rays, X-rays, and ultraviolet rays) spanning a wavelength region of 0.03–300 nm, are filtered by the atmosphere above our planet. The other wavelengths enter our atmosphere. The radiation that reaches the earth's surface is now subjected to reflection by the atmosphere, the clouds, and the earth's surface. The total solar radiation that is reflected amounts to about 30%. The balance of 70% of incoming solar radiation is absorbed by the atmosphere, clouds, land, and oceans. Table 1.1 gives the estimated contributions by the different entities toward reflection and absorption. However, solar energy powers the life on the earth solely by absorption. Almost all of the short wavelength radiation coming from the sun (ultraviolet light) is absorbed by the ozone layer in the stratosphere. This absorption is very important as it protects life on the earth.

Figure 1.4 shows the path for the greenhouse effect and the solar radiation that is emitted and the radiation reaching the earth. Note that only part of the solar radiation reaches the earth.

TABLE 1.1 Pathways for the dissipation of solar radiation

	Reflection (%)
Atmosphere	6
Clouds	20
Surface	4
Absorption (%)	
Atmosphere	16
Clouds	3
Land and oceans	51

Source: http://en.wikipedia.org/wiki/Greenhouse_effect.

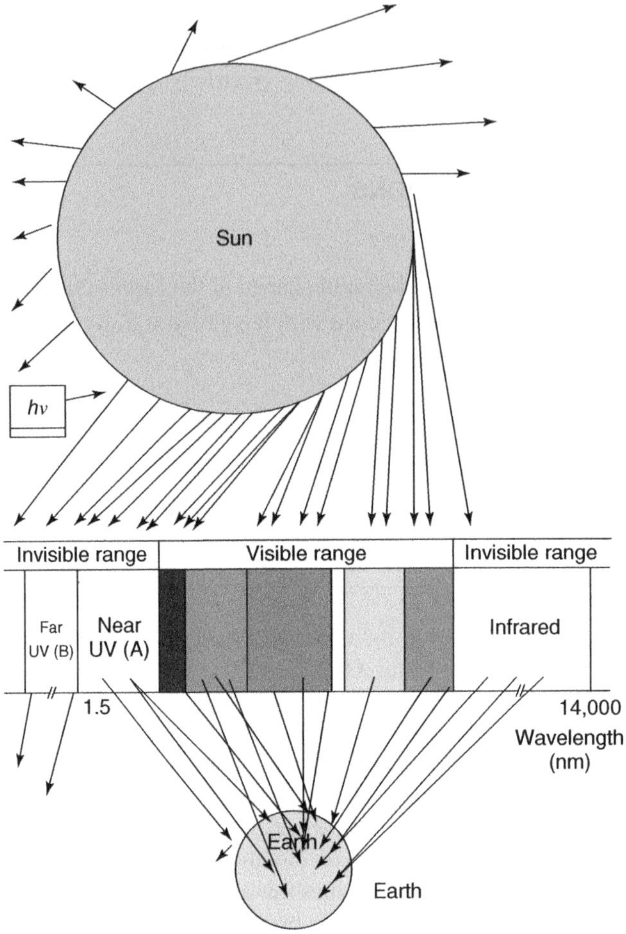

Figure 1.4 Solar radiation.

1.3.1 What Is the Effect of Solar Radiation Reaching the Earth?

The solar radiation reaching the earth heats the surface. This heating effect can be calculated from the radius of the earth (R) [0.635×10^7 m], solar constant (S) that describes the average amount of radiation that earth receives from the sun [1.37 kW/m^2], and Albedo (A) [the fraction of the radiation reflected by the planet] of the earth as given by equation 1.1.

$$\text{Amount of solar radiation reaching the earth's surface} = \pi R^2 S(1 - A) \qquad (1.1)$$

This radiation heats up the earth to an effective temperature, T_e. The photons of the wavelengths shown by arrows pointing to earth in Figure 1.4 reach the entire surface and are not localized. If the earth emits radiation as a blackbody, the infrared radiation emitted will follow Stefan–Boltzmann law, according to which

$$\text{Amount of radiation reemitted by earth} = 4\pi R^2 k T_e^4 \qquad (1.2)$$

where k = Stefan–Boltzmann constant.

Equating (1.1) with (1.2)

$$T_e = [S(1 - A)/4k]^{1/4} \tag{1.3}$$

By substituting the constants in equation 1.3, it is possible to estimate the effective temperature T_e. The earth's temperature based on this equilibrium model would be about $T_e = 253$ K(−20°C). This is not a suitable condition for life as it would be a frozen world. However, this situation does not exist because of the greenhouse effect and we have an average temperature on the earth of about 288 K (15°C).

1.3.2 How the Temperature Is Kept Higher Than the Equilibrium Model

We have considered in the earlier discussions that the sun's radiation is a blackbody radiation that reaches the earth. The temperature of the sun is much higher than the earth (5880 K vs 288 K), and also the earth's surface (earth diameter $= 1.27 \times 10^7$ m and surface area 0.51×10^{15} m^2) is much smaller than the sun (diameter $= 1.39 \times 10^9$ m and surface area $= 0.609 \times 10^{19}$ m^2). Due to these factors, Wien's displacement law proposes that the wavelength of radiation emitted by the earth should be longer than the one coming from the sun. It is typically in the infrared region of 1000 nm. The sun's radiation that reaches the earth is a visible wavelength of about 500 nm. As the earth radiates infrared radiation, it is absorbed by molecules in the atmosphere, typically molecules such as carbon dioxide, water vapor, nitrous oxide, ozone, and methane. These molecules have the capability to absorb the infrared radiation and reemit it to keep the earth's temperature higher than predicted by the equilibrium model. The molecules absorbing the earth's radiation are called greenhouse gases and the process is known as the greenhouse effect. In other words, the greenhouse effect is a process of absorption of infrared radiation emitted from the earth by the greenhouse gases. Thus, most of the thermal radiation of the earth does not escape and is contained in the atmosphere. Only about 6% of the total radiation from the earth escapes into space. For infrared radiation to be absorbed, the molecule should have a permanent dipole moment or asymmetric stretching or bending that can cause a temporary dipole moment. In the atmosphere, nitrogen and oxygen molecules are available in high concentrations and do not contribute to infrared absorption; this is due to the fact that these molecules do not have permanent dipole moment. The molecules such as water vapor, nitrous oxide, ozone, and methane have permanent dipole moment with the exception of carbon dioxide that possesses temporary dipole moment.

1.3.3 Quality of Life

The quality of our living depends on the environment we have around us. The effective temperature, T_e, is one of the deciding factors. If the atmosphere around us has a higher carbon dioxide level, then it will absorb more of the radiation emitted by the earth and reradiate it to the earth. This results in higher T_e on the earth and consequently in global warming. If global warming continues to take place, then a stage might be reached when our existence is threatened. Here we could compare the greenhouse effect of other planets. Venus is rich in carbon dioxide and hence it causes the greenhouse effect on its surface, where the temperature is such that a metal like lead can melt. On the other hand, Mars has very small amounts of greenhouse gases and hence produces a minimum greenhouse effect.

The carbon dioxide level in the earth's atmosphere has increased due to heavy industrialization from the original value of 313 ppm in 1960 to the present value of 375 ppm.

The average temperature of the earth has increased by 0.5°C. This has been discussed as global warming in several scientific meetings. If this trend were to continue, then after a very long time, the effective temperature may not be tolerable for our living. At this stage, increased water evaporation will take place that will affect the quality and quantity of drinking water. It may cause higher rainfalls that may result in flooding. Another possible concern is in a rising sea level that can also cause flooding of the land. Increased temperature may cause spread of infectious diseases, forest fires, and demand for more air conditioning for our living (Table 1.2).

Figure 1.5 gives the carbon dioxide level before Christ (BC) and expected level in 2015.

1.3.4 The Ecosphere

Based on our current understanding of the greenhouse effect, it is desirable to examine the ecosphere, which is not only made up of the environment but also includes all the living things. It extends from the stratosphere to the deep abyss of ocean, with several interacting entities. We may divide the ecosphere into local ecosystems. Among these ecosystems within ecosystems, there may be interactions that will affect the atmosphere. With increasing industrialization (producing more carbon dioxide that is let into the atmosphere) and deforestation (absence of photosynthesis resulting in more carbon dioxide in the atmosphere) in the ecosphere, more of the greenhouse gases will surround us that would result in increasing the temperature on the earth. Another problem that we face is the destruction of the ozone layer (this layer filters ultraviolet rays from reaching the earth) by the

TABLE 1.2 Polluted air: causes and remedies

Type of pollution	Gases involved	Sources	Remedies
Universal problem Greenhouse gas effect	CO_2, CFC, CH_4, N_2O, and O_3	Fuel combustion Forest fires Volcano eruption Chlorofluorocarbons (CFC) Volatile organic compounds Peroxy acetyl nitrate	Lesser usage of fossil fuels Using nonpolluting fuels Forest conservation Stopping volcanoes (???)
Acid rain	Sulfur and nitrogen oxides, ammonia and hydrochloric acid vapors	Caused by combustion of fuels and industrial gases Chemical pulping used in paper industries	Using gas absorbers for desulfuration and denitration Efficient combustion
Ozone layer	Fluorocarbons, CH_3CCl_3, CCl_4, O_3	Decomposition of O_3 with Cl in UV light	Substitution and collection of CFO
Local problem-smog	SO_2, HCl, CO, sulfuric acid mist	Industrial waste gas Combustion of coal	Smoke treatment
Photochemical smog	NO_x, SO_x, nonmethane HC	Combustion of fuel	Same as acid rain

CFC, chlorofluorocarbons; CFO, chlorofluoro oxides; ???, questionable.

References: Energy and air pollution, World energy outlook special report, International Energy Agency, OECD/IEA, 2016; S. Cole, G. Ellen (14 December 2015). "New NASA Satellite Maps Show Human Fingerprint on Global Air Quality." *NASA*. Retrieved 14 December 2015.

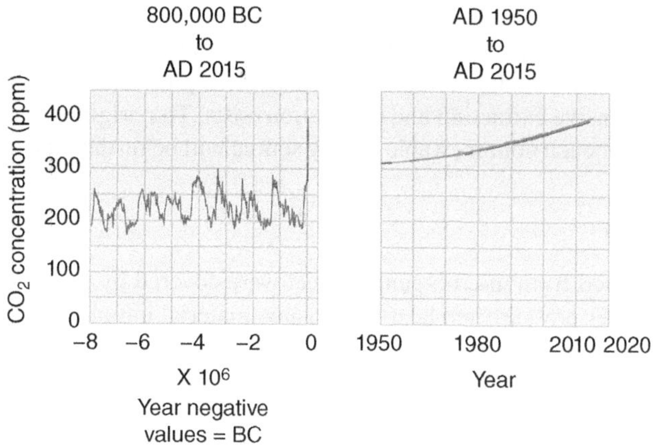

Figure 1.5 Global atmospheric concentrations of carbon dioxide over time. [www.epa.gov/climatechange/indicators.]

fluorocarbons that will allow shorter wavelength radiation from the sun to penetrate through the layer and will have significant interaction with our ecosystem. This may bring about destruction of plants, animals, and humans living on our planet. Although the physical and chemical processes involved in greenhouse effect suggests caution in our industrialization and deforestation, it may be several millions of years before the effective temperature can reach the limit of destruction of life on our planet due to the above-mentioned causes.

The greenhouse gas effect has not been accepted by some scientists. The skeptics consider it as a normal and natural process on the planet that will not affect our living. However, those who accept the greenhouse effect tend to think that it can be reduced by controlling industrial and automobile exhausts. This step would reduce the carbon dioxide level in the atmosphere. Several countries are enforcing the automobile emission controls to a very low level (0–2%). Reductions can also be achieved by use of fuels that do not produce greenhouse gases. In this context, hydrogen technology plays a role as it is used as a fuel to power cars (as fuel cells) and in home heating. The "Kyoto Protocol" enforces that industrialized countries should bring down the emissions of greenhouse gases by 5% by 2010 as compared to 1990.

Toward the end of 2015—at the Climate conference in Paris—195 countries adopted the first universal global climate deal, also known as Paris Agreement. This document describes sets of global actions to avoid dangerous climate change by limiting global warming to 1.5°C.

1.4 NONRENEWABLE ENERGY RESOURCES

Energy sources are nonrenewable when they are depleted in a foreseeable time, typically within a couple of hundred years. Nonrenewable energy is abundantly used and makes up most of our energy resources. Among the nonrenewable sources, the most prevalent are fossil fuels, which formed more than 200–300 million years ago, either from plants, resulting in coal, or from microorganisms, leading to petroleum and gas. Another nonrenewable supply of energy is available nowadays as nuclear fuel, based on the fission of heavy nuclei such as uranium-235. Besides these conventional sources, there are also oil sand and natural gas hydrates, which, however, require more complex extraction methods such as hydraulic fracturing (fracking) and are not commercially used in a significant quantity yet because of their relatively expensive extraction and production costs. The origin, production, use, and specific problems for each nonrenewable energy source will be further discussed as follows.

1.4.1 Petroleum

Petroleum was formed from microorganisms that were covered by sand and silt below the earth's surface. Under pressure and heat, the organic material initially formed waxy solids, so-called kerogen, a precursor to gaseous and liquid organic compounds. Crude oil, or simply oil, as petroleum is also called, is typically a yellow/black viscous liquid that contains gaseous, liquid, and solid hydrocarbons. We will use terms *oil* and *petroleum* interchangeably. In refineries, the petroleum is separated into purer fractions by a distillation process. The details on the fractionation of oil are described in Section 2.6.3.1. Figure 1.6 and Table 1.3 show the regions and major countries and their oil reserves.

Typically, oil is obtained by drilling into reservoirs beneath the earth's surface, including from platforms in the oceans. The oil reserves of the world amounted to 148 Gt in 2004. The area of the world in which the most oil is found is the Middle East, which has about 65% of the world reserves. Most of the world's oil is produced by the following countries: Saudi Arabia, Iran, Iraq, Kuwait, and the United Arab Emirates. These and other countries are members of the OPEC (Organization of Petroleum Exporting Countries). Other major oil-producing non-OPEC countries are Russia and Mexico. If nonconventional oil reserves are included, Canada would possess significant reserves based on oil sand. Figure 1.7 indicates how oil production has increased within the last decades.

If we continue to consume petroleum at the current rate, we should run out of this fossil fuel in the next 40–50 years by most estimates. Over half the petroleum used by Americans comes from countries other than the United States. Most of the oil is consumed for gasoline

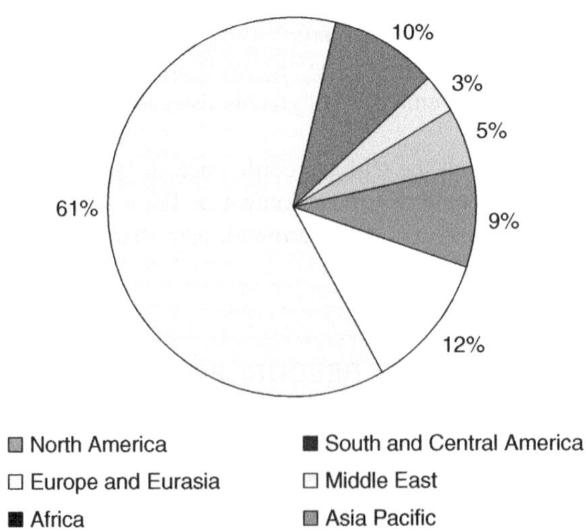

Figure 1.6 Oil reserves in the world in 2005. [*Source*: Based on data from BP, available at http://www.bp.com/productlanding.]

TABLE 1.3 Oil reserves by country in billions of barrels

Africa and the Middle East			North America, Central America, and South America			Europe, Asia, and Oceania		
	Low estimate	High estimate		Low estimate	High estimate		Low estimate	High estimate
Africa			North America			Western Europe		
Algeria	11.4	11.8	Canada	16.5	178.0	United Kingdom	4.1	4.5
Libya	33.6	39.1	Mexico	12.9	14.8			
Nigeria	35.3	35.9	United States	21.3	29.3	Norway	7.7	8.0
Total	100.8	113.8	Total	50.7	222.1	Total	16.2	17.3
Middle East			Central and South America			Eastern Europe and Former USSR		
Iran[a]	125.8	132.7	Brazil	10.6	11.2	Russia	60	72.4
Iraq[a]	115	115	Venezuela	52.4	361.2	Kazakhstan	9	39.6
Kuwait[a]	48	101.5	Total	76	401.1	Total	79.2	121.9
Qatar	15.2	15.2						
Saudi Arabia[a]	261.9	264.3						
UAE[a]	69.9	97.8						
Total	657.3	733.9						
TOTAL WORLD RESERVES: 1016.4–1650.7						Asia and Oceania		
						China	15.4	16.0
						Australia	1.5	4
						India	4.9	5.6
						Indonesia	4.3	4.3
						Total	36.2	39.8

[a]This reserve number cannot be verified.

Source: Wikipedia 2007 (orig.: EIA) http://wikipedia.com, secondary source: http:www.eia.doe.gov/, last accessed 6/1/2007.

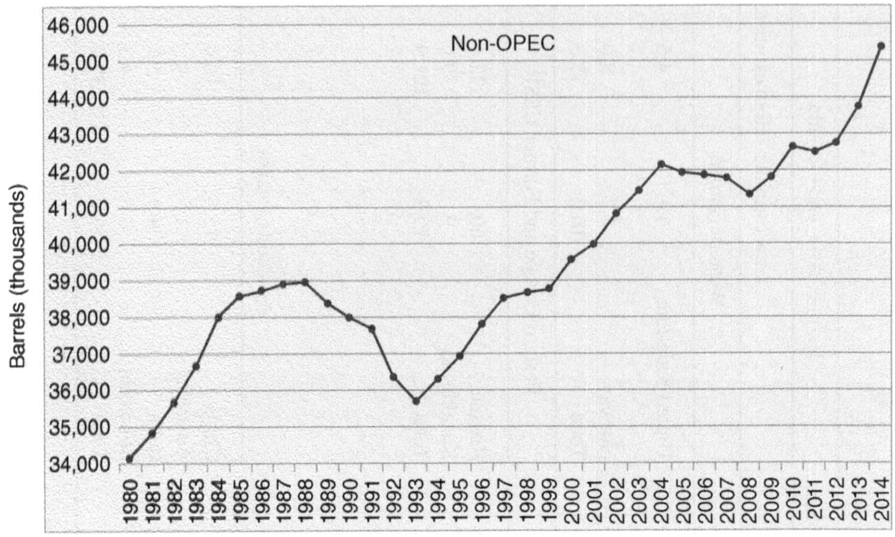

Figure 1.7 World petroleum production. [*Source*: Energy Information Administration/Annual Energy Review 2005 (http://www.eia.doe.gov/emeu/aer/pdf/pages/sec_10.pdf).]

production, and more than half for fuels altogether, including diesel oil, heating oil, and airplane fuel. However, oil is also used to make more valuable products such as plastics, rubber, and medicines.

1.4.2 Natural Gas

Because natural gas is formed by similar anaerobic processes as oil, involving the decay of microorganisms several million years ago, most natural gas reservoirs are geographically close to oil reservoirs (conventional natural gas). Natural gas that occurs by itself in bedrock is distinguished as unconventional natural gas. A substantial amount of natural gas is methane, with amounts ranging between 50% and 90%, with the remainder consisting of hydrocarbons with 2–4 carbons. Table 1.4 shows the countries' natural gas reserves. Natural gas is frequently transported through pipelines. Alternatively, it is liquefied at low temperatures (−160°C) and used as LNG (liquefied natural gas) and transported in containers. Due to its lower viscosity, LNG can be more easily transported and processed than oil. It is mostly used as fuel in industry and residential homes. Similar to oil, it is also used effectively as precursor for the production of plastics and pharmaceuticals. Since natural gas is a purer mixture of hydrocarbons and contains less by-products than oil, it produces less environmentally problematic products when burned. For example, the level of nitrogen oxides or particulate matter is negligible. Another advantage of natural gas is that it is cheaper than oil. Similar to most volatile hydrocarbons, natural gas is highly flammable and therefore potentially hazardous. Natural gas itself does not have a characteristic smell. For detection in case of leakage, sulfur-containing compounds such as mercaptans are added to it. These additives can be easily detected by their garlic-like smell and thus indicate whether the gas is present. The world's natural gas reserves are forecasted to last about 60 more years—slightly longer than our oil reserves.

TABLE 1.4 World natural gas reserves: world proven natural gas reserves by country, 2005 and 2006

	2005	2006	% change 06/05
North America	7,420.0	7,590.0	2.3
Canada	1,633.0	1,665.0	2.0
United States	5,787.0	5,925.0	2.4
Latin America	7,312.0	7,716.0	5.5
Argentina	439.0	415.0	−5.5
Bolivia	740.0	740.0	0.0
Mexico	408.0	388.0	−4.9
Trinidad and Tobago	530.0	530.0	0.0
Venezuela	4,315.0	4,708.0	9.1
Latin America others	880.0	935.0	6.2
Eastern Europe	58,878.0	58,890.0	0.0
Romania	628.0	628.0	0.0
Former USSR	58,099.0	58,113.0	0.0
Eastern Europe others	151.0	149.0	−1.3
Western Europe	5,561.0	5,396.0	−3.0
Germany	257.0	255.0	−0.8
Netherlands	1,387.0	1,347.0	−2.9
Norway	3,007.0	2,892.0	−3.8
United Kingdom	481.0	481.0	0.0
Western Europe others	429.0	421.0	−1.9
Middle East	72,834.0	72,319.0	−0.7
Iran, I.R.	27,580.0	26,850.0	−2.6
Iraq	3,170.0	3,170.0	0.0
Kuwait	1,572.0	1,572.0	0.0
Oman	995.0	980.0	−1.5
Qatar	25,636.0	25,636.0	0.0
Saudi Arabia	6,900.0	7,154.0	3.7
UAE	6,060.0	6,040.0	−0.3
Middle East others	921.0	917.0	−0.4
Africa	14,132.0	14,165.0	0.2
Algeria	4,504.0	4,504.0	0.0
Angola	270.0	270.0	0.0
Egypt	1,895.0	1,940.0	2.4
Libya, S.P.A.J.	1,491.0	1,420.0	−4.8
Nigeria	5,152.0	5,210.0	1.1
Africa others	821.0	821.0	0.0
Asia and Pacific	14,928.0	14,824.0	−0.7
Australia	2,605.0	2,605.0	0.0
Bangladesh	436.0	435.0	−0.2
China	2,449.0	2,449.0	0.0
India	1,101.0	1,075.0	−2.4
Indonesia	2,769.0	2,659.0	−4.0
Malaysia	2,480.0	2,480.0	0.0
Myanmar	538.0	538.0	0.0
Pakistan	798.0	798.0	0.0
Papua New Guinea	428.0	435.0	1.6
Asia and Pacific others	1,324.0	1,350.0	2.0
Total world	181,065.0	180,899.0	−0.1
OPEC	89,419.0	89,193.0	−0.3
OPEC percentage	49.4	49.3	

Source: http://www.opec.org/library/Annual%20Statistical%20Bulletin/interactive/FileZ/Main.htm see then also: Oil and gas data, T34 World proven natural gas reserves by country, 1980–2006, last accessed: 6/1/2007.

1.4.3 Coal

Coal is the most abundant fossil fuel and is used to produce most (approximately 60%) of the world's electricity. It was formed about 300 million years ago (earlier than petroleum) from highly compressed residues of plants. Chemically, coal consists mainly of carbon, ash, which represents silicates and metals, and some sulfur. Some metals in coal are toxic, and sulfur is not desired, since it leads during combustion to sulfur oxides, which are hazardous in the environment. Coal appears in nature in different grades. At high pressure and low moisture anthracite is formed, which has, with 95%, the highest carbon content. Bituminous coal, which is relatively soft, contains 60–80% carbon. Subbituminous coal has a lower carbon content of about 40%; however, its lowest sulfur content makes it useful. At lower pressure and higher moisture lignite is formed, which only contains 25% carbon. Coal was obtained traditionally by underground mining, but more recently there is a trend to surface mining. The latter leads to less human casualties, however, causing more changes to the landscape and environment. Table 1.5 displays the world's reserves of coal.

TABLE 1.5 Proved recoverable coal reserves at end-2006 (million tons)

Country	Bituminous and anthracite	Subbituminous and lignite	Total	Share
United States	111,338	135,305	246,643	27.1
Russia	49,088	107,922	157,010	17.3
China	62,200	52,300	114,500	12.6
India	90,085	2,360	92,445	10.2
Australia	38,600	39,900	78,500	8.6
South Africa	48,750	0	48,750	5.4
Ukraine	16,274	17,879	34,153	3.8
Kazakhstan	28,151	3,128	31,279	3.4
Poland	14,000	0	14,000	1.5
Brazil	0	10,113	10,113	1.1
Germany	183	6,556	6,739	0.7
Colombia	6,230	381	6,611	0.7
Canada	3,471	3,107	6,578	0.7
Czech Republic	2,094	3,458	5,552	0.6
Indonesia	740	4,228	4,968	0.5
Turkey	278	3,908	4,186	0.5
Greece	0	3,900	3,900	0.4
Hungary	198	3,159	3,357	0.4
Pakistan	0	3,050	3,050	0.3
Bulgaria	4	2,183	2,187	0.2
Thailand	0	1,354	1,354	0.1
North Korea	300	300	600	0.1
New Zealand	33	538	571	0.1
Spain	200	330	530	0.1
Zimbabwe	502	0	502	0.1
Romania	22	472	494	0.1
Venezuela	479	0	479	0.1
Total	478,771	430,293	909,064	100.0

Source: http://www.wikipidia.com. Secondary source: http://www.eia.doe.gov/oiaf/aeo/supplement/pdf/suptab_114.pdf, last accessed: 4/1/2008.

Typically 80–90% of the coal is used in the country of origin, which is due to its relatively difficult transportation compared to oil and natural gas. Because of the environmentally hazardous gases formed during combustion, coal scrubbers or filters must be used before letting the gaseous product into the atmosphere. The expected time for the world's coal reserves to be depleted is within 200 years, by far the longest among the nonrenewable fuels.

Besides electricity, coal can be converted to gases as "coal gas" for similar uses as natural gas. It can be also converted to syngas, a mixture of CO and H_2. As the price of oil and natural gas increases, this option becomes more attractive. For the same reason, coal is liquefied to substitute for oil (see, e.g., the Fischer–Tropsch process under Organic Chemistry, Section 2.6.3.2, p. 110). As coke, coal is an important ingredient in the production of pig iron and steel. Fine carbon is used in plastics and rubber as a reinforcing agent.

1.4.4 Nuclear Energy

So far, the only form of nuclear energy that has been practically used is nuclear energy from the fission of heavy nuclei, such as uranium-235 or plutonium-239. Plutonium-239 is used mainly in breeder reactors, which have the advantage that they produce fissionable material. However, these types of reactors are politically unpopular because of their higher safety issues. In the past, France has built many breeder reactors, whereas the United States or Germany did not.

Per kilogram of fuel, nuclear fission produces about 10,000 times the amount of energy as coal. However, nuclear reactors involve several disadvantages, which have made them less popular as an energy source. The major issue is that the fissionable isotopes are radioactive and would cause disease if they would be released into the atmosphere, which would occur in the event of an accident. The worst accident so far happened in Chernobyl in the former Soviet Union in 1989, when the building holding the nuclear reactor exploded. Furthermore, even in the absence of an accident, the spent nuclear fuel rods still remain radioactive for several hundred thousand years. They must be carefully disposed. Currently, spent nuclear fuel rods are kept in containers below water pools. In the United States, they are to be eventually placed between aquifers under the Yucca Mountains, in Nevada. In addition, regular objects close to the nuclear core of the reactor such as instruments or clothes become radioactive and have to be treated with caution. The required safety and waste disposal measures make nuclear fuel less economical than simply based on its theoretical efficiency.

To obtain fissionable uranium-235, it has to be enriched from the more abundant uranium-238 as found in minerals, such as pitchblende. Table 1.6 shows the world's uranium sources. About 16% of the world's electricity is obtained from nuclear power plants in about 30 countries. Figure 1.8 shows the status of commercial nuclear power plants of different countries. A comparison with other methods of electricity generation is illustrated in Figures 1.9 and 1.10. The world's reserves of uranium are expected to last for about 50 years.

In research labs, the fusion of hydrogen isotopes has been successful, but "cold fusion" at moderate temperatures remains a dream of scientists because it would be the most environmentally friendly source of energy, with resources that would last us more than 100 million years. Though attempted many times to date, fusion of nuclei at close to ambient temperatures could not be carried out successfully.

TABLE 1.6 Known recoverable resources of uranium

	Tonnes U	Percentage of world
Australia	1,074,000	30
Kazakhstan	622,000	17
Canada	439,000	12
South Africa	298,000	8
Namibia	213,000	6
Brazil	143,000	4
Russian Fed.	158,000	4
USA	102,000	3
Uzbekistan	93,000	3
World total	3,537,000	

Source: Reasonably Assured Resources plus Inferred Resources, to US\$80 kg^{-1} U, 1/1/03, from OECD NEA & IAEA, Uranium 2003: Resources, Production and Demand, updated 2005. Available at http://www.world-nuclear.org/.

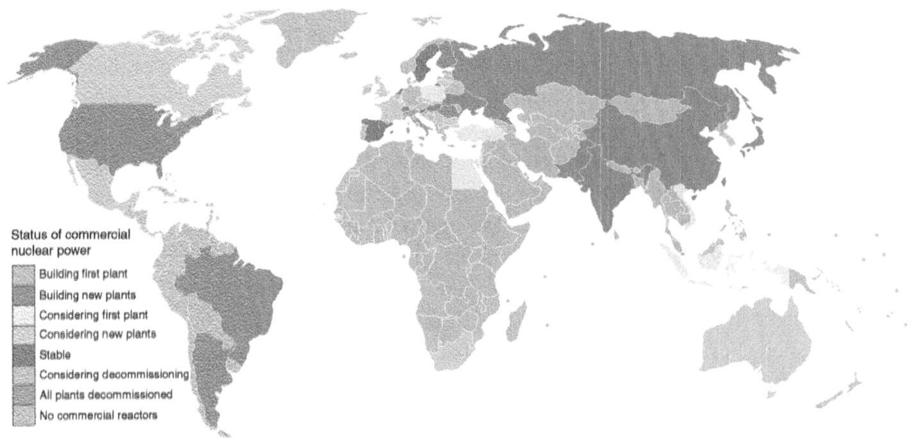

Figure 1.8 The world's nuclear power. [*Source*: https://en.wikipedia.org/wiki/Nuclear_power_by_country. Used under CC BY SA 3.0 https://creativecommons.org/licenses/by-sa/3.0/.]

1.4.5 Outlook

Both oil and natural gas are used as precursors for hydrogen as fuel (see also Section 2.6). Coal is currently the fastest growing energy source in the world. The major issue with all nonrenewable energy sources is that they create environmentally harmful and dangerous products. Even when the toxic side products such as oxides of nitrogen or sulfur are removed, the fossil fuels are the major contributors to an increase in the emission of greenhouse gases, such as CO_2 and CH_4, and thus lead to global warming. One issue with hydrogen as a renewable energy source is that currently most sources of hydrogen production involve nonrenewable resources, for example, syngas from coal or conversion of natural gas by so-called steam reforming: $CH_4 + H_2O \rightarrow CO + 3H_2$. Figure 1.11 shows how rapidly the price of petroleum has been increasing in recent decades. In 2008, the price of oil reached as high as \$145 per barrel.

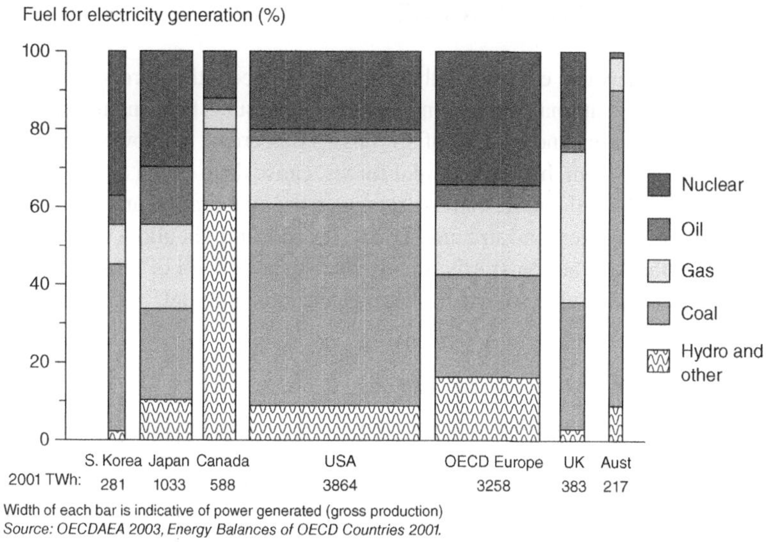

Figure 1.9 Electricity cost by different methods.

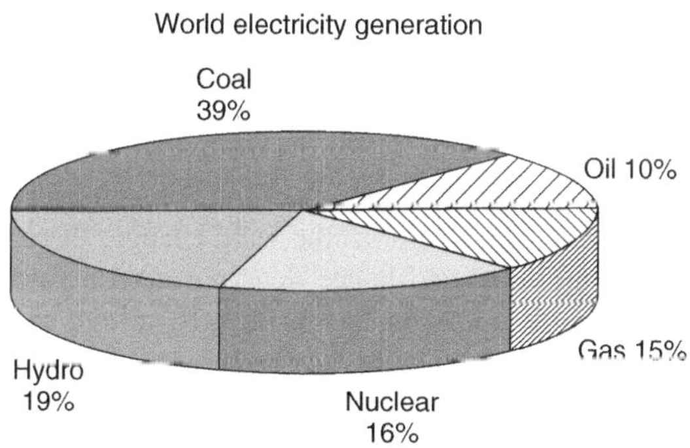

Figure 1.10 Different methods of generation of electricity (2003). [*Source*: OECD Factbook 2006-ISBN 92-64-03561-3-© OECD 2006.]

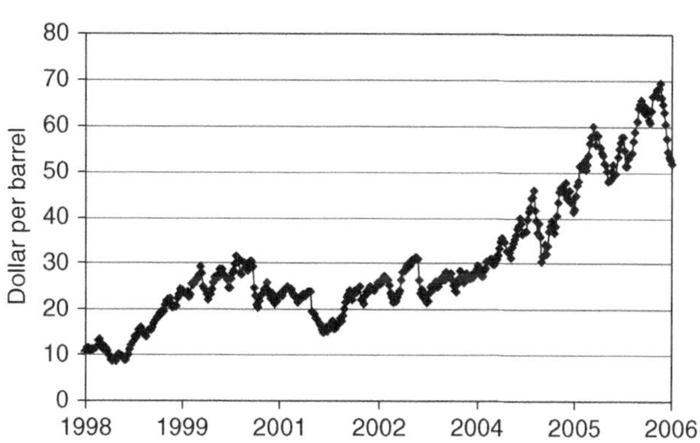

Figure 1.11 Change in oil price in the United States during 1998–2006. [*Source*: Figure constructed from data from the EIA, available at http://tonto.eia.doe.gov/dnav/pet/pet_pri_wco_k_w.htm.]

1.5 RENEWABLE ENERGY SOURCES

The source for the renewable energies is the sun and is linked either directly or indirectly to the power of the earth's internal and external changes. The sun's heat and the earth's surface temperatures cause heating and cooling of air masses that create the powerful winds. Those winds, along with heat from the sun and tidal forces, cause deep ocean currents and surface waves. The combined wind and the sun's heat cause the evaporation and precipitation that result in flowing rivers, lakes, and streams. The water and sunlight allow vegetation to grow as organic matter that can be transformed fuel. Once captured, all of these energies can be converted for use as renewable sources of heat, electricity, and fuel.

1.5.1 Wind Energy

Wind is moving air that causes the uneven heating and cooling on the earth's surface due to the absorption of heat by the earth. In this process, the land surfaces heat faster than water. The warmer air over land rises, resulting in the cooler air over the water moving out. This is a perennial process. A never-ending cycle of moving air causing atmospheric wind can be harnessed to produce large amounts of electricity. For achieving this, wind turbines are used. These turbines absorb the kinetic energy by turning aerodynamic blades. Typically, two or three blades are mounted on a shaft, forming a rotor. As wind flows over the blades, the low-pressure air is forced to lift and pull, turning the rotor. The rotor is connected to a drive shaft that spins a generator, which converts mechanical energy into electricity. One wind turbine can produce 1.5–4 million kilowatt hours (kWh) of electricity in a year (U.S. Department of Energy, 2005).

Upwind turbines have a wind vane that measures and communicates the direction of the wind to a yaw drive that reorients the rotor as the wind changes direction. The downwind turbines are reoriented by the wind and hence does not need a yaw drive. The rotor is computer-controlled in the startup at wind speeds of 8–16 miles/h and shut down at wind speeds over 65 miles/h. Some wind power systems combine with solar energy sources and hydrogen storage systems to store excess energy, with the added bonus of pure water as a by-product of its process.

Wind farms are clusters of turbines that produce electricity that is carried to a power grid for sale to utility companies. Wind power also provides *distributed energy*, which means it is on a small scale, close to the user and available immediately. In Europe, many individual homeowners share wind resources from small cooperative operations.

Offshore wind farms are clusters of turbines mounted on a floating platform, either close to shore or out in deeper water where they are less visible from shore. They have an average of 3-MW capacity. The electricity that is generated is transported via an undersea cable to an onshore power grid. Using this method, the public concern over the extensive use of land for turbine installation has been overcome.

Giant offshore turbine installations are increasing, particularly in Europe. Siemens in Germany has developed direct-drive offshore turbine with 8-MW power rating, utilizing extended rotor diameter of 154 m [19]. Germany is developing a huge 5-MW turbine that has 200-ft-long blades and stands 400 ft high (National Geographic, 2005). The government restrictions require that the turbines be installed at least 3 nautical miles from shore, which creates logistical problems and higher costs.

Wind power currently supplies 2.5% of the world's electricity and yet it is the fastest growing alternative energy source [23]. However due to unpredictable weather conditions the typical wind turbine efficiency of 25% of nominal power is estimated.

1.5.2 Solar Energy

Solar energy is the most effective and stable source of renewable energy on our planet. The radiation is emitted from the sun that is operating at a temperature of 6000 K; the wavelengths of radiation lie in the range of visible and near infrared. Using solar energy it has been shown that electricity, hot water, heating, and cooling for dwellings could be produced. Solar technologies that involve electricity have been designated as photovoltaic and ones using thermal energy have been called the thermal systems.

Photovoltaic (solar cell) systems convert sunlight directly into electricity. A solar cell is based on semiconductors that absorb the sunlight. The electrochemical processes in the photovoltaic systems are described in more detail in Section 1.5.2.3.

The thermal system operates by absorbing the sun's radiation; it converts the stored heat into electricity for hot water preparation, building heating and cooling, and energy generation.

As shown in Ref. [6], the solar radiation budget is determined by the difference between the absorbed solar energy flux and the outgoing long wavelength radiation at the top of the atmosphere. These two components interact in a very complex way with most of the atmospheric processes tending to balance each other, thereby maintaining average constant climatic conditions. On a smaller temporal and spatial scale, this balance is disrupted, and the regional results of the radiation budget then contribute to atmospheric and oceanic circulations. In the context of climate perturbations, modifications in vegetation or in the formation of ice on the earth's surface can cause greenhouse gases such as carbon dioxide to increase. This will have adverse effects. As for interannual variations, the climatic anomalies such as the El Nino are likely to occur. This highlights the role of cloudiness in the radiation balance and explains why understanding of the interactions between clouds and radiation is a major goal of climate research (Figures 1.12 and 1.13). Accordingly, calculations at a conservative average of only 200 W/m^2, the net yearly solar energy input to the planet corresponds to 2.22×10^{11} GW-year, equivalent to 7.577×10^{20} BTU, 757,700 Quads, or 1.29×10^{14} petroleum barrels/year: this amount is equivalent to 353,000 million barrels /day. The actual petroleum world consumption is around 70 million barrels/day. Therefore, the solar available energy power is 5000 times the total energy power derived

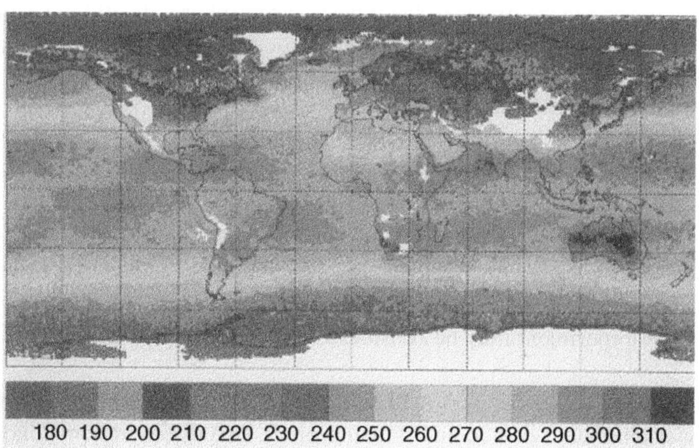

Figure 1.12 Thermal radiation contour. Outgoing thermal radiation (W/m^2) at the top of the atmosphere in January 1988 (7:30), calculated without clouds.

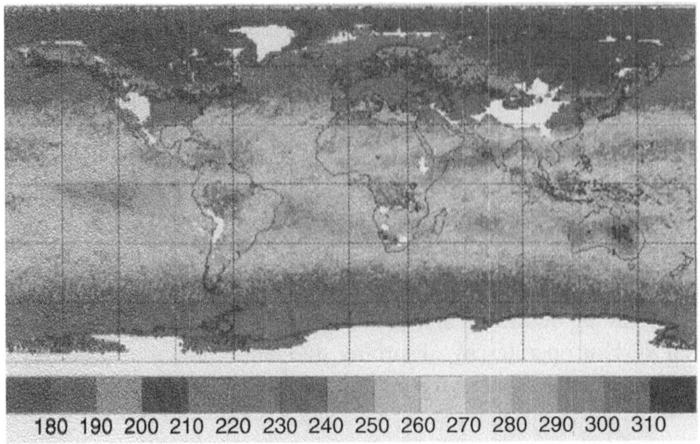

Figure 1.13 Thermal radiation contour. Outgoing thermal radiation (W/m^2) at the top of the atmosphere in January 1988 (7:30), with cloud cover. The cloud cover radically changes this zonal distribution by preventing a part of the thermal radiation, up to 60 W/m^2.

from actual world oil produced combustion (counted at 100% efficiency). Even at an energy transformation efficiency of only 10% and covering the energy collectors with only 1% of land surface, solar energy would provide about six times the actual oil used equivalent.

Solar energy is a sustainable, nonpolluting source of energy; however, implementation of current technology based on relatively lower efficiency of solar cells requires occupations of substantial earth area located in desert and sunny lands that are often very far from the main regions of human habitants. In addition, the influence of such extensive land use on the ecosphere requires further investigation.

1.5.2.1 Solar Spectrum

The wavelengths of the electromagnetic spectrum of solar energy range from 0.2 to 2.5 μm (http://www.esru.strath.ac.uk/Courseware/Class-16110/). The solar spectrum is shown in Figure 1.14. Note that one angstrom (1 Å) corresponds to 10^{-8} cm or 10^{-10} m. The sun emits radiation in the ultraviolet, visible, and near-infrared regions. Due to the existence of ozone layer in the stratosphere, the ultraviolet photons are absorbed by this region, with the result that we observe only the visible radiation in the region of 4000–8000 Å on earth.

1.5.2.2 How Do We Convert the Solar Radiation to Electricity?

During the 19th century, the race was on for understanding the composition and structure of matter. This race led us to our current understanding of atomic structure—namely that atoms contain electrons, protons, and neutrons. During the course of this race, Albert Einstein discovered a phenomenon called the photoelectric effect. He focused a beam of photons onto a low work function metal such as selenium metal and collected the electrons ejected out of the metal. This experiment may be considered as the starting point for the development of solar cells.

Solar cells are of two types: photovoltaic and photoelectrochemical. The photovoltaic is completely a solid-state device and is made up of two different semiconductors. The photoelectrochemical solar cell uses a semiconductor in a liquid electrolyte. These two types are discussed as follows.

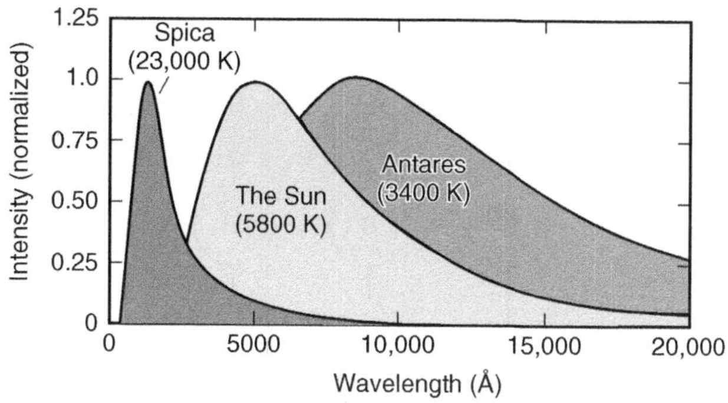

Figure 1.14 Solar spectrum. [*Source*: Reproduced with permission from http://csep10.phys.utk.edu/astr162/lect/sun/spectrum.html.]

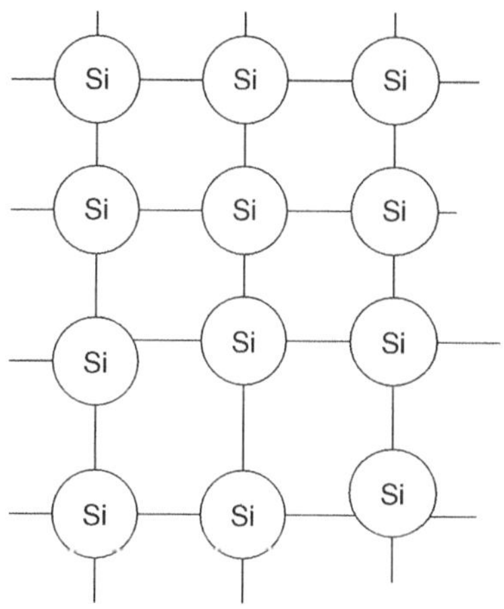

Figure 1.15 Intrinsic semiconductor.

1.5.2.3 Photovoltaic Cells Materials such as silicon, germanium, and compounds of II–V and II–VI elements in the periodic table exhibit a property of semiconduction and are called semiconductors. These semiconductors have electrical conductivity lower than metals and higher than insulators. Silicon is an ideal one for our discussion as it is a cheap material for fabrication of photovoltaic cells. When pure silicon is doped with an acceptor element such as boron, it becomes a positively charged (p-type) semiconductor. On the other hand, when it is doped with a donor element such as phosphorus, it becomes a negatively charged (n-type) semiconductor. We will take the example of silicon. Here, silicon has a valency of 4, and each silicon atom is bonded to four other silicon atoms as shown in Figure 1.15. When it is doped with boron that has a valency of 3, silicon is replaced with boron as shown in Figure 1.16.

Boron does not have the necessary number of electrons to form bonds with silicon. Hence, it leaves a hole, and such a semiconductor is a p-type semiconductor or boron-doped silicon. Any atom that has a deficient number of electrons will produce this effect. For example, instead of boron, aluminum, indium, or gallium can be used. Donor levels are in the range of 1 in 10^6 atoms/cm^3 or less. The situation is different when silicon

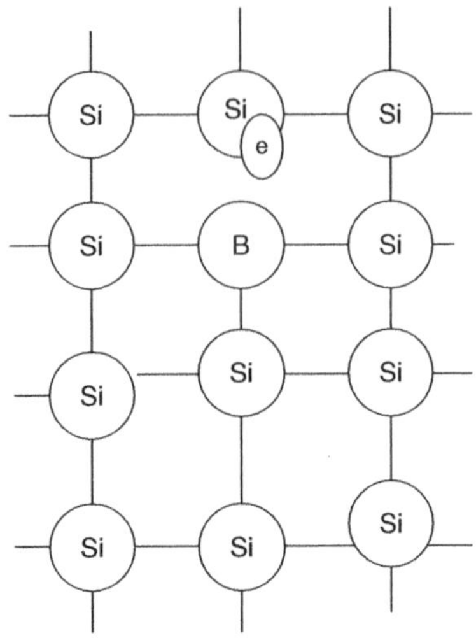

Figure 1.16 Boron doped semiconductor.

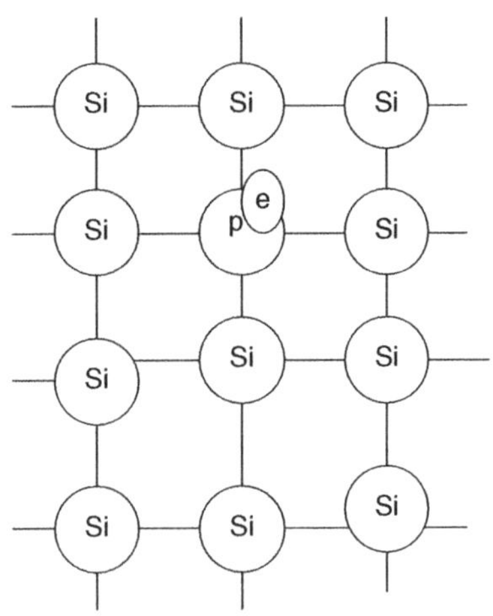

Figure 1.17 Phosphorus-doped silicon.

is phosphorus doped. Phosphorus has a valency of 5 and replacement of a silicon by phosphorus results in the structure depicted in Figure 1.17. One valence electron of phosphorus is free and is available for conduction. The phosphorus atom is negatively charged and hence is defined as the n-doped semiconductor or phosphorus-doped silicon. The illustrations in the figures indicate how the n- and p-type semiconductors are formed. The doping levels will decide the number of phosphorus atoms in silicon or boron atoms in silicon. The doping can be done to a high level to make a semiconductor into a metal.

Let us consider a case of combining an n-type semiconductor and p-type semiconductor as shown in Figure 1.18. It is called p–n diode. This diode is connected to a battery supply where a positive terminal of the battery is connected to the p-type silicon and negative terminal to the n-type silicon. Under the above conditions (called the forward bias), p-Si

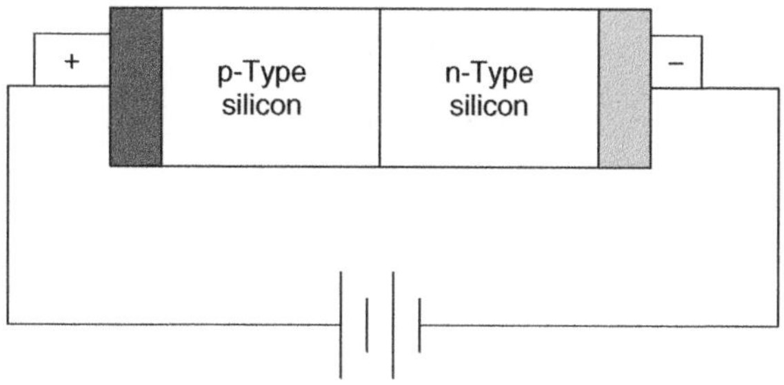

Figure 1.18 p–n diode with forward bias.

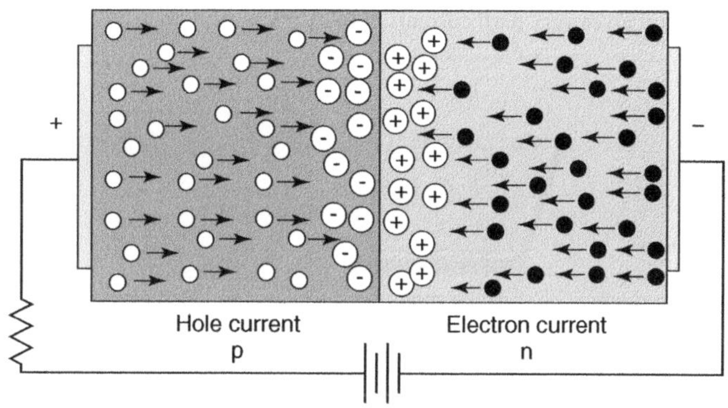

Figure 1.19 Electron and hole flow in the forward-bias condition. [*Source*: Reproduced with permission from Wikipedia.]

is made more positive such that the electron movement occurs from the left to right. An electron can move across the junction and is a downhill process. An electron can fill in the hole in this biasing condition. The conduction direction for the electron is from right to left. Under reverse-bias conditions, the p-silicon is made negative such that it is harder for the electrons to move across the junction. The flow of electrons and holes in the forward-bias condition is shown in Figure 1.19. White circles are holes and black circles are electrons in the diagram. When an electron crosses over the junction, electron–hole recombination occurs, shown in the figure by a circle with minus sign. The current that flows at different applied voltages can be monitored, and Figure 1.20 is a typical curve for the forward-bias conditions.

1.5.2.4 Solar Cells A solar cell is based on the photovoltaic effect that was first discovered by Alexendre Edmond Becquerel in 1839. The first solar cell was constructed in 1883 using selenium and gold. It produced an efficiency of 1%. Subsequently, solar cells were constructed using a p-type semiconductor and an n-type semiconductor as discussed in the previous section. When solar radiation falls on the p–n device, charge carriers are generated that results in the conversion of solar photons into electricity. The band gap of the semiconductor absorbing the radiation is an important factor for this conversion. Silicon is ideal as its band gap energy is about 0.60 eV and a large part of the solar radiation will be effective in producing electricity.

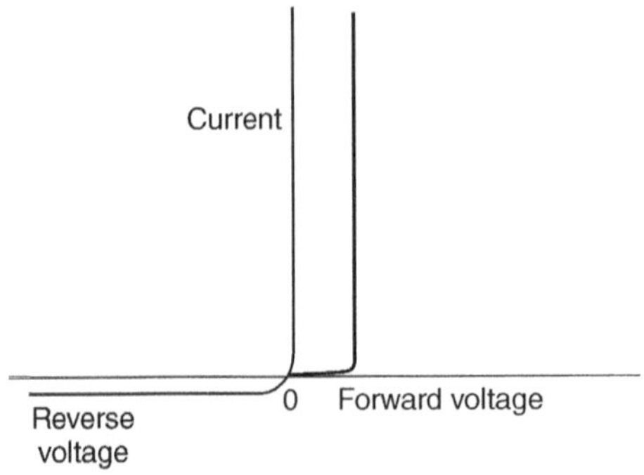

Figure 1.20 Current flow under forward- and reverse-bias conditions. For silicon, the current flow starts at about 0.5 V and gives very high currents at 0.7 V.

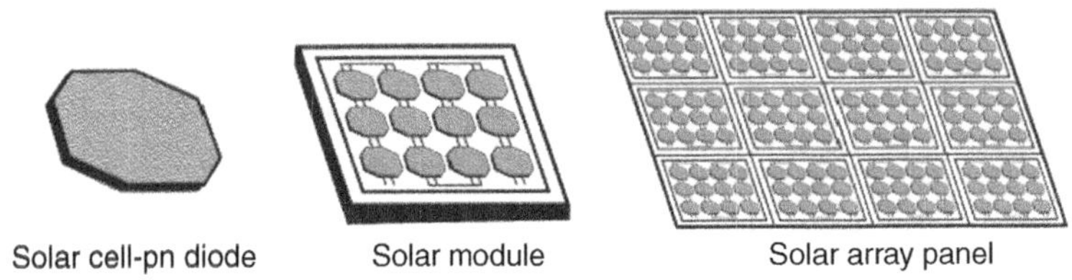

Figure 1.21 Solar panels used for conversion of sunlight to direct current (DC electricity).

A solar cell is primarily converting the incident photons to electricity. Figure 1.21 shows a single solar cell, a solar module made of several solar cells connected together, and a solar array panel where several modules are connected together. There are three generations of solar cells developed so far. Each of the generations has been aimed at improving the conversion efficiency. In the first generation of solar cells, the efficiency (defined as photons to electricity) was about 5–6% using silicon. This was increased to 30% in the most efficient multiple junction solar cells in the third generation. The first generation is made of a large area single-layer p–n junction diode; it is useful in producing electricity from the incident radiation. In the second-generation solar cells, multiple layers of p–n junction diodes were developed. With these solar cells, longer wavelength radiations are also utilized, resulting in higher conversion efficiency. With the third-generation solar cells, a light-absorbing material is added onto the solar cell. These light-absorbing materials are dyes, organic polymers, and quantum dots. Today, solar cells can give an efficiency of about 30%. The voltage generated by a photovoltaic cell is given by the following equation:

$$\text{Voltage } (V) \text{ (in V)} = \{\{kT\}/q\}\ln\{I_{\text{Ph}}/I_{\text{drak}}\} \tag{1.4}$$

where k is the Boltzmann constant, T is the absolute temperature, q is the charge, I_{Ph} is the current that flows in solar radiation, and I_{dark} is the current that flows when no solar radiation is falling on the voltaic cell. The values of k and q are 1.35×10^{-23} J/K and 1.60×10^{-19} C, respectively, with semiconductors such as gallium arsenide or indium selenide.

CRITICAL THINKING QUESTION

A solar photovoltaic cell produces a photocurrent of 0.4 uA when operated at 1000 K. In the dark, the photovoltaic cell gave a current of 1 nA. Calculate the voltage that can be obtained from the photovoltaic cell?

In the problem, the following data is given:

$$T = 1000\,\mathrm{K}$$
$$I_{\mathrm{ph}} = 0.4\,\mathrm{uA} = 0.4 \times 10^{-6}\,\mathrm{A}$$
$$I_{\mathrm{dark}} = 1\,\mathrm{nA} = 1 \times 10^{-9}\,\mathrm{A}$$

The constants k and q are available in the handbooks. The values are inserted in equation 1.4

$$\text{Voltage } V(\text{in V}) = [\{1.35 \times 10^{-23}\,\mathrm{J/K} \times 1000\,\mathrm{K}\}/1.60 \times 10^{-19}\,\mathrm{C}] \times \ln\{0.4 \times 10^{-6}\,\mathrm{A}/1 \times 10^{-9}\,\mathrm{A}\} = 0.505\,\mathrm{V}$$

Ans. 0.505 V

New materials for solar cells including perovskite solar cells.

1.5.2.4.1 First-Generation Solar Cells Solar cells operate on the principle of absorption of radiation from the sun and converting it to electricity. The working description of the solar cell is the absorption of radiation by a suitable semiconductor that has band gap energy equal to the visible radiation from the sun. The radiation excites the electrons from the valence band to the conduction band. The chosen semiconductors for solar cells are given in Figure 1.23. The practical solar cell that was introduced is shown in Figure 1.22. This uses silicon wafer with antireflection layer, titanium. The arrow in the figure points to the electron (yellow dot).

The yellow layer above aluminum represents p-type silicon and the wavy yellow line on the top represents n-type silicon. The contacting metal of the n-type silicon is shown in the circle with an arrow pointing to it.

1.5.2.4.2 Second-Generation Solar Cells Following the introduction of silicon solar cell, thin film technology using new semiconductors have emerged. Gallium arsenide and other semiconductors of group 3 and group 5 have been introduced, which gave 20% solar conversion efficiencies. Thin film solar technology provided a limited stability and hence a limited marketing value. The second-generation solar cells are shown in Figure 1.23.

Figure 1.23 shows iron sulfide (FeS_2), copper indium selenide ($CuInSe_2$), copper indium sulfide ($CuInS_2$), cadmium telluride (CdTe), gallium arsenide (GaAs), amorphous silicon (a-Si), and crystalline silicon (c-Si). The thicknesses of the films used in each of the solar cells are given on the bar chart. When the nanoworld started in the late 20th century, many

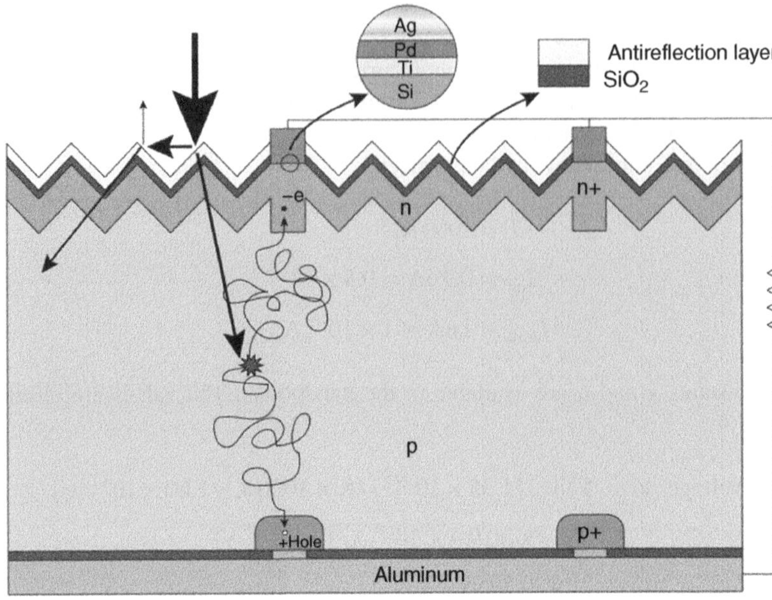

Figure 1.22 First-generation solar cell. [*Source*: Taken from https://en.wikipedia.org/wiki/Solar_cell.]

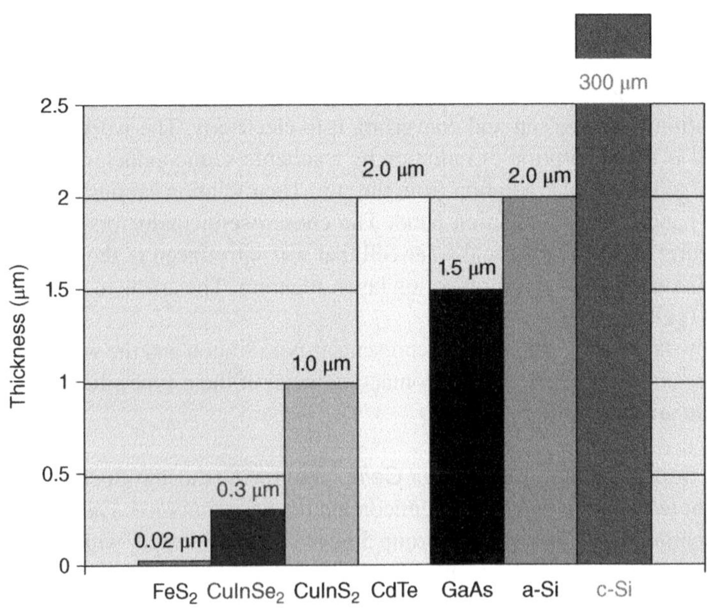

Figure 1.23 Second-generation solar cell materials with solar cell size marked on the bar chart. [*Source*: Badawy (2015) [24]. Reproduced with permission of Elsevier.]

new solar cells were developed. They are classified as (a) dye-sensitized solar cells, (b) quantum dot solar cells, and (c) organic solar cells.

Three ways of solar energy conversion by using semiconductor nanocrystals have been proposed:

1. Metal–semiconductor photovoltaic cell (also called Schottky junction photovoltaic cell)
2. Polymer–semiconductor hybrid solar cells
3. Quantum dot sensitized solar cells.

The new approaches are aimed at size-controlled energy levels for the conversion. A review of all the solar cells suggests that harvesting photons from the sun at greater efficiency has been the goal. The single crystal solar cells that were introduced in the beginning (called first-generation solar cells) had about 15% power efficiency. The single crystal manufacturing involves technology at a high cost. This led to the use of polycrystalline silicon solar cells that brought the cost down but at the sacrifice of power efficiency, which was around 5%.

1.5.2.4.3 Third-Generation Solar Cells The thrust now is on the use of quantum dots that can deliver high efficiency at low cost. With these solar cells, an efficiency limit of 32% is projected. Around 2013, solar cells having perovskite structured compound was introduced. An example of perovskite material is methylammonium lead halide, $CH_3NH_3PbX_3$ [methylammonium cation ($CH_3NH_3{}^+$) is surrounded by PbX6 octahedral] where X is a halogen such as Cl, Br, or I. Another perovskite is $CH_3NH_3SnX_3$. The efficiency of this solar cell is expected to reach 22.1%. The manufacturing cost is low compared to other solar cells and may hit the market soon.

1.5.2.5 Photoelectrochemical Solar Cells The sun's radiation has been successfully used to (a) generate electricity using photoelectrolytic cells, (b) produce a good fuel such as hydrogen, (c) produce chemicals that are less expensive, and (d) convert plant carbohydrates into more useful liquid fuel such as alcohol. These methods are depicted diagrammatically in Figure 1.24, where the methodology of utilization of solar energy is described in column 1, and the material absorbing the solar radiation is shown in column 2. The third column gives a pictorial representation of the method. The interface that produces the output in the method is shown under the column "interface." In the last column, the nature of the output upon solar radiation striking the interface is given. For example, the solar radiation falling on a semiconductor that is placed inside an electrolytic solution is electrical and hence this method is called the photoelectrolytic cell. This method is analogous to the solar cell we discussed in Section 1.5.2.3. In all the methods other than the photoelectrolytic method, the output is chemical. For example, we could decompose water to hydrogen gas, which is called photoelectrosynthesis. The photocatalytic and photobiological methods are shown in Figure 1.24. In these methods, chemicals absorb the solar radiation and transfer the energy until it is stored in stable molecules A few details regarding these methods are discussed in the following sections.

1.5.2.6 Photoelectrolytic Cells In the photoelectrolytic pathway to generate electricity from sunlight, two types of devices have been developed: photovoltaic (liquid-junction solar cells that differ from solid-state solar cells) and photogalvanic devices.

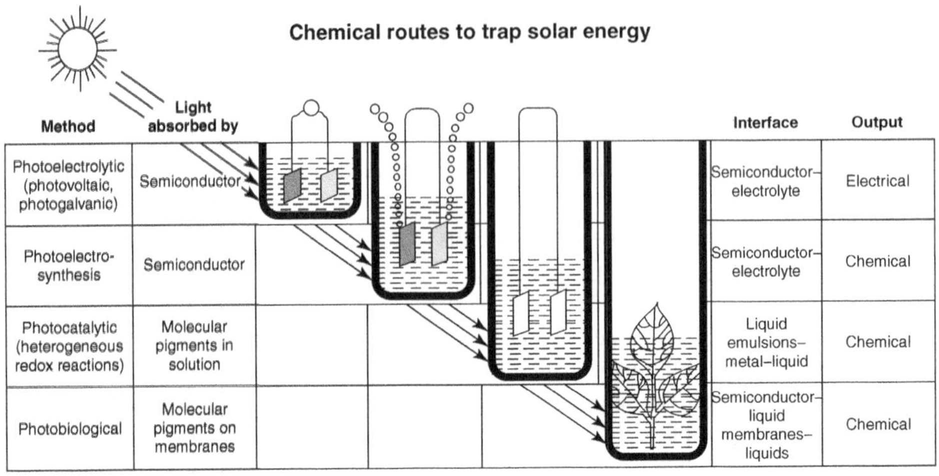

Figure 1.24 Chemical routes for solar energy.

With the liquid-junction photovoltaic cell, the device is made up of a semiconductor electrode and a metal electrode. When solar radiation strikes the semiconductor, electrons are excited from the valence band to the conduction band leaving a vacancy, called the hole, in the valence band. The electrons in the conduction band move through the external circuit resulting in the flow of current. The hole that has been created by the absorption of radiation oxidizes a species in solution. In general, in aqueous solutions, oxidation of water to oxygen occurs at this electrode. Simultaneously at the counter electrode a reduction process occurs. The flow of electrons that occurs upon excitation of the semiconductor is the photocurrent.

In a normal semiconductor, energy gradients do not exist and hence the excited electron recombines with the hole so fast that it is very difficult to bring about the separation of the charges (electron and hole). Hence it is very difficult to drive chemical reactions on its surface. This situation changes dramatically when it is placed in an electrolytic solution. The charge separation occurs because of the space-charge region underneath the surface with the electrolyte (see Figure 1.25) that is present in this situation, promoting the chemical reactions to occur on the surface of the semiconductor. Note that in Figure 1.25 the conduction and valence bands are bent near the electrode-solution region (this is called band bending). The atoms of the semiconductor near the electrolyte solution have higher energies relative to the bulk of the semiconductor. These cases are shown for n- and p-type semiconductors in Figure 1.25. The electrons and holes on the surface can react with different redox systems. This, however, is based on the energetics of the redox reactions.

1.5.2.7 Photoelectrosynthetic Cells The decomposition of water using solar radiation is a difficult process. We know from our practical experience that leaving a bucket of water in sunlight does not result in the splitting of it into hydrogen and oxygen. However, the situation changes when the bucket of water contains a semiconductor such as n-type titanium dioxide. The water decomposes to hydrogen and oxygen. This occurs due to absorption of solar radiation by the semiconductor, resulting in the electron and hole generation. This decomposition of water to hydrogen and oxygen was reported in 1970 by Fujishima and Honda, who collected 11 L of hydrogen on a good Japanese summer day

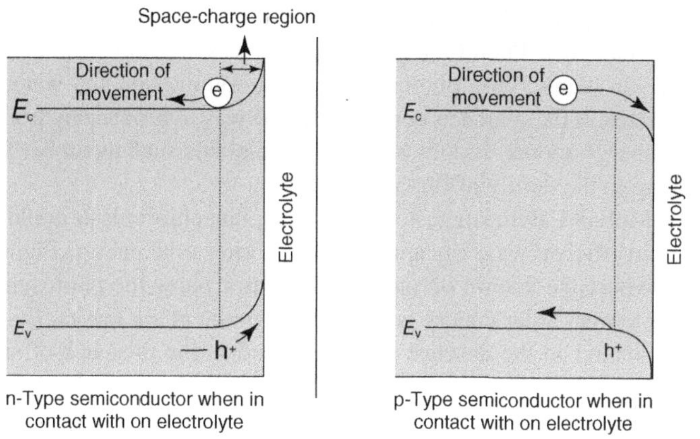

Figure 1.25 Space charge regions when n- or p-type semiconductors are in contact with an electrolyte. E_v and E_c are the valence and conduction bands energy levels.

using a single crystal of n-titanium dioxide electrode and a platinum counter electrode. This electrosynthetic cell worked with an efficiency of 0.4%. This result is quite impressive as the titanium dioxide absorbed only 4% of the solar radiation that is incident on it.

Photoelectrosynthetic cell is a device that drives an uphill chemical reaction using solar energy photons so that solar energy is stored in the chemical; uphill meaning requires external energy to be given to the system for decomposition. In the aforementioned example, water decomposition to hydrogen and oxygen is an uphill process—solar energy is able to drive it using a semiconductor. The first step is a process of absorption of solar energy of appropriate wavelength being absorbed by the semiconductor (n-TiO_2 absorbs radiation with energies greater than about 3.2 eV-bandgap energy of the semiconductor)

$$\text{n-TiO}_2 + h\nu \rightarrow \text{n-TiO}_2 + \text{h}^+ + \text{e}^- \qquad (1.5)$$

where h^+ is a hole and e is the electron. This is followed by

$$2\text{H}_2\text{O} + 4\text{h}^+ \rightarrow \text{O}_2 + 4\text{H}^+ \qquad (1.6)$$

and

$$2\text{H}^+ + 2\text{e}^- \rightarrow \text{H}_2(\text{g}) \qquad (1.7)$$

Over the years, several improvements were made in the materials used for solar energy absorption. Using silicon dioxide or magnesium oxide doped iron oxide as the electrode for solar energy absorption, and sodium sulfate or sodium hydroxide as the electrolyte, 4 L of hydrogen per hour could be produced for a square meter of the electrode. Recently, nanosized semiconductor particles have been used in photoelectrosynthetic cells.

1.5.2.8 *Photocatalytic Cells* Solar energy can be successfully utilized for driving a chemical reaction using a catalyst. By this method, several useful chemicals can be produced at a faster rate. Ammonia synthesis from nitrogen and hydrogen is one example. Using zinc-doped p-gallium phosphide, this reaction occurs spontaneously with the help of photons. The production of aromatic compounds at semiconductor particles and at colloidal particles is an interesting example of this type.

1.5.2.9 Photobiological Systems Photosynthesis that occurs in plants every day is an example of this category. Photobiological systems seek to mimic the complex photosynthetic processes in plants. With photosynthesis, carbon dioxide and water combine to produce carbohydrates. If the synthesis continues all the way to produce hydrocarbons, then it can be a useful energy source. It does occur in some plants such as rubber tree (*Hevea*). These plants belong to the family of Euphorbiaceae.

Nobel laureate Melvin Calvin suggested that green plant chloroplasts could be used as a model to understand efficient ways of capturing and storing solar energy. Figure 1.26 gives the energy cycle, where the arrows on the left of the tree show the photosynthesis in the tree producing the leaves. Solar energy is now stored in the green leaves. Bacterial action on decaying plant matter in the absence of oxygen and in the presence of silt and water produces coal. This occurs after a long time and millions of years of bacterial action under pressure. The first step in this decay is the formation of peat-compressed plant matter that contains twigs and leaves. The subsequent step is the formation of brown coal or lignite, followed by the formation of bituminous coal. This coal is used in the thermal power reactors to produce electricity and that is shown on the right side of Figure 1.26. In this energy cycle, solar energy is ultimately converted into electricity.

1.5.2.10 Solar Heater Solar energy is used for heating water for domestic and industrial purposes. Swimming pools in several countries are heated by solar heaters. This requires the usage of solar thermal collectors. In several European and Asian countries, a very high percentage (up to 75%) of domestic hot water supply comes by this method.

1.5.2.11 Solar Cooker A thermally insulated box that traps solar energy has been successfully used for cooking food. In India, chapatis (Indian bread) are made in desert areas using this type of box. In the developed countries it is used for pasteurization and fruit canning.

1.5.2.12 Solar Pond Differential salt concentration causes density gradients that restrict the heat exchange by natural convection. By filling a pond with three layers of

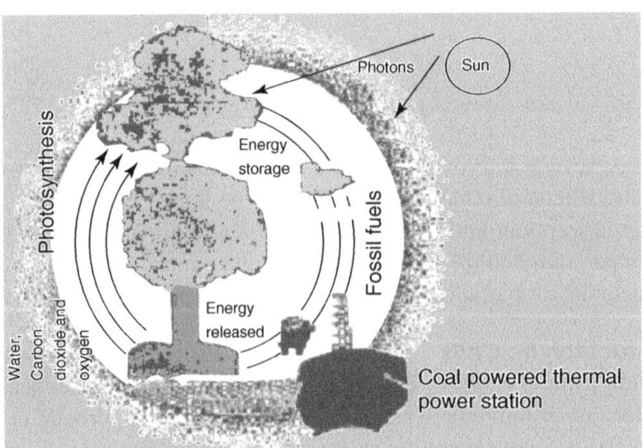

Figure 1.26 Photosynthetic energy cycles.

water containing three different concentrations of salt, with the top layer having the lowest concentration and the bottom layer having the highest concentration of salt, solar energy is trapped in the bottom layers. This approach is particularly useful for rural areas for heating buildings or generating electricity.

1.5.2.13 Solar Energy for Splitting Water In recent years, experiments have been carried out to split water into hydrogen by concentrating sunlight through several stages and focusing with fiber optics. This method is discussed in detail in Chapter 2.

1.5.3 Geothermal Energy

Subterranean planetary activity continually generates and stores an enormous energy resource as heat. Geothermal energy from the earth's internal heat can be extracted and used to generate electricity for heating homes and businesses. Geothermal energy can be found in the earth's crust, or lithosphere, in tectonically active regions. The most accessible regions of the lithosphere are thin or have been disrupted by recent (10 million years or less) volcanic or earthquake activity. There are four types of geothermal resources: hydrothermal systems, geo-pressured zones, hot and dry rocks, and magma from the earth's core.

Hydrothermal systems are found where groundwater has seeped down into the earth along the fault lines and becomes heated by hot rock. The high-pressure conditions that are found deep below the earth's surface can heat and store hot water at temperatures well above the boiling point of water on the surface. As heated groundwater moves down through rock fractures, it eventually pools in reservoirs, where it may rise back to the surface through natural convection processes. These hydrothermal reservoirs have temperatures ranging from 250 to 600 °F and are the source of the earth's natural hot springs and geysers (Figure 1.27). Hot water vaporizes into steam as it nears the lower pressures at the surface. Steam is separated and used to power turbines that generate electricity. The hot water component, which has a lower temperature than the steam, is processed through a binary plant to generate electricity (Figure 1.28).

Geo-pressured zones are subsurface areas where saltwater, or brine, becomes trapped between layers of hot, impermeable rock. The brine reaches temperatures of 200–400 °F and becomes very highly pressurized. This reservoir of heat and hydraulic pressure can be tapped to generate electricity. Sometimes, dissolved methane can be found in large quantities within the brine, increasing the potential energy resource.

Figure 1.27 Geothermal energy hot springs in Bridgeport, CA. [*Source*: Reproduced with permission from Wikipedia.]

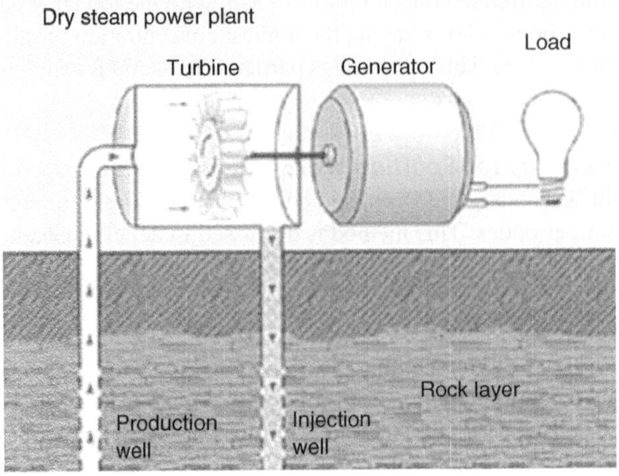

Figure 1.28 Geothermal energy. [*Source*: Reproduced with permission of National Renewable Energy Laboratory (NREL).]

Hot and dry rocks can be found everywhere among the lithosphere, making it a potentially plentiful geothermal resource on a global scale. However, this type of solid rock is found at depths greater than 2 miles beneath the earth's surface and where there is no liquid to carry the heat. To tap this energy, man-made wells are drilled into which water is pumped under high pressure to create fracture networks among the heated rocks. The circulating water absorbs the heat from the rock and creates hydrothermal reservoirs. Once the water reaches high enough temperatures, it can be pumped back to the surface and used to generate electricity.

Magma or molten rock from the earth's mantle and lower crust is what generates the volcanic activity. Most magma originates at depths of 20 miles or more, but in some global regions—near volcanoes and mid-ocean ridges—significant amounts of magma may also be found closer to the surface. These magma reservoirs, or calderas, can be tapped for their geothermal energy. Extracting the energy from magma requires well-drilling equipment that can reach to great depths—with some drilling locations at the bottom of the ocean—and withstand temperatures above 2000°F. The technology of this magnitude is still in development. Typically, water is pumped into the well to solidify the magma and acts as a heat exchanger to further generate electricity. The other possibility is to mix water directly with the iron oxide in the magma to generate hydrogen. The collected hydrogen by-product can be burned to generate energy.

Geothermal heat pump systems make use of the constancy of temperature found in the shallow ground, less than 10 ft below the surface. These systems are for direct-use applications and can be used to generate heating, air conditioning, and hot water in buildings. The pumps have three basic parts: the ground heat exchanger, the heat pump unit, and ductwork to deliver the heated or cooled air. The heat exchanger is a loop of pipe that is buried in shallow ground near the building. A mixture of water and antifreeze is circulated through the pipe, either absorbing or releasing heat within the ground. In the winter, the pump draws heat from the warmer ground and transfers it to the building, and in the summer it removes heated air from the building and releases it within the ground.

Geothermal energy is a clean, efficient (see Figures 1.27 and 1.28), and abundant heat source for small, end-user applications, such as district heating systems, space heating, industrial heating, greenhouses, agriculture and livestock farms, aquaculture, seawater distilleries, and organic drying facilities.

1.5.4 Biomass Energy

Any organic material made from plants or animals is considered biomass. Energy from the sun is stored in plants via photosynthesis in the form of chemical energy. The plant-eating animals absorb this chemical energy. When biomass is burned, the stored chemical energy can be released. It can also be converted to other usable energy forms, such as methane gas, ethanol, biodiesel fuel, and biogases. The conversion processes used are anaerobic digestion, gasification, and fermentation that take place in biorefineries, similar to oil refineries and petrochemical plants. However, the biomass industry worldwide uses a diverse scale of conversion processes that are dependent upon the variety of feedstock available in a particular region.

Biomass energy can be produced from wood, food crops (Figure 1.29), grasses, forestry by-products (such as sawdust), agricultural by-products (such as peanut hulls), manure, and other organic municipal solid wastes (MSW). Wood is currently the most common form of biomass used.

There are three types of biomass energy applications: biopower, biofuels, and bioproducts. Biopower is the heat and electricity generated by burning biomass directly or converting it into biogases (methane) or liquid fuels that burn. Raw wood is used for fuel, as well as wood fuel pellets that have a greater energy density. Pellets are made by compressing

Figure 1.29 Corn field for ethanol production. [*Source*: Reproduced from: http://upload.wikimedia .org/wikipedia/en/0/0a/Maize_ear.jpg.]

sawdust, wood shavings, and paper into small cylinders that have low moisture content. In addition to compressed wood waste, fuel pellets can be made from agricultural waste such as ground peanut hulls, straw, corn, and rice husks. They are easily stored and transported, plus they are a clean, efficient way of utilizing the by-products from local forestry and agricultural industries. Solid-waste landfills are full of potential biopower. As organic matter decomposes, a mixture of methane gas and carbon dioxide is released. The methane gas can be recovered and used to produce electricity. In addition, the mineral-rich residue from the gasification process is a sludge that can be collected and sold as fertilizer. Biofuels are liquid fuels, such as ethanol and biodiesel, which are used for transportation. They can be used on their own or blended with gasoline and diesel fuel. Biofuels burn cleaner and produce fewer air pollutants. Ethanol is produced in the fermentation process that extracts the glucose from corn, wheat, rice, potatoes, beets, and even yard waste. However, woody crops (such as poplar and willow trees) and switchgrass are emerging as new dedicated crops for making ethanol. Gasoline engines can run on a 15% ethanol to 85% gasoline mixture, but fuels that have higher ethanol contents need special vehicles. Biodiesel fuel is distilled from vegetable oils and animal fats that are recycled from restaurant waste grease and sometimes from plant sap. Diesel engines can use biodiesel fuels as a safe, biodegradable fuel that reduces emissions. Biofuel efficiency is currently less than 1%, due to the energy that is expended to grow, fertilize, and harvest the vegetation that makes it.

The use of biomass energy offers many economical and ecological benefits. Most importantly, it has the potential to reduce greenhouse gas emissions in significant amounts because biomass is carbon-neutral. This is because the carbon dioxide released when biomass is burned or converted is equal to the amount of carbon dioxide it absorbed during photosynthesis. Plant tissue also has almost no sulfur content, thereby reducing acid rain when used. Agriculture and forestry industries are supported and enhanced by the use of biomass energies because the feedstocks for power generation come from residues from crops and wood. Biomass energy optimizes land use because feedstock crops can be cultivated on land that is not suitable for food crops and would otherwise be unused. Bioproducts can replace the traditional ones made from petroleum. And biomass energy recycles unused material, which significantly supports the waste management—particularly in large, urban cities such as New York.

1.5.5 Hydropower Energy

Of all the renewable energies that generate electricity, hydropower is the most widely used. Hydroelectric power plants harness the mechanical energy in moving water and located on or near a dynamic water source (see Figure 1.30). The water's flow or fall determines the amount of available energy. As the water rushes through a pipe intake, its pressure pushes against turbine blades that spin an electrical generator. The electricity is then transmitted through power lines to a grid. In a storage system, the water remains in a reservoir created by a dam and then is released when electricity demand is high. Hydropower is clean, with no waste products that pollute the water or the air. It does impact the environment when dams interrupt the natural habitat; for instance, fish trying to swim upstream to spawn.

1.5.6 Ocean Energy

The ocean's abundant energy can be captured in the tides, the waves, the deep-ocean currents, and in the warmer surface waters in tropical locales [3]. Tidal energy systems harness

Figure 1.30 Hydroelectric power generators at Nagarjuna dam and hydroelectric plant, India. [*Source*: Reproduced with permission from Wikipedia.]

the power in the tides caused by the sun–moon's gravitational pull and the earth's rotational pull. High- and low-tide pulls can cause near-shore water levels to vary up to 40 ft. A tidal range of 10 ft is needed in an area that has an inlet in order to produce energy that is economically feasible. Tidal energy plants have a dam, or tidal barrage, that stretches across an inlet. The barrage has gates that allow ocean water to fill into a tidal basin during the incoming high tides and to empty on the outgoing tide through a turbine placed within the basin. Tidal fences are another option that can be installed in channels. They are vertical-axis turbines mounted in a fence that forces all tidal water to pass through the turbines. The tidal turbines generate electricity using the same technology as wind turbines. Wave energy systems harness the power in the waves as the wind blows them over the ocean's surface. The world's coastlines are ideal sites to capture wave energy. As waves approach shore, they are "focused" into a narrow channel, which increases their energy power before they are forced through turbines. Ocean thermal energy conversion (OTEC) is a system that extracts the sun's heat from the surface water to produce electricity. In tropical waters, the surface temperature can be 40 or more degrees warmer than the deeper water. These systems are not economical because the electricity that is produced must be transported to land.

Currently, at least 48 countries are involved in renewable energy development and utilization. Due to current economic conditions, the need for renewable energy has never been greater and has demonstrated strong growth within the last decade. In 2015, global investments in renewable have reached over US$2859 billion [22]. More than 1.7 million people are directly employed by the industry and the 160 GW of installed renewables represents 4% of global capacity [8].

As shown in Bloomberg's New Energy Outlook, "Cheaper coal and cheaper gas will not derail the transformation and decarbonization of the world's power systems. By 2040, zero-emission energy sources will make up 60% of installed capacity. Wind and solar will account for 64% of the 8.6 TW of new power generating capacity added worldwide over the next 25 years, and for almost 60% of the $11.4 trillion invested [20]."

1.6 ENERGY STORAGE

The biggest disadvantage of electric power is that electricity cannot be economically stored directly, but it can be stored in other forms that consume a lot of energy for conversion and subsequent distribution. By storing the power during off-peak periods and releasing it at peak times, coinciding with periods of peak consumer demand, energy storage can transform this spontaneous power into schedulable, high-value products. Energy storage in the form of hydrogen can be more effective in comparison to other available energy storage options, which are described as follows.

1.6.1 Pumped-Hydro Storage

Pumped-hydro facilities consist of two large reservoirs; one located at a low level and the other situated at a higher elevation. During off-peak hours, water is pumped from the lower to the upper reservoir, where it is stored. To generate electricity, the water is then released back down to the lower reservoir, passing through hydraulic turbines and generating electrical power. Known applications are limited to powerful energy plants but require substantial real estate for installation.

1.6.2 Compressed Air Energy Storage

Compressed air energy storage systems use off-peak power to pressurize air into reservoirs, which is then released during peak daytime hours to be used in gas turbines for power production. Existing facilities are sized in the range of several hundred megawatts. In a gas turbine, roughly two thirds of the energy produced is used to pressurize the air. The idea is to use low-cost power from an off-peak load facility in place of the more expensive gas turbine-produced power to compress the air for combustion. Since facilities have no need for air compressors tied to the turbines, they can produce two to three times as much power as conventional gas turbines for the same amount of fuel [11]. No technical information or economical data are available for pneumatic energy storing from small sources of renewable energy.

1.6.3 Flow Batteries

Flow batteries, also known as regenerative fuel cells, are capable of storing and releasing energy through a reversible electrochemical reaction between two redox solutions. These are called dissolved redox electrolyte systems. Designs exist around the use of zinc bromide ($ZnBr_2$), vanadium bromide (VBr), and sodium bromide (NaBr) as the electrolytes. Charging of the facility occurs when electrical energy from the grid is converted into potential chemical energy. Release of the chemical energy occurs within an electrochemical cell, with a separate compartment for each electrolyte, physically separated by an ion-exchange membrane. The technology is a closed-loop cycle, so there is no discharge of the regenerative electrolyte solutions from the facility. The scale of the facility is based primarily on the size of the electrolytic tanks.

1.6.4 Batteries

A number of battery technologies exist for use as utility-scale energy storage facilities. Primarily, these installations have been lead-acid, but other battery technologies such as sodium sulfide (Na_2S) and lithium ion are quickly becoming commercially available. All batteries are electrochemical cells. During discharge, ions from the anode are released into the solution and oxides are deposited on the cathode. Reversing the electrical charge through the system recharges the battery. When the cell is being recharged, the chemical reactions are reversed, restoring the battery to its original condition. Cost, environmental issues related to used batteries disposal, and limited number of recharge cycles make this approach not very valuable.

1.6.5 Superconducting Magnetic Energy Storage

Superconducting magnetic energy storage systems store energy in the magnetic field created by the flow of direct current in a coil of cryogenically cooled superconducting material. The system includes a superconducting coil, a power conditioning system, a cryogenically cooled refrigerator, and a cryostat/vacuum vessel. They are highly efficient at storing electricity (>95%), and provide both real and reactive power. These facilities are used to provide grid stability in a distribution system and power quality at manufacturing plants requiring ultraclean power, such as microchip fabrication facilities. This technology is still in the development stage and not cost-effective for small distribution generation systems.

1.6.6 Supercapacitors

Based on a study the U.S. National Renewable Energy Laboratory conducted in 1997 on a system in Deering, Alaska, the largest fuel saving with supercapacitors (the high energy density electrochemical device, typically on the order of a thousand times greater than an electrolytic capacitor) came from relatively short-term storage. Evidence from this test indicated that a storage capability of 10 min reduced the fuel use by 18%, the diesel running time by 19%, and the number of diesel starts by 44%. We believe that short time storage capability making use of supercapacitors is a not very attractive option for wind energy storage. However, applications such as peak shaving measure, especially into electric propulsion systems, are becoming an everyday practice.

1.6.7 Flywheels

A flywheel energy storage system works by accelerating a rotor to a very high speed and maintaining the energy in the system as inertial energy. Advanced composite materials are used for the rotor to lower its weight while allowing for the extremely high speeds; energy is stored in the rotor in proportion to its momentum. The flywheel releases the energy by reversing the process and using the motor as a generator. As the flywheel releases its stored energy, the flywheel's rotor slows until it is discharged. In application as energy storage for industrial use, flywheels are limited to shaving peak power for relatively a short term due to substantial time needed for power charge and discharge. Other limitations are related to safety and cost consideration.

1.6.8 Hydrogen Storage

An attractive method of producing hydrogen gas is through electrolysis of water using a dedicated electrolyzer. The generated hydrogen gas can be coupled to a fuel cell system to produce electricity. However, most often it is preferable to store hydrogen gas and transport it to far-off places for subsequent utilization in fuel cells. Hydrogen and oxygen from an electrolyzer may be stored separately in pressure tanks or other hydrogen storage media, such as metal hydrides. Hence the generated hydrogen is an energy carrier that can be transported to any location of use and discharged on set schedule at the required time. The economic benefits of storing the energy in the form of hydrogen are namely emissions reduction, efficiency of utilization, improvement in the operation conditions, and relative simplicity of energy transportation to the required site (see Section 3.33).

1.7 ENERGY ETHICS

There are many examples showing that human civilization in its need for energy has had a tremendous impact on our planet. In a few generations, the supply of fossil fuels, which were created in over several hundred million years ago, is predicted to be depleted. The emission of SO_2 into the atmosphere following the combustion of coal and oil is more than from natural sources such as the decomposition of dimethyl sulfide from oceans. The release of NO from the combustion of fossil fuels and biomass is now probably larger than natural sources creating photochemical smog in urbanized areas. “Greenhouse” gases, such as CO_2 and CH_4, that are often associated with energy production, are contributing to the global warming process. These and other impacts of human activities on the earth’s ecosystem have led to the discussion for defining the current geological period as the “Anthropocene Epoch.”

An energy policy is needed that will promote conservation of energy to minimize the impact on our planet and enable the transition toward sustainable management of the environment at every level of the society—from individual, family, local, national, to international. Individuals, as consumers and members of society, must recognize that there are connections between our consumption choices and the rest of the world. Consumers need to purchase more energy-efficient appliances, homes, and vehicles for long-distance single-person trips. Choosing hybrid vehicles with better energy efficiency, locally grown products, installing solar photovoltaic panels, and purchasing green power will help meet future energy needs of society and preserve our environment. Many consumers help save energy by recycling and purchasing recycled materials rather than buying new products.

Governmental policies on taxes, subsidies, incentives, and standards need to be in place to encourage both energy efficiency for automobiles, factories, appliances and homes, and improvement of the environment. In some countries, excellent bike infrastructure and public transportation together with high taxes for vehicle registration have led to the use of bike, subway, or bus rather than car for transportation. The governments with higher energy prices have lower energy needs per capita than countries with lower energy prices.

To ensure the sustainability of human civilization on our planet, society has to concentrate on improving energy efficiency by utilizing alternative sources of energy; one of them being hydrogen technology. By switching to renewable energy sources and hydrogen technology, we would be able to contribute to sustainable management of the environment.

PROBLEMS

Section 1.3

1. What is the wavelength region of infrared radiation?
2. Name the greenhouse effect gases?
3. Atmosphere contains nitrogen besides carbon dioxide and oxygen. Why are they not greenhouse effect gases?
4. What wavelengths of sun's radiation region the earth? What happens to other wavelengths?
5. What are the different sources for the greenhouse effect gases?
6. What are the various ways the greenhouse gas contents can be reduced in the atmosphere?
7. How do aerosols contribute to the greenhouse gas effect?
8. Compare the relative contributions of greenhouse effect gases arising from natural gas, oil, and coal.
9. Discuss how clouds play a role in the global warming.
10. What is Kyoto agreement?
11. Explain how the temperature of the earth is kept higher than the equilibrium model.
12. Calculate the fraction of the radiation reflected by the earth if the earth's temperature changed from the original value of 14–15°C.

Section 1.5

1. What is the difference between photovoltaic cell and photoelectrochemical cell?
2. Describe the different types of harvesting solar energy.
3. Describe how we could split water by photosynthetic cell.
4. Explain how n-type and p-type semiconductors are produced.

MULTIPLE CHOICE QUESTIONS

1. Of the following gases, which is NOT a greenhouse gas?
   (a) CO_2
   (b) CH_4
   (c) O_3
   (d) N_2O
2. Oil and natural gas are
   (a) fossil fuels
   (b) nonfossil fuels

3. When fossil fuels are burnt,
   (a) they produce O_2
   (b) they produce CO_2
   (c) nothing happens
   (d) they produce inert gases
4. When solar radiation reaches the earth, it contains
   (a) X-rays
   (b) γ-rays
   (c) visible wavelengths
   (d) short wavelength ultraviolet light
5. When more and more carbon dioxide goes into atmosphere,
   (a) more heat is trapped
   (b) it decomposes
   (c) helps grow more plants
   (d) generates ozone
6. How much of solar radiation is dissipated by clouds?
   (a) 100%
   (b) 50%
   (c) 3%
   (d) 90%
7. How much of solar radiation is reflected by earth?
   (a) 6%
   (b) 100%
   (c) 80%
   (d) None
8. Based on solar model, what is the predicted temperature of the earth?
   (a) 100°C
   (b) 50°C
   (c) −20°C
   (d) 15°C
9. How many years did it take for oil to form from microorganisms?
   (a) 200,000
   (b) 2,000,000
   (c) 20,000,000
   (d) 200,000,000
10. Which is the purest form of coal?
   (a) Anthracite
   (b) Bituminous coal
   (c) Subbituminous coal
   (d) Lignite

11. Which countries have the largest reserves of the following nonrenewable energy sources: oil, natural gas, coal, and uranium, respectively?
    (a) Canada, United States, Mexico, and South Africa
    (b) Norway, India, China, and Brazil
    (c) Saudi Arabia, Russia, United States, and Australia
    (d) Egypt, Turkey, Germany, and Japan

BIBLIOGRAPHY

1. http://www.cnn.com/2003/ALLPOLITICS/02/06/bush-energy/index.html (accessed 25 May 2017).
2. US Government Energy Information Administration. http://www.eia.doe.gov/, http://wilcoxen.maxwell.insightworks.com/pages/804.html (accessed 25 May 2017).
3. What is tidal energy? http://www.oceanenergycouncil.com/ocean-energy/tidal-energy (accessed September 2017).
4. World Energy Projections: Resources, Consumption and CO2 Emission http://www.combusem.com/ENERGY1.HTM (accessed September 2017).
5. Clerici, A. (2004) World Energy Council Survey of Energy Resources, Presentation, www.eia.doe.gov (accessed 25 May 2017).
6. Oil & Gas Journal, Vol. 102, No. 47, Dec. 20, 2004, last accessed 4/1/2008 from http://www.ogj.com/currentissue/index.cfm?p = 7&v = 102&i = 47 (accessed 25 May 2017).
7. http://www.bp.com/home.do?categoryId = 1 (accessed 25 May 2017).
8. Martinot, E. (2006) Global revolution—a status report on renewable energy worldwide by Renewable Energy World MARCH/APRIL, Also can be accessed from http://ngm.nationalgeographic.com/ngm/0508/feature1/text2.html (accessed 25 May 2017).
9. Crutzen, P.J. http://en.wikipedia.org/wiki/ (accessed 25 May 2017).
10. Rubbia, C. (2000) The Future of Energy, 18th IAEA Fusion Energy Conference, Sorrento, Italy, October 4, 2000
11. Baxter, R. (2002) Energy storage: enabling a future for renewables? *Renewable Energy World*, July–August 2002.
12. Chen, I.Y.D., Santhanam, K.S.V., and Bard, A.J. (1981) Solution redox couple for electrochemical energy storage. *Journal of the Electrochemical Society*, **128**, 1460.
13. Chen, I.Y.D., Santhanam, K.S.V., and Bard, A.J. (1982) Solution redox couples for electrochemical energy storage II. *Journal of the Electrochemical Society*, **129**, 61.
14. Yang, Z., Zhang, J., Kintner-Meyer, M.C. *et al.* (2011) Electrochemical energy storage for green grid. *Chemical Reviews*, **111** (5), 3577–3613.
15. Weber, A.Z., Mench, M.M., Meyers, J.P. *et al.* (2011) Redox flow batteries, a review. *Journal of Applied Electrochemistry*, **41**, 1137.
16. Protich, Z., Wong, P., and Santhanam, K.S.V. (2016) A new graphene composite with a high coulombic efficiency. *Journal of Power Sources*, **332**, 337.
17. https://www.hydrogen.energy.gov/ (accessed 25 May 2017).
18. https://ec.europa.eu/clima/policies/international/negotiations/paris/index_en.htm (accessed 25 May 2017).
19. http://www.windpowermonthly.com/10-biggest-turbines (accessed 25 May 2017).
20. https://www.bloomberg.com/company/new-energy-outlook/ (accessed 25 May 2017).
21. http://energystorage.org/energy-storage/energy-storage-technologies (accessed 25 May 2017).

22. Max-Planck-Institute for Chemistry Division of Atmospheric Chemistry http://www.mpic.de/en/employees/honors-and-awards/the-anthropocene.html; http://fs-unep-centre.org/sites/default/files/publications/globaltrendsinrenewableenergyinvestment2016lowres (accessed 25 May 2017).
23. https://www.iea.org/topics/renewables/subtopics/wind/ (accessed 09 June 2017).
24. Badawy, W.A. (2015) A review on solar cells from Si-single crystals to porous materials and quantum dots. *Journal of Advanced Research*, **6**, 123–132.
25. Kalisman, P., Nakibli, Y., and Amirav, L. (2016) Perfect photon-to hydrogen conversion efficiency. *Nano Letters*, **16**, 1776.
26. D&C Dec 4, 2015: NY Sets Policy on Energy Use.
27. D&C Dec. 14, 2015: Kerry Predicts Climate Deal will Transform Economies.
28. D&C April 4, 2015: 2015 Tucson Fuel Cell Stands Out in Small Field.
29. D&C Sept 9, 2016: Lockheed Martin Opens Waste-to-Energy Plant.
30. C&E News July 13, 2015: Fight Over Future Fuels: EPA's proposed Renewable Fuel Standard displaces biotech firms, oil interests, and biofuel makers.

CHAPTER 2

Chemistry Background

2.1 REVERSIBLE REACTIONS AND CHEMICAL EQUILIBRIUM

Chemical reactions are of two types: irreversible and reversible. Let us write a general chemical reaction

$$A + B \rightarrow C + D \tag{2.1}$$

where A and B are called the reactants. C and D are called the products. After the reaction has proceeded for appreciable time, both the reactants (A and B) are completely used up in the above reaction. Only C and D are available in the medium. These two entities are incapable of combining to give the reactants. This type of reaction is called an irreversible reaction, where complete conversion of the reactants to products occurs. An example of this type of reaction is

$$SnO_2 + 4\,Li \rightarrow 2Li_2O + Sn \tag{2.2}$$

However, if a reaction occurs as

$$X + Y \leftrightarrows F + Z \tag{2.3}$$

where X and Y are the reactants and F and Z are the products, here, even after a long time, the reactants (X and Y) will be present along with the products (F and Z). This can be viewed as though X and Y react to produce products F and Z (forward reaction) and the reverse reaction of F and Z reacting to produce X and Y will also occur. This type of reaction is called a reversible reaction. In reversible reactions, only a portion of reactants is converted to products after a certain length of time. A well-known example of the reversible reaction is the synthesis of ethyl acetate from acetic acid and ethyl alcohol:

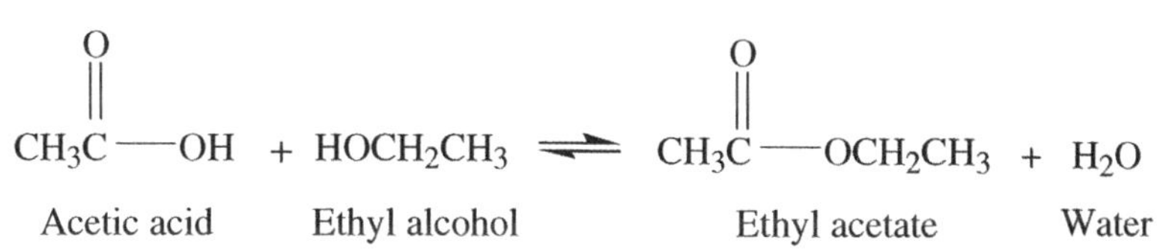

(2.4)

Introduction to Hydrogen Technology, Second Edition.
K.S.V. Santhanam, Roman J. Press, Massoud J. Miri, Alla V. Bailey, and Gerald A. Takacs

Book Companion site : www.wiley.com/go/santhanam/hydrogentech_2e

Another common example given is the Haber–Bosch process, in which hydrogen and nitrogen combine to form ammonia:

$$3H_2(g) + N_2(g) \leftrightarrows 2NH_3(g) \tag{2.5}$$

The double arrows show that the reaction is reversible.

When the concentrations of the reactants and products have no net change over time, this is defined as a state of chemical equilibrium. This state occurs when the forward chemical reaction proceeds at the same rate as the reverse reaction. At the point when the synthesis of ethyl acetate achieves equilibrium, the reaction vessel contains all four substances—acetic acid, ethyl alcohol, ethyl acetate, and water—and the rates of reaction of esterification (forward) and reaction of hydrolysis of ester (reverse) are equal. In the Haber–Bosch process, equilibrium is reached when the rate of production of ammonia equals its rate of decomposition. The following pictures (Figures 2.1 and 2.2) demonstrate conditions for the general case of chemical equilibrium.

At a particular level, equilibrium is dynamic; at the same time, some molecules of reactants form products and some molecules of products form reactants. A common example

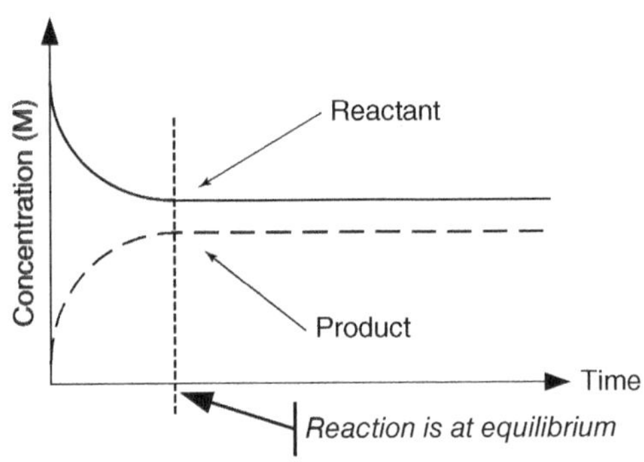

Figure 2.1 Conversion of a reactant and yield of a product with time. At equilibrium, the concentration of reactant and product remain constant.

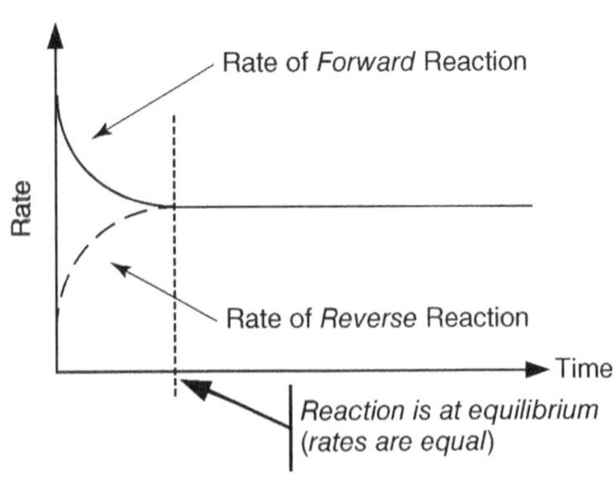

Figure 2.2 The rates of the forward and reverse reactions with time. At equilibrium, the rates of the forward and reverse reactions become the same.

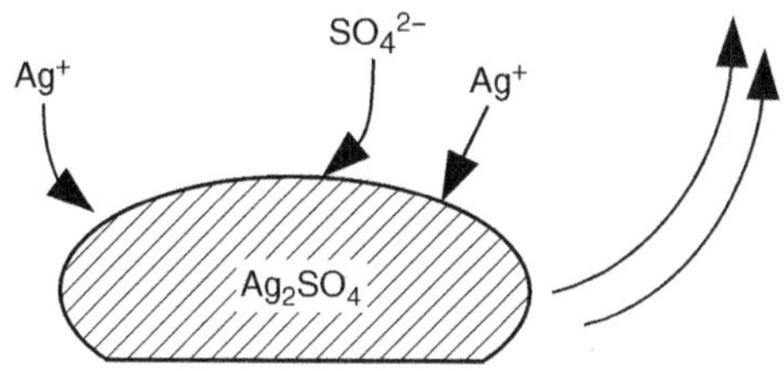

Figure 2.3 Equilibrium conditions of dissolving of silver sulfate in water.

of a dynamic equilibrium is dissolving of silver sulfate (Ag_2SO_4) in water:

$$Ag_2SO_4(s) \leftrightarrows 2Ag^+(aq) + SO_4^{2-}(aq) \tag{2.6}$$

At equilibrium, as much silver sulfate dissolves (forward arrow) as Ag^+ and SO_4^{2-} ions combine to form a precipitate (backward arrow) (Figure 2.3).

2.1.1 Equilibrium Equations and Equilibrium Constants

Every reversible reaction has a characteristic equilibrium constant (K or K_{eq}), given by an equilibrium equation. Let us consider a general case of a reversible chemical reaction:

$$a\text{A} + b\text{B} + \cdots \leftrightarrows c\text{C} + d\text{D} + \cdots \tag{2.7}$$

where A, B, ... are the reactants; C, D, ... are the products; and a, b, c, and d are the coefficients in the balanced equation.

The equilibrium constant K is the number obtained by multiplying the equilibrium concentrations of the products and dividing by the equilibrium concentrations of the reactants, with the concentration of each substance raised to a power equal to its coefficient in the balanced equation. An equilibrium expression for the general case of chemical reactions is

$$K = \frac{[\text{C}]^c \cdot [\text{D}]^d}{[\text{A}]^a \cdot [\text{B}]^b} \tag{2.8}$$

where K = equilibrium constant, $[\text{A}]^a$ and $[\text{B}]^b$ = reactant A and B concentration raised to a power equal to coefficients, and $[\text{C}]^c$ and $[\text{D}]^d$ = products C and D concentration raised to a power equal to coefficients.

In an equilibrium constant expression, the following are observed:

- The concentrations of the chemical species can be expressed as molarity (in aqueous solutions), K is designated K_c, or partial pressures (for gases), K is designated K_p.
- All concentrations are equilibrium values.
- The units of equilibrium constant K will depend on the number of moles of reactants and products.
- K value varies with temperature.
- Pure condensed phases such as solids, liquids, or solvents never appear in an equilibrium expression because their concentrations are constant.

To evaluate the equilibrium constant K, we insert the equilibrium concentrations into the equilibrium expression. For example, using the experimental data, molar concentrations for the reactants and product [CO], $[H_2]$, and $[CH_3OH]$ we can find the value of K_c for the reaction

$$CO(g) + 2H_2(g) \leftrightarrows CH_3OH(g) \tag{2.9}$$

$$K_c = \frac{[CH_3OH]}{[CO] \cdot [H_2]^2} \tag{2.10}$$

Proceeding in the same way, the value of K for the reaction, where one of the reactants is in solid state and its concentration does not appear in the equilibrium constant expression:

$$2H_2(g) + 2C(s) \leftrightarrows H_2C = CH_2(g) \tag{2.11}$$

$$K_c = \frac{[H_2 = CH_2]}{[H_2]^2} \tag{2.12}$$

When the reactants and products in the chemical reactions are gases, we can formulate the equilibrium expression in terms of partial pressures instead of molar concentration. For the Haber process, for example, $3H_2(g) + N_2(g) \leftrightarrows 2NH_3(g)$. In an equilibrium mixture of gases at 500°C, the partial pressure of H_2 is 0.928 atm, the partial pressure of N_2 is 0.432 atm, and that of NH_3 is $2{:}24 \times 10^{-3}$ atm. The equilibrium constant for this reaction is

$$K_p = \frac{(p_{NH_3})^2}{(p_{N_2})(p_{H_2})^3} = \frac{(2.24 \times 10^{-3})^2}{(0.432)(0.928)^3} = 1.45 \times 10^{-5} \tag{2.13}$$

Note, for the same reaction, the numerical value of K_c is generally different from K_p, so it is very important to indicate which of these equilibrium constants we are using. For the Haber process, for example, the values K_p and K_c at 25°C are 5.8×10^5 and 3.5×10^8, respectively. In general,

$$K_p = K_c(RT)^{\Delta n} \tag{2.14}$$

where R is the gas constant (0.082057 L atm/K mol), T is the temperature, and Δn is the difference between the total moles of gaseous product and the total moles of gaseous reactant. The value Δn for ammonia synthesis is

$$\Delta n = 2 - 4 = -2 \tag{2.15}$$

What is the meaning of the equilibrium constant, K? The value of K indicates the position of a reaction at equilibrium: which reaction—forward or reverse—is favorable? Knowing the value of K, we can determine how much reactant or product will be formed. The meaning of the equilibrium constants can be summarized in the following way:

1. If K is very much larger than 1, the reaction goes essentially to completion. For example, $K = 150$ means that a relatively high amount of products are present at equilibrium.
2. If K is very much smaller than 1, the reaction does not proceed forward. For example, $K = 10^{-30}$ means that only reactants are present at equilibrium.

3. If K equals 1, neither a forward nor a reverse reaction is favored; both reactants and products present at equilibrium.

As an example, let us consider the combustion reaction of hydrogen:

$$2H_2(g) + O_2(g) \leftrightarrows 2H_2O(g) \tag{2.16}$$

The equilibrium-constant expression is

$$K_p = \frac{p^2_{[H_2O]}}{p^2_{[H_2]}p_{[O_2]}} = 1.35 \times 10^{80} \tag{2.17}$$

The value of K_p indicates that there should be infinitesimal amounts of hydrogen and oxygen gases at equilibrium. The equilibrium mixture should be nearly pure water.

2.1.2 Le Chatelier's Principle: The Effect of Changing Conditions on Equilibria

At equilibrium, the system is in a state of balance; the forward and reverse processes are occurring at equal rates. What happens to this dynamic state of balance if we disturb the system, changing conditions such as temperature, pressure, or the concentration of a reactant or product? We can understand these effects in terms of a principle first put forward by Henry-Louis Le Chatelier (1850–1936), a French industrial chemist. Le Chatelier's principle can be stated as follows: If a system at equilibrium is disturbed by a change in temperature, pressure, or the concentration of one of the components, the system will shift its equilibrium position so as to counteract the effect of the disturbance. We will use the Le Chatelier's principle to discuss how a system at equilibrium responds to three major changes in external conditions: (a) adding or removing a reactant or product, (b) changing the pressure, and (c) changing the temperature.

2.1.2.1 Change in Reactant or Product Concentrations What will happen if the concentration of one of the reactants or products changes? Let us consider the general case of chemical reactions and the first case—increasing the concentration of the reactant A.

$$aA + bB + \cdots \leftrightarrows cC + dD + \cdots \tag{2.18}$$

According to Le Chatelier's principle, the reaction will relieve the "stress" and consume the added reactant A. As a result, the forward reaction will be favored so the value of the products will increase; the concentration of the reactant B will be lower because the reactant B will react with the added reactant A. However, the reverse reaction will also speed up until the new equilibrium state is established with the same value of equilibrium constant K. We can make the same conclusion considering the algebraic equation describing K:

...then this increases

This remains constant → $K = \dfrac{[C]^c \cdot [D]^d}{[A]^a \cdot [B]^b}$ (2.19)

If this increases... ...then this decreases...

Following in the same manner, we can see that removing the product (or products) would also cause a shift in the reaction to the right, producing more products, whereas adding C or D to the system at equilibrium would cause the reaction to shift to the left, reducing the greater C or D concentrations. In both cases, again, the value of K remains unchanged:

If this decreases (removed)

$$\text{This remains constant} \rightarrow K = \frac{[C]^c \cdot [D]^d}{[A]^a \cdot [B]^b} \tag{2.20}$$

...then this decreases (consumes)

Decreasing the concentration of one of the reactants will cause the reverse reaction to be favored:

...then this decreases

$$\text{This remains constant} \rightarrow K = \frac{[C]^c \cdot [D]^d}{[A]^a \cdot [B]^b} \tag{2.21}$$

If this decreases...

The Haber reaction $3H_2(g) + N_2(g) \leftrightarrows 2NH_3(g)$ is a good example to demonstrate the factors that might be varied to increase the yield of ammonia. Adding more reactants, N_2 and H_2, to an equilibrium mixture of N_2, H_2, and NH_3, or removing NH_3 from the equilibrium mixture causes the reaction to form more NH_3. In the industrial production of ammonia, the product is continuously removed by selectively liquefying it so that the reaction is driven close to completion.

2.1.2.2 *Effect of Pressure Change*

Le Chatelier's principle indicates that the system where one or more of the substances involved is a gas will respond to an increase in pressure by shifting the equilibrium in the direction that decreases the total number of gas molecules and thus decreases the pressure. Conversely, decreasing the pressure will favor the reaction that produces more gas molecules.

As an example, let us consider the ammonia synthesis equilibrium:

$$\underbrace{N_2(g) + 3H_2(g)}_{\text{4 mol of gas}} \rightleftarrows \underbrace{2NH_3(g)}_{\text{2 mol of gas}} \tag{2.22}$$

In this case, 4 mol of gas is converted to 2 mol of gas. What happens if we increase the total pressure of equilibrium mixture by decreasing, for example, the volume? According to Le Chatelier's principle, we expect that increasing the pressure will favor the forward reaction, formation of ammonia, because it will decrease the total number of reactants and thus decrease the pressure relieving the stress. Figure 2.4 represents the industrial Haber

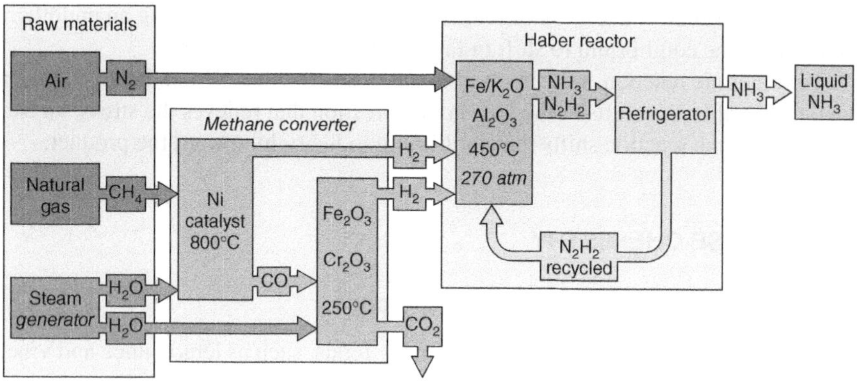

Figure 2.4 The Haber process for ammonia synthesis. [*Source*: Olmsted and Williams (2006) [103]. Reproduced with permission of John Wiley & Sons.]

process for ammonia synthesis where the effect of pressure change and change in reactant and product concentrations are used to run the reaction to essential completion.

Considering the formation reaction of hydrogen fluoride from hydrogen and fluorine, when the sum of the moles of reactants and the sum of moles of products are equal, we can predict the following:

$$\underbrace{H_2(g) + F_2(g)}_{\text{2 mol of gas}} \rightleftarrows \underbrace{2HF(g)}_{\text{2 mol of gas}} \tag{2.23}$$

and the equilibrium will not change when the pressure of the system is increased or decreased.

2.1.2.3 *Effect of Temperature Changes* We can apply Le Chatelier's principle to deduce rules for the temperature dependence of equilibrium. A simple way to do this is to think of heat as a reactant or product that increases or decreases the equilibrium.

We can consider heat as a reactant in an endothermic reaction (heat is absorbed) and heat as a product in an exothermic reaction (heat is released).

a) Endothermic reaction:

$$a \cdot \mathrm{A} + b \cdot \mathrm{B} + \mathrm{Heat} \leftrightarrows c \cdot \mathrm{C} + d \cdot \mathrm{D} \quad (\Delta H > 0) \tag{2.24}$$

Example:

$$N_2O_4(g) \leftrightarrows 2NO_2(g) \quad (\Delta H = 58\ \mathrm{kJ/mol}) \tag{2.25}$$

b) Exothermic reaction:

$$a \cdot \mathrm{A} + b \cdot \mathrm{B} \leftrightarrows c \cdot \mathrm{C} + d \cdot \mathrm{D} + \mathrm{Heat} \quad (\Delta H < 0) \tag{2.26}$$

Example:

$$Cl_2(g) + H_2(g) \leftrightarrows 2HCl(g) \quad (\Delta H = -44\ \mathrm{kcal/mol}) \tag{2.27}$$

In an endothermic reaction, an increase in temperature at equilibrium occurs as we add a reactant that causes the equilibrium to shift to the right to favor the product formation, the direction that consumes the excess of reactant (heat). Conversely, cooling an endothermic reaction causes the equilibrium to shift in favor of the reverse reaction.

In an exothermic reaction, we can observe the opposite effect. When heat is added as a product, the equilibrium shifts to the left, in the direction that relieves the stress, an excess of heat. Cooling the reaction shifts the equilibrium to the right, toward the product.

2.2 ACID–BASE CHEMISTRY

2.2.1 A Brief Review: The Arrhenius Concept

Since ancient times, people have known that many foods, such as lemon juice and vinegar, taste sour because they are acids; and some foods taste bitter and feel slippery because they are bases. People have long been familiar with some properties of acids and bases: dissolving certain metals with formation of hydrogen and reacting with a number of organic dyes. But what makes a substance behave as an acid or a base? It was not before the 19th century that two useful definitions of acids and bases were introduced.

The early definitions of acids and bases are based on a suggestion that, on being dissolved in water, they are separated into their individual ions. The Swedish scientist Svante Arrhenius defined acids as substances that release hydrogen ions into water. For example, hydrochloric acid dissolves in water, releasing hydrogen ion (proton) and chloride anion:

$$HCl(aq) \rightarrow H^+(aq) + Cl^-(aq) \tag{2.28}$$

Bases were defined as substances that dissolve in water to release hydroxide ions, OH^-, into water solution. Sodium hydroxide, NaOH, is a typical Arrhenius's base:

$$NaOH(aq) \rightarrow Na^+(aq) + OH^-(aq) \tag{2.29}$$

With Arrhenius's theory, a number of things were explained. Among them are why all acids have similar properties to each other and why all bases are similar, and why acids and bases react to counteract each other (we will discuss the reaction of neutralization in the following section). However, this theory has limits. For example, it does not explain why some substances, such as ammonia, can act like a base even though they do not contain hydroxide ions.

Before we explore definitions of acids and bases, we have to note two important things. The first is why we call hydrogen ion a proton. The proton is the ion that remains when a hydrogen atom loses its single electron (Figure 2.5). The second is that a proton does not exist in water in this simple form; a proton interacts with the lone pair of electrons of a water molecule to form a hydrated hydrogen ion, which is called a hydronium ion H_3O^+:

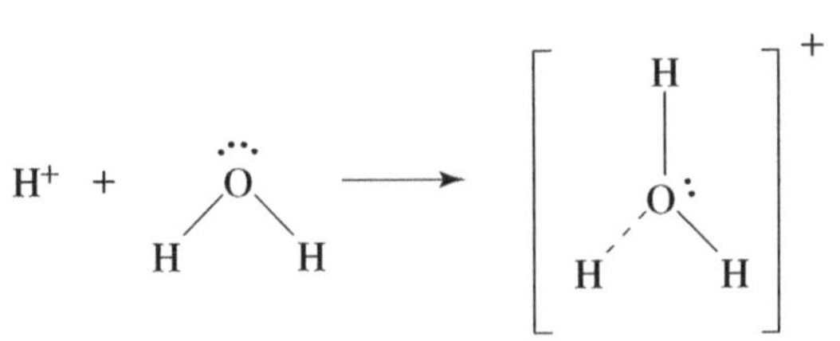

(2.30)

Hydronium ion H_3O^+

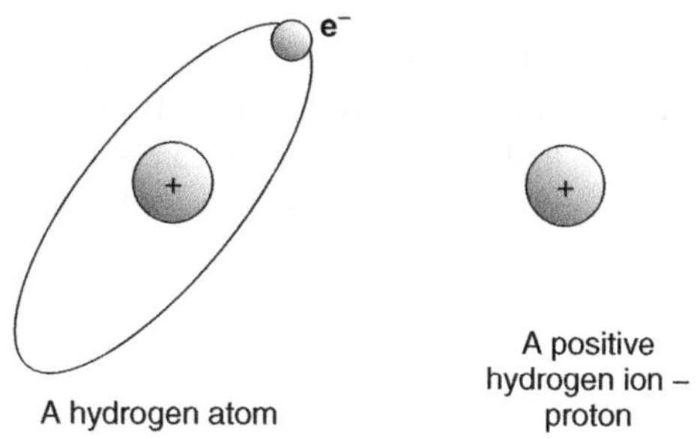

Figure 2.5 A hydrogen atom and a positive hydrogen ion (a proton).

Knowing this, we will, however, often use the proton for simplicity and convenience as a species to represent acidity in aqueous solutions.

2.2.2 Brønsted–Lowry Acids and Bases, Proton-Transfer Reactions

The more comprehensive and general definition of acid and base was independently proposed in 1923 by the Danish chemist Johannes Brønsted and the English chemist Thomas Lowry. A Brønsted–Lowry acid is any substance that is able to give a proton to another molecule or ion; an acid is a proton donor. A Brønsted–Lowry base is any substance that can accept H^+ from an acid. For example, in the reaction between hydrogen chloride (HCl) and ammonia (NH_3), HCl acts as an acid donating a proton, and ammonia acts as a base accepting the proton from the acid. This reaction can run not only in water but also in the gas phase to form solid NH_4Cl:

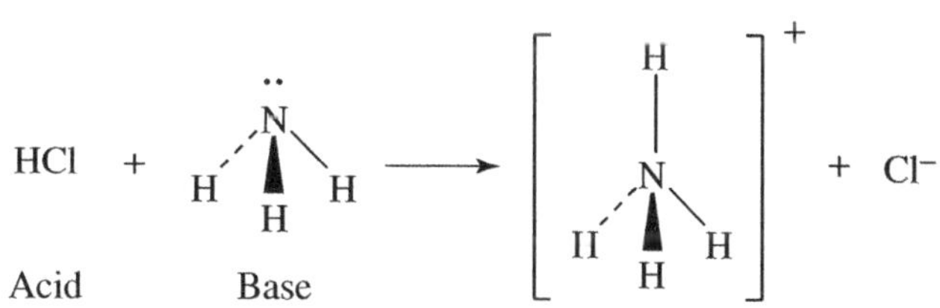

(2.31)

Note that the theory also explains behavior of compounds in water. For example, nitric acid, defined as Arrhenius acid, acts also as a Brønsted–Lowry acid, donating a proton to a molecule of water. A molecule of water is an acceptor of a proton, so water is a Brønsted–Lowry base:

$$HNO_3(aq) + H_2O(l) \rightarrow NO_3^-(aq) + H_3O^+(aq) \tag{2.32}$$

According to Brønsted–Lowry, ammonia in water is a base (proton acceptor) and water is an acid (proton donor):

$$NH_3(aq) + H_2O(l) \leftrightarrows NH_4^+(aq) + OH^-(aq) \tag{2.33}$$

We can also consider ammonia as an Arrhenius base because adding it to water leads to an increase in the concentration of OH^-.

As we see, a Brønsted–Lowry acid and a base always work together to transfer a proton. In other words, a substance can act as an acid only if another substance acts as a base, and this process is called a proton-transfer reaction. To be a Brønsted–Lowry acid, a species (molecule or ion) should have a hydrogen atom, which can be lost as an H^+; in contrast, to be a Brønsted–Lowry base, a species must have a lone pair of electrons and be able to form a coordinate covalent bond, without which it cannot accept H^+ from an acid.

Some molecules or ions can function as an acid or a base depending on the circumstances. For example, water is neither an Arrhenius acid nor an Arrhenius base because it does not donate an appreciable concentration of proton or hydroxide ions to a solution; however, in some reactions water can behave either as Brønsted–Lowry acid or base. For example, water acts as an acid with ammonia (2.33) and as a base with nitric acid (2.32). Substances such as water that can act as an acid and as a base are called *amphoteric*.

Another example of amphoteric species is the hydrogen phosphate anions:

$$HPO_4^{2-}(aq) + H_2O(l) \leftrightarrows H_3O^+(aq) + PO_4^{-3}(aq) \quad (2.34)$$

$$HPO_4^{2-}(aq) + H_2O(l) \leftrightarrows H_2PO_4^-(aq) + OH^-(aq) \quad (2.35)$$

A large number of Brønsted–Lowry acids and bases are known. Acids that are capable of donating one proton are called monoprotic acids. Well-known monoprotic acids are hydrochloric acid (HCl), nitric acid (HNO_3), and acetic acid (CH_3COOH). Acids that are capable of donating more than one proton are called polyprotic acids. For example, sulfuric acid H_2SO_4 is a diprotic acid and phosphoric acid H_3PO_4 is a triprotic acid.

2.2.3 Relative Strengths of Acids and Bases, Equilibrium Constants for Acids and Bases

Acids differ in their ability to donate a proton, and bases differ in their ability to accept a proton or donate a hydroxide ion to a water solution. Table 2.1 indicates relative strengths of acids and bases.

Strong acids and bases exist in an aqueous solution entirely as ions. In other words, they are completely 100% ionized, or dissociated. Strong acids and bases are strong electrolytes due to the fact that the presence of ions can be indicated by the ability of solutions to conduct electricity. There are just a few strong acids in existence: six monoprotic acids, HCl, HBr, HI, HNO_3, $HClO_3$, and $HClO_4$, and one diprotic acid, H_2SO_4. For example, nitric acid usually completely ionizes in water; therefore, we do not use equilibrium arrows; the reaction lies completely to the right forming ions:

$$HNO_3(aq) + H_2O(l) \rightarrow NO_3^-(aq) + H_3O^+(aq) \quad (2.36)$$

In general, dissociation of a "strong" acid can be described as follows:

$$HA(aq) \rightarrow H^+(aq) + A^-(aq) \quad (2.37)$$

The situation with the diprotic acid H_2SO_4 is more complex: sulfuric acid is only a strong acid in regard to its capability of donating its first proton:

$$H_2SO_4(aq) + H_2O(l) \leftrightarrows HSO_4^-(aq) + H_3O^+(aq) \quad (2.38)$$

TABLE 2.1 Acid dissociation constants, K_a, for some acids (25°C)

Acid	Conjugate base	Equilibrium reaction	Strength of base K_a	K_b
Sulfuric (H_2SO_4)	HSO_4^-	$H_2SO_4(aq) + H_2O(l) \rightleftarrows H_3O^+(aq) + HSO_4^-(aq)$	Large	Very small
Hydronium ion (H_3O^+)	H_2O	$H_3O^+(aq) + H_2O(l) \rightleftarrows H_2O(l) + H_3O^+(aq)$	1.0	1.0×10^{-14}
Hydrogen sulfate ion (HSO_4^-)	SO_4^{2-}	$HSO_4^-(aq) + H_2O(l) \rightleftarrows H_3O^+(aq) + SO_4^{2-}(aq)$	1.2×10^{-2}	8.3×10^{-13}
Phosphoric acid (H_3PO_4)	$H_2PO_4^-$	$H_3PO_4(aq) + H_2O(l) \rightleftarrows H_3O^+(aq) + H_2PO_4^-(aq)$	7.5×10^{-3}	1.3×10^{-12}
Hydrofluoric (HF)	F^-	$HF(aq) + H_2O(l) \rightleftarrows H_3O^+(aq) + F^-(aq)$	6.8×10^{-4}	1.4×10^{-11}
Formic (HCOOH)	$HCOO^-$	$HCOOH(aq) + H_2O(l) \rightleftarrows H_3O^+(aq) + HCOO^-(aq)$	1.8×10^{-4}	5.6×10^{-11}
Acetic (CH_3COOH)	CH_3COO^-	$CH_3COOH(aq) + H_2O(l) \rightleftarrows H_3O^+(aq) + CH_3COO^-(aq)$	1.8×10^{-5}	5.6×10^{-10}
Carbonic (H_2CO_3)	HCO_3^-	$H_2CO_3(aq) + H_2O(l) \rightleftarrows H_3O^+(aq) + HCO_3^-(aq)$	4.3×10^{-7}	2.4×10^{-8}
Dihydrogen phosphate ion ($H_2PO_4^-$)	HPO_4^{2-}	$H_2PO_4^-(aq) + H_2O(l) \rightleftarrows H_3O^+(aq) + HPO_4^{2-}(aq)$	6.2×10^{-8}	1.6×10^{-7}
Ammonium ion (NH_4^+)	NH_3	$NH_4^+(aq) + H_2O(l) \rightleftarrows H_3O^+(aq) + NH_3(aq)$	5.6×10^{-10}	1.8×10^{-5}
Hydrocyanic (HCN)	CN^-	$HCN(aq) + H_2O(l) \rightleftarrows H_3O^+(aq) + CN^-(aq)$	4.9×10^{-10}	2.5×10^{-5}
Hydrogen phosphate ion (HPO_4^{2-})	PO_4^{3-}	$HPO_4^{2-}(aq) + H_2O(l) \rightleftarrows H_3O^+(aq) + PO_4^{-3}(aq)$	3.6×10^{-13}	2.8×10^{-2}
Water (H_2O)	OH^-	$H_2O(l) + H_2O(l) \rightleftarrows H_3O^+(aq) + OH^-(aq)$	1.0×10^{-14}	1.0
Hydrogen (H_2)	H^-	$H_2(g) + H_2O(l) \rightleftarrows H_3O^+(aq) + H^-(aq)$	Very small	Large
Ethanol (C_2H_5OH)	$C_2H_5O^-$	$C_6H_5OH(aq) + H_2O(l) \rightleftarrows H_3O^+(aq) + C_2H_5O^-(aq)$	Very small	Large
		Strength of acid		

The second proton can be donated to the solution as well, but not completely, and as a result the hydrogen sulfate ion HSO_4^- gives up a proton with a great difficulty by representing the "weak" acid. An acid is defined as a weak acid if it ionizes to a lower extent (between 1% and 5%); therefore, most of the species (molecules or polyatomic ions) will stay nonionized in water. All organic acids, such as acetic acid (CH_3OOH) and propanoic acid (CH_3CH_2COOH), and many inorganic acids, such as nitrous acid (HNO_2) and carbonic acid (H_2CO_3), are weak acids.

In a similar way, the bases are strong if they are completely ionized in water into metal cations and hydroxide ions. They usually accept a proton easily and hold it tightly (it is said that they have a high affinity for H^+). All monohydroxic and three dihydroxic bases, $Ca(OH)_2$, $Sr(OH)_2$, and $Ba(OH)_2$, are the strong bases. For example, $Ba(OH)_2$ ionizes in water completely,

$$Ba(OH)_2(aq) \rightarrow Ba^{2+}(aq) + 2(OH)^-(aq) \tag{2.39}$$

In general, dissociation of a strong base can be described follows:

$$M(OH)_n(aq) \rightarrow M^{n+}(aq) + nOH^-(aq) \quad (2.40)$$

A weak base is a base that has only a slight affinity for H^+ and holds it weakly. An example of a weak base is ammonia. Usually this reaction is described by using a double arrow, indicating dissolving ammonia in water as equilibrium:

$$NH_3(aq) + H_2O(l) \leftrightarrows NH_4^+(aq) + OH^-(aq) \quad (2.41)$$

There are a few ways to estimate the relative strengths of acids and bases, especially the weak acids and bases. One is to describe the reaction of weak acids and bases in water using an equilibrium equation, and then to find the equilibrium constant that in turn will show the relative concentrations of ions at equilibrium. For the general reaction

$$HA\ (aq) + H_2O\ (l) \leftrightarrows H_3O^+(aq) + A^-(aq) \quad (2.42)$$

the equilibrium constant is equal to

$$K = \frac{[H_3O^+] \cdot [A^-]}{[HA] \cdot [H_2O]} \quad (2.43)$$

As water is acting as a solvent, its concentration will be constant and has no effect on the equilibrium. Based on this conclusion, we can put water concentration and value of the K together and get a new constant called the acid dissociation constant K_a:

$$K_a = K[H_2O] = \frac{[H_3O^+] \cdot [A^-]}{[HA]} \quad (2.44a)$$

Similarly, we can write the same general equilibrium expression for a weak base B in water:

$$B(aq) + H_2O\ (l) \leftrightarrows BH^+(aq) + OH^-(aq) \quad (2.44b)$$

$$K_b = \frac{[BH^+] \cdot [OH^-]}{[B]} \quad (2.44c)$$

Table 2.1 shows K_a values for some weak acids and K_b for some weak bases. The table illustrates several important conclusions:

- The larger value of K_a constitutes a stronger acid.
- Strong acids have K_a values much greater than 1, which means that dissociation is favored. For strong acids, such as HCl or HNO_3, the K_a values are so large that it is not useful to measure them.
- Weak acids have K_a values much less than 1, which means that dissociation is not favored.
- Removing the second proton from a polyprotic acid is always more difficult than the first one and, as a result, the K_a becomes lower.

- Most organic acids with carboxylic group COOH have K_a with the magnitude less than 10^{-3}.
- As the strength of an acid increases (larger K_a), the strength of its conjugate base decreases (smaller K_b).

2.2.4 Acid–Base Equilibrium, Conjugate Acid–Base Pairs, General Trends in Acid–Base Reactions

In each of the chemical equations we have written so far, we can see an important consequence of the Brønsted–Lowry definitions: the products of an acid–base reaction are themselves acids and bases. For example, consider the reaction of an H-A acid that donates a proton to the base B: and therefore H-A is a Brønsted–Lowry acid and B: is a Brønsted–Lowry base:

$$\text{H-A} + \text{B} :\leftrightarrows: \text{A}^- + \text{B}^+\text{H}$$

$$\text{Acid} \quad \text{Base} \quad \text{Base} \quad \text{Acid} \tag{2.45}$$

Reaction between products is also possible; the product B^+H can donate a proton to the $:A^-$, so B^+H acts as the Brønsted–Lowry acid and: A^- acts as the Brønsted–Lowry base. That is, those acid–base reactions are reversible; in any acid–base equilibrium, both the forward reaction and the reverse reaction involve proton transfer. A pair of acid and base constituents such as H-A and: A^- or B^+H and B: that differ only in the presence or absence of one H^+ are called *a conjugate acid–base pair*.

Every reaction between Brønsted–Lowry acid and base has two sets of conjugate acid–base pairs. For example, consider the following reactions and those in Table 2.1.

Conjugate pair 1

$$PO_4^{3-}(aq) + H_2O(l) \leftrightarrows HPO_4^{2-}(aq) + OH^-(aq) \tag{2.46}$$

Conjugate pair 2

Conjugate pair 1

$$H_2S(aq) + H_2O(l) \leftrightarrows H_3O^+(aq) + HS^-(aq) \tag{2.47a}$$

Conjugate pair 2

Acid Base Conj. acid Conj. base

It is possible to predict the direction of proton-transfer reactions considering the relative strength of individual bases or acids. For example, we can think of the reactions as being governed by the relative abilities of the base on the right side of the equation and the left side of the equation accepting the protons. Consider, for example, a proton transfer reaction between an acid HA and water:

Forward reaction ⟶

Brønsted–Lowry bases

$$\underset{\text{Acid}}{HA(aq)} + \underset{\text{Base}}{H_2O(l)} \rightleftharpoons \underset{\text{Conj. acid}}{H_3O^+(aq)} + \underset{\text{Conj. base}}{A^-(aq)} \tag{2.47b}$$

Brønsted–Lowry acids

⟵ Reverse reaction

If the water molecules (the base in the forward reaction) are a stronger base than A^- (the conjugate base of HA), then H_2O will abstract the proton from HA to produce H_3O^+ and A^-. In this case, the equilibrium will dominate to the right. This describes the behavior of a strong acid in water. For example, nitric acid is completely ionized in water, forming H_3O^+ and NO_3^- ions, because H_2O is a stronger base than NO_3^-, so H_2O accepts the proton to form hydronium ion:

Brønsted–Lowry acids

$$\underset{\text{Acid}}{HNO_3(aq)} + \underset{\text{Stronger base}}{H_2O(l)} \rightarrow \underset{\text{Acid}}{H_3O^+(aq)} + \underset{\text{Weaker base}}{NO_3^-(aq)} \tag{2.48}$$

Brønsted–Lowry bases

If A^- is a stronger base than H_2O, the equilibrium will dominate to the left. This describes the behavior of a weak acid in water. For example, when acetic acid CH_3COOH dissolves in water, the solution consists mainly of CH_3COOH molecules with a low concentration of H_3O^+ and CH_3COO^- ions. It occurs because the CH_3COO^- ion is a stronger base than water and more easily accepts a proton from H_3O^+.

Brønsted–Lowry bases

$$CH_3COOH(aq) + \underset{\text{Weaker base}}{H_2O(l)} \rightleftharpoons H_3O^+(aq) + \underset{\text{Stronger base}}{CH_3COO^-(aq)} \tag{2.49a}$$

Brønsted–Lowry acids

For the following proton-transfer reaction, we determine that the CO_3^{2-} represents the stronger base than SO_4^{2-} (the conjugate base of HSO_4^-), and therefore the equilibrium will lie predominantly to the right, favoring products:

Brønsted–Lowry bases

$$HSO_4^-(aq) + CO_3^{2-}(aq) \rightleftarrows SO_4^{2-}(aq) + HCO_3^-(aq) \quad (2.49b)$$

Stronger base — Weaker base

Brønsted–Lowry acids

As a result, we can conclude that in every acid–base reaction the position of equilibrium favors transfer of the proton to the stronger base.

2.2.5 Dissociation of Water and pH Scale

The process of autoionization of water was discovered many years ago by Friedrich Kohlrausch (1840–1910) who found that very pure water still conducts electricity to a very small extent because of very low concentrations of H_3O^+ and OH^- ions. Water autoionization reaction is a result donating proton from one molecule of water to another,

$$2H_2O\ (l) \leftrightarrows H_3O^+(aq) + OH^-(aq) \quad (2.50)$$

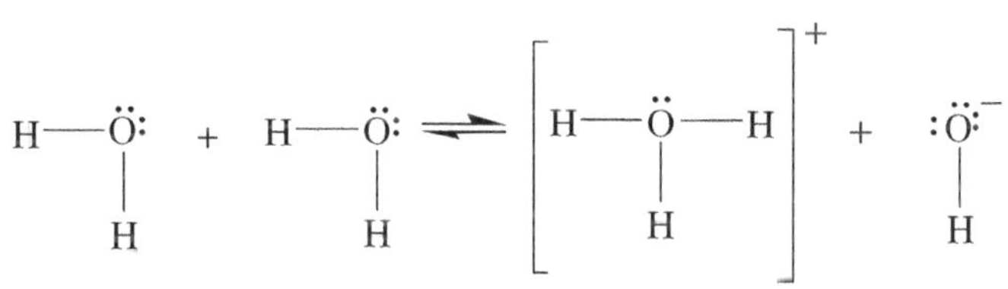

(2.51)

The autoionization of water demonstrates its ability to act as either a Brønsted–Lowry acid or Brønsted–Lowry base; therefore, we determine that water is amphoteric. The autoionization of water is an equilibrium process that lies far to the left side. In fact, in pure water at room temperature, only 2 out of every 10^9 molecules are ionized at any instant. The equilibrium-constant expression for autoionization is

$$K = \frac{[H_3O^+] \cdot [OH^-]}{[H_2O]^2} \quad (2.52)$$

The concentration of water as a solvent (pure liquid) can be considered a constant and excluded from the equilibrium-constant expression. We use symbol K_w, to denote the new equilibrium constant, called the ion-product constant or ionization constant for water:

$$K_w = [H_3O^+][OH^-] \quad (2.53)$$

Using pure water electrical conductivity measurements, we define that $[H_3O^+] = [OH^-] = 1.0 \times 10^{-7}$ M at 25°C; therefore,

$$K_w = [H_3O^+][OH^-] = 1.0 \times 10^{-14} \quad (2.54)$$

This equation is very important because it can be applied to all aqueous solutions, not just to pure water. Since the product of $[H_3O^+]$ and $[OH^-]$ is always constant $(1{:}0 \times 10^{-14})$ for any solution, we can calculate $[H_3O^+]$ or $[OH^-]$ if we know one of the constituents:

$$[H_3O^+] = 1.0 \times 10^{-14}/[OH^-] \quad \text{and} \quad [OH^-] = 1.0 \times 10^{-14}/[H_3O^+]$$

According to the value of $[H_3O^+]$ and $[OH^-]$, the aqueous solutions would be classified as neutral, acidic, or basic if the following conditions are satisfied:

- In a neutral solution, $[H_3O^+] = 1.0 \times 10^{-7}$ M and $[OH^-] = 1.0 \times 10^{-7}$ M.
- In an acidic solution, $[H_3O^+] > [OH^-]$; $[H_3O^+] > 1.0 \times 10^{-7}$ M and $[OH^-] < 1.0 \times 10^{-7}$ M.
- In a basic solution, $[H_3O^+] < [OH^-]$; $[H_3O^+] < 1.0 \times 10^{-7}$ M and $[OH^-] > 1.0 \times 10^{-7}$ M.

There is an easier way to express and compare H_3O^+ concentration using pH scale. The pH of a solution could be defined as the negative of the base-10 logarithm for the hydronium concentration:

$$\mathrm{pH} = -\log[H_3O^+] \tag{2.55}$$

We also can obtain another useful equation if we use the negative logarithms of both sides in the expression $K_w = [H_3O^+][OH^-] = 1.0 \times 10^{-14}$:

$$\mathrm{p}K_w = \mathrm{pH} + \mathrm{pOH} = 14 \text{ at } 25^\circ\mathrm{C} \tag{2.56}$$

Thus, for aqueous solutions, the conclusion is as follows:

- In a neutral solution, pH = 7 and $[H_3O^+] = 1.0 \times 10^{-7}$ M.
- In an acidic solution, pH < 7 and $[H_3O^+] > 1.0 \times 10^{-7}$ M.
- In a basic solution, pH > 7 and $[H_3O^+] < 1.0 \times 10^{-7}$ M.

Figure 2.6 demonstrates the pH scale and the values of some common substances.

2.2.6 Buffer Solutions

Buffers are extremely important in all areas of science. Biochemists are particularly concerned with buffers because the proper functioning of any biological system depends on the pH.

For the reaction described as follows,

$$\underset{\text{Acid}}{HA(aq)} + \underset{\text{Base}}{H_2O(l)} \leftrightarrows \underset{\text{Conj. acid}}{H_3O^+(aq)} + \underset{\text{Conj. base}}{A^-(aq)} \tag{2.57}$$

we can calculate

$$K_a = \frac{[H_3O^+][A^-]}{[HA]} \tag{2.58}$$

and

$$[H_3O^+] = K_a\frac{[HA]}{[A^-]} \tag{2.59}$$

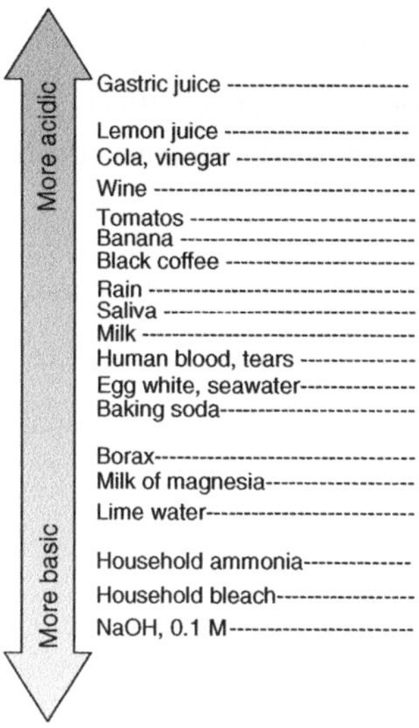

$[H^+]$ (M)	pH	pOH	$[OH^-]$ (M)
←$1(1\times10^{-0})$	0.0	14.0	$1X10^{-14}$
←1×10^{-1}	1.0	13.0	$1X10^{-13}$
←1×10^{-2}	2.0	12.0	$1X10^{-12}$
←1×10^{-3}	3.0	11.0	$1X10^{-11}$
←1×10^{-4}	4.0	10.0	$1X10^{-10}$
←1×10^{-5}	5.0	9.0	$1X10^{-9}$
←1×10^{-6}	6.0	8.0	$1X10^{-8}$
←1×10^{-7}	7.0	7.0	$1X10^{-7}$
←1×10^{-8}	8.0	6.0	$1X10^{-6}$
←1×10^{-9}	9.0	5.0	$1X10^{-5}$
←1×10^{-10}	10.0	4.0	$1X10^{-4}$
←1×10^{-11}	11.0	3.0	$1X10^{-3}$
←1×10^{-12}	12.0	2.0	$1X10^{-2}$
←1×10^{-13}	13.0	1.0	$1X10^{-1}$
←1×10^{-14}	14.0	0.0	$1(1X10^{0})$

Figure 2.6 H^+ concentrations and pH values of some common substances at 25°C.

The pH also could be determined if we use the following equation:

$$pH = pK_a + \log\frac{[A^-]}{[HA]} \tag{2.60}$$

Equation 2.60 is the central equation for buffers, named the "Henderson–Hasselbalch equation." It shows the pH of a solution, the ratio of the concentrations of conjugate acid and base, and the pK_a for the acid. If a solution is prepared from the weak base B and its conjugate acid BH^+, the analogous equation will look like this:

$$pH = pK_a + \log\frac{[B]}{[BH^+]_{\leftarrow pK_a \text{ applies to this acid}}} \tag{2.61}$$

As we see, in an acidic buffer, $[H_3O^+]$ and pH depend on the ratio [HA] to $[A^-]$. One typical example of a buffer system is an acetic acid/acetate, for example, CH_3COOH/CH_3COONa; this is the acetic acid–acetate ion buffer:

$$CH_3COOH\ (aq) + H_2O(l) \leftrightarrows H_3O^+(aq) + CH_3COO^-(aq) \tag{2.62}$$

$$[H_3O^+] = K_a\frac{[CH_3COOH]}{[CH_3COO^-]} \tag{2.63}$$

If a small amount of OH^- is added to the system (e.g., NaOH), the acid component (free acetic acid) will neutralize the added OH^-. If an acid (e.g., HCl) is added to the system, the base component (the basic acetate, which is the conjugate base) neutralizes the added H_3O^+. In both cases, the pH remains relatively constant.

Other common examples of buffer systems are as follows:

- HF/KF, hydrogen fluoride/fluoride (an acidic buffer)
- $NaHCO_3/Na_2CO_3$, sodium bicarbonate/carbonate (an acidic buffer)
- $H_2PO_4^-/HPO_4^{2-}$, dihydrogen phosphate/hydrogen phosphate (an acidic buffer)
- NH_3/NH_4Cl, ammonia/ammonium salt (a basic buffer).

There are some important conditions required for the solution to work as a buffer:

- The ratio $[HA]/[A^-]$ should be close to 1.
- The values of [HA] and $[A^-]$ in the buffer solution should be approximately 10 times as high as the amounts of either acid or base added.

2.3 CHEMICAL THERMODYNAMICS

2.3.1 Introduction

Chemical thermodynamics is concerned with the transformation of various forms of energy and heat within chemical systems. Changes in thermodynamic quantities will be calculated as a system converts from one state to another. One example is the heat evolved or absorbed in a chemical reaction. Other quantities, such as entropy and Gibbs free energy, provide useful information about the spontaneity of a chemical change. Specific examples are given for use in the chemistry of hydrogen technology.

2.3.2 Thermodynamic System and Changes in State

A thermodynamic system is that part of the universe whose properties are under investigation. The system is confined to a definite place in space by the boundaries that separate it from the rest of the universe, the surroundings. The properties of a system are those physical quantities that are perceived by the senses or by experimental methods of investigation. The properties of a gas include pressure (P), volume (V), temperature (T, in K), and number of moles (n) that are related to one another through an equation of state, such as the ideal gas equation

$$PV = nRT \tag{2.64}$$

where R is the ideal gas constant. A system is in a definite state when each of its properties has a definite value or number. A system undergoes a change in its state when it goes from a specified initial state or set of numerical values (P_i, V_i, T_i, n_i) to a specified final state with a different set of numerical values (P_f, V_f, T_f, n_f).

2.3.3 Work and Heat

Work, W, is any quantity that flows across the boundary of a system during a change in state and is completely convertible into the lowering of a weight ($w = mg$ where m is the mass and g is the gravitational acceleration) in the surroundings. Work is positive when it is done by the surroundings on the system, hence, lowering a weight in the surroundings by a distance h.

$$W = mgh \tag{2.65}$$

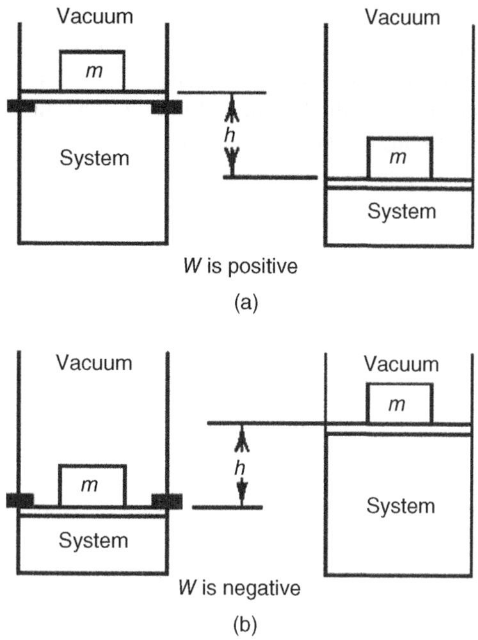

Figure 2.7 (a) When the stoppers are pulled out, the mass in the surroundings compresses the system and does a positive amount of work on the system. (b) When the stoppers are pulled out, the system lifts a mass in the surroundings and the system does work on the surroundings.

Figure 2.7 shows examples of a change in the state of a system when (a) work is done on a gaseous system containing hydrogen (positive work) and (b) when the system does work on the surroundings by lifting a mass in the surroundings (negative work). Often, the pressure of the surroundings provides the force per unit area (mg/A) to decrease the volume of the system ($\Delta V = hA$) when a positive amount of work is done on the system.

$$W = -P\Delta V \tag{2.66}$$

Heat, q, is a quantity that flows across the boundary of a system during a change in state because of a difference in temperature between the system and its surroundings and flows from a point of higher to a point of lower temperature. Heat is positive when it has flowed from the surroundings to the system.

The amounts of heat and work transferred to the system depend on the pathways of the processes used during the change in state.

2.3.4 Internal Energy and the First Law of Thermodynamics

If heat and work both flow from the surroundings to the system, then the system experiences a change in state and an increase in its internal energy (E).

$$\Delta E = q + W \tag{2.67}$$

In contrast to heat and work, the change in internal energy is equal to the difference between the internal energy of the final and initial states ($\Delta E = E_f - E_i$). Internal energy is said to be a thermodynamic function and is independent of the pathway during the change in state.

Equation 2.67 is a statement of the first law of thermodynamics. All of the energy that has flown from the surroundings is now in the system. The surroundings lost an amount of energy ΔE that has been gained by the system. The first law is also referred to as the principle of conservation of energy where the energy of the universe is a constant. Energy has not been created or destroyed but has been transferred from the surroundings to the system with the total energy of the universe being the same before and after the change.

2.3.5 Heat of Reaction at Constant Volume and Constant Pressure

If the volume of a system is kept constant, as in a bomb calorimeter, which is a strong metal vessel with a screw-down top and a valve arrangement through with the gases may be introduced, no work is done as the reaction occurs since there is no change in volume. The heat evolved for a reaction that gives off energy (exothermic) at constant volume (q_v) in a bomb calorimeter may be determined from the increase in temperature (ΔT) and the heat capacity of the system (C_v), which is the amount of heat required to change the temperature of the system by 1°C. The heat evolved will then be directly related to the change in internal energy, which is the heat of the reaction at constant volume.

$$\Delta E = q_v = C_v \Delta T \tag{2.68}$$

Many chemical reactions are often carried out open to the atmosphere, at constant atmospheric pressure. Therefore, the measured heat of reaction is at constant pressure and not at constant volume.

$$\Delta E = q_p - P\Delta V \quad \text{or} \quad q_p = \Delta E + P\Delta V \tag{2.69}$$

The heat of reaction at constant pressure is defined as an enthalpy change where enthalpy (H) is a thermodynamic function independent of the pathway and defined as

$$H = E + PV \tag{2.70}$$

The heat of reaction, ΔH, is determined from the temperature change and the heat capacity of the system at constant pressure (C_p).

$$\Delta H = q_p = C_p \Delta T \tag{2.71}$$

The heat given off under standard state conditions for the combustion of gaseous hydrogen to form liquid and gaseous water has been measured, and their values are shown as follows. The negative signs for the heats of reaction indicate exothermic reactions when heat is given off.

$$H_2(g) + 1/2O_2(g) \rightarrow H_2O(l) \quad \Delta H^\circ = -285.83 \text{ kJ/mol} \tag{2.72}$$

$$H_2(g) + 1/2O_2(g) \rightarrow H_2O(g) \quad \Delta H^\circ = -241.82 \text{ kJ/mol} \tag{2.73}$$

The standard state of a liquid or a solid substance is the pure substance, while for a gas it is the ideal gas at a pressure of 1 bar and 298 K. The superscript indicates that all species are in their standard state. Some heat capacities per mole of a chemical at standard conditions are tabulated in Table 2.2.

TABLE 2.2 Chemical thermodynamic properties at 298.15 K and 1 bar

Substance	ΔH_f° (kJ/mol)	ΔG_f° (kJ/mol)	$\overline{S}^\circ$ (J/K mol)	$\overline{C}_p^\circ$ (J/K mol)
O(g)	249.170	231.731	161.055	21.912
O_2(g)	0	0	205.138	29.355
O_3(g)	142.7	163.2	238.93	39.20
H(g)	217.965	203.247	114.713	20.784
H^+(g)	1536.202			
H^+(ao)	0	0	0	0
H_2(g)	0	0	130.684	28.824
OH(g)	38.95	34.23	183.745	29.886
OH^-(ao)	−229.994	−157.244	−10.75	−148.5
H_2O(l)	−285.830	−237.129	69.91	75.291
H_2O(g)	−241.818	−228.572	188.825	33.577
H_2O_2(l)	−187.78	−120.35	109.6	89.1
He(g)	0	0	126.150	20.786
Ne(g)	0	0	146.328	20.786
Ar(g)	0	0	154.843	20.786
Kr(g)	0	0	164.082	20.786
Xe(g)	0	0	169.683	20.786
F(g)	78.99	61.91	158.754	22.744
F^-(ao)	−332.63	−278.79	−13.8	−106.7
F_2(g)	0	0	202.78	31.30
HF(g)	−271.1	−273.2	173.779	29.133
Cl(g)	121.679	105.680	165.198	21.840
Cl^-(ao)	−167.159	−131.228	56.5	−136.4
Cl_2(g)	0	0	223.066	33.907
ClO_4^-(ao)	−129.33	−8.52	182.0	
HCl(g)	−92.307	−95.299	186.908	29.12
HCl(ai)	−167.159	−131.228	56.5	−136.4
HCl in $100H_2O$	−165.925			
HCl in $200H_2O$	−166.272			
Br(g)	111.884	82.396	175.022	20.786
Br^-(ao)	−121.55	−103.96	82.4	−141.8
Br_2(l)	0	0	152.231	75.689
Br_2(g)	30.907	3.110	245.463	36.02
HBr(g)	−36.40	−53.45	198.695	29.142
I(g)	106.838	70.250	180.791	20.786
I^-(ao)	−55.19	−51.57	111.3	−142.3
I_2(cr)	0	0	116.135	54.438
I_2(g)	62.438	19.317	260.69	36.90
HI(g)	26.48	1.70	206.594	29.158
S(rhombic)	0	0	31.80	22.64
S(monoclinic)	0.33	0.1	32.6	23.6
S(g)	278.805	238.250	167.821	23.673
S_2(g)	128.37	79.30	228.18	32.47
S^{2-}(ao)	33.1	85.8	−14.6	
SO_2(g)	−296.830	−300.194	248.22	39.87

(continued overleaf)

TABLE 2.2 ***(Continued)***

Substance	ΔH_f° (kJ/mol)	ΔG_f° (kJ/mol)	$\overline{S}^\circ$ (J/K mol)	$\overline{C}_p^\circ$ (J/K mol)
SO_3(g)	−395.72	−371.06	256.76	50.67
SO_4^{2-}(ao)	−909.27	−744.53	2.01	−293
HS^-(ai)	−17.6	12.08	62.8	
H_2S(g)	−20.63	−33.56	205.79	34.23
H_2SO_4(l)	−813.989	−690.003	156.904	138.91
H_2SO_4(ai)	−909.27	−744.53	20.1	−293
N(g)	472.704	455.563	153.298	20.786
N_2(g)	0	0	191.61	29.125
NO(g)	90.25	86.57	210.761	29.844
NO_2(g)	33.18	51.31	240.06	37.20
NO_3^-(ao)	−205.0	−108.74	146.4	−86.6
N_2O(g)	82.05	104.20	219.85	38.45
N_2O_4(l)	−19.50	97.54	209.2	142.7
N_2O_4(g)	9.16	97.89	304.29	77.28
NH_3(g)	−46.11	−16.45	192.45	35.06
NH_3(ao)	−80.29	−26.50	111.3	
NH_4^+(ao)	−132.51	−79.31	113.4	79.9
HNO_3(l)	−174.10	−80.71	155.60	109.87
HNO_3(ai)	−207.36	−111.25	146.4	−86.6
NH_4OH(ao)	−366.121	−263.65	181.2	
P(s, white)	0	0	41.09	23.840
P(g)	314.64	278.25	163.193	20.786
P_2(g)	144.3	103.7	218.129	32.05
P_4(g)	58.91	24.44	279.98	67.15
PCl_3(g)	−287.0	−267.8	311.78	71.84
PCl_5(g)	−374.9	−305.0	364.58	112.8
C(graphite)	0	0	5.74	8.527
C(diamond)	1.895	2.900	2.377	6.113
C(g)	716.682	671.257	158.096	20.838
C_2(g)	0	−0.0330	144.960	29.196
CO(g)	−110.525	−137.168	197.674	29.142
CO_2(g)	−393.509	−394.359	213.74	37.11
CO_2(ao)	−413.80	−385.98	117.6	
CO_3^{2-}(ag)	−677.14	−527.81	−56.9	
CH(g)	595.8			
CH_2(g)	392.0			
CH_3(g)	138.9			
CH_4(g)	−74.81	−50.72	186.264	35.309
C_2H_2(g)	226.73	209.20	200.94	43.93
C_2H_4(g)	52.26	68.15	219.56	43.56
C_2H_6(g)	−84.68	−32.82	229.60	52.63
HCO_3^-(ao)	−691.99	−586.77	91.2	
HCHO(g)	−117	−113	218.77	35.40
HCO_2H(l)	−424.72	−316.35	128.95	99.04
H_2CO_3(ao)	−699.65	−623.08	187.4	
CH_3OH(l)	−238.66	−166.27	126.8	81.6

TABLE 2.2 (*Continued*)

Substance	ΔH_f° (kJ/mol)	ΔG_f° (kJ/mol)	$\overline{S}^\circ$ (J/K mol)	$\overline{C}_p^\circ$ (J/K mol)
CH_3OH(g)	−200.66	−161.96	239.81	43.89
$CH_3CO_2^-$(ao)	−486.01	−369.31	86.6	−6.3
C_2H_4O(l, ethylene oxide)	−77.82	−11.76	153.85	87.95
CH_3CHO(l)	−192.30	128.12	160.2	
CH_3CO_2H(l)	−484.5	−389.9	159.8	124.3
CH_3CO_2H(ao)	−485.76	−396.46	178.7	
C_2H_5OH(l)	−277.69	−174.78	160.7	111.46
C_2H_5OH(g)	−235.10	−168.49	282.70	65.44
$(CH_3)_2O$(g)	−184.05	−112.59	266.38	64.39
C_3H_6(g, propene)	20.42	62.78	267.05	63.89
C_3H_6(g, cyclopropane)	53.30	104.45	237.55	55.94
C_3H_8(g, propane)	−103.89	−23.38	270.02	73.51
C_4H_8(g, 1-butene)	−0.13	71.39	305.71	85.65
C_4H_8(g, 2-butane, *cis*)	−6.99	65.95	300.94	78.91
C_4H_8(g, 2-butane, *trans*)	−11.17	63.06	296.59	87.82
C_4H_{10}(g, butane)	−126.15	−17.03	310.23	97.45
C_4H_{10}(g, isobutane)	−134.52	−20.76	294.75	96.82
C_6H_6(g)	82.93	129.72	269.31	81.67
C_6H_{12}(g, cyclohexane)	−123.14	31.91	298.35	106.27
C_6H_{14}(g, hexane)	−167.19	−0.07	388.51	143.09
C_7H_8(g, toluene)	50.00	122.10	320.77	103.64
C_8H_8(g, styrene)	147.22	213.89	345.21	122.09
C_8H_{10}(g, ethylbenzene)	29.79	130.70	360.56	128.41
C_8H_{10}(g, octane)	−208.45	16.64	466.84	188.87
Si(s)	0	0	18.83	20.00
SiO_2(s, α)	−910.94	−856.64	41.84	44.43
Sn(s, white)	0	0	51.55	26.99
Sn^{2+}(ao)	−8.8	−27.2	−17	
SnO(s)	−285.8	−256.9	56.5	44.31
SnO_2(s)	−580.7	−519.6	52.3	52.59
Pb(s)	0	0	64.81	26.44
Pb^{2+}(ao)	−1.7	−24.43	10.5	
PbO(s, yellow)	−217.32	−187.89	68.70	45.77
PbO_2(s)	−277.4	−217.33	68.6	64.64
Al(s)	0	0	28.33	24.35
Al(g)	326.4	285.7	164.54	21.38
Al_2O_3(s, α)	−1675.7	−1582.3	50.92	79.04
$AlCl_3$(s)	−704.2	−628.8	110.67	91.84
Zn(s)	0	0	41.63	25.40
Zn^{2+}(ao)	−153.89	−147.06	−112.1	46
ZnO(s)	−348.28	−318.30	43.64	40.25
Cd(s, γ)	0	0	51.76	25.98
Cd^{2+}(ao)	−75.90	−77.612	−73.2	
CdO(s)	−258.2	−228.4	54.8	43.43
$CdSO_4 \cdot \frac{8}{3}H_2O$(s)	−1729.4	−1465.141	229.630	213.26
Hg(l)	0	0	76.02	27.983
Hg(g)	61.317	31.820	174.96	20.786

(*continued overleaf*)

TABLE 2.2 (*Continued*)

Substance	ΔH_f° (kJ/mol)	ΔG_f° (kJ/mol)	$\overline{S}^\circ$ (J/K mol)	$\overline{C}_p^\circ$ (J/K mol)
Hg^{2+}(ao)	171.1	164.40	−32.2	
HgO(s, red)	−90.83	−58.539	70.29	44.06
Hg_2Cl_2(s)	−265.22	−210.745	192.5	102
Cu(s)	0	0	33.150	24.435
Cu^+(ao)	71.67	49.98	40.6	
Cu^{2+}(ao)	64.77	65.49	−99.6	
Ag(s)	0	0	42.55	25.351
Ag^+(ao)	105.579	77.107	72.68	21.8
Ag_2O(s)	−31.05	−11.20	121.3	65.86
AgCl(s)	−127.068	−109.789	96.2	50.79
Fe(s)	0	0	27.28	25.10
Fe^{2+}(ao)	−89.1	−78.90	−137.7	
Fe^{3+}(ao)	−48.5	−4.7	−315.9	
Fe_2O_3(s, hematite)	−824.2	−742.2	87.40	103.85
Fe_3O_4(s, magnetite)	−1118.4	−1015.4	146.4	143.43
Ti(s)	0	0	30.63	25.02
TiO_2(s)	−939.7	−884.5	49.92	55.48
U(s)	0	0	50.21	27.665
UO_2(s)	−1084.9	−1031.7	77.03	63.60
UO_2^{2+}(ao)	−1019.6	−953.5	−97.5	
UO_3(s, γ)	−1223.8	−1145.9	96.11	81.67
Mg(s)	0	0	32.68	24.89
Mg(s)	147.70	113.10	148.650	20.786
Mg^{2+}(ao)	−466.85	−454.8	−138.1	
MgO(s)	−601.70	−569.43	26.94	37.15
$MgCl_2$(ao)	−801.15	−717.1	−25.1	
Ca(s)	0	0	41.42	25.31
Ca(g)	178.2	144.3	154.884	20.786
Ca^{2+}(ao)	−542.83	−553.58	−53.1	
CaO(s)	−635.09	−604.03	39.75	42.80
$CaCl_2$(ai)	−877.13	−816.01	59.8	
$CaCO_3$(calcite)	−1206.92	−1128.79	92.9	81.88
$CaCO_3$(aragonite)	−1207.13	−1127.75	88.7	81.2
Li(s)	0	0	29.12	24.77
Li^+(ao)	−278.49	−293.31	13.4	68.6
Na(s)	0	0	51.21	28.24
Na^+(ao)	−240.12	−261.905	59.0	46.4
NaOH(s)	−425.609	−379.494	64.455	59.54
NaOH(ai)	−470.114	−419.150	48.1	−102.1
NaOH in $100H_2O$	−469.646			
NaOH in $200H_2O$	−469.608			
NaCl(s)	−411.153	−384.138	72.13	50.50
NaCl(ai)	−407.27	−393.133	115.5	−90.0
NaCl in $100H_2O$	−407.066			
NaCl in $200H_2O$	406.923			
K(s)	0	0	64.18	29.58
K^+(ao)	−252.38	−283.27	102.5	21.8
KOH(s)	−424.764	−379.08	78.9	64.9

TABLE 2.2 (*Continued*)

Substance	ΔH_f° (kJ/mol)	ΔG_f° (kJ/mol)	$\overline{S}^\circ$ (J/K mol)	$\overline{C}_p^\circ$ (J/K mol)
KOH(ai)	−482.37	−440.50	91.6	−126.8
KOH in $100H_2O$	−481.637			
KOH in $200H_2O$	−481.742			
KCl(s)	−436.747	−409.14	82.59	51.30
KCl in $100H_2O$	−419.191			
Rb(s)	0	0	76.78	10.148
Rb^+(ao)	−251.17	−283.98	121.50	
Ca(s)	0	0	85.23	32.17
Cs^+(ao)	−258:28	−292:02	133:05	−10:5

[a]The values are from The NBS Tables of Chemical Thermodynamic Properties (1982). The standard state pressure is 1 bar (0.1 MPa). The compounds are in the order of elements used in these tables. For the elements represented here, the order is O, H, He, F, Cl, Br, I, S, N, P, C, Pb, Al, Zn, Cd, Hg, Cu, Ag, Fe, Ti, Mg, Ca, Li, Na, K, Rb, and Cs. The standard state for a strong electrolyte in aqueous solution is the ideal solution at unit mean molality (unit activity). The thermodynamic properties of the completely dissociated electrolyte are designated by ai. The thermodynamic properties of undissociated molecules in water are designated by ao. The properties of organic substances with more than two carbon atoms are from Stull, D.R., Westrum, E.F., and Sinke, G.C. (1969) *The Chemical Thermodynamics of Organic Compounds*, Wiley, Hoboken, NJ. The NBS Tables of Chemical Thermodynamic Properties have been published as a supplement to Volume II (1982) of the *Journal of Physical and Chemical Reference Data* and may be ordered from the American Chemical Society, 1155 Sixteenth St., NW, Washington, DC 20036. The conversion to the new standard state pressure is described by Freeman, R.D. (1985) *J. Chem. Educ.*, 62: 681.

2.3.6 Standard Enthalpies and Heats of Formation

The heat of reaction may be viewed as being equal to the enthalpy of the products minus the enthalpy of the reactants, where $\overline{H}^\circ$ refers to the enthalpy per mole at standard conditions. Therefore, the heats of reaction for equations 2.72 and 2.73 are equal to

$$\Delta H^\circ(2.72) = \overline{H}^\circ(H_2O(l)) - \overline{H}^\circ(H_2(g)) - 1/2\overline{H}^\circ(O_2(g))$$

$$\Delta H^\circ(2.73) = \overline{H}^\circ(H_2O(g)) - \overline{H}^\circ(H_2(g)) - 1/2\overline{H}^\circ(O_2(g))$$

Reactions (2.72) and (2.73) are also called formation reactions since one mole of the product compound is formed from the reactant elements in their most stable form and phase at 1 bar pressure and 298 K. By convention, elements at these conditions are assigned an enthalpy per mole of zero. Thus, the heats of formation, ΔH_f, for $H_2(g)$, $O_2(g)$, $H_2O(l)$, and $H_2O(g)$ are 0, 0, −285.83, and −241.82 kJ/mol, respectively, under standard conditions.

Table 2.2 displays some chemical thermodynamic properties including standard enthalpies per mole or heats of formation that are useful in calculating heats of reaction at standard conditions of 1 bar pressure at 298 K.

2.3.7 Hess's Law

Standard enthalpies of individual processes can be combined to obtain the enthalpy of another process. This application of the first law is called Hess's law, which states that the standard enthalpy change of an overall reaction is the sum of the individual reactions into which a reaction may be divided. The basis of the law is the path independence of the heat of reaction and, therefore, the reactant and product chemicals are only important in calculating the enthalpy and not the intermediates. For example, the heat of vaporization, ΔH_{vap}, for $H_2O(l) \rightarrow H_2O(g)$ may be calculated from thermodynamic tables using

$$\Delta H^\circ_{vap} = \Delta H^\circ_f(H_2O(g)) - \Delta H^\circ_f(H_2O(l))$$

explaining $\Delta_f H = 1H_f$

Hess's law provides an alternate means for obtaining the heat of vaporization using the heats of reaction for equations 2.72 and 2.73.

$$\begin{array}{rl} H_2O(l) \rightarrow H_2(g) + 1/2O_2(g) & \Delta H^\circ = +285.83\ \text{kJ/mol} \\ H_2(g) + 1/2O_2(g) \rightarrow H_2O(g) & \Delta H^\circ = -241.82\ \text{kJ/mol} \\ \hline H_2O(l) \rightarrow H_2O(g) & \Delta H^\circ_{vap} = +44.01\ \text{kJ/mol} \end{array}$$

The first reaction above has been written where $H_2O(l)$ is a reactant and is now endothermic by 285.83 kcal/mol while equation 2.72 was exothermic by that amount. Because the change in enthalpy is path independent, the enthalpy of gaseous hydrogen and oxygen intermediates are not needed to determine the heat of vaporization of water.

Before pipelines were built to deliver natural gas, towns and cities contained plants that produced a fuel known as "town gas," which contained hydrogen that was formed by passing steam over red-hot charcoal.

$$C(s) + H_2O(g) \rightarrow CO(g) + H_2(g) \qquad (2.74)$$

Hess's law may be employed to calculate the heat of reaction for equation 2.74 using measured heats of oxidation for C(s), CO(g), and $H_2(g)$.

$$\begin{array}{rl} H_2O(g) \rightarrow H_2(g) + 1/2O_2(g) & \Delta H^\circ = +241.82\ \text{kJ/mol} \\ C(s) + 1/2O_2(g) \rightarrow CO(g) & \Delta H^\circ = -110.53\ \text{kJ/mol} \\ \hline C(s) + H_2O(g) \rightarrow CO(g) + H_2(g) & \Delta H^\circ = 131.29\ \text{kJ/mol} \end{array}$$

or

$$\begin{array}{rl} H_2O(g) \rightarrow H_2(g) + 1/2O_2(g) & \Delta H^\circ = +241.82\ \text{kJ/mol} \\ CO_2(g) \rightarrow CO(g) + 1/2O_2(g) & \Delta H^\circ = +282.98\ \text{kJ/mol} \\ C(s) + O_2(g) \rightarrow CO_2(g) & \Delta H^\circ = -393.51\ \text{kJ/mol} \\ \hline C(s) + H_2O(g) \rightarrow CO(g) + H_2(g) & \Delta H^\circ = 131.29\ \text{kJ/mol} \end{array}$$

Reaction (2.74) is endothermic by 131.29 kJ/mol and has the same value whether $O_2(g)$ or $O_2(g)$ and $CO_2(g)$ are the intermediates in the Hess's law calculations since the heat of reaction does not depend on the pathway taken to obtain the products.

2.3.8 Second Law, Changes in Entropy, and the Carnot Cycle

When faced with the problem of determining whether a particular reaction will occur, the first consideration often is the amount of available energy from the reaction. Hot steam was obviously required to overcome the endothermicity of reaction (2.74) to produce gaseous CO and H_2 from coal. However, a number of endothermic reactions do occur spontaneously. Energy alone is not sufficient to decide whether a reaction will occur, and this is where the thermodynamic function entropy (S) becomes important. In addition, entropy will provide information about the limitation of converting thermal energy into useful work.

The second law of thermodynamics defines small differential changes in entropy for a reversible process, such as equilibrium, as

$$dS = dq_{rev}/T \tag{2.75}$$

where dq_{rev} is the heat absorbed in a reversible process carried out at temperature T. Changes in entropies for reversible phase changes, such as melting (s → l), vaporization (l → g), and sublimation (s → g), are easily determined using the positive changes in enthalpies at the constant temperature for melting, boiling, and sublimation.

When there are temperature changes within a phase, dq_{rev} is expressed in terms of the heat capacity at constant pressure of 1 bar, C_p° and the differential change in temperature, dT.

$$dq_{rev} = C_p^\circ dT \quad \text{and} \quad dS^\circ = C_p^\circ dT/T \tag{2.76}$$

Integration of the change in entropy equation 2.76 over small temperature changes where the heat capacity is a constant independent of temperature gives the change in entropy from the initial to final temperature within the phase.

$$\Delta S^\circ = C_p^\circ \ln(T_f/T_i) \tag{2.77}$$

Using these equations, the calculated values for the changes in entropy are found to be positive for the phase changes (s → l, l → g, s → g) and for increases in temperature within a phase.

Historically, entropy was first introduced with the performance of the simplest reversible heat engine called the Carnot cycle, which investigated the principles governing the transformation of thermal energy into mechanical energy. The reversible Carnot cycle gave the highest fraction for the conversion of thermal energy into useful work, called the efficiency, ε, where T_h is the temperature of the high-temperature reservoir and T_l is the temperature of the low-temperature surroundings.

$$\varepsilon = (T_h - T_l)/T_h \tag{2.78}$$

This efficiency equation will be helpful when evaluating the conversion of thermal energy from energy sources, such as solar and nuclear, into the production of hydrogen fuel.

Irreversible heat engines, which have friction in the moving parts and produce an effect on the surroundings, have much lower efficiencies of conversion. Equation 2.78 represents the maximum efficiency for converting thermal energy into useful work.

2.3.9 Third Law of Thermodynamics and Calculations of Entropy Changes for Chemical Reactions

The third law of thermodynamics states that the absolute value of entropy of a pure, perfectly crystalline substance is zero at the absolute zero of temperature. A pure, perfectly crystalline substance is completely ordered and, therefore, any imperfection in the substance will introduce disorder and randomness within the system. An alternate way of stating the third law is that it is impossible to reach absolute zero temperature where there is perfect order. An increase in temperature of the crystalline substance will increase the energy within the substance and increase disorder through its molecular motions. When the temperature reaches the point for a phase change, molecules in the new phase will have more randomness and hence experience an increase in disorder. Entropy is a measure of the disorder or randomness of the system.

Assuming the third law of thermodynamics, the entropy per mole for a chemical ($\overline{S}°$) may be determined at the standard conditions of 1 bar pressure by using equations 2.75 and 2.76 to calculate the entropy increase for both the temperature rise within each phase and at the phase changes needed to reach 298 K. These calculated entropies per mole are called standard molar entropies and some values are tabulated in Table 2.2. Standard molar entropies are useful in calculating the change in entropy for a chemical reaction. The following equation illustrates the calculation for determining the change in entropy at 298 K for the formation of "town gas," reaction (2.74). The change in entropy is equal to the entropy of the products minus the entropy of the reactants.

$$\Delta S° = \overline{S}°(H_2(g)) + \overline{S}°(CO(g)) - \overline{S}°(H_2O(g)) - \overline{S}°(C(s)) \tag{2.79}$$

The change in entropy for reaction (2.74) may be predicted to be positive, since there is more disorder in the product state (2 mol of gas) compared to the reactant state (1 mol of solid and 1 mol of gas). The standard molar entropies for gaseous hydrogen, carbon monoxide, water, and solid carbon graphite from Table 2.2 are 130.684, 197.674, 188.825, and 5.74 J/K mol, respectively, which results in a change in entropy for reaction (2.74) of 133.793 J/K mol.

2.3.10 Gibbs Free Energy

The thermodynamic function Gibbs free energy, G, is defined in terms of enthalpy, temperature, and entropy

$$G = H - TS \tag{2.80}$$

If the temperature is constant, then the change in Gibbs free energy is made up of a change in enthalpy and a term containing a change in entropy.

$$\Delta G = \Delta H - T\Delta S \tag{2.81}$$

Changes in ΔG are a measure of the tendency for a reaction to take place spontaneously. The exothermicity of a chemical reaction is associated with unstable reactants that exchange energy to acquire a more stable state as products. Superimposed on this is the tendency of the system to achieve maximum entropy by becoming more disordered or having more freedom in the product state. The system is trying to satisfy these two opposing tendencies toward minimum energy (maximum chemical stability) and maximum entropy (maximum disorder). ΔG indicates the balance between these two factors and is a measure of the tendency to take place. For a spontaneous change, ΔG is negative and this is most optimally achieved for an exothermic reaction and a reaction that has more disorder in the products than in the reactants. Often the change in enthalpy and entropy oppose one another; however, for a reaction to be thermodynamically favorable it must have a negative ΔG. For example, the chemical equation to form "town gas," reaction (2.74), was endothermic by 131.29 kJ/mol and had a change in entropy of 133.793 J/K mol. Therefore, the change in ΔG at 298 K for reaction (2.74)

$$\Delta G^\circ = 131.29 - (298)(133.793)/1000 = 91.4\ \mathrm{kJ/mol} \tag{2.82}$$

is positive, which indicates that reaction (2.74) is not spontaneous at room temperature under standard conditions. Hot steam was therefore needed to make reaction (2.74) spontaneous at a temperature higher than room temperature.

Changes in Gibbs free energy for formation reactions, $\Delta G_f{}^\circ$, given in Table 2.2 may also be used to calculate ΔG for a chemical reaction, illustrated as follows, for reaction (2.74):

$$\begin{aligned}\Delta G^\circ &= \Delta G_f^\circ(H_2(g)) + \Delta G_f^\circ(CO(g)) - \Delta G_f^\circ(H_2O(g)) - \Delta G_f^\circ(C(s))\\ &= 0 + (-137.168) - (-228.578) - 0 = 91.4\ \mathrm{kJ/mol}\end{aligned}$$

2.3.11 Relationship between Thermodynamics and Chemical Equilibrium

If the chemical reaction is not at standard conditions for gaseous reactants and products, ΔG is then obtained using equation 2.83 where the reaction quotient, Q_p, is similar to the algebraic expression for the equilibrium constant but instead includes the pressures of reactants and products that are not at equilibrium.

$$\Delta G = \Delta G^\circ + RT \ln Q_p \tag{2.83}$$

For production of hydrogen gas by reaction (2.74), the reaction quotient is

$$Q_p = (PH_2)(PCO)/(PH_2O) \tag{2.84}$$

Solids, such as C(s), and liquids do not appear in the reaction quotient, but only gases. Therefore, if there is a very high initial pressure of gaseous water compared to the initial pressures of gaseous H_2 and CO, the reaction quotient in equation 2.84 is a very small fraction. Since the natural logarithm (ln) of a fraction is a negative number, the $RT \ln Q_p$ term in equation 2.83 may make ΔG negative and result in a spontaneous reaction even though ΔG° is positive. As the reaction proceeds, the pressure of gaseous reactant water will decrease and the pressure of the gaseous hydrogen and carbon monoxide products will increase. The reaction will continue until the magnitude of the $RT \ln Q_p$ is equal to but opposite in sign to the ΔG° term in equation 2.83. At that point, $\Delta G = 0$ and there will be chemical equilibrium. At chemical equilibrium, Q_p becomes K_p, the equilibrium constant, where the pressures are now the equilibrium pressures for the chemical reaction. Therefore, at equilibrium equation 2.83 becomes

$$\Delta G^\circ = -RT \ln K_p \tag{2.85}$$

Equilibrium constants may be calculated from standard Gibbs free energy changes, $\Delta G°$, which are obtained from thermochemical tables.

$$K_p = e^{-\Delta G°/(RT)} \tag{2.86}$$

For the production of hydrogen by the "town gas," equation 2.74, the equilibrium constant at room temperature becomes

$$K_p = e^{-91.4\ \mathrm{kJ}/(8.314\ \mathrm{J/K})(1/1000\ \mathrm{kJ/J})(298\ \mathrm{K})} = 9.6 \times 10^{-17}$$

The low value for the equilibrium constant indicates a low pressure of products relative to the reactants in reaction (2.74) at room temperature.

The convenience of natural gas led to the development of commercially available technologies to form hydrogen from light hydrocarbons containing mostly methane by steam reforming. The first step in the steam reforming of methane is the reaction with steam to produce a synthetic gas, called syngas, which is a mixture containing carbon monoxide and hydrogen.

$$CH_4(g) + H_2O(g) \rightarrow CO(g) + 3H_2(g) \tag{2.87}$$

The change in Gibbs free energy for reaction (2.87) at room temperature may be computed to be

$$\begin{aligned}\Delta G° &= 3\Delta G_f°(H_2(g)) + \Delta G_f°(CO(g)) - \Delta G_f°(H_2O(g)) - \Delta G_f°(CH_4(g))\\ &= 0 + (-137.168) - (-228.572) - (-50.72) = 142.124\ \mathrm{kJ/mol}\end{aligned}$$

which corresponds to the equilibrium constant

$$K_p = e^{-142.124\ \mathrm{kJ}/(8.314\ \mathrm{J/K})(1/1000\ \mathrm{kJ/J})(298\ \mathrm{K})} = 1.2 \times 10^{-25}$$

Reaction (2.87) is very unfavorable thermodynamically at room temperature under standard conditions. Reaction (2.87) occurs with the combustion of fossil fuels at elevated temperatures (700–925°C) and the use of a catalyst that is protected from corrosion by a desulfurization process of the feedstock. With the current high energy prices for fossil fuels, the convenience of the use of natural gas is being questioned and technologies are being developed to produce coal-fuel power plants with a target of zero emission, hydrogen production and carbon dioxide sequestration capabilities (see Chapter 4).

2.3.12 Temperature Dependence of the Heat of Reaction

Since many reactions, such as the production of "town gas" from coal (2.74) and syngas (2.87), occur at temperatures other than room temperature, there is often a need to calculate the heat of reaction at different temperatures.

The heat of reaction at constant pressure, $\Delta H°$, is the difference between the enthalpy of the products and the enthalpy of the reactants.

$$\Delta H° = H°_{\mathrm{products}} - H°_{\mathrm{reactants}} \tag{2.88}$$

The change in the heat of reaction with temperature is given by the derivative of equation 2.88 and equals the change in the heat capacity for the chemical reaction, ΔC_p°, as defined in equation 2.89.

$$d(\Delta H^\circ)/dT = d(H^\circ_{\text{products}})/dT - d(H^\circ_{\text{reactants}})/dT = (C^\circ_{p\ \text{products}}) - (C^\circ_{p\ \text{reactants}}) = \Delta C^\circ_p \quad (2.89)$$

Writing equation 2.89 in the differential form, we have

$$d(\Delta H^\circ) = (\Delta C_p^\circ)dT \quad (2.90)$$

Integrating equation 2.90 from a fixed temperature T_0 to another temperature T and rearranging, we obtain

$$\Delta H_T^\circ = \Delta H_{T_0}^\circ + \int_{T_0}^{T} (\Delta C_p^\circ)dT \quad (2.91)$$

As shown in Table 2.3, the heat capacities per mole, $\overline{C}_p^\circ$, are commonly expressed in a power series as function of temperature.

$$\overline{C}_p^\circ = \alpha + \beta T + \gamma T^2 + \delta T^3 \quad (2.92)$$

For the syngas reaction (2.86), the change in heat capacity for the chemical reaction becomes

$$\begin{aligned}
\Delta C_p^\circ &= 3\overline{C}_p^\circ(H_2(g)) + \overline{C}_p^\circ(CO(g)) - \overline{C}_p^\circ(H_2O(g)) - \overline{C}_p^\circ(CH_4(g)) \\
&= 3(29.088 - 0.192 \times 10^{-2}T + 0.400 \times 10^{-5}T^2 - 0.870 \times 10^{-9}T^3) \\
&\quad + (28.142 + 0.167 \times 10^{-2}T + 0.537 \times 10^{-5}T^2 - 2.221 \times 10^{-9}T^3) \\
&\quad - (32.218 + 0.192 \times 10^{-2}T + 1.055 \times 10^{-5}T^2 - 3.593 \times 10^{-9}T^3) \\
&\quad - (19.875 + 5.021 \times 10^{-2}T + 1.268 \times 10^{-5}T^2 - 11.004 \times 10^{-9}T^3) \\
&= 63.313 - 5.622 \times 10^{-2}T - 0.586 \times 10^{-5}T^2 + 9.766 \times 10^{-9}T^3
\end{aligned}$$

TABLE 2.3 Molar heat capacities at constant pressure as a function of temperature from 300 to 1800 K: $\overline{C}_p^\circ = \alpha + \beta T + \gamma T^2 + \delta T^3$

	α (J/K mol)	β 10^{-2} (J/K^2 mol)	γ 10^{-5} (J/K^3 mol)	δ 10^{-9} (J/K^4 mol)
$N_2(g)$	28.883	−0.157	0.808	−2.871
$O_2(g)$	25.460	1.519	−0.715	1.311
$H_2(g)$	29.088	−0.192	0.400	−0.870
$CO(g)$	28.142	0.167	0.537	−2.221
$CO_2(g)$	22.243	5.977	−3.499	7.464
$H_2O(g)$	32.218	0.192	1.055	−3.593
$NH_3(g)$	24.619	3.75	0.138	—
$CH_4(g)$	19.875	5.021	1.268	−11.004

Source: S.I. Sandles, Chemical and Engineering Thermodynamics 3rd ed. (1999) Reproduced with permission of John Wiley & Sons.

The heat of reaction at 1000 K may be determined from the following equation:

$$\Delta H^\circ_{1000\ \mathrm{K}} = \Delta H^\circ_{298\ \mathrm{K}} + \int_{298}^{1000} (63.313 - 5.622 \times 10^{-2}T - 0.586 \times 10^{-5}T^2 + 9.766 \times 10^{-9}T^3)\mathrm{d}T$$

where the heat of reaction at 298 K is calculated to be endothermic by 206.103 kJ/mol from the heats of formation given in Table 2.2.

$$\begin{aligned}\Delta H^\circ_{1000\ \mathrm{K}} &= 206{,}103 + 63.313T]|_{298}^{1000} - 5.622 \times 10^{-2}T^2/2]|_{298}^{1000} \\ &\quad - 0.586 \times 10^{-5}T^3/3]|_{298}^{1000} + 9.766 \times 10^{-9}T^4/4]|_{298}^{1000} \\ &= 206{,}103 + 63.313(1000 - 298) - 2.811 \times 10^{-2}(1000^2 - 298^2) \\ &\quad - 0.1953 \times 10^{-5}(1000^3 - 298^3) + 2.4415 \times 10^{-9}(1000^4 - 298^4)\ \mathrm{J/mol} \\ \Delta H^\circ_{1000\ \mathrm{K}} &= 225.5\ \mathrm{kJ/mol}\end{aligned}$$

2.3.13 Temperature Dependence of the Change in Entropy and Equilibrium Constant for a Chemical Reaction

One method for determining the equilibrium constant at temperatures other than room temperature is to first obtain the change in entropy for the reaction. Then, using the heat of reaction at the desired temperature and equation 2.81, the standard free energy change, ΔG°, for the chemical reaction may be calculated. Next, the equilibrium constant at the desired temperature may be obtained from equation 2.86.

The equation for determining the change in entropy for a chemical reaction, ΔS°, is derived similarly to how the temperature dependence of the heat of reaction was determined.

$$\Delta S^\circ = S^\circ_{\mathrm{products}} - S^\circ_{\mathrm{reactants}}$$

Employing equation 2.76, equation 2.93 is obtained

$$\Delta S^\circ_T = \Delta\ S^\circ_T + \int_{To}^{T} (\Delta C^\circ_p/T)\mathrm{d}T \tag{2.93}$$

The entropy change at 1000 K for the syngas reaction (2.86) may be determined from the following equation:

$$\Delta S^\circ_{1000\ \mathrm{K}} = \Delta S^\circ_{298\ \mathrm{K}} + \int_{298}^{1000} (63.313/T - 5.622 \times 10^{-2} - 0.586 \times 10^{-5}T + 9.766 \times 10^{-9}T^2)\mathrm{d}T$$

where the entropy change at 298 K is calculated to be 214.6 J K/mol from the standard molar entropies given in Table 2.2.

$$\begin{aligned}\Delta S^\circ_{1000\ \mathrm{K}} &= 214.6 + 63.313\ \ln T]|_{298}^{1000} - 5.622 \times 10^{-2}T]|_{298}^{1000} \\ &\quad - 0.586 \times 10^{-5}T^2/2]|_{298}^{1000} + 9.766 \times 10^{-9}T^3/3]|_{298}^{1000}\end{aligned}$$

$$= 214.6 + 63.313 \ln(1000/298) - 5.622 \times 10^{-2}(1000 - 298)$$
$$- 0.293 \times 10^{-5}(1000^2 - 298^2) + 3.255 \times 10^{-9}(1000^3 - 298^3)\text{J/K mol}$$
$$= 252.3 \text{ J/K mol}$$

$$\Delta G^{\circ}_{1000\text{ K}} = \Delta H^{\circ}_{1000\text{ K}} - 1000\Delta S^{\circ}_{1000\text{ K}} = (225.5) - (1000)(252.3)/1000$$
$$= -26.8 \text{ kJ/mol}$$

$$K_p = e^{-(-26.8 \text{ kJ}/(8.314 \text{ J/K})(1/1000 \text{ kJ/J})(1000 \text{ K}))} = 25.1$$

Studying the reaction (2.87) at 1000 K has decreased the standard free energy from 142.124 to −26.8 kJ/mol, which has resulted in an equilibrium constant of 25.1 compared to 1.2×10^{-25} at 298 K. The higher equilibrium constant at 1000 K favors the formation of the hydrogen product.

Alternatively, the equilibrium constant at any temperature may be determined from a double integration with the first integration calculating the temperature dependence for ΔH° (2.91) and the second using the following equation:

$$\ln K_{pT} = \ln K_{p298} + \int_{298}^{T} (\Delta H^{\circ}/(RT^2))\mathrm{d}T \tag{2.94}$$

2.3.14 Gibbs Free Energy Change, Nernst Equation, and Equilibrium Constant for an Electrochemical Reaction

When reversible work beyond volume expansion, for example, electric work (W_{el}), is done on a system, then

$$\Delta G = W_{el} = -nFE \tag{2.95}$$

where n is the number of moles of electrons transferred in the electrochemical reaction, F is the Faraday (96,485 C/mol of electrons), and E is the electromotive force of the cell in volts. Similarly, the standard free energy change for the electrochemical reaction is written in terms of E° the standard electromotive force for the cell.

$$\Delta G^{\circ} = -nFE^{\circ} \tag{2.96}$$

The relationship between E and E° is determined using equation 2.79, which includes the reaction quotient, Q_p.

$$-nFE = -nFE^{\circ} + RT \ln Q_p \tag{2.97}$$

Equation 2.97 is called the Nernst equation and is usually written as

$$E = E^{\circ} - \frac{RT}{nF} \ln Q_p \tag{2.98}$$

When the temperature is 298 K, $R = 8.314$ J/K mol and the logarithm is expressed in base 10, the Nernst equation becomes

$$E = E^{\circ} - \frac{0.059}{n} \log_{10} Q_p \tag{2.99}$$

Since $\Delta G = 0$ at equilibrium, equation 2.95 gives $E = 0$ for the equilibrium mixture. At equilibrium, the reaction quotient is the equilibrium constant, K, and (2.98) becomes

$$E^\circ = \frac{RT}{nF} \ln K \tag{2.100}$$

By measuring the standard electromotive force of an electrochemical cell, the equilibrium constant for the cell electrochemical reaction may be obtained from equation 2.101.

$$K = e^{nFE^\circ/(RT)} \tag{2.101}$$

2.4 CHEMICAL KINETICS

Chemical kinetics deals with how fast reactions occur, that is, the rates of the reactions, and factors that influence the reaction rates. The time for chemical reactions to go to completion can range from relatively slow, for example, hours for some organic reactions, to about 10^{-6} s (i.e., microseconds) for acid/base neutralizations, and some reactions can be as fast as 10^{-15} s (i.e., femtoseconds). The rates of chemical reactions are of course greatly important in the chemical industry. As in any industry "time is money," and often it is critical that a product is desired in a certain time. We can also use chemical kinetics to predict how long it takes to reach a critical concentration, for example, how long a pharmaceutical drug will be effective. Ultimately, using chemical kinetics we can also understand how a chemical reaction occurs at the more detailed, molecular level.

A reaction does not have to be strongly exothermic to be fast, or vice versa. Also, the change in free energy, introduced in Section 2.3, is not related to the rate of a reaction. In fact, there is *no direct correlation between the thermodynamics of a reaction and its kinetics*. Some reactions are highly spontaneous but proceed very slowly. For example, if hydrogen and chlorine are placed in a flask, nothing will occur. Only if some energy is provided, for example, in form of an igniting spark, will the reaction begin and hydrogen chloride will form with an outburst of energy, observable as heat and light [$H_2 + Cl_2 \rightarrow 2HCl$]. Other reactions such as the oxidation of nitrogen oxide to nitrogen dioxide occur fast but only with a small release of energy [$2NO + O_2 \rightarrow 2NO_2$]. While in thermodynamics only state functions of the reactants and products and ultimately the free energy change of the reaction are relevant to predict the spontaneity of a reaction, the *reaction pathway* of how exactly the reactant(s) are converted to product(s) is subject to the reaction's kinetics. The thermodynamic data only depend on the type of reactants or products; however, as we will see for a given reaction different reaction pathways are possible.

We will first discuss kinetics at the *macroscopic* level, and later gradually approach the kinetics at the *microscopic* level. The former relates to the determination of rates as it is done with relatively simple tools in the laboratory. If we have a way to observe when a reaction product has formed, for example, indicated by a change in color, and we can measure the time with a stopwatch, we can determine the rate at the macroscopic level. The determination of kinetics at the microscopic or molecular level, usually requires more sophisticated techniques, such as spectroscopy, since we need more detailed experimental evidence about the steps by which the reaction proceeds, which is typically much more complex than shown in the balanced chemical equation.

Four major factors can influence the rate of a chemical reaction:

1. The rate does depend on the concentration of reactant(s). If the amount of a reactant is larger in a given volume, the reaction will be generally faster. This is why, for example, food exposed to air spoils rapidly as it reacts with the oxygen in the air. However, vacuum-packed food spoils more slowly (the vacuum is not perfect and the concentration of oxygen is only lower but not zero).
2. A reaction usually occurs more rapidly at a higher *temperature*. That is why we boil eggs (i.e., $T = 100°C$), rather than let them ferment slowly.
3. A substance that acts as a catalyst could be present, which speeds up a reaction. A naturally present catalyst is chlorophyll in green leaves, which helps to produce glucose and oxygen from carbon dioxide and water, an otherwise very slow reaction.
4. Finally, the surface area of a reactant can be critical. That is why, once in a while, explosions occur in wheat silos because fine wheat dust reacts very fast with the oxygen in the air by combustion.

2.4.1 The Rate of a Chemical Reaction

The rate of a chemical reaction is typically abbreviated as rate, and occasionally represented by the symbols r or R. Its unit is expressed as concentration per time. Concentration is the amount of a substance in a volume, and in chemistry most often molarity, $M = \text{mol/L}$, is used as unit for the concentration. The unit of time depends on how fast the reaction actually is; we prefer to use seconds (s) for fast reactions, and larger time units for slower reactions, that is, minutes (min), hours (h), days (d), or years (a). Rates always have positive values.

Figure 2.8 shows for the decomposition of hydrogen iodide: $2HI(g) \rightarrow H_2(g) + I_2(g)$, how the concentrations of the reactant and the products change with time. Reactant concentrations always decrease, whereas product concentrations increase as the reaction proceeds. Furthermore, we recognize that the changes are not monotonous, but that the curves are steeper at the beginning with the high reactant concentration than as time goes on and the reactant concentration decreases. We also can observe that hydrogen and iodine increase at the same rate, but the hydrogen iodide decreases more rapidly, that is, twice as fast

The simplest type of rate that can be determined is the average rate. It is based on the change in concentration for a chosen *time interval*.

For a reactant A:

$$\text{Average rate of A} = \frac{[A]_2 - [A]_1}{t_2 - t_1} = -\frac{\Delta[A]}{\Delta t} \tag{2.102}$$

$[A]_1$ or $[A]_2$ are the concentrations of A at the times t_1 and t_2, respectively. Because the reactant's concentration is decreasing as the reaction proceeds, the change in the concentration of A or $\Delta[A]$ is negative. Therefore, for reactants a minus sign is added in front of $\Delta[A]$ to turn the rate itself into a positive number.

For a product B, its average rate is defined as

$$\text{Average rate of B} = \frac{[B]_2 - [B]_1}{t_2 - t_1} = \frac{\Delta[B]}{\Delta t} \tag{2.103}$$

Since products always are formed in the reaction, the value of $\Delta[B]$ is positive; no minus sign is required to obtain a positive rate for products.

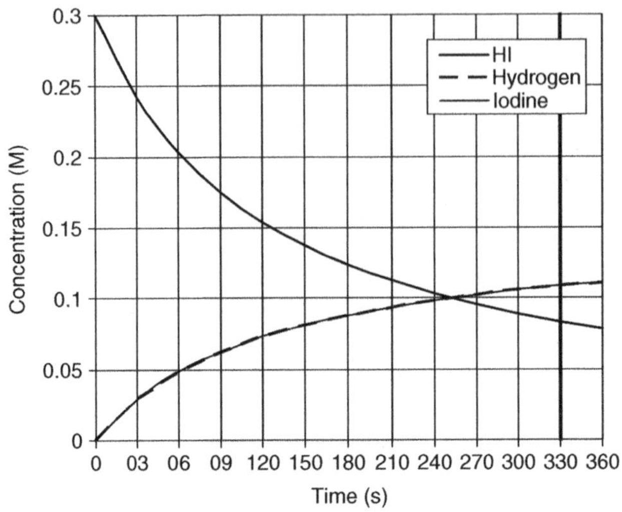

Figure 2.8 Concentration versus time for the decomposition of hydrogen iodide into the elements.

For the decomposition of hydrogen iodide, shown earlier, assume that a 1-L flask was filled with 0.300 mol of HI, and that the following experimental data were obtained:

Time (s)	[HI] (M)
0	0.3001
30	0.2425
60	0.2034
90	0.1752
120	0.1538
150	0.1371
180	0.1240
210	0.1127
240	0.1034
270	0.0956
300	0.0888
330	0.0830
360	0.0779

We can then determine the average rates for the disappearance of HI between (a) 0 and 60 s, (b) 90 and 210 s, and (c) 270 and 300 s as follows:

a) $\text{rate}_{\text{ave},0-60\text{ s}} = -(0.2034 - 0.3001)\text{M}/(60 - 0)\text{s} = 1.61 \times 10^{-3}\ \text{M/s}$

b) $\text{rate}_{\text{ave},90-210\text{ s}} = -(0.1127 - 0.1752)\text{M}/(210 - 90)\text{s} = 5.21 \times 10^{-4}\ \text{M/s}$

c) $\text{rate}_{\text{ave},240-330\text{ s}} = -(0.0830 - 0.1034)\text{M}/(330 - 240)\text{s} = 2.27 \times 10^{-4}\ \text{M/s}$

Figure 2.9 shows the three different average rates for the reaction mentioned earlier. Again, we can confirm, this time in a quantified manner, that the average rate changes with time, and that it gets more accurate, as we choose a smaller time interval.

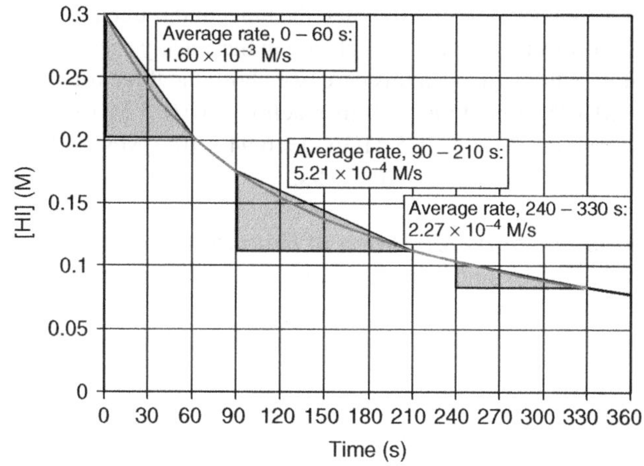

Figure 2.9 Three different average rates for the decomposition of hydrogen iodide described in the example on average rates.

2.4.1.1 Initial and Instantaneous Rates Using the types of curves in the concentration/time diagram for a reactant, as shown in Figures 2.8 and 2.9, we can define more specific rates for a given reaction. The rate at the very beginning ($t=0$) of the reaction is called the initial rate and can be determined by the slope of tangent at the very beginning of the reaction or $t=0$. The instantaneous rate is the rate at a selected *point in time* (not a time interval) during the reaction, and can, for example, graphically be determined by the slope of tangent at that specific time. These types of rates are illustrated in Figure 2.10. The advantage of the instantaneous rate over the average rate is that the accuracy of the former rate does not depend on the selected time interval.

In Figure 2.10, it is shown how we can, using graphical determination, estimate (a) the initial rate and (b) the instantaneous rates at 120 and at 270 s into the reaction (in M/h).

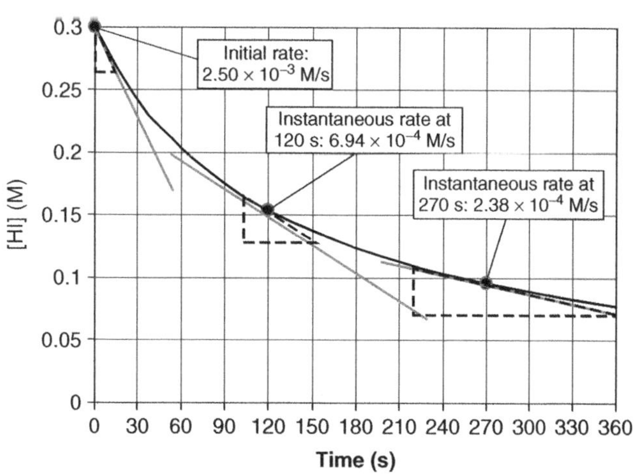

Figure 2.10 Different types of rates based on the decrease of the concentration of a reactant with time. $R_{Init.}$: initial rate; $R_{Inst.}$: instantaneous rate.

The initial rate is found by constructing first the tangent through the time zero and using it as hypotenuse to form a rectangular triangle along the *x* and *y* axes and then determining the slope by dividing *y* through *x*. (In this case, we chose the three triangles with exactly a decrease of 0.025 M.) The initial rate for this reaction is (0.025 M/10 s) = 2.5×10^{-3} M/s. The instantaneous rate at 120 s: (0.025 M/36 s) = 6.94×10^{-4} M/s, and the instantaneous rate at 270 s: (0.025 M/105 s) = 2.38×10^{-4} M/s.

2.4.1.2 Relationships between Rates of Reactants and Products

For a reaction with a relatively simple stoichiometry, as with the decomposition of hydrogen iodide into the elements, a detailed formula for conversions is hardly required. Based on the balanced chemical equation, $2HI \rightarrow H_2 + I_2$, with the given stoichiometric coefficients, we can simply conclude that the decomposition of hydrogen iodide is twice as fast as the formation of each of the two elements formed as products. This also means that the rate of the formation of hydrogen and the rate of the formation of iodine are equal.

Another relatively simple example for producing hydrogen is the "syngas" reaction:

$$CH_4(g) + H_2O(g) \rightarrow CO(g) + 3H_2(g)$$

How much faster is hydrogen formed than carbon monoxide? The answer is evident: by 3, or hydrogen is formed three times as fast as carbon monoxide.

For chemical reactions that are stoichiometrically more complex, we rather should determine the relationships between rates of participants in the reaction using the following approach. For a balanced reaction in the general form,

$$a\text{A} + b\text{B} \rightarrow c\text{C} + d\text{D}$$

we can write

$$\text{rate} = -\frac{1}{a}\frac{\Delta[\text{A}]}{\Delta t} = -\frac{1}{b}\frac{\Delta[\text{B}]}{\Delta t} = \frac{1}{c}\frac{\Delta[\text{C}]}{\Delta t} = \frac{1}{d}\frac{\Delta[\text{D}]}{\Delta t} \tag{2.104}$$

and we can now determine the relationship between the rates of reactants and products.

Consider, for example, the combustion of butane according to

$$2C_4H_{10}(g) + 13O_2(g) \rightarrow 8CO_2(g) + 10H_2O(g)$$

What will the rate of the formation of carbon dioxide be (in M/s), if the rate for the disappearance of oxygen is 0.450 M/s? If we use equation 2.104 and rearrange it to obtain the rate of carbon dioxide, we obtain

$$-\frac{1}{13}\frac{\Delta[O_2]}{\Delta t} = \frac{1}{8}\frac{\Delta[CO_2]}{\Delta t}$$

Then,

$$\text{rate}(CO_2) = \frac{\Delta[CO_2]}{\Delta t} = 8 \times \left(-\frac{1}{13}\frac{\Delta[O_2]}{\Delta t}\right) - \frac{8}{13} \times (-0.450\ \text{M/s})$$

$$= \underline{0.277\ \text{M/s}}$$

2.4.2 The Rate Law

For a given chemical reaction, we can experimentally determine its rate law. This is not to be confused with a natural law, such as the first law of thermodynamics. The rate law is an equation for the rate of a chemical reaction.

For the general formula of a chemical reaction as

$$aA + bB \rightarrow cC + dD$$

the *general form of the rate expression* is

$$\text{rate} = k \cdot [A]^x \cdot [B]^y \qquad (2.105)$$

where k is the "rate constant," x is called the "reaction order of A," and y is the "reaction order of B." The exponents x and y are not necessarily the same as the stoichiometric coefficients a and b. The sum $(x+y)$ is the "reaction order of the whole reaction," also called more commonly just the "reaction order." The unit of the rate constant depends on the overall reaction order of a reaction.

Individual orders of reactions are more typically positive; however, they do not have to be integers. Some reactants do not affect the concentration and thus have a reaction order of zero. For the rare case of an "inhibitor," the reaction order can even be negative, which means that increasing this type of reactant actually decreases the rate.

2.4.3 Determination of the Rate Law: The Method of the Initial Rates

The purpose of the method of the initial rates is to determine the rate expression for a certain reaction using a set of *experimental* data. We initially determine the reaction orders of all reactants and then can determine the rate constant. Knowing those quantities, we can calculate or predict the rate for any set of reactant concentrations we may be interested in.

Let us assume that three reactants are involved in a chemical reaction: $2A + B + C \rightarrow D + 2E$. To be able to solve a problem using the method of the initial rates, we are usually provided with sufficient experimental data, usually in form of a short table, as shown in the following problem. Each row in the table represents data for one experiment, showing the concentrations of the reactants and the resulting initial rate. As a principle that is generally true in science and engineering, the simplest way to determine the effect, or in kinetic terms the order, of a reactant, let us say A, is to use an experiment in which the concentration of that particular reactant, that is [A], changes, whereas the concentrations of the other reactants do not, that is, the [B] and [C], must stay constant. In the same fashion, one can determine the order of the next reactant, that is, B. For its determination, one needs to pick experiments in which [B] changes but [A] and [C] do not. The same principle then is also applied to the third or, if necessary, more reactants.

We have the following three tasks:

1. Determine the specific rate expression for the following reaction.
2. Calculate the rate constant for this reaction.
3. Determine the rate when, for example, $[A] = 5.10 \times 10^{-2}$ M, $[B] = 0.142$ M, and $[C] = 0.365$ M.

For the reaction $2A + B + C \rightarrow D + 2E$, the following kinetic data were obtained:

Exp. no.	[A] (M)	[B] (M)	[C] (M)	Initial rate* (M/min)
1	5.42×10^{-3}	2.41×10^{-2}	0.268	2.73×10^{-6}
2	5.42×10^{-3}	2.41×10^{-2}	0.536	2.72×10^{-6}
3	1.08×10^{-2}	2.41×10^{-2}	0.268	5.46×10^{-6}
4	5.42×10^{-3}	4.82×10^{-2}	0.536	1.09×10^{-5}

*These initial rates have been measured experimentally and therefore contain some measurement error.

Solution:

1. Determination of the rate expression for this reaction.
   By comparing experiments 1 and 3, one can find the order of A as x:

$$\frac{\text{rate}_3}{\text{rate}_1} = \frac{5.46 \times 10^{-6}}{2.73 \times 10^{-6}} = 2 = \left(\frac{[A]_3}{[A]_1}\right)^x = \left(\frac{1.08 \times 10^{-2}}{5.42 \times 10^{-3}}\right) = 2^x$$

   Since $2^x = 2$, x must be 1.
   By comparing experiments 2 and 4, y can be determined:

$$\frac{\text{rate}_4}{\text{rate}_2} = \frac{1.09 \times 10^{-5}}{2.73 \times 10^{-6}} = 4 = \left(\frac{[B]_4}{[B]_2}\right)^y = \left(\frac{4.82 \times 10^{-2}}{2.41 \times 10^{-2}}\right) = 2^y$$

   Because $2^y = 4$, y must be 2.
   By comparing experiments 1 and 2, z can be determined:

$$\frac{\text{rate}_2}{\text{rate}_1} = \frac{2.72 \times 10^{-6}}{2.73 \times 10^{-6}} \approx 1 = \left(\frac{[C]_2}{[C]_1}\right)^z = \left(\frac{0.536}{0.268}\right)^z = 2^z$$

   Because $2^z = 1$, z must be 0.
   The rate expression for this specific reaction (at constant temperature) then is $\text{rate} = k \cdot [A][B]^2$.
   If we are not sure that the reaction order necessarily will be an integer, we can use, for example, to determine x, the following formula:

$$x = \ln(\text{rate}_3/\text{rate}_1)/\{\ln([A]_3/[A]_1)\}.$$

2. Determination of k
   Using any of the three experiments, one can then determine k, for example, with values of Exp. No. 1:

$$k = 2.73 \times 10^{-6}\ \text{M/min} \times \frac{1}{5.42 \times 10^{-3}\ \text{M}} \times \frac{1}{(2.41 \times 10^{-2}\text{M})^2}$$

$$= 0.867\ \text{M}^{-2}\ \text{min}^{-1}$$

3. Determination of the rate for $[A] = 5.10 \times 10^{-2}\text{M}$, $[B] = 0.142$ M, and $[C] = 0.365$ M.

$$\text{rate} = 0.867\ \text{M}^{-2}\text{min}^{-1} \times 5.10 \times 10^{-2}\ \text{M} \times (0.142\ \text{M})^2 = 8.92 \times 10^{-4}\ \text{M/min}$$

2.4.4 Integrated Rate Expressions

2.4.4.1 *First-Order Reactions* A reaction for which the rate is only proportional to one reactant concentration, that is, rate $= k \cdot [A]$, is called a first-order reaction. Using equation 2.102 for very small or infinitesimally small changes in the concentration with time, we obtain

$$-\frac{d[A]}{dt} = k \cdot [A] \tag{2.106}$$

It is more convenient to use the integrated form of the rate expression of a first-order reaction:

$$\ln \frac{[A]_{\circ}}{[A]_t} = k \cdot t \tag{2.107a}$$

whereby $[A]_{\circ}$ is the original reactant concentration and $[A]_t$ the concentration at a certain time t. Because of the ratio of concentrations in equation 2.107a, any property that is proportional to concentration, such as pressure, can be used to determine the first-order rate constants.

Equation 2.107a is useful to determine the time it takes to obtain a certain concentration of a reactant (e.g., at what time 75% of reactant has been converted to product). To determine the concentration obtained after a given time ($[A]_t$) or the original concentration ($[A]_{\circ}$), it is more helpful to use the algebraic form of this equation:

$$\ln [A]_t - \ln [A]_{\circ} = -k \cdot t \tag{2.107b}$$

We consider the time at which *half* of the original material has reacted, the so-called half-life. For a first-order reaction, we can convert equation 2.107a for this special case:

$$t_{1/2} = \ln \frac{1}{1/2} \times \frac{1}{k} = \frac{0.693}{k} \tag{2.108}$$

The rate constants of all first-order reactions have the unit: 1/time unit.

The decomposition of all radioactive materials follows a first order rate law. Figure 2.11 shows this characteristic pattern.

Though all radioactive materials decay by this pattern, the specific half-lives for each radioactive isotope vary significantly. For example, the uranium 235 isotope has a half-life of approximately 5 billion years, whereas that of strontium 90 is about 30 years, and that of meta-stable technetium 99 is only 6 h. In essence, ^{99m}Tc mostly decays into non harmful products within a week because of which it is used as a tracer in medical diagnosis. However, uranium 235 used as fuel in nuclear power plants stays around "forever," which is why the spent uranium fuel rods need to be carefully disposed.

Because of this first-order decay pattern, we can use radioactive substances as "internal clocks," which tell us how old a substance is. At the beginning, when the radioactive substance has just been formed, 100% of it is radioactive, but as time goes on some of the material will decay and will not be radioactive anymore, and only the remaining portion will stay radioactive. By measuring how much of the substance radiates ($=[A]_t$), we then can determine how old it is, that is, t. This method is called radioactive isotope dating.

Of the three isotopes of hydrogen, regular hydrogen (^{1}H), deuterium (^{2}H), and tritium (^{3}H), only tritium is radioactive. Tritium emits β-particles as it decays into helium-3 with a

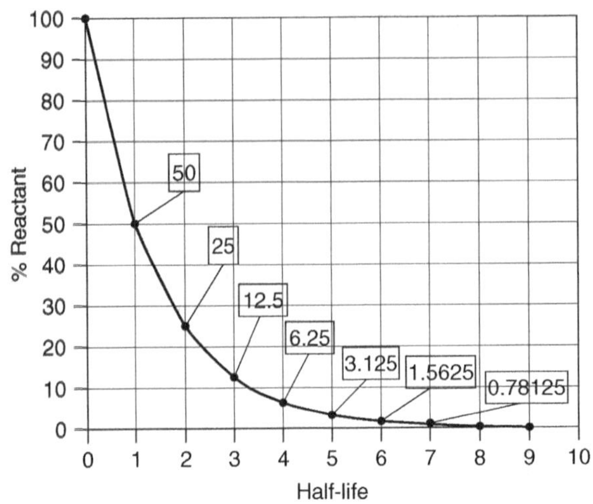

Figure 2.11 General pattern for radioactive decay and first-order reactions. In the case of radioactive decay, the reactant is a radioactive isotope emitting subatomic particles, for example, α, β, and/or γ-rays.

half-life of 12.3 years. The amount of tritium is quite small (less than $1/10^{18}$ compared to regular hydrogen). However, its presence can be measured by β-particle emission with good instrumentation such as a Geiger counter. Imagine you have a bottle of wine, and would like to know its age. As long as the grapes the wine was made from were connected to the vine, its tritium concentration remained at a constant level since it was always supplied through the ground water. Once the grapes were harvested and thus cut off from the ground water, however, the amount of radioactive tritium starts to decrease.

Assume that a sample of the wine you took has a tritium concentration that is 23% of that of fresh grapes. How old is the wine in years? First, we use equation 2.108 to determine k:

$$k = 0.693/12.3\ a = 0.0563 \cdot a.$$

Then we use equation 2.107a to determine t:

$$t = \ln \frac{1}{0.23} \times \frac{1}{0.0563} = 26.1\ a.$$

So this wine is a little more than 26 years old.

Another good example for a chemical reaction that is of first order is the decomposition of hydrogen peroxide:

$$2H_2O_2(aq) \rightarrow 2H_2O(l) + O_2(g)$$

The first-order rate expression then is

$$\frac{-d[H_2O_2]}{dt} = k \cdot [H_2O_2]$$

Figure 2.12 illustrates this specific case of this first-order reaction. The decomposition of the hydrogen peroxide in neutral aqueous solution can have a half-life of about 10 h.

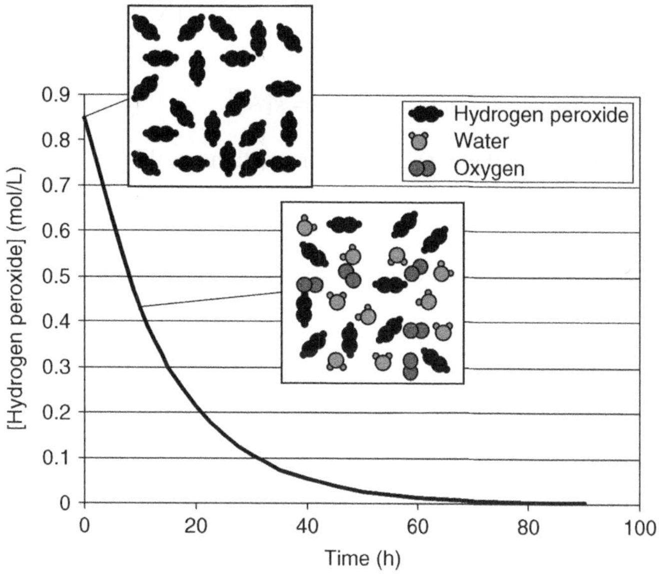

Figure 2.12 Decomposition of hydrogen peroxide as an example of a reaction based on a first-order rate law. The first "snapshot" is taken at the beginning ($t = 0$) and the second after one half-life ($t_{1/2} = 10$ h), representing the reaction mixture at elevated temperatures when all three substances are in the gaseous state.

Hydrogen peroxide decomposes much faster in the presence of a catalyst, as we will discuss later in this section.

Let us solve a problem related to the decomposition of phosphine, a first-order reaction, according to

$$4\,PH_3(g) \rightarrow P_4(g) + 6\,H_2(g).$$

(a) Find the rate constant of this reaction, if the pressure of phosphine dropped from originally 0.825–0.660 atm in 47.5 s?

(b) Determine the pressure after 80 s? Using equation 2.107a,

Solution:

(a)

$$k = \ln \frac{0.825}{0.660} \times \frac{1}{47.5\ \mathrm{s}} = 4.70 \times 10^{-3}\ \mathrm{s}^{-1}$$

(b)

$$\ln p(\mathrm{A}) = -4.70 \times 10^{-3}\ \mathrm{s}^{-1} \times\ 80\ \mathrm{s} + \ln 0.825 = -0.568$$

$$p(\mathrm{A}) = 0.567\ \mathrm{atm}$$

2.4.4.2 Second-Order Reactions Some chemical reactions follow a second-order rate law, which means that the overall reaction rate is two. In more typical examples, two reactants have each a reaction order of one, that is, rate = k[A][B]. If we have only one reactant with a reaction order of two, we can write

$$\text{rate} = \text{d[A]}/\text{d}t = k[\text{A}]^2 \tag{2.109}$$

The integrated rate law for the latter second-order reaction has this form:

$$1/[\text{A}]_t = k \cdot t + 1/[\text{A}]_0 \tag{2.110}$$

The unit of the rate constant of second-order reactions is L/mol × 1/time unit, for example, 1/(M s).

The half-life for a second-order reaction is defined as

$$t_{1/2} = \frac{1}{k[\text{A}]_0} \tag{2.111}$$

Notice that in contrast to the half-life expression for a first-order rate law, here the half-life is also dependent of the original reactant concentration.

Consider a reaction: A reaction is second order according to 2 X(g) → Y(g)

(a) What is the rate constant of this reaction, if it takes 5.50 s for the concentration of X to drop from 0:612 to 0:124 M?
(b) What is the half-life (in seconds) under these circumstances?

Solution:

a) We convert equation 2.108 to resolve for the rate constant;

$$k = \left(\frac{1}{[\text{A}]_t} - \frac{1}{[\text{A}]_\circ}\right)\frac{1}{t} = \left(\frac{1}{0.124\,\text{M}} - \frac{1}{0.612\,\text{M}}\right)\frac{1}{5.50\,\text{s}} = 1.17\,\text{M}^{-1}\text{s}^{-1}$$

b) The half-life can be found by plugging the given value for $[\text{A}]_\circ$ and the calculated value of k.

$$t_{1/2} = \frac{1}{k[\text{A}]_\circ} = \frac{1}{1.17\ \text{M}^{-1}\text{s}^{-1} \times 0.612\ \text{M}} = 1.40\ \text{s}$$

2.4.4.3 *Zero-Order Reactions* Zero-order reactions do not occur as frequently as first- and second-order reactions. Often they involve reactions at heterogeneous surfaces, for example, the decomposition of dinitrogen oxide into the elements on platinum catalyst ($2N_2O(g) \rightarrow 2N_2(g) + O_2(g)$).

In zero-order reactions, the following relations must be true:

$$\text{Rate} = -\frac{\text{d}[A]}{\text{d}t} = k[\text{A}]_\circ = k \tag{2.112}$$

and integrated: $[\text{A}] = [\text{A}]_\circ - kt$. This means that the rate for zero-order reactions is independent of the concentration of the reactant. However, in reality, some reactant A must be present to observe this independence.

In Figure 2.13, three diagrams are shown that show the change of concentration with time for the zero-, first-, and second-order reactions. In the case of zero-order reactions, no mathematical conversion is necessary, since equation 2.111 already represents a straight-line plot. For the first- and second-order reactions, the straight line plots are given in analogy to equations 2.107b and 2.110.

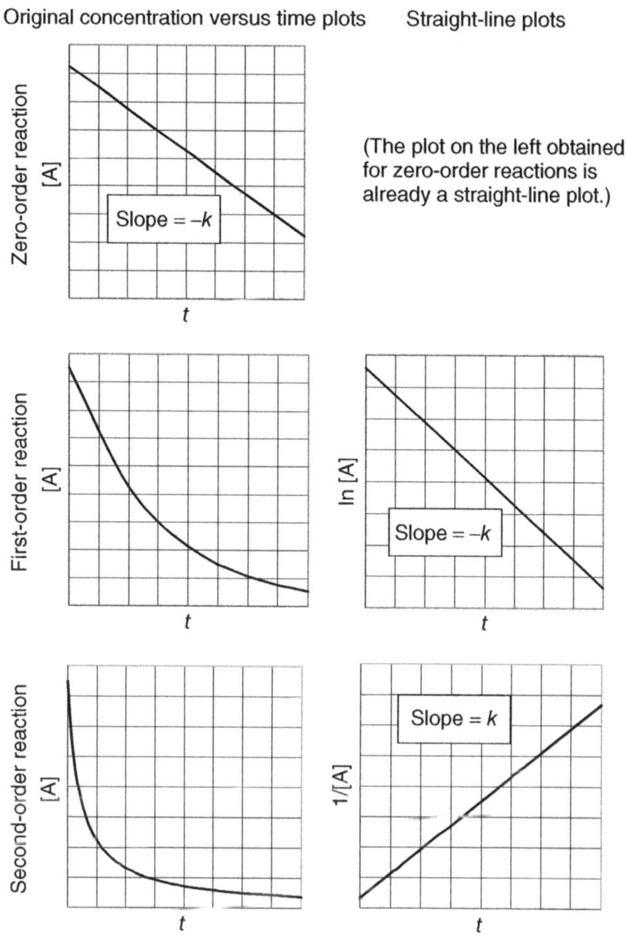

Figure 2.13 Concentration versus time dependence for the zero-, first-, and second-order reactions with straight-line plots.

2.4.5 Collision Theory

We have learned that the faster a typical reaction, the higher are the reactant concentrations. This is because at higher reactant concentrations, more collisions between reactant molecules take place. However, there are further requirements for a chemical reaction to proceed successfully, that is, to lead to the formation of the products:

1. The velocities of the reactant molecules must be sufficiently high, that is, they must possess a minimum kinetic energy, E_{kin}, for the particular reaction.
2. The molecules must also collide with the right geometric orientation.

These two conditions are illustrated in Figure 2.14.

The minimum kinetic energy level molecules or atoms need to successfully convert to product is also called activation energy, E_a. In Figure 2.15, the activation energy is indicated as part of an energy/reaction coordinate plot.

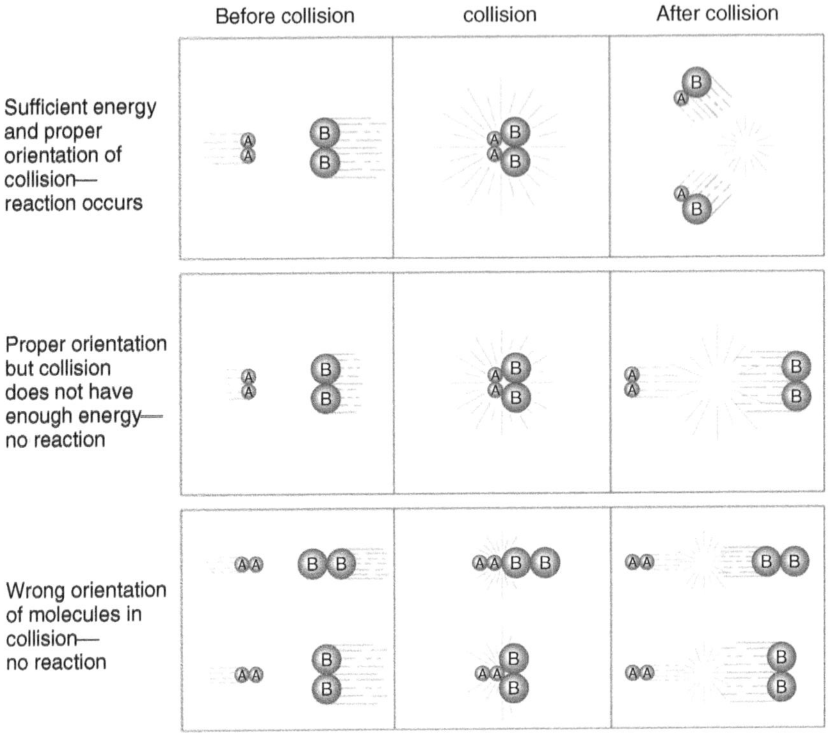

Figure 2.14 Different scenarios for the collision of reactant molecules. [*Source*: Malone (2004) [104]. Reproduced with permission of John Wiley & Sons.]

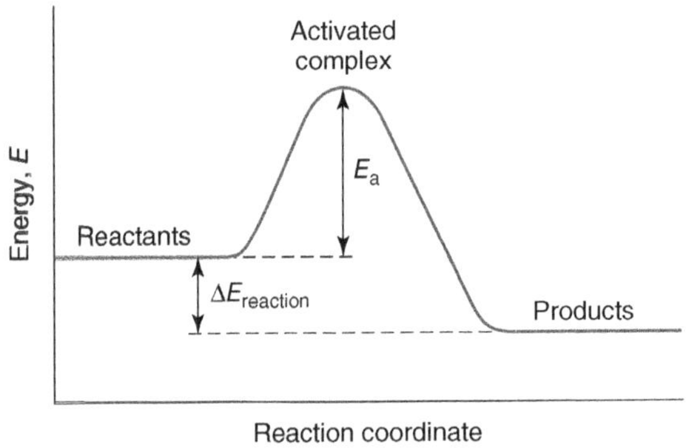

Figure 2.15 Energy/reaction coordinate diagram for the example of an exothermic reaction (uncatalyzed). The reaction kinetics is related to how the reactants are converted to the products, that is, the reaction pathway. Instead of activated complex, the term transition state is also frequently used. [*Source*: Olmsted and Williams (2006) [103]. Reproduced with permission of John Wiley & Sons.]

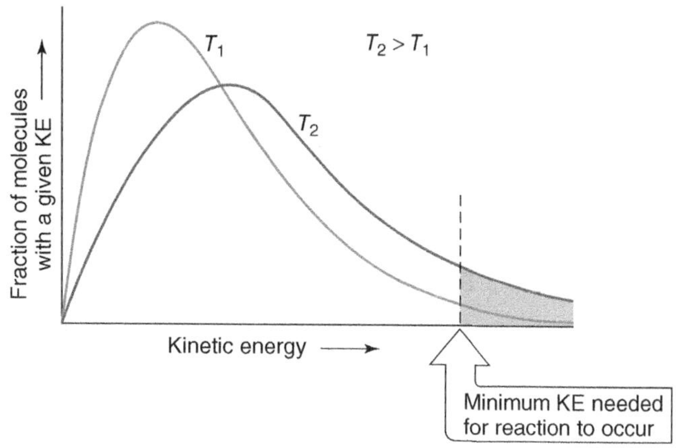

Figure 2.16 Distribution of kinetic energies of molecules at two different temperatures $T_1 < T_2$. The colored areas below curves correspond to number of molecules that react. [*Source*: Malone (2004) [104]. Reproduced with permission of John Wiley & Sons.]

2.4.6 The Effect of Temperature on the Rate

With an increase in temperature, the rate of a chemical reaction typically goes up. As the temperature is raised, not all reactant molecules will convert to product, however. Rather, with increasing temperature the number of reactant molecules that have a kinetic energy higher than the activation energy also increases, and only these molecules will be able to convert to product. The kinetic energy of the molecules depends on their velocity according to $E_{kin} = mv^2/2$. The velocity of the molecules and therefore their energy are statistically distributed. The distribution for their different kinetic energies is plotted in Figure 2.16.

2.4.6.1 Dependence of Rate Constant on Temperature The relation between the rate constant and the temperature for a specific reaction is called the Arrhenius equation:

$$k = Ae^{\frac{-E_a}{R \cdot T}} \tag{2.113}$$

Equation 2.113 expresses that as the temperature T increases, the rate constant k increases. Realize that as T increases, the exponent of the base of the natural logarithm, e, also increases due to the negative sign. E_a, the activation energy, is constant for a given reaction. A is the frequency factor, a constant and geometry-dependent term, and R is the ideal gas constant, used as 8.314 J/(mol K).

2.4.6.2 Comparison of Rates at Two Temperatures Assume the activation energy for a reaction is 84.0 kJ/mol. How much faster will this reaction occur if one raises the temperature from 25 to 35°C?

What we need to do is to compare the rates at the two different temperatures as represented by the rate constants, determined as k_2/k_1, whereby we chose T_1 to stand for the lower temperature, that is, 25°C, and T_2 for the higher temperature, respectively. We then obtain

$$\ln\frac{k_2}{k_1} = \frac{84.0\,\text{kJ/mol}}{8.314\,\text{J/(mol} \times \text{K)} \times 10^{-3}\text{kJ/J}}\left(\frac{1}{298.15\text{K}} - \frac{1}{308.15\text{K}}\right) = 1.10$$

Then

$$\frac{k_2}{k_1} = 3.00$$

So the reaction will be three times faster. Many natural chemical reactions have a lower activation energy close to $E_a = 53$ kJ/mol. For these reactions, an increase of 10°C in temperature doubles the rate of the reaction.

2.4.7 The Mechanism of a Reaction

The mechanism of a chemical reaction describes the detailed events by which a chemical reaction proceeds. We can also consider this as kinetics at the *microscopic* level. We learned earlier that, for the determination of the abovementioned rate laws (at the *macroscopic* level), relatively simple lab equipment suffices. We can measure when a reaction has been essentially completed by using, for example, a color indicator and determining the time of the reaction with a stopwatch. However, to determine the mechanism of many chemical reactions, more sophisticated instrumentation is required. To study the mechanism of a reaction one needs to determine the actual pathway of the reaction and find empirically which chemical species are formed during the reaction ("activated complex" as shown in Figure 2.15, also called "transition state").

One experimental issue is that generally the chemical species that are formed during the reaction are very short-lived and therefore cannot be isolated or captured. If one waits too long, all one obtains is (are) the expected product(s), without any clue of how it was formed from the reactant(s). Therefore, ideally instrumentation is required that can take "snapshots" of the transition state which in some cases is possible by using spectroscopic methods or lasers. Such characterizations require the so-called *in situ* measurements, which mean the measurements have to take place at the location of the reaction and right during the formation of new species.

The chemical reactions, which take place at the most detailed, that is, microscopic, level are termed *elementary reactions*. We use the term *molecularity* to label the reaction orders at the microscopic level. Three examples of molecularities are given with their corresponding elemental reactions in the following paragraph. If the molecularity of the reaction is known, its rate law can be predicted. Since the elementary reaction represents exactly which molecules react, it is straightforward to determine the reaction orders or molecularities. However, realize that only for elementary reactions can the rate law be directly predicted from the stoichiometry of the reaction, and this is not true for the more commonly used chemical equations at the macroscopic level mentioned earlier in this section on kinetics.

Elemental reaction	Molecularity
$A \rightarrow B$	Unimolecular
$A + B \rightarrow C$	Bimolecular
$A + B + C \rightarrow D$	Trimolecular

Most chemical reactions are *bimolecular*, in agreement with our collision model mentioned above, that is, two molecules collide forming product(s). *Unimolecular* reactions, involving decompositions of single molecules, are quite rare. The chances for three molecules to meet at the right time and place, *trimolecular* reactions, are also quite low.

Even reactions that macroscopically are of first order often involve bimolecular steps. Once the mechanism of a reaction has been revealed, and its molecularities are known, the orders for the macroscopic reaction also can be determined; however, the reverse, that is, the determination of elementary steps from macroscopic kinetics, is not possible.

If a reaction is composed of several elementary steps, only the slowest of these will be rate determining. Consider the following example, in which the rate law for the net reaction is asked for:

1.	$A + B \rightarrow C$	(slow)
2.	$C + D \rightarrow E$	(fast)
3.	$E + D \rightarrow F + G$	(fast)
Net reaction:	$A + 2D \rightarrow F + G$	

Solution:

It would be wrong to apply only the net reaction as the basis for the rate law, that is, rate = $k \cdot [A] \cdot [D]^2$ is incorrect. Since only the first step is slow and therefore rate determining, we must use it to determine the correct rate law:

$$\text{Correct rate law:}\quad \text{rate} = k \cdot [A] \cdot [B]$$

2.4.7.1 Determining the Mechanism for a Reaction with an Initial Fast Step Involving Equilibrium In certain cases, a mechanism may exist with a known rate law that appears to involve more than two steps, including an initial fast step involving an equilibrium. For example, consider the decomposition of dinitrogen pentoxide according to

$$2N_2O_5(g) \rightarrow 4NO_2(g) + O_2(g)$$

We can assume the following steps as part of the reaction mechanism:

$$N_2O_5 \underset{k_{-1}}{\overset{k_1}{\rightleftharpoons}} NO_2 + NO_3 \quad \text{(fast, equilibrium)}$$

$$NO_2 + NO_3 \overset{k_2}{\rightarrow} NO + NO_2 + O_2 \quad \text{(slow)}$$

$$NO_3 + NO \overset{k_3}{\rightarrow} 2\ NO_2 \quad \text{(fast)}$$

Experimentally, the rate law was determined as

$$\text{Rate} = k\ [N_2O_5]$$

However, based on the slowest step the rate expression should be

$$\text{Rate} = k_2[NO_2]\ [NO_3]$$

The intermediate NO_3 does not appear in the overall equation. In order to resolve this issue, we need to find a relation that allows substitution for this intermediate with the substances in the familiar overall equation.

At the equilibrium, the rates of the forward and the reverse equations must be equal:

$$\text{Rate}_{\text{forward}} = k_1[\text{N}_2\text{O}_5] = \text{Rate}_{\text{reverse}} = k_{-1}[\text{NO}_2][\text{NO}_3]$$

Now we can rearrange this to express

$$[\text{NO}_3] = k_1[\text{N}_2\text{O}_5]/k_{-1}[\text{NO}_2]$$

If we now substitute for $[NO_3]$ in the rate expression for the rate-determining step, we obtain

$$\text{Rate} = k_2[\text{NO}_2][\text{NO}_3] = k_2[\text{NO}_2]k_1[\text{N}_2\text{O}_5]/k_{-1}[\text{NO}_2]$$

Finally, we can place all constants in the front and obtain

$$\text{Rate} = k_2k_1/k_{-1}[\text{N}_2\text{O}_5]$$

This is the same equation that was found to be the experimentally determined rate law, so in this case the mechanism is in agreement and the slow step is still rate determining.

2.4.8 The Influence of a Catalyst on the Rate of a Reaction

A catalyst influences the reaction rate in the following manner:

- It speeds up the chemical reaction, both in the forward and reverse directions.
- It causes a lowering of the activation energy of the chemical reaction.
- It changes the pathway of the chemical reaction.
- It is neither consumed, nor produced in the net reaction.

In Figure 2.17, the effect of a catalyst on the activation energy is illustrated.

A catalyst has no influence on the equilibrium's position according to Le Chatelier's principle, nor does it influence the value of the equilibrium constant as a change in temperature does. However, when a catalyst is added to system at the beginning of a reaction, it takes less time to reach the state of equilibrium.

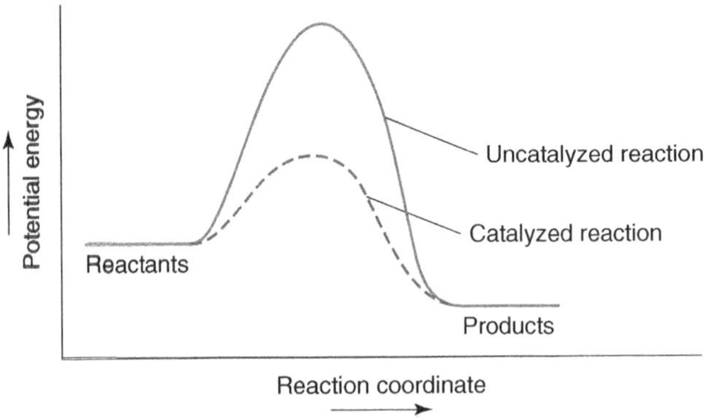

Figure 2.17 The effect of a catalyst on a given reaction: it lowers the activation energy of that reaction. [*Source*: Malone (2004) [104]. Reproduced with permission of John Wiley & Sons.]

A general scheme involving a catalyst and an intermediate is given here:

1.	$A + B \rightarrow C + D$
2.	$C + E \rightarrow F + B$
Net reaction:	$A + E \rightarrow D + F$

In this net reaction, B is a catalyst because it reacts in the first step and then is produced in the second step; thus, overall it is not consumed. In contrast, C is an intermediate, because it is the product of the first step but then reused as a reactant in the second step.

For example, the decomposition of hydrogen peroxide can be catalyzed by many substances. The following reaction scheme shows one such decomposition:

1.	$H_2O_2(aq) + I^-(aq) \rightarrow OI^-(aq) + H_2O(l)$
2.	$H_2O_2(aq) + OI^-(aq) \rightarrow I^-(aq) + H_2O(l) + O_2(g)$
Net:	$2H_2O_2(aq) \rightarrow 2H_2O(l) + O_2(g)$

(a) What is the catalyst and (b) what is an intermediate?

The *solution*: Neither the iodide, I^-, nor the hypoiodide, OI^-, appear in the net reaction.

a) I^- is the catalyst because it is consumed in the first step but formed in the second step.

b) OI^- is a product in the first step but a reactant in the second step. It is the intermediate.

2.4.8.1 *Types of Catalysts*

1. *Homogeneous catalyst*: A homogeneous catalyst is present in the same physical phase (gas, liquid, or solid) as the reactants. *Example*: Chlorine radicals that catalyze ozone decomposition.
2. *Heterogeneous catalyst*: A heterogeneous catalyst is located in a different phase than the reactants. *Example*: Rh/Pt in a catalytic converter used in automobiles (converting CO into CO_2, and NO, NO_2 into N_2).
3. *Enzymes*: Catalysts occurring in biological systems are called enzymes. Their shape is very important for their function, termed also "key/lock" model. In Figure 2.18, it is shown how the enzyme sucrase splits sucrose, a disaccharide consisting of glucose and fructose.

A substance that also affects the rate of a chemical reaction, but in the opposite direction compared to a catalyst, that is, it *slows down a reaction by increasing the activation energy*, is called an *inhibitor*. Inhibitors are used industrially to avoid undesired reactions. For example, antioxidants are added to food or polymers to retard reactions leading to their deterioration.

2.4.8.2 *Heterogeneous Catalysis* One industrially important example of a heterogeneous catalysis is applied in the so-called Haber process for the synthesis of ammonia

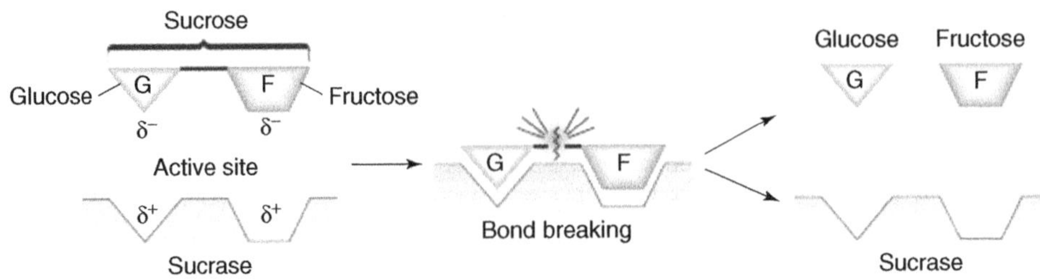

Figure 2.18 In this scheme, it is shown how the shape of the enzyme (here sucrase) fits with that of a substrate (here, sucrose) to catalyze the reaction, splitting of the substrate into smaller compounds as products (here glucose and fructose). [*Source*: Malone (2004) [104]. Reproduced with permission of John Wiley & Sons.]

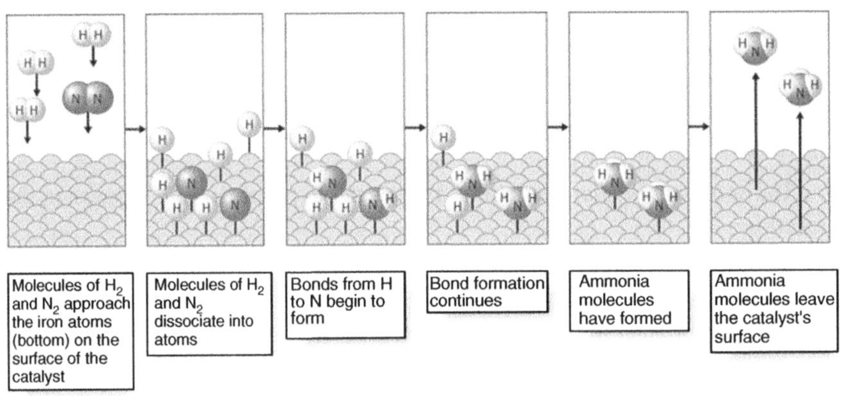

Figure 2.19 Heterogeneous formation of ammonia from hydrogen and nitrogen. [*Source*: Brady (2004) [105]. Reproduced with permission of John Wiley & Sons.]

from the elements (Figure 2.19). The presence of gaseous reactants and product and the solid catalyst makes this an example of a heterogeneous catalysis.

$$3H_2 + N_2 \xrightarrow{Fe} 2NH_3$$

For this process, Le Chatelier's principle (discussed earlier) is applied to favor product formation. The ammonia is removed from the system by liquefying it in a condenser. A relatively high pressure of 200 atm is required because the product has half the volume of the reactants based on the stoichiometric coefficients in the balanced equation. Although the process is exothermic, the process is run at approximately 400°C to keep the rate of the reaction high. To further increase the rate, a heterogeneous catalyst is added—iron. Without the iron, the probability of the reactant molecules to meet with the right orientation is relatively small.

The reaction shown in Figure 2.19 goes through six different stages:

1. Initially, each of the gaseous reactant molecules, hydrogen and nitrogen, become physically approach the metal surface, also called *adsorption*.
2. The molecules separate into atoms.

3. The atoms *migrate* toward each other.
4. New N–H bonds are formed.
5. The formation of ammonia molecules is completed.
6. Finally, the ammonia molecules leave the surface and go into the gas phase, also called *desorption*.

Though the formation of ammonia involves a heterogeneous catalysis involving hydrogen, it is frequently considered an example of a *nitrogen fixation*.

2.4.9 A Closer Look at Hydrogenations: Catalysis Involving Addition of Hydrogen

Most hydrogenations, that is, the addition of hydrogen to a substance, are catalyzed reactions. Hydrogenations are mostly used to convert organic compounds into different forms, for example, ethylene, according to

$$CH_2 = CH_2 + H - H \rightarrow CH_3 - CH_3$$

Without the presence of a catalyst, hardly any hydrogenation occurs below 500°C. The majority of hydrogenation catalysts contain late transition metals such as platinum, palladium, rhodium or ruthenium, or nickel. The latter is nonprecious and therefore used for more economical processes. These heterogeneous catalysts exist frequently as pure forms of the transition metals. As in the case of the described ammonia synthesis, the catalyst initially adsorbs the gaseous compounds, that is, the hydrogen and the organic compound, and helps in the faster formation of new bonds on the surface.

Due to the hydrogen being gaseous, such reactions need to be carried out in high-pressure steel reactors, also termed *autoclaves*, at pressures between 100 and 3000 psi. For example, vegetable oils are converted to margarine by adding hydrogen to double bonds in the longer hydrocarbon chains in the presence of a nickel catalyst. This leads to solidification of the product, so it can be spread on bread.

Nickel is often used in industry in the form of the so-called Raney catalyst. This catalyst is prepared from an alloy of nickel and aluminum by adding sodium hydroxide. The base dissolves the aluminum and what remains is a porous and therefore highly active nickel catalyst. For hydrogenations in which the purity of the product is more relevant, pure platinum is used as catalyst. For the selective hydrogenation of alkynes to alkanes, a palladium catalyst, which is poisoned deliberately with $CaCO_3$, is most preferable, called the Lindlar catalyst. The conversion of alkynes into alkenes is typically done with pure palladium as catalyst (Adam's catalyst).

In contrast to the above-mentioned type of heterogeneous hydrogenation, many modern hydrogenation catalysts are homogeneous. They represent metalorganic compounds, which contain a central atom, such as platinum, palladium, or rhodium, which is coordinatively bonded to other atoms such as chlorine and typically to organic groups, called *ligands*. Because of the organic ligands, these compounds are soluble in organic solvents, in contrast to the pure metals or their inorganic salts. One of the earlier discovered and efficient hydrogenation catalysts is the Wilkinson catalyst, $(Ph_3P)_3RhCl$, called chlorotris(triphenylphosphine)rhodium(I), shown in Figure 2.20. This catalyst is frequently used for the preparation of alkanes from alkenes.

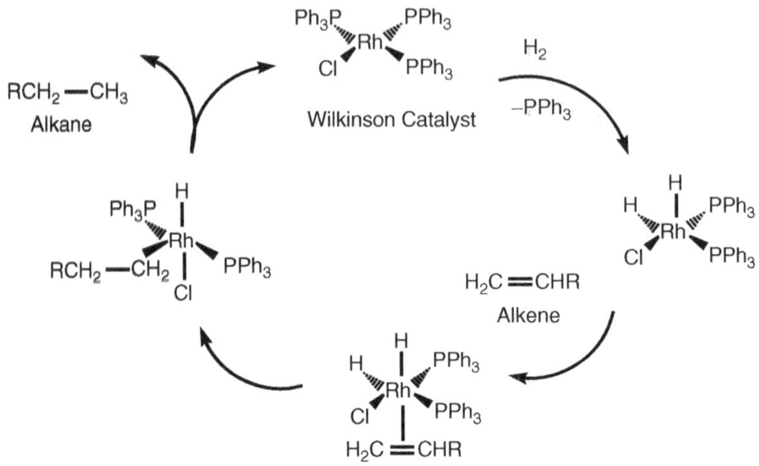

Figure 2.20 The Wilkinson catalyst—an example of a homogeneous catalyst.

The Wilkinson catalyst can be prepared by reacting rhodium(III) chloride trihydrate with triphenylphosphine in ethanol. Once activated, the central metal atom in the catalyst forms new coordinative bonds, whereby the square planar complex forms an octahedral structure. The mechanism of this homogeneous hydrogenation involves an oxidative addition of hydrogen, that is, the metal actually loses two electrons as it adds the two hydrogen atoms while releasing one of the triphenyl phosphine groups. The two hydrogen atoms are then added to the alkene in two subsequent steps.

There are several advantages of homogeneous versus heterogeneous hydrogenation. A homogeneous catalyst is active not only at its surface but also on a molecular level, and therefore a larger quantity of the catalyst is efficient, resulting in higher rates. The homogeneous complexes can be designed to obtain higher yields of hydrogenated product. In cases where the stereochemistry of the product is important, such as for chiral compounds, homogeneous catalysts are also more versatile. On the other hand, heterogeneous catalysts are easier to separate from the final product. This disadvantage of a homogeneous hydrogenation catalyst can be overcome by distributing it on an inert carrier or support. Furthermore, heterogeneous catalysts are generally easier to prepare.

Very important catalysts for the production of hydrocarbons from hydrogen and carbon monoxide are iron and cobalt catalysts used in the Fischer–Tropsch syntheses. These products are the basic materials for many other organic compounds, which are used for numerous important applications, such as the production of medicines, detergents, and plastics. Heterogeneous catalysts are mostly used for this reaction; however, recent work with homogeneous catalysts has also been proven possible. For more details on these hydrogenations, see Section 2.6.

Hydrogenation catalysts are also used for the refining of crude oil to remove sulfur. These processes are collectively also termed *hydrodesulfurization*. In these processes, heterogeneous catalysts composed of molybdenum and cobalt oxides on alumina are applied to convert the sulfur into hydrogen sulfide, which can be easily removed from the product of interest. For example,

$$RSH + H_2 \rightarrow RH + H_2S$$

2.4.10 Summary

To summarize this section, we conclude that the rate of a given reaction can be expressed through the rate law for that reaction. The rate law always expresses the rate as the rate constant, which, except for zero-order reactions, is multiplied with the concentration(s) of reactant(s) using exponents for each reactant's order. Therefore, the rate constant allows one to determine the rate at any concentration of reactant(s), as long as the temperature has not been changed or a catalyst has not been added. An increase in temperature usually leads to an increase in the rate constant, governed by the Arrhenius equation, however, not to a change in the activation energy of a reaction. The only method we discussed qualitatively that even causes a change of the activation energy for a given reaction is the addition of a catalyst.

2.5 ELECTROCHEMISTRY (OXIDATION–REDUCTION REACTIONS)

Electrochemistry deals with oxidation–reduction reactions. Oxidation is a process of removal of electron/electrons and reduction is a complimentary process of addition of electron/electrons. Oxidation reactions occur at the anode and reduction reactions occur at the cathode. Taking the example of electrolysis of water, at the anode water is oxidized to oxygen. This process involves the removal of electrons. Equation 2.114 gives the number of electrons released.

$$2\,H_2O \rightarrow O_2 + 4\,H^+ + 4\,e^- \tag{2.114}$$

or

$$H_2O \rightarrow 1/2O_2 + 2\,H^+ + 2e^- \tag{2.115}$$

Equation 2.114 states that 2 mol of water are oxidized to 1 mol of oxygen molecule, releasing four electrons. Equation 2.115 represents that 1 mol of water is oxidized to a half mole of oxygen molecule releasing two electrons. The latter statement is helpful in the context of thermodynamics. An oxidation–reduction reaction is often abbreviated as a "redox" reaction.

The question of how many electrons are involved in any oxidation–reduction reaction can be easily answered by following the rules discussed in Chapter 5, which are specifically oriented toward the hydrogen technology. A redox reaction, however complex it may be, can be solved for the number of electrons by using those ion-electron rules. Table 2.4 gives a few representative redox reactions.

The number of electrons required for the reduction as shown in Table 2.4 would be predictable based on the charge neutralization basis. Some redox reactions proceed through an intermediate step involving a smaller number of electrons. For example, $Cu^{2+} + e \rightarrow Cu^+$ is also a viable reaction. However, it is reactive and hence may not live in solution under normal conditions. It could be stabilized by forming complexes.

2.5.1 Types of Electrochemical Cells

Two types of electrochemical cells are possible. In the first type, the free energy change in the chemical reaction is manifested as the electrical energy. In the second type, electrical energy is used to bring about chemical reactions at the electrodes. The first type of cell is called a galvanic cell and the second type is an electrolytic cell.

TABLE 2.4 Table of redox reactions at 25°C

$Cu^{2+}(aq) + 2e = Cu(s)$	0.340 V
$Zn^{2+}(aq) + 2e = Zn(s)$	−0.762 V
$Pb^{2+}(aq) + 2e = Pb(s)$	−0.125 V
$Ag^{+}(aq) + e = Ag(s)$	0.799 V
$Na^{+}(aq) + e = Na(s)$	−2.714 V
$K^{+}(aq) + e = K(s)$	−2.925 V
$Li^{+}(aq) + e = Li(s)$	−3.045 V
$Al^{3+}(aq) + 3\ e = Al(s)$	−1.670 V
$Sn^{2+}(aq) + 2\ e = Sn(s)$	−0.136 V
$Cl_2(g) + 2\ e = 2\ Cl^{-}(aq)$	1.358 V

Source: Bard and Faulkner (2001) [108]. Reproduced with permission of John Wiley & Sons.

A large number of galvanic cells are available today. The Daniel cell was one of the earliest ones to be developed using Zn metal acting as anode in 1 M zinc sulfate solution and Cu cathode in a solution of 1 M copper sulfate. At anode, Zn metal is oxidized and zinc ion is formed. Copper ion is reduced at the Cu cathode. The two reactions are written as

$$Zn(s) \rightarrow Zn^{2+}(aq) + 2e \tag{2.116}$$

$$Cu^{2+}(aq) + 2e \rightarrow Cu(s) \tag{2.117}$$

The overall reaction is sum of the above two reactions:

$$Zn(s) + Cu^{2+}(aq) \rightarrow Zn^{2+}(aq) + Cu(s) \tag{2.118}$$

CRITICAL THINKING QUESTION

What is the difference between a Galvanic cell and electrolytic cell?

Galvanic cell is one in which free energy change is manifested as an output voltage; for example, Li battery. All batteries fall into this category.

Electrolytic cell is an example of driving the reactions in the desired direction. Electrolysis of water to produce hydrogen and oxygen is an example of an electrolytic cell. Here energy is expended to produce chemicals.

The galvanic cell develops a potential of 1.10 V. This voltage is a function of concentration of the electrolytes, temperature, and agitation. Other examples of galvanic cells are batteries that we use in everyday life.

CRITICAL THINKING QUESTION

1 Place a Cu metal plate in a solution of $ZnSO_4$ solution. In another case, place a Zn metal in a solution of $CuSO_4$. Predict which one will undergo reaction.

Answer:

$Cu\ (s) + ZnSO_4\ (aq) = $ No reaction

Zn (s) + $CuSO_4$ (aq) = will show a spontaneous deposition of copper on zinc. Copper ion is reduced on zinc surface. For reduction of copper ion, electrons are required. These electrons come from zinc metal. We can write the reaction as

$$Zn\,(s) = Zn^{2+}(aq) + 2e\ (Zn)$$

$$\underline{Cu^{2+}(aq) + 2e\ (Zn) = Cu\,(s)}$$

$$\underline{Zn\,(s) + Cu^{2+}(aq) = Zn^{2+}(aq) + Cu\,(s)}$$

Examine the $CuSO_4$ solution. It contains only Cu^{2+} and $SO_4{}^{2-}$. No undissociated $CuSO_4$ is present. It is a strong electrolyte. By placing zinc metal in this solution,

$$Zn(s) + Cu^{2+}(aq) + SO_4{}^{2-}(aq) = Zn^{2+}(aq) + Cu(s) + SO_4{}^{2-}$$

Examine the reactant side and product side of the above equation you notice $SO_4{}^{2-}$(aq) is present on both sides. So it has not taken part in the reaction and hence it is called the spectator ion. So virtually we can write

$$Zn(s) + Cu^{2+}(aq) = Zn^{2+}(aq) + Cu(s)$$

The above equation is called the net reaction.

Why Cu(s) placed in $ZnSO_4$ shows no reaction is because the Gibbs free energy change for zinc to be deposited on Cu is positive. It is an uphill reaction. The above illustrations are useful in understanding the operation of Daniel cell (Figure 2.21).

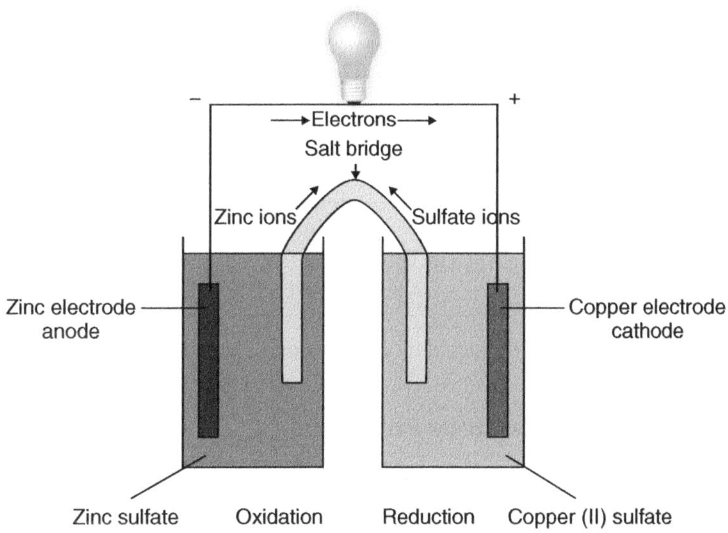

Figure 2.21 Daniel cell.

2.5.2 Electrode Potentials

A metal that is placed in a solution of its ions develops a potential. For example, zinc metal placed in a solution of $ZnSO_4$ develops a potential called a single electrode potential. Similarly, Cu metal in a solution of $CuSO_4$ develops a potential. Although a metal develops a potential in a solution, it is not possible to measure the single electrode potentials; viewed in simpler term, a multimeter has two input leads. If one lead is connected and the other one is hanging loose, the meter will not show any reading. In order for the meter to respond, it is necessary to connect the other lead to another cell developing a potential. In the previous example of a Daniel cell, Zn metal and Cu metal are connected to the two leads of the meter to give a reading. The meter reads the difference in the two potentials. It does not give the value of each single electrode potential. The single electrode potential is measured with respect to a common electrode such as hydrogen electrode. Please note that this also is not a true value. However, using a common electrode potential concept, it has been possible to predict the voltages of galvanic cells and free energy of reactions. A hydrogen electrode is called a reference electrode and, using this reference, all the single electrode potential values have been measured.

2.5.3 Hydrogen Electrode

A hydrogen electrode consists of a platinum black immersed in a solution of 1 M HCl with hydrogen gas at 1 atm pressure passing over it. Platinum black is formed using H_2PtCl_6. The potential of the platinum surface depends on the H^+ ion concentration of the solution and on partial pressure of the hydrogen gas. The electrode potential is zero at all temperatures. A symbolic representation of the hydrogen electrode is shown as follows:

$$\mathrm{Pt, H_2}(p = 1\ \mathrm{atm})/\mathrm{H^+}(1\ \mathrm{M}) \tag{2.119}$$

When this electrode is used as anode, oxidation of hydrogen occurs releasing two electrons. Used as cathode, reduction of hydrogen ion occurs to produce hydrogen molecule. The potential of this electrode is taken arbitrarily as zero.

Other reference electrodes have been generated whose potentials are measured with reference to hydrogen electrode. The following list gives the secondary reference electrodes and their potentials:

Electrode	Representation	Potential[a]
Saturated calomel electrode (SCE)	$Hg/HgCl_2$, saturated KCl	0.2415 V
Silver/silver chloride electrode (Ag/AgCl)	Ag (s)/AgCl(s)	0.2230 V
Calomel electrode (NCE)	$Hg/HgCl_2$, 1 M KCl	0.2682 V

[a]Potentials are with reference to normal hydrogen electrode at 25°C.

2.5.4 Measurement of Electrode Potentials

A normal or standard hydrogen electrode is a hypothetical standard reference for measurement of electrode potentials. However, it is not convenient for use in the laboratory and, hence, a calibrated reference electrode such as the ones discussed in Section 2.5.3

is used in determining the potentials. These reference electrodes are easy to handle and can be commercially bought at nominal cost. For measurement of electrode potentials, a cell is constructed by combining the reference electrode with the one whose potential is to be determined. The electromotive force (EMF) of such a cell is measured. From the known value of the reference electrode, the potential of the unknown system is determined using

$$\text{Cell EMF} = E_c - E_a \tag{2.120}$$

E_c is the cathode potential and E_a is the anode potential. If the potential of the unknown system is measured using 1 M solution, then the electrode potential of such a system is called standard electrode potential. For example, Zn electrode in a solution of 1 M $ZnSO_4$ will give a potential that will be referred to as the standard electrode potential. Similarly, all systems whose potentials are measured using 1 M solutions will be listed as standard electrode potentials. The standard potential

a) is dependent on temperature T;
b) is a relative quantity with respect to a common reference such as normal hydrogen electrode;
c) will not change sign by contacting metal; and
d) measures the driving force for the half cell reaction.

If measurements of electrode potentials are done at any other concentration, then the potential is shifted as given by the following equation established by Nernst:

$$E = E^\circ - \{RT/nF\}\ln\{a_{red}/a_{ox}\} \tag{2.121}$$

Here, E° is the standard electrode potential, a_{ox} and a_{red} are activities of oxidized and reduced species in solution, n is the number of electrons transferred, and F is the Faraday Let us take, for example, a redox reaction:

$$AgCl(s) + e \rightarrow Ag(s) + Cl^-(aq) \tag{2.122}$$

The standard potential for this half-cell reaction is 0.220 V. For reaching this potential, it is necessary to use 1 M NaCl solution. However, if we use a smaller concentration of NaCl, then the Nernst potential is given by

$$E = E^\circ_{Ag/AgCl} - \{RT/F\}\ln\{a_{Ag}a^-_{Cl}/a_{AgCl}\} \tag{2.123}$$

Taking activity of solid components as unity at 298 K, the above equation is simplified to

$$E = E^\circ_{Ag/AgCl} - 0.0592\log a^-_{Cl} \tag{2.124}$$

$$= 0.220 - 0.0592\log C^-_{Cl}\cdot f^-_{Cl} \tag{2.125}$$

where $a = C \times f$ is used in arriving at equation 2.125. f is the activity coefficient of the ion and C is the concentration.

2.5.5 Cell Voltages

The availability of standard potentials enables the calculation of cell voltages. In Section 2.5.1, it was shown that the Daniel cell gives a voltage of about 1.10 V. Let us calculate the cell voltage that would be expected from the Daniel cell using the standard potential. The standard potentials of the redox couples involved are

$$Zn^{2+}(aq) + 2e \rightarrow Zn(s) \quad E^\circ = -0.762\ V$$

$$Cu^{2+}(aq) + 2e \rightarrow Cu(s) \quad E^\circ = 0.340\ V$$

Since zinc is anode and Cu is the cathode (see Section 2.5.1),

$$E = E_c - E_a$$

$$E = 0.340 + 0.762 = 1.102\ V$$

This value is in agreement with the experimentally determined voltage for the Daniel cell. Using this approach, it is possible to calculate the voltages and construct batteries or fuel cells that are used in practical life. This approach of predicting the fuel cell voltage is discussed in detail in Chapter 5.

2.6 ORGANIC CHEMISTRY

2.6.1 Organic Chemistry and Organic Compounds

Organic chemistry deals with the chemistry of organic compounds, which always contain carbon atoms. The oxides of carbon and carbonates traditionally are not considered organic. Organic chemistry is separated from inorganic chemistry, that is, the chemistry of all the other compounds, which are not organic, because there are so many compounds of the former category. By now, we have identified more than 10 million organic compounds compared to less than a million compounds that are inorganic. Based on the average amount of compounds per element, compounds of carbon, which is only 1 of 109 known elements, are favored in nature by more than 1000:1 compared to the average number of compounds per other element. Naturally, there are also many chemical reactions involving organic substances. The separation into these two areas of chemistry therefore is mainly organizational, and does not mean the fundamental nature of organic chemistry would be different from inorganic chemistry. We also use the same type of molecular structure formulas, called Lewis formulas, which are used for inorganic compounds, such as water or sulfuric acid, and also throughout organic chemistry.

Organic chemistry is important in our everyday lives. Dyes, perfumes, medicines, pesticides, and plastics all contain organic compounds. Our food contains organic substances. An orange, for example, contains several hundred such compounds. All forms of life, that is, plants, animals, and humans, must contain organic compounds. Therefore, the ingredients of most of our fuels, predominantly crude oil and coal, are organic compounds, because they were formed millions of years ago from either microorganisms or plants, respectively. There is a separate discipline in chemistry, called biochemistry, which deals exclusively with the chemistry of life. However, many modern, synthetic materials are made of polymers, that is, plastics, rubber, and fibers, which also are mostly organic. We will discuss these further in the subsequent section on polymers (2.7).

2.6.1.1 The Uniqueness of Carbon Carbon is unique. It hardly forms ionic bonds but mostly covalent bonds due to its four valence electrons. Carbon atoms can form strong covalent bonds to their own kind and hydrogen. As a result, carbon can form linear, branched, or cyclic molecules, as shown in Figure 2.22. It is because of the covalency of the carbon bonds to the other atoms that organic compounds typically differ in their macroscopic properties from compounds with polar or ionic bonds. While most inorganic compounds are metal salts with ionic bonds, organic compounds take the form of gases, liquids, and solids. With increasing molar mass of organic compounds, their melting and boiling points increase. Organic substances are typically soluble in each other because of their similar nonpolar nature, while they do not typically dissolve in water. Therefore, organic compounds are at best weak electrolytes. Almost all organic compounds are combustible, and many can be used as fuel, which is not the case in inorganic compounds.

Figure 2.23 gives an overview of the classification of organic compounds. If only carbon and hydrogen are present in a compound, it is called a *hydrocarbon*. If the hydrocarbon contains only single bonds, it is called an *alkane*. However, carbon atoms can also form multiple bonds between each other. If a compound contains at least one double bond between two carbon atoms, it is called an *alkene*. If a hydrocarbon contains a triple bond, it is named an *alkyne*. Notice that in these singular forms the third to last letter, a, e, or y, is critical to determine to which subgroup the hydrocarbon belongs. Compounds containing only single carbon–carbon bonds are saturated, while those containing multiple bonds are unsaturated. In addition, carbon atoms can form rings consisting of single bonds, called (ali-)*cyclic* compounds, and they can form unsaturated rings, called *aromatic* compounds. In their individual representatives, aromatic compounds carry similar to the alkenes an e as the third letter from the end, for example, benzene (see its structure in Table 2.5).

Organic compounds do not include only hydrocarbons but also substances that contain, in addition to atoms of carbon and hydrogen, atoms of other main group elements, also

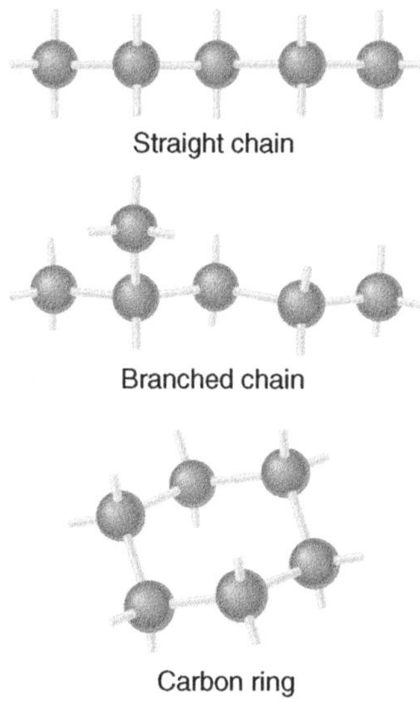

Figure 2.22 Carbon atoms can form straight and branched chains and rings. [*Source*: Brady (2004) [105]. Reproduced with permission of John Wiley & Sons.]

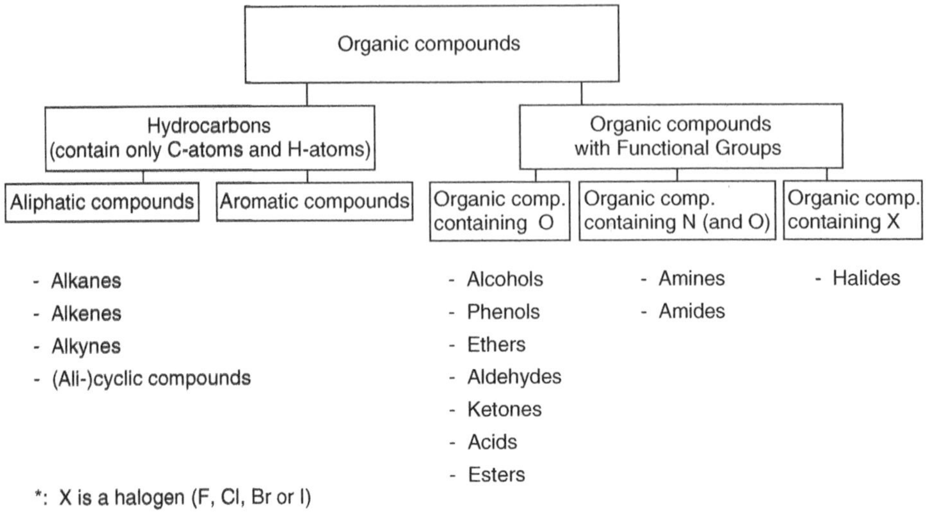

Figure 2.23 Classification of organic compounds. Hydrocarbons are discussed in more detail in Sections 2.6.2–2.6.6 and organic compounds with functional groups in Sections 2.6.7.1–2.6.7.7.

called heteroatoms. Sometimes, these latter types of substances are also called organic compounds containing *functional groups* because of their higher reactivity compared to hydrocarbons (however, in a sense, multiple bonds also are functional because, as we will see later, they make alkenes more reactive than the alkanes). The most dominant atom next to carbon and hydrogen in organic compounds is oxygen, present in ethers, alcohols, aldehydes, ketones, carbocylic acids, and esters. Nitrogen atoms are part of amines and amides, and organic halides. Several organic compounds contain silicon, phosphorous, or sulfur, but will not be included in our discussion.

2.6.2 Hydrocarbons

2.6.2.1 The Main Representatives of Hydrocarbons and Their Shapes In Table 2.5, the names and formulas of representative hydrocarbons and their shapes as ball-and-stick models are given. Notice that the additional structural formulas are simpler to draw because they represent projections of the three-dimensional ball-and-stick models.

The shapes of these hydrocarbons and related organic compounds are related to the hybridizations of the carbon atoms they contain. These hybridizations are described in most textbooks under chemical bonding and we will not further explain them other than to remind the reader that in all alkanes the carbons undergo sp^3-hybridization with tetrahedron angles between C–H bonds; in alkenes and benzene sp^2-hybrids are formed, resulting in C–H bond angles of 120°; and in alkynes containing sp-hybrids, those bond angles are 180°. In the multiple bonds, the remaining p-orbitals form π bonds. Therefore, double or triple bonds cannot be rotated freely around the bond axis, as is the case for a single bond. This has important consequences regarding the numbers of isomers for unsaturated compounds versus saturated compounds, discussed later.

The condensed formulas are even more efficient because in them no C–H bonds are shown, since they only appear in one simple form. For alicyclic and aromatic hydrocarbons it is very common to even omit the element symbols for hydrogen for efficiency. However, keep in mind that they are really present, and have to be included, for example, to calculate

TABLE 2.5 Names, structures, and formulas for representative hydrocarbons

Class	Representative[a]	Ball-and-stick model	Structural formula	Condensed formula	Molecular formula
Alkanes	Methane	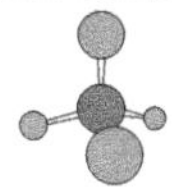	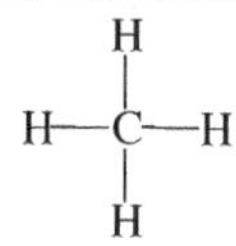	CH_4	CH_4
	Ethane	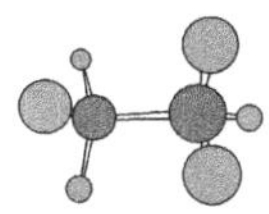	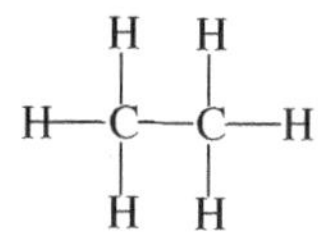	CH_3-CH_3	C_2H_6
Alkenes	Ethylene (Ethene)	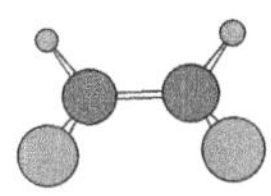	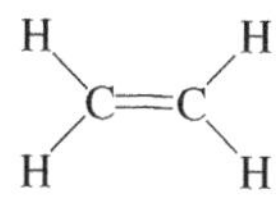	$CH_2{=}CH_2$	C_2H_6
Alkynes	Acetylene (ethyne)	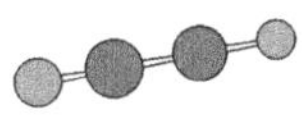	H—C≡C—H	CH≡CH	C_2H_2
Alicyclic compounds	Cyclohexane	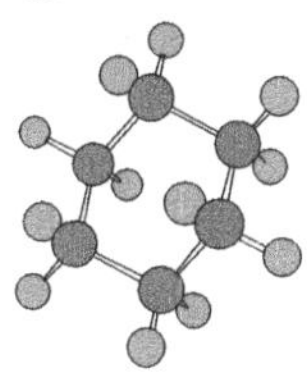	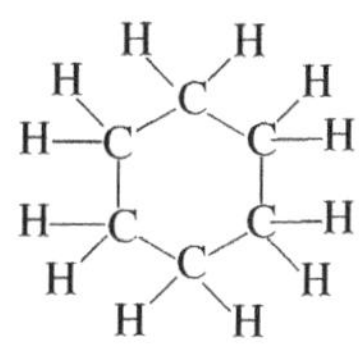	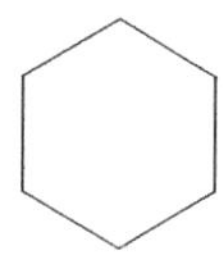	C_6H_{12}
Aromatic compounds	Benzene	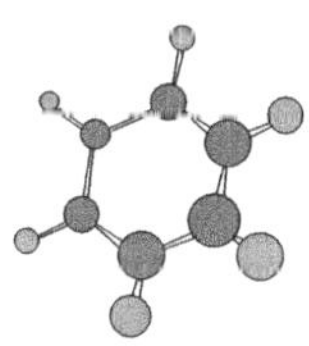	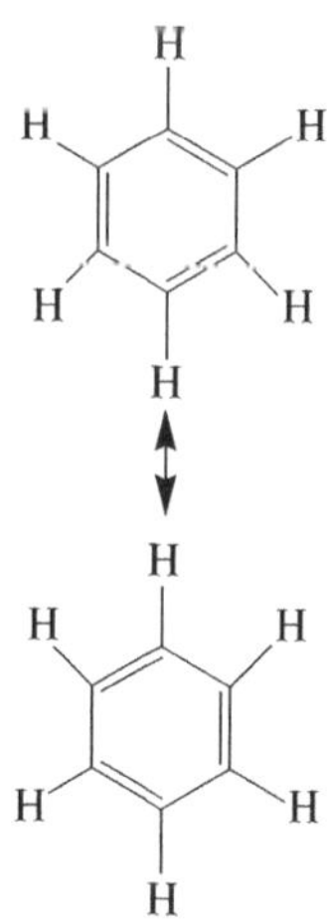	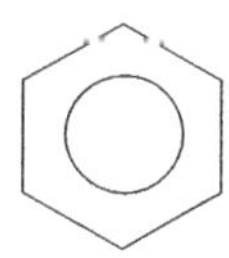	C_6H_6

Gray-colored atoms are carbon atoms and lighter gray atoms are hydrogen atoms.
[a]If IUPAC names are different from common names, they are given in brackets.

the molar mass of the molecule (in case of cyclohexane the molar mass is = 84 g/mol and for benzene 78 g/mol).

2.6.3 Alkanes

In Table 2.6, the common names of the alkanes with up to 10 carbons are listed, including some of their isomers. All alkanes have the general formula C_nH_{2n+2}. The names for the first

TABLE 2.6 Alkanes with up to 10 C-atoms and some of their branched isomers

Number of C-atoms in alkane	Name of alkane	Molecular formula general: C_nH_{2n+2}	Number of possible isomers, including linear alkane	Name of related isomer(s)	Condensed structure of branched isomer
1	Methane	CH_4	1	None	None
2	Ethane	C_2H_6	1	None	None
3	Propane	C_3H_8	1	None	None
4	Butane	C_4H_{10}	2	Isobutane	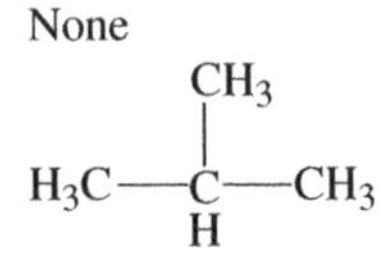
5	Pentane	C_5H_{12}	3	Isopentane	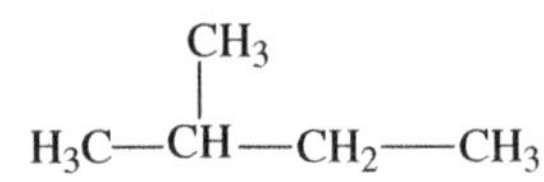
				Neopentane	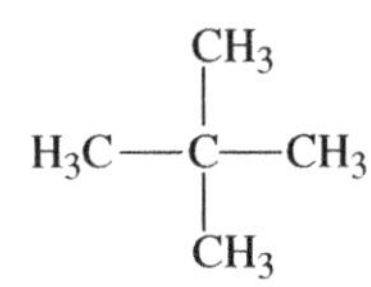
6	Hexane	C_6H_{14}	5	(IUPAC names preferred)	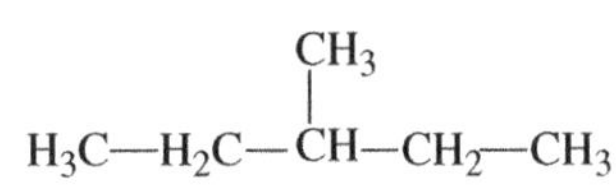
7	Heptane	C_7H_{16}	9	(IUPAC names preferred)	—
8	Octane	C_8H_{18}	18	(IUPAC names preferred)	—
9	Nonane	C_9H_{20}	35	(IUPAC names preferred)	—
10	Decane	$C_{10}H_{22}$	75	(IUPAC names preferred)	—

four alkanes, that is, with up to four carbon atoms, are solely based on historical reasons. Starting with five carbons we use Greek numbers, that is, penta for five, then hexa for six, and so on. In the older nomenclature, the unbranched alkanes were called *n*-alkanes, with *n* standing for normal; however, they can be distinguished by just carrying the plain alkane name, for example, butane, whereas only their isomers require differentiation, for example, isobutane. Remember that the isomers of a substance have one and the same molecular formula but different spatial structures. In Figure 2.24, three representations of the compound pentene are shown, demonstrating that the shape can change due to single bond rotation.

2.6.3.1 Sources and Properties of Alkanes Alkanes can be easily obtained from crude oil by fractionated distillation, which implies that the oils' ingredients are carefully separated by their boiling points. The scheme of the fractionation of crude oil is shown in Figure 2.25. Because alkanes have only single bonds, they cannot undergo too many chemical reactions and are used mainly for a few important applications. Many of them are used as fuels, including gases released from the oil. The liquid alkanes with a few carbon

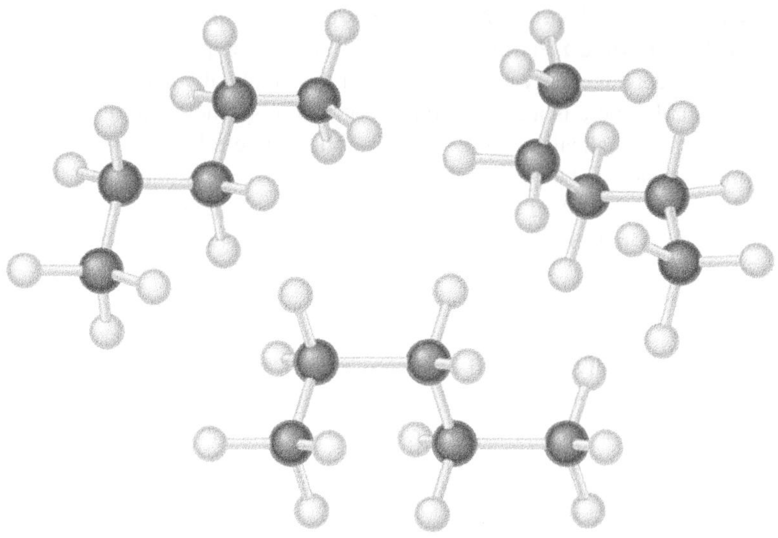

Figure 2.24 Three representations of pentane. [*Source*: Brady (2004) [105]. Reproduced with permission of John Wiley & Sons.]

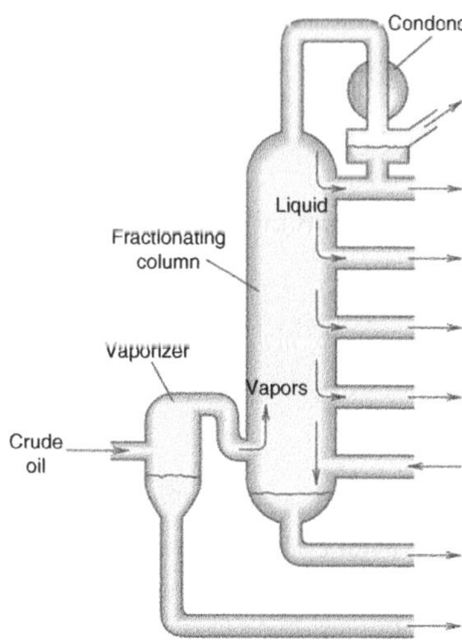

Fraction	Approximate number of carbons	Approximate boiling range (°C)	Major uses
Gases	1–5	0–80 (collected in this range)	Home-heating, cooking fuel, and factory use
Petroleum ethers	5–7	30–110	Solvents
Gasoline	6–12	30–200	Automobile fuel
Kerosene	12–24	175–275	Jet fuel, some home heating, portable stoves and lamps
Gas oil	18+	250–400	Heating oil, diesel fuel
Steam			
Residue 2	19+	300+	Lubricants, paraffin wax, petroleum jelly
Residue 1	—	—	Asphalt, pitch, petroleum, coke (paving, coating, and structural uses)

Figure 2.25 Alkanes obtained from the fractionation of crude oil. [*Source*: Malone (2004) [104]. Reproduced with permission of John Wiley & Sons.]

atoms can be used as solvents in the chemical industry. As the number of carbon atoms reaches 20, the alkanes become viscous enough to be used as lubricants in engines and as semisolid materials such as paraffin oil or waxes used in cosmetics. Whatever cannot be boiled in the initial heating of the crude oil mostly becomes asphalt for the paving of our roads and roofs.

2.6.3.2 *Syntheses of Alkanes from Hydrogen and Carbon Monoxide: Syngas and the Fischer–Tropsch Process* As described earlier, petroleum or oil is the most important fuel to meet our energy needs. Whenever the supply of crude oil drops or its price increases remarkably, there is a need to replace it with a synthetic alternative or produce feedstock products synthetically. At the same time, cleaner petroleum products, which have a lower sulfur content, can be produced. A very efficient method to achieve the synthesis of hydrocarbons is the so-called Fischer–Tropsch process. This method was developed in the early 20th century in Germany, which has plenty of coal but no oil. It is represented by the following reaction:

$$(2n + 1)H_2 + nCO \rightarrow C_nH_{2n+2} + nH_2O$$

The reaction is typically catalyzed by iron or cobalt and run at temperatures between 200 and 350°C and elevated pressure (e.g., 25 bar). Its reactants can be produced by the partial combustion of methane. For this reason, untransportable or "stranded" natural gas can be applied. A more frequently used source for the two reactants is syngas, which is the mixture of hydrogen and carbon monoxide.

2.6.3.3 *Syngas: A Source for Hydrogen* It is the mixture of the reactants for the Fischer–Tropsch process; however, it does not contain any organic compounds. We will discuss here briefly what syngas is and how it is made. Syngas consists of a mixture of varying levels of hydrogen, carbon monoxide, and carbon dioxide. It is formed by the gasification of coal or simpler hydrocarbon products.

$$C(s) + O_2(g) \rightarrow CO_2(g)$$

$$CO_2(g) + C(s) \rightarrow 2\,CO(g)$$

$$C(g) + H_2O(g) \rightarrow CO(g) + H_2(g)$$

It can be also produced via the steam reforming reaction from natural gas:

$$CH_4(g) + H_2O(g) \rightarrow CO(g) + 3\,H_2(g)$$

Syngas is used primarily as fuel, for the industrial synthesis of hydrogen.

2.6.4 Alkyl and Aryl Groups

As can be seen in Table 2.6, the number of hydrocarbons even containing only 10 carbon atoms can get very large due to the many isomers. The latter cannot be distinguished by trivial names, as are used for the simpler unbranched representatives, because it would require too much memorization and would be clearly inefficient. Primarily for this reason, names for *groups* that are bonded to larger structures in organic molecules have been introduced. These are generally represented as -R and their names are derived from the basic common names of the alkanes already introduced above. For example, by removing a hydrogen atom from ethane, we obtain the group ethyl, changing the ending from -ane to -yl. A list of the most frequently used alkyl and two aryl groups is given in Table 2.7.

Note that the names for the groups are never used by themselves. There is no substance called methyl or another called ethyl. The reason the first branched alkane is isobutane, while the first isomeric alkyl group is already isopropyl is that the latter as a group has a bond to a potential, and yet undefined partner atom. In other words, if you would remove

TABLE 2.7 Names and structures of alkyl and aryl groups

Number of C-atoms	Name of group	Molecular formula	Condensed structural formula of group and its isomeric group(s)
1. Alkyl groups			
1	Methyl-	CH_3–	H_3C–
2	Ethyl-	C_2H_5–	H_3C–CH_2–
3	Propyl-	C_3H_7–	H_3C–CH_2–CH_2–
3	Isopropyl-	*i*-C_3H_7–	CH_3 | H_3C—CH—
4	Butyl-	C_4H_9–	H_3C–CH_2–CH_2–CH_2–
4	Isobutyl-	*i*-C_4H_9-	CH_3 | H_3C—CH—CH_2—
4	*sec*-Butyl-	*s*-C_4H_9–	CH_3 | H_3C—CH_2—CH—
4	*tert*-Butyl-	*t*-C_4H_9–	CH_3 | H_3C—C— | CH_3
2. Aryl groups			
6	Phenyl-	C_6H_5–	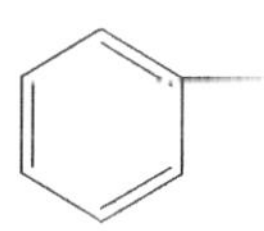
7	Benzyl-	C_6H_5-CH_2–	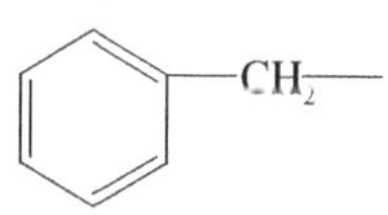

from isobutane the methyl group at its end, which is further away from the branch, you obtain isopropyl. Two examples of compounds containing an isopropyl group are isopropyl alcohol and isopropyl chloride.

Because of its four valence electrons, a carbon atom must always form four bonds. If these are four single bonds, the following four possibilities can arise, which are shown as follows. If the carbon is at the end of the chain, it forms a methyl group (including the hydrogens), and it is called a primary carbon atom. If it is between two other carbon atoms in a chain, it forms a methylene group and itself becomes a secondary carbon. If the carbon is at a branching point, it forms a methine group and it becomes a tertiary carbon. Finally, if a carbon should be connected to four other carbon atoms, it becomes a quaternary carbon and, since no hydrogen is connected to in this case, it is not considered a group.

CH_3-	—CH_2-	—CH— (with one further bond below)	—C— (with further bonds above and below)
Methyl-group	Methylene-group	Methine-group	
Primary C-atom	Secondary C-atom	Tertiary C-atom	Quaternary C-atom

2.6.5 The Systematic Names of Hydrocarbons and Introduction to the IUPAC Nomenclature

Knowing now how we name alkanes and the smaller alkyl groups, we can construct the names of thousands of more complex alkanes. This systematic nomenclature (i.e., procedure of naming) was developed by the International Union for Pure and Applied Chemistry and therefore is also called IUPAC nomenclature. One of their goals was to introduce a completely unambiguous system that would allow scientists to differentiate between all the millions of organic compounds. In the IUPAC nomenclature, an organic compound is subdivided into structural elements, which compose the name. To be efficient, these structural elements should be as large as possible.

For some of the very simple hydrocarbons, such as the simple alkanes, there is no need for further simplification and the common chemical name will be the same as the IUPAC name, for example, hexane. However, branched alkane isomers are much easier to name using the IUPAC nomenclature. We will discuss the names of alkanes followed by the names of alkenes to learn the governing principles for the IUPAC nomenclature, which can be applied also to other organic compounds containing other functionalities or heteroatoms. Instead of alkyl or alkyl groups, we could also generally use the term *substituent*, which also could include, for example, a chlorine atom.

2.6.5.1 *IUPAC Rules to Name Alkanes*

1. *Identify the longest continuous chain in the molecule, and use it for the parent name, placing it at the end of the name to be completed.* If there is a choice between continuous chains use the one with the most branches. Try to avoid using names of higher alkyl groups unless necessary. Keep in mind that the single bonds in a continuous chain can rotate and may form corners in the two-dimensional projection.
2. *Determine all alkyl groups, which branch off from the continuous chain, in the compound, and place their names in front of the parent name.* Assign each group the smallest possible number, to indicate the *position* of this group at the continuous chain (the particular C-atom in the longest chain the group is attached to). Start numbering the carbons in the continuous chain at the end with more branches.
3. *If several groups with different lengths or type are present, name them in alphabetical order.* For example, ethyl comes before methyl in the name. Prefixes in italics such as *sec*- or *tert*- are neglected, but iso-, neo-, or cyclo must be considered in this step, for example, as in 5-*tert*-butyl-6-isopropyl-3-methyl dodecane.
4. *If several groups of the same type are present, use the prefixes di, tri, tetra, and so on. If more than one branch of the same type is linked to a main-chain carbon, the positioning number is used multiple times, separated by a comma.* These prefixes for numbers are not considered relevant in the alphabetical ordering. For example, as in 2,3,3-trimethyl-4-propyl-nonane.

2.6.5.2 *Examples of IUPAC Names of Alkanes*

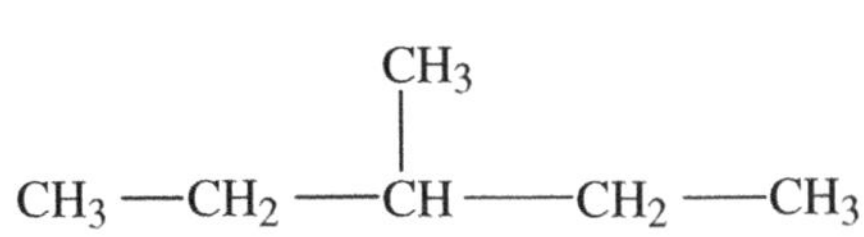

3-Methyl-pentane

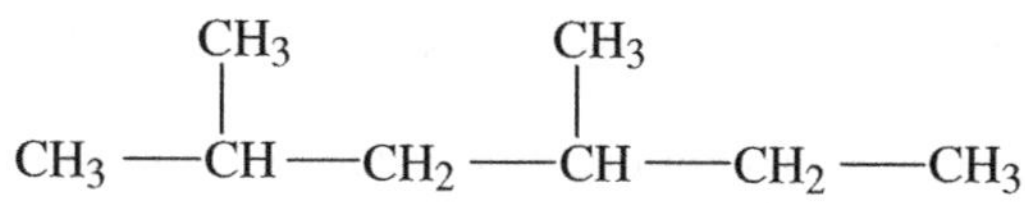

2,4-Dimethylhexane

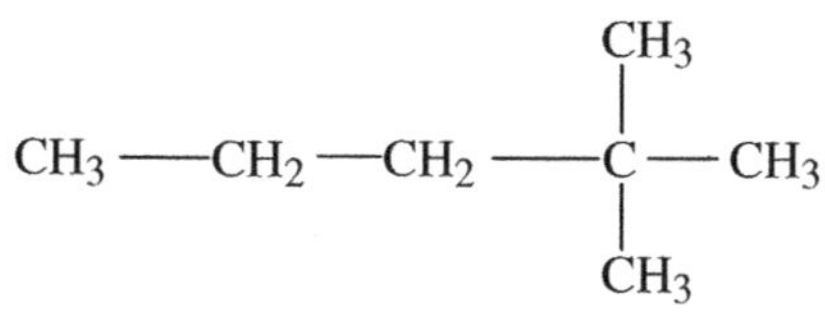

2,2-Dimethylpentane

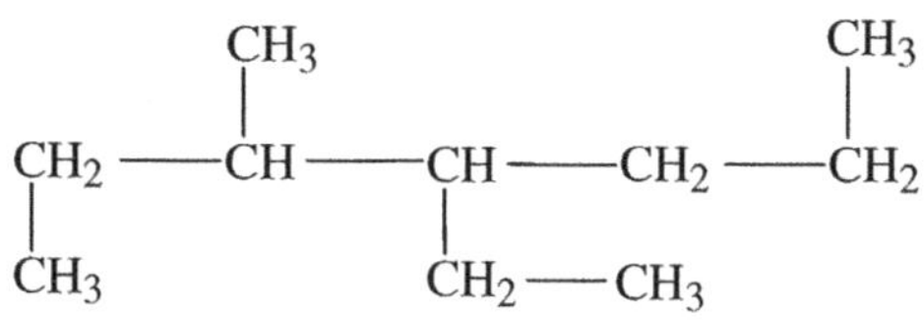

4-Ethyl-3-methylheptane

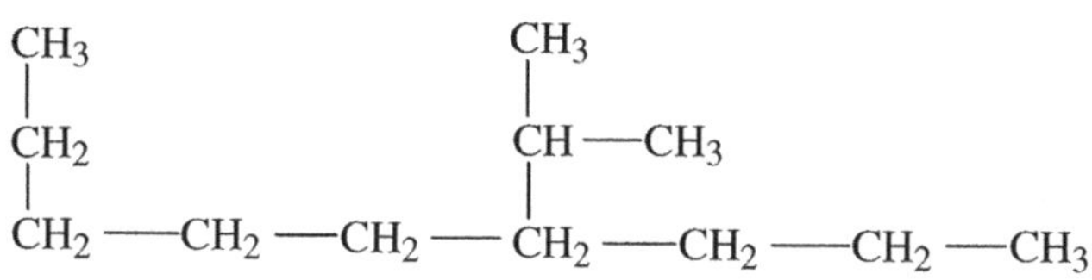

4-Isopropylnonane

2.6.6 Alkenes

Alkenes are produced in refineries by "cracking" longer alkanes, resulting in the formation of double bonds, for example, butane can be split into ethane and ethylene. Because their double bond(s) are reactive, alkenes can undergo many reactions, particularly addition reactions, some of which are shown in Figure 2.26. The example of ethylene shows that a

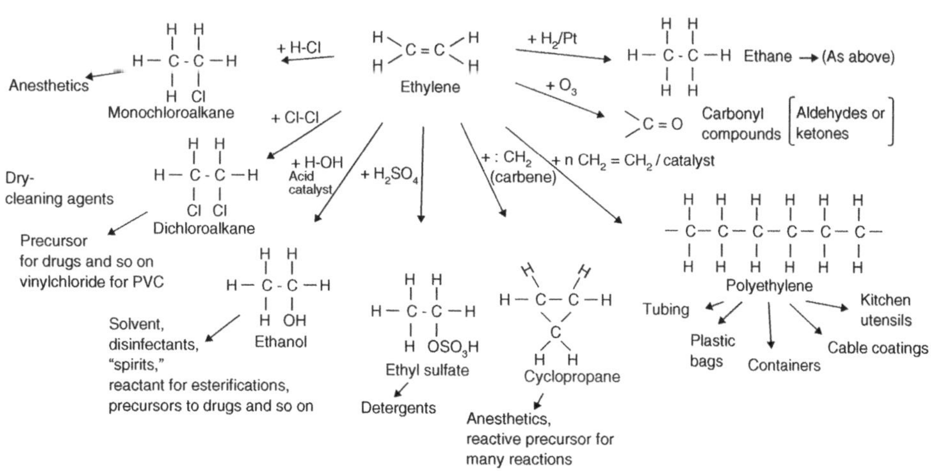

Figure 2.26 Reactions of ethylene as an example of an alkene and the products that can be formed. [*Source*: Miri (2006) [51]. Reproduced with permission of Pearson.]

variety of commercially important products can be obtained, including precursors for other materials such as pharmaceuticals.

We will discuss in the following sections the naming of one more category of organic substances, the alkenes. This is to learn how the IUPAC nomenclature works in principle for all other organic compounds other than alkanes. Because there are so many organic compounds, in the naming process the functional groups have to be ranked by importance, that is, prioritized. As a general principle, in IUPAC nomenclature the first letters of an organic compound are identical to those of the alkane with the same number of carbons, followed by an ending that indicates the functional group. For example, the IUPAC name of any compound that contains four carbon atoms always starts with the initial letters *eth-*. But the examples given for the alkenes illustrate this principle further.

2.6.6.1 IUPAC Names of Alkenes The IUPAC names of simple alkenes differ from their common names. The IUPAC name of the simplest alkene, which contains two carbon atoms, is ethene (instead of the common name ethylene). The first letters indicate as a general principle the number of carbons present in the compound, while the ending indicates that it is an alkene. In the same manner, an alkene with three carbons is called propene (and not propylene as the common name), and so on.

To name more complex alkenes, similar rules are used as for the naming of branched alkanes mentioned earlier. The double bonds have priority over the branches, however, which means that the double bond(s) must be part of the parent name, and that it(they) is(are) numbered with the smallest number possible. Therefore, for hydrocarbons containing (a) double bond(s) the following rules are applied:

1. The longest continuous chain must contain the double bond(s), and use it for the parent name having now the ending *-ene*. (e.g., a continuous chain with eight carbons and a double bond is called an octene.)
2. Identify the position of each double bond, using the smallest number possible to indicate where each double bond starts. (e.g., 3-octene has a double bond between the third and fourth carbon.)
3. If the continuous chain contains more than one double bond, use *before* the -ene ending di- tri-, and so on to indicate their number. (For example, 1,3-butadiene has two double bonds at each end of the molecule.)

2.6.6.2 Geometric Isomerism Because a carbon–carbon double bond cannot freely rotate, in contrast to a carbon–carbon single bond, another form of isomerism can occur than that of the isomerism due to branching mentioned earlier. This type of isomerism is commonly referred to as *cis/trans* isomerism. If there are substituents (i.e., atoms other than hydrogen or groups of atoms) on each of the two carbon atoms they can be oriented in the *cis* or in the *trans* position.

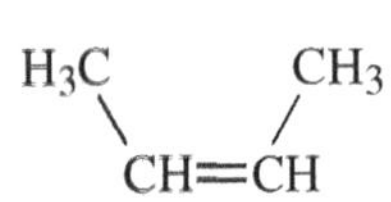

cis-2-Butene

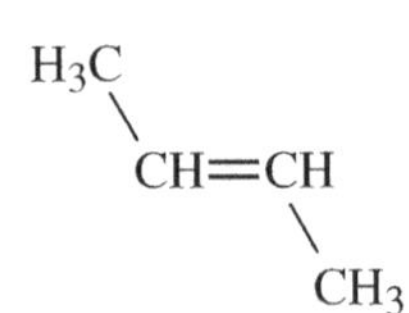

trans-2-Butene

If the substituents are on the same side of the double bond, the prefix *cis* is used; if they are on opposite sides, *trans* is used. To better understand this, you can draw a dotted line through the double bond's longer axis, as shown in the diagram above.

2.6.6.3 Examples for IUPAC-Naming Alkenes

1-Butene

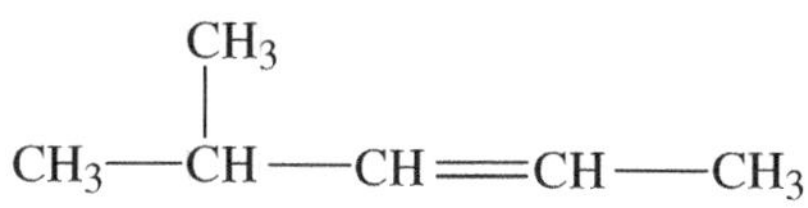

4-Methyl-2-pentene

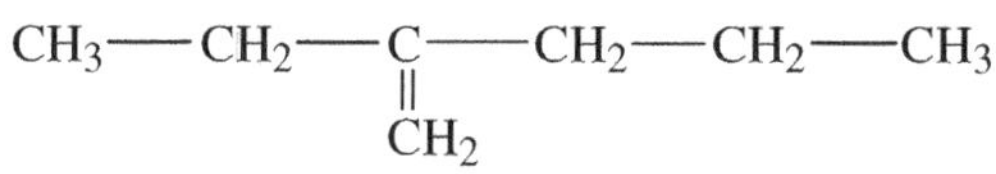

2-Ethyl-1-pentene

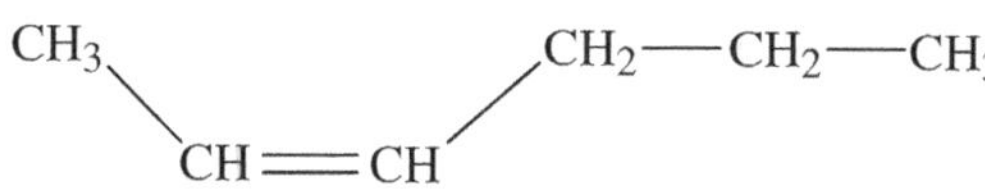

cis-2-Hexene

In alkynes, two C-atoms, which are triple bonded, have each only one substituent left to bond with (a C-atom can only have a maximum of four bonds). Therefore, there are not that many compounds in this category. The simplest alkyne has the IUPAC name ethyne and is very reactive due to its two π bonds. It is therefore used in many chemical reactions allowing the products to bear a double bond. It is also used in welding due to the high energy that is released when it combusts by reaction with oxygen.

2.6.6.4 *Aromatic Compounds* Compounds containing aromatic rings, such as benzene, are more stable than alkenes. Therefore, it takes much higher energies to add atoms to an aromatic ring and to convert the bonds in the ring to single bonds. Instead of addition reactions, benzene and its derivatives rather undergo substitutions, that is, the hydrogen atoms at the ring are replaced by atoms or groups of atoms. Toluene is an important intermediate and solvent in organic syntheses. Toluene is preferred in many processes because it is less carcinogenic than benzene. We will discuss other aromatic systems with heteroatoms in the following section. The structures of three aromatic hydrocarbons are shown as follows:

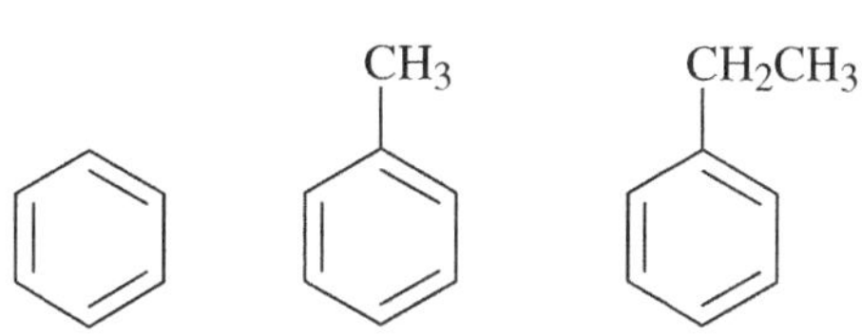

Benzene Toluene Ethylbenzene

Aromatic rings can also be attached forming *condensed* aromatic compounds. An example is naphtalene, which gives mothballs their characteristic scent. Other condensated aromatic rings appear in the form of anthracene or phenanthrene, which are used as dyes and precursors of synthetic drugs.

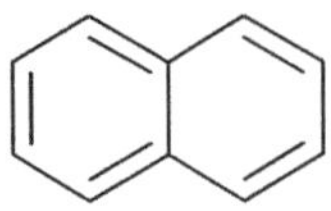

Naphtalene

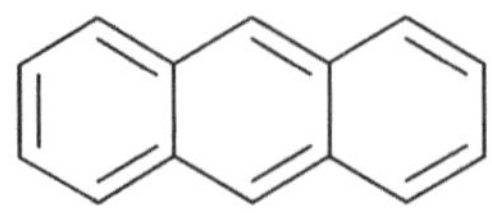

Anthracene

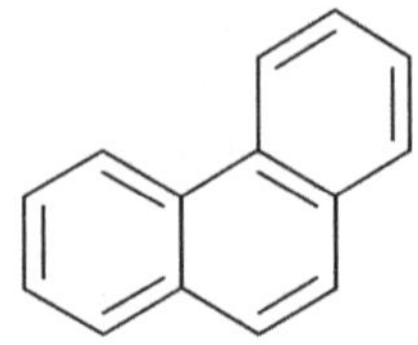

Phenanthrene

2.6.7 Organic Compounds Containing Heteroatoms

Many organic compounds used daily contain other elements besides carbon and hydrogen, called heteroatoms. They are present in medicines such as aspirin, in soda drinks as citric acid, or in coffee and tea as caffeine. An overview of these compounds is given in Table 2.8, including typical representative for each class. (The classes are given in order

TABLE 2.8 Overview of organic compounds with functional groups

Class name	General formula	Name of representative compound	Structure of representative compound
Organic halide	R-X X = F, Cl, Br, I	Chloroform *Trichloromethane (IUPAC)*	$CHCl_3$: H—C(Cl)(Cl)—Cl
Ether	R—O—R'	Diethyl ether, "Ether" *3-Oxapentane (IUPAC)*	C_2H_5—O—C_2H_5
Amine	R—N(H(R″))(H)R′	Methyl amine *Aminomethane (IUPAC)*	H_3C—N(H)—H
Phenol	Ar—OH	Phenol	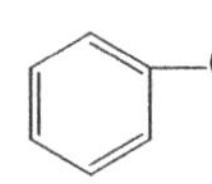—OH
Alcohol	R—OH	Ethyl alcohol, "Alcohol," and Ethanol *(IUPAC)*	C_2H_5—OH
Ketone	R—C(=O)—R′	Acetone *Propanone (IUPAC)*	H_3C—C(=O)—CH_3
Aldehyde	R—C(=O)—H	Formaldehyde, "Formalin" *Methanal (IUPAC)*	H—C(=O)—H
Carboxylic acid	R—C(=O)—OH	Acetic acid, *Ethanoic acid (IUPAC)*	H_3C—C(=O)—OH
Ester	R—C(=O)—O—R′	Ethyl acetate *Ethylethanoate (IUPAC)*	H_3C—C(=O)—O—C_2H_5
Amide	R—C(=O)—N(H)—H(R)	Acetamide	H_3C—C(=O)—N(H)—H(R)

Some IUPAC names are less preferred and shown in italics. If no IUPAC name is shown, the common name and IUPAC name are identical.

of their priority by IUPAC rules, with organic halides having the lowest and amides the highest priority.) We will only briefly mention some highlights on each of these classes.

2.6.7.1 *Organic Halides* Organic halides contain halogen atoms, that is, fluorine, chlorine, bromine, or iodine atoms. If the only heteroatom they contain is a halogen, they are also generally called halohydrocarbons. The simplest chloroalkane is methyl chloride (IUPAC name: chloromethane). If two hydrogens of methane are substituted with chlorine atoms, dichloromethane is formed. Better known is chloroform (other common name: methyltrichloride; IUPAC name besides chloroform: trichloromethane). Chloroform was used for many years as an anesthetic for surgeries. However, it is carcinogenic and has been replaced with halothane (IUPAC name: 2-bromo-2-chloro-1,1,1-trifluoroethane), which contains both chlorine and fluorine atoms. If all four hydrogen atoms in methane are replaced with chlorine, carbon tetrachloride (Carbontet, IUPAC name: tetrachloromethane) is obtained, which is used frequently as a solvent in organic synthesis. The structures of some halides are given as follows:

CH_3-Cl

Methyl chloride

$Cl-CH_2-Cl$

Methylene chloride

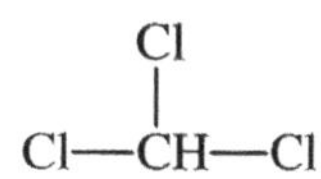

Chloroform

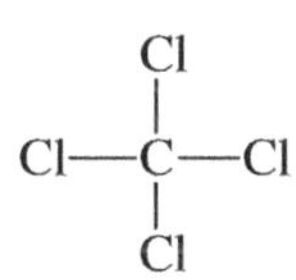

Carbon tetrachloride

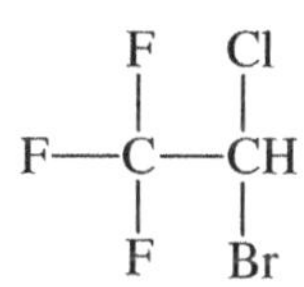

Halothane

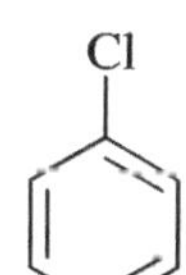

Chlorobenzene

Because chlorofluorocarbons (CFCs) appeared to be stable or inert compounds, they were used until the 1980s as refrigerants and propellants in air-conditioners and aerosol sprays. One example of a CFC is dichlorofluoromethane. However, they disintegrate in the higher atmosphere producing chlorine radicals, which catalytically destroy ozone and have led to a depletion of the ozone layer, which protects us from excessive UV radiation. They are now being phased out, and replaced by hydrochlorofluorocarbons (HCFCs), which decompose in the troposphere causing no harm to the ozone.

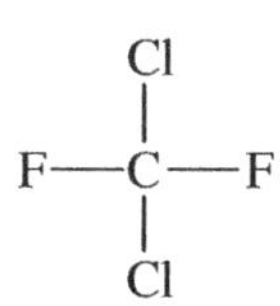

Dichlorodifluoromethane
— an example of a chlorofluorocarbon

2.6.7.2 *Alcohols, Ethers, and Phenol* All alcohols contain at least one hydroxyl group, –OH. They typically have higher boiling points in comparison to alkanes with the same number of carbon atoms because the molecules are attracted by hydrogen bonding. The simplest alcohol is methyl alcohol (IUPAC name: methanol). Methanol is toxic even at low levels and causes blindness and potentially death if ingested. It is sometimes an ingredient of improperly prepared alcoholic beverages. It is an important chemical in industry. Fuel cells can be based on the combustion of methanol.

The alcohol commonly in beverages is ethyl alcohol (IUPAC: ethanol). It is naturally obtained by the fermentation of glucose in sugars or starch, whereby yeast acts as an enzyme.

$$\underset{\text{Glucose}}{C_6H_{12}O_6} \xrightarrow{\text{Yeast}} \underset{\text{Ethanol}}{2\,CH_3CH_2OH} + 2CO_2$$

At high concentrations in the human body, ethanol acts as a depressant and can be toxic. Industrial alcohol is degenerated, that is a poison is added to it that is difficult to remove, so it cannot be used as a beverage. The government controls this to be able to impose taxes on the beverages. In beer and wine, alcohol content does not exceed about 15%. However, by distillation, the alcohol concentration can be increased to higher levels (i.e., 50% or 100% proof) in a variety of liquors. Ethanol is an important industrial solvent and reactant. In recent years, ethanol has received attention as an alternative fuel. It is already mixed with gasoline to improve fuel efficiency of cars (up to 10%). However, there is also a trend to use ethanol as a biofuel from crops.

Rubbing alcohol has the common name isopropyl alcohol. It is sold to consumers to be used as a disinfectant or for other medical purposes. It got its name from being rubbed on the bodies of patients to act as a circulation stimulant, since it evaporates easily and thereby cools the skin. The simplest compound that contains two alcoholic groups is ethylene glycol (IUPAC name: 1,2-ethane-diol). It is used as antifreeze in cars due to its low freezing point of −13°C. A polyol containing three hydroxylic groups is glycerol (also called glycerine, IUPAC name: 1,2,3-propane-triol). It is used as lubricant in medicines, food, and machines.

$$\underset{\text{Methanol}}{CH_3-OH} \qquad \underset{\text{Ethanol}}{CH_3-CH_2-OH} \qquad \underset{\text{Isopropanol}}{CH_3-\overset{\overset{\displaystyle CH_3}{|}}{CH}-OH}$$

$$\underset{\text{Ethylene glycol}}{\underset{OH}{\underset{|}{CH_2}}-\underset{OH}{\underset{|}{CH_2}}} \qquad \underset{\text{Glycerol}}{\underset{OH}{\underset{|}{CH_2}}-\underset{OH}{\underset{|}{CH}}-\underset{OH}{\underset{|}{CH_2}}}$$

Phenols, similar to alcohols, have hydroxyl groups, which are attached to benzene rings. However, they differ in their properties and reactivities from alcohols. Because of the aromatic ring strongly bonding to the oxygen, the hydrogen is only weakly bonded to the oxygen and can be donated more easily, that is, making phenols slightly acidic. Resorcinol is a phenol that is used in disinfectants for sore throats.

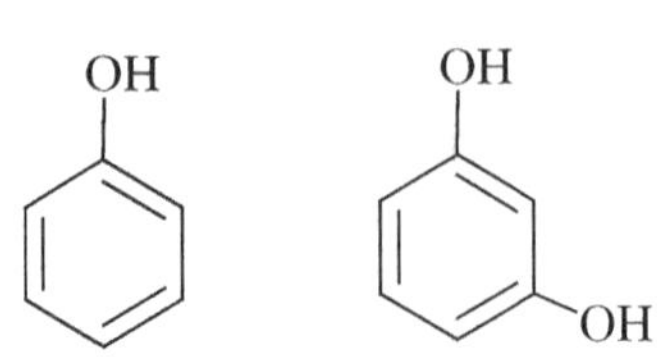

Phenol Resorcinol

All ethers possess a characteristic group of atoms—a single-bonded oxygen atom between two carbon atoms. Their boiling points are lower than those of the alcohols with the same number of carbon atoms because they cannot form hydrogen bonding. They are therefore extremely flammable liquids. Their simplest representative is dimethyl ether (IUPAC name: methoxymethane). However, commonly we refer to diethyl ether (IUPAC name: ethoxyethane) simply as "ether" because it was used many years ago for many applications such as an anesthetic for surgeries. It is, however, not used anymore for this purpose due to its high flammability. Ether is used frequently as a solvent and also as a reagent in chemical reactions.

$$CH_3 - O - CH_3 \qquad CH_3 - CH_2 - O - CH_2 - CH_3$$

Dimethylether Diethylether

2.6.7.3 *Carbonyl Compounds* All carbonyl compounds contain a carbon to which an oxygen is double bonded. This makes the carbon relatively electropositive. The compounds that fall into this category are aldehydes, ketones, carboxylic acids, esters, and amides.

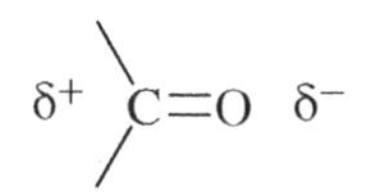

The carbonyl-group

2.6.7.4 *Aldehydes and Ketones* In an aldehyde, the carbon atom of the carbonyl group is connected to carbon atom on one side and a hydrogen atom on the other side. The simplest aldehyde is formaldehyde (IUPAC name: methanal), which is an exception because the carbonyl C-atom is connected to two H-atoms. It is a gas at room temperature but dissolves well in water. Its aqueous solution, called formalin, is a disinfectant and is used to preserve body parts in anatomy. It is also a reactant for the production of thermoset polymers. The next higher aldehyde is acetaldehyde (IUPAC name: ethanal). Benzaldehyde is an aromatic aldehyde, which smells like bitter almonds. Aldehyde occurs in many fruits and other plants, giving them their characteristic flavor.

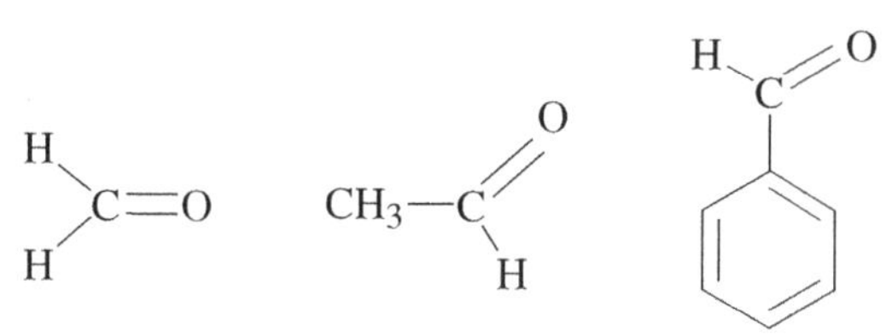

Formaldehyde Acetaldehyde Benzaldehyde

Aldehydes can be produced from the oxidation of alcohols, for example,

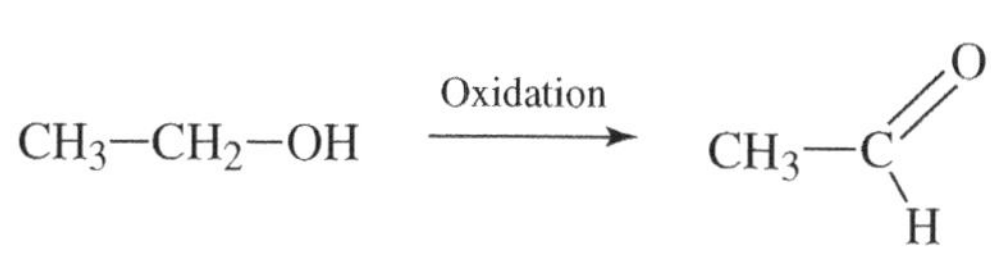

Ethanol Acetaldehyde

In ketones, the carbon atom of the carbonyl compound is connected to two carbon atoms on each side. The simplest example is acetone (dimethylketone, IUPAC name: propanone). It is a liquid at room temperature and an excellent solvent, and an ingredient of nail polish removers. Methylethylketone (IUPAC name: 2-butanone) is an important ingredient in paints and varnishes, and quickly evaporates. Acetophenone (methylphenylketone, IUPAC name: 1-phenylethanone) is an example of an aromatic ketone used in fragrances and chewing gum.

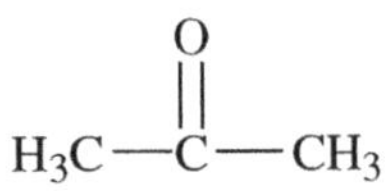

Acetone

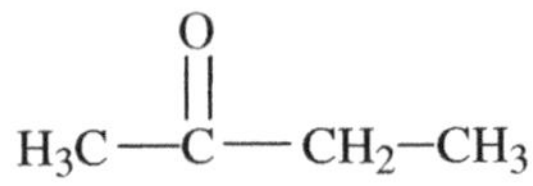

Methylethylketone

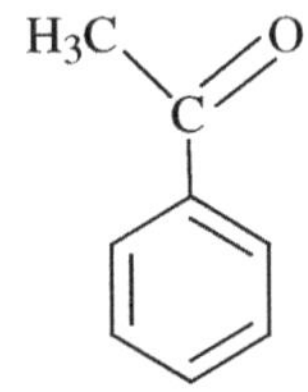

Acetophenone

2.6.7.5 Carboxylic Acids and Esters An organic acid contains carboxylic group where the carbon of the carbonyl group is also bonded to a hydroxyl group, which weakens the bond to the hydrogen atom easing proton formation.

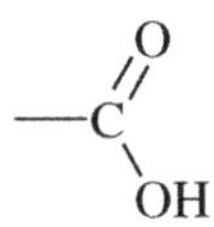

A carboxylic group

All organic acids are weak acids. The simplest carboxylic acid is formic acid (IUPAC name: methanoic acid). Its common name has its roots in that it was found in the secretion of ant stings (*formica*, Latin for ant). Formic acid is a liquid and easily dissolves in water. It is used as an antibacterial agent. A commonly used representative of organic acids is acetic acid (IUPAC name: ethanoic acid). In diluted form it becomes vinegar. Higher aliphatic acids typically have unpleasant smells, such as butyric acid (IUPAC name: butanoic acid), which got its name from rancid butter and has an odor resembling sweat. Benzoic acid is the simplest aromatic carboxylic acid and is used as a food preservative. Oxalic acid is the simplest dicarboxylic acid. It interferes in the metabolism of humans and an excess of it leads to kidney stones.

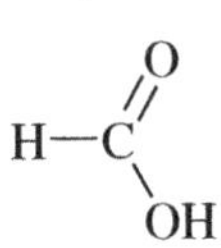

Formic acid

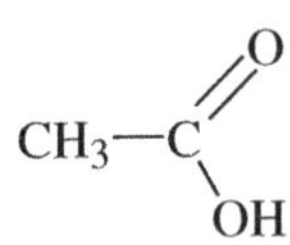

Acetic acid

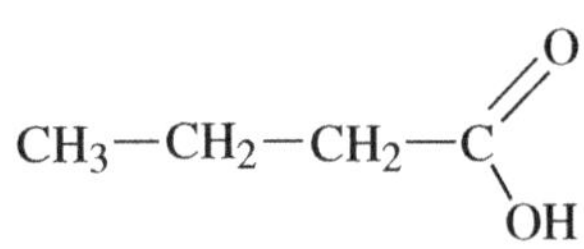

Butyric acid

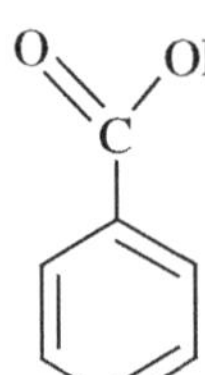

Benzoic acid

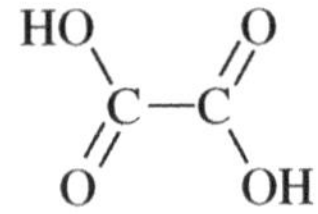

Oxalic acid

Acids can be formed by oxidation of aldehydes, for example,

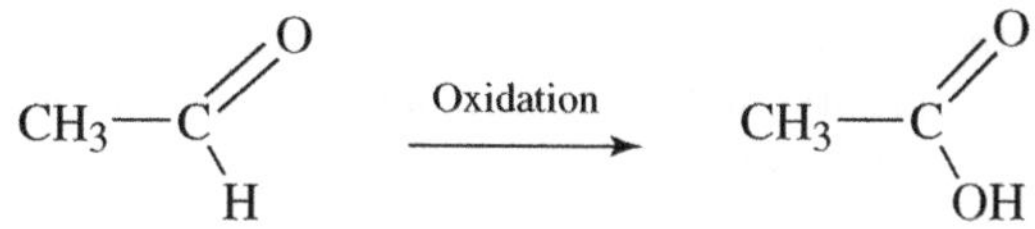

Acetaldehyde Acetic acid

An ester group contains a carbonyl group, in which the carbon is also connected to an etheric oxygen:

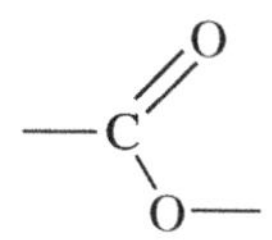

An ester-group

Organic acids react with alcohols to form esters, also called esterification. For example,

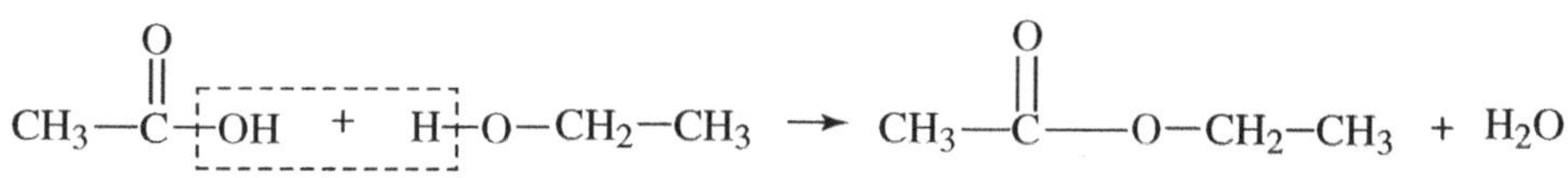

Acetic acid Ethanol Ethyl acetate

In contrast to the acids they are formed from, esters typically have very pleasant odors. Ethyl acetate (common names: acetic acid ethyl ester, ethyl ethanoate) is a liquid, which like most esters is used as fragrance. While ethyl formate has a rum-like smell, ethyl butanoate smells like pineapple.

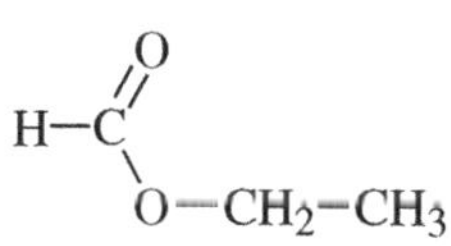

Ethyl formate

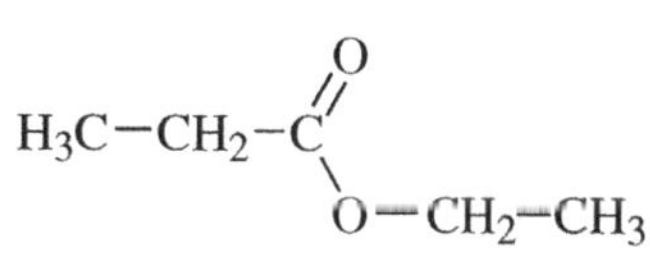

Ethyl butanoate

2.6.7.6 Amines Amines are organic derivatives of the weak inorganic base ammonia, NH_3. Therefore, they are also slightly basic, and, similar to ammonia, amines can be protonated forming organic ammonium ions:

$$CH_3 - NH_2(l) + H_2O(l) \rightleftharpoons CH_3 - NH_3^+(aq) + OH^-(aq)$$

Depending on the number of carbon atoms attached to the nitrogen in the amine and the remaining hydrogen atoms, one distinguishes the following: primary amines, which are only connected to one carbon atom; secondary amines, connected to two carbon atoms; and tertiary amines, in which the nitrogen has all its available valences connected to carbon atoms. Many amines have unpleasant fish-like odors, some even smelling like decaying animals, because of which one is named cadaverine (1,5-diaminopentane, IUPAC name: pentane-1,5-diamine). They are used in many medicines, giving them a bitter taste.

Amphetamine (or benzedrine, IUPAC name: 1-phenylpropane-2-amine) is known as a stimulant. Amines are also used in the production of polyamides and polyurethane. The simplest amine is methylamine. An aromatic amine is aniline, which has been produced many years in the dye industry, and is a precursor to the synthetic color mauve. Some examples of amines are shown as follows:

$$CH_3{-}NH_2$$

Methylamine
(a primary amine)

$$CH_3{-}CH_2{-}NH{-}CH_2{-}CH_3$$

Diethylamine
(a secondary amine)

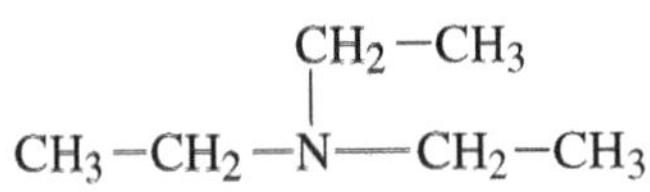

Triethylamine
(a tertiary amine)

$$NH_2{-}CH_2{-}CH_2{-}CH_2{-}CH_2{-}CH_2{-}NH_2$$

Cadaverine

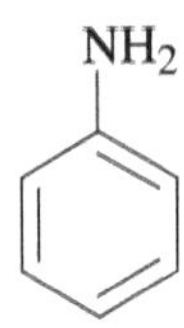

Aniline

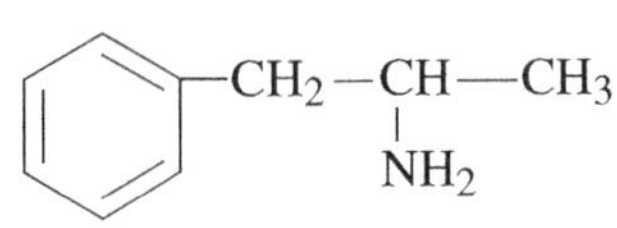

Amphetamine

2.6.7.7 *Amides* In the amide group, a carbonyl group is combined with an amino group:

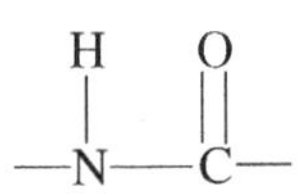

The amide group

Amides are weaker bases than amines and quite water soluble. Formamide (methanamide) and acetamide (ethanamide) are water-soluble liquids, while benzamide is a solid. These are important precursors for pharmaceutical drugs.

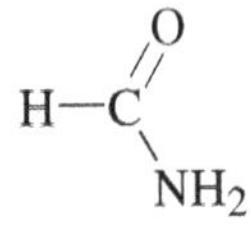

Formamide

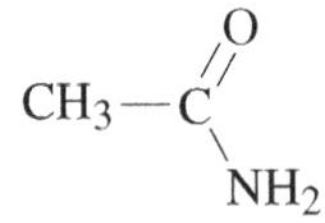

Acetamide

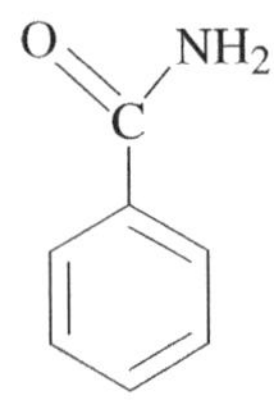

Benzamide

Amides are formed in the neutralizations of amines as weak bases, with carboxylic acids, for example,

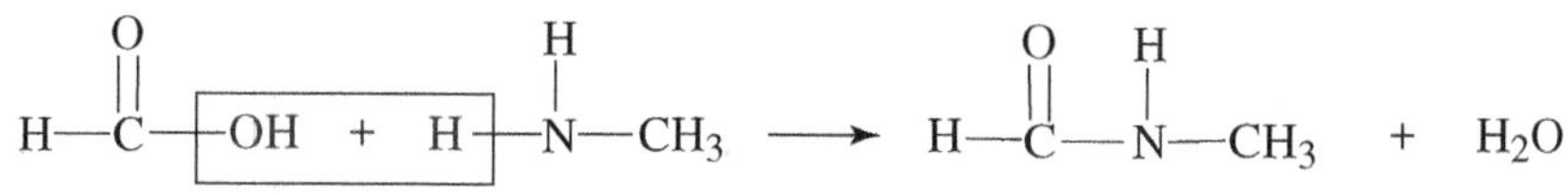

Formic acid Methylamine *N*-Methyl-formamide

2.6.8 Organic Reactions Involving Hydrogen

In Chapters 3 and 4, we will consider physical and chemical properties of hydrogen, including chemical reactions for hydrogen production. In this chapter, we look at the organic reactions involving hydrogen as a reactant or catalyst.

2.6.8.1 Hydrogen as a Reducing Agent It is known that an organic compound has been reduced if the reaction increases the number of C–H bonds or decreases the number of C–O, C–N, or C–X bonds (where X denotes a halogen). Hydrogen is well known as a common reducing agent. Let us now take a look at some examples of oxidation–reduction reactions that take place on carbon and other atoms of various classes of organic compounds.

2.6.8.1.1 Addition of Hydrogen to Alkenes Hydrogen adds to the double bond of an alkene to form an alkane in the presence of a transition metal catalyst such as platinum, palladium, or nickel. Without the catalyst, the activation energy of the reaction is too high for the reaction to occur. The reaction of addition of hydrogen in the presence of catalyst is called catalytic hydrogenation. The reaction is heterogeneous because the catalysts are used in the form of a dispersion of metals or metal oxides on charcoal (Pt/C, Pd/C). After reaction, the catalyst can be easily removed by filtration.

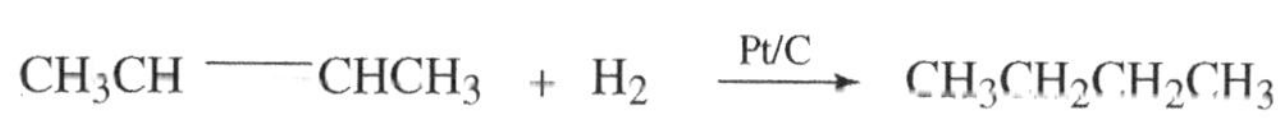

2-Butene Butane

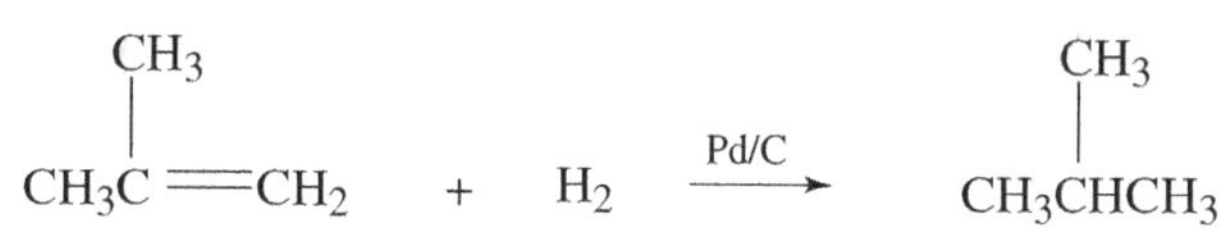

2-Methylpropene 2-Methylpropane

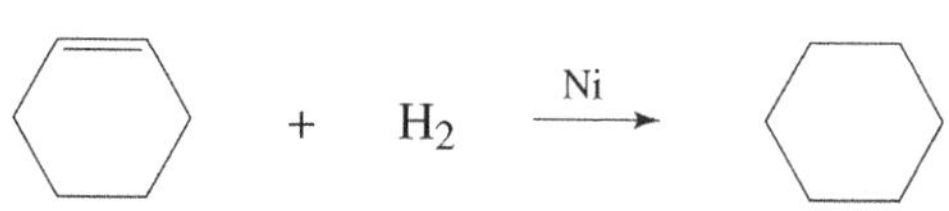

Cyclohexene Cyclohexane

The mechanism of catalytic hydrogenation is not completely understood; however, it is known that the reaction goes through adsorption of hydrogen on the surface of the metal and formation of a complex between alkene and metal by overlapping its p orbitals with vacant orbitals of the metal. This results in breaking the π bond of the alkene and σ bond of H_2 and formation of the new C–H σ bond to produce an alkane (Figure 2.27). This is why

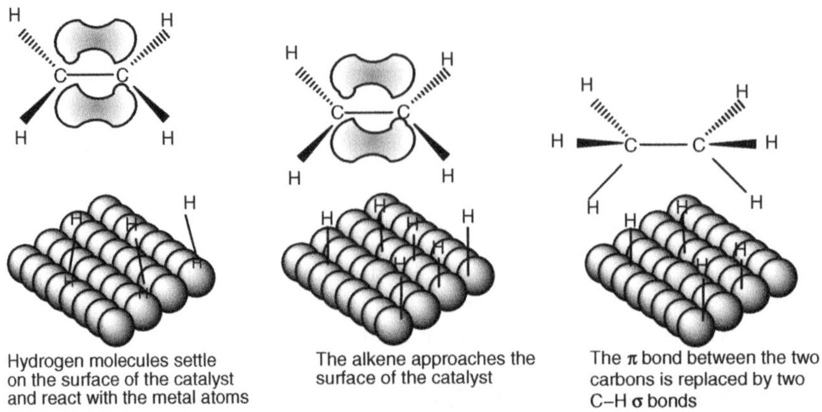

Figure 2.27 Catalytic hydrogenation of an alkene.

the catalytic reduction is stereoselective; because the two hydrogen atoms add from a solid surface, they add with syn stereochemistry—with formation of *cis*-isomers. For example, addition of hydrogen to 1,2-diethylcyclohexene yields *cis*-1,2-diethylcyclohexane:

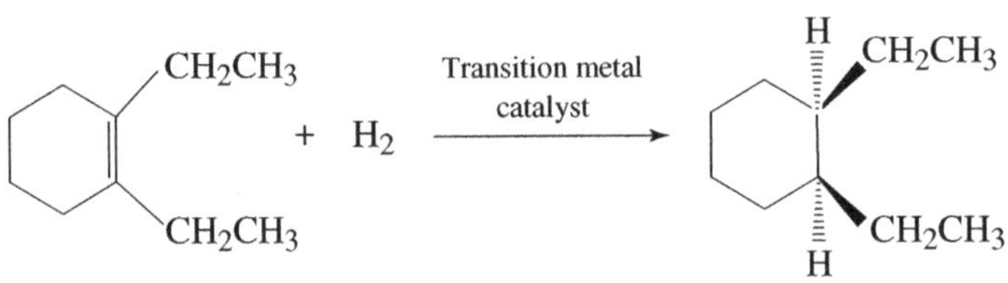

1,2-Diethylcyclohexene   *cis*-1,2-Diethylcyclohexene

Hydrogenation is exothermic, releasing about 80–120 kJ (20–30 kcal) of heat per mole of hydrogen. This means that head releases when a weaker π bond has been converted to a stronger σ bond. The higher the energy level of the alkene (the less stable alkene), the higher is the heat of hydrogenation. The energy level of the alkene depends on its structure:

- The degree of substitution of the carbon–carbon double bond: the higher the number of substitutes, the lower the heat of reaction. The heat decreases, for example, in the raw ethylene (no substituents) → propene (one substituent) → butene (two substituents).
- If the alkene is in form of *cis*- or *trans*-isomer: the heat of reduction of a *trans* alkene is lower than that of a *cis*-alkene.

2.6.8.1.2 Addition of Hydrogen to Alkynes Similar to the π bond of an alkene, the two π bonds of an alkyne readily undergo addition reactions; in the presence of a metal catalyst such as platinum, palladium, or nickel, hydrogen adds to alkyne in the same manner that it adds to an alkene, reducing it to alkane:

$$CH_2CH_3C{\equiv}CH \xrightarrow{H_2,Pt/C} CH_3CH_2CH{=}CH_2 \xrightarrow{H_2,Pt/C} CH_3CH_2CH_2CH_3$$

1-Butyne (alkyne)   1-Butene (alkene)   Butane (alkane)

The reaction takes place in two stages, with an alkene intermediate. The reaction can be stopped at the alkene stage if not efficient, but partially deactivated metal catalyst (Lindlar catalyst) is used.

The catalytic hydrogenation of alkynes is similar to the hydrogenation of alkenes because only *syn* addition of hydrogen occurs; the alkyne sits on the surface of the metal catalyst, and the hydrogen atoms are delivered to the triple bond from the surface of catalyst.

2.6.8.1.3 Reduction of Nitrogen- and Oxygen-Containing Groups Catalytic hydrogenation can also be used to reduce carbon–nitrogen double and triple bonds or nitrogen–oxygen double bond. The products of the reaction are amines:

$$CH_3CH_2CH{=}NCH_3 + H_2 \xrightarrow{Pd/C} CH_3CH_2CH_2NHCH_3$$

Methylpropylamine

$$CH_3CH_2CH_2C{\equiv}N + 2H_2 \xrightarrow{Pd/C} CH_3CH_2CH_2CH_2NH_2$$

Butylamine

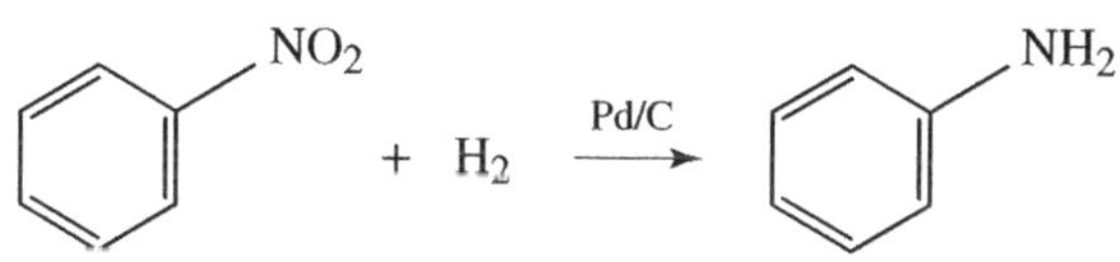

Nitrobenzene Aminobenzene

Reduction of aldehydes and ketones occurs in the presence of Raney nickel as the metal catalyst. Raney nickel is finely dispersed nickel with adsorbed hydrogen. Pt and Rh catalysts are also used for hydrogenation of ketones and aldehydes. Aldehydes are reduced to primary alcohols, and ketones are reduced to secondary alcohols:

$$CH_3CH_2CH_2\overset{\overset{O}{\|}}{C}H + H_2 \xrightarrow{\text{Raney Ni}} CH_3CH_2CH_2CH_2OH$$

Butanal (an aldehyde) — (primary alcohol)

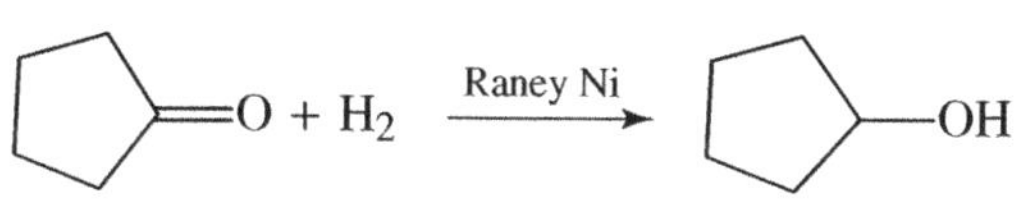

Pentanon (ketone) — Pentanol (secondary alcohol)

The metal catalysts are not selective, which means that if an aldehyde or ketone has an isolated carbon–carbon double bond, both functional groups undergo the addition of hydrogen under these conditions:

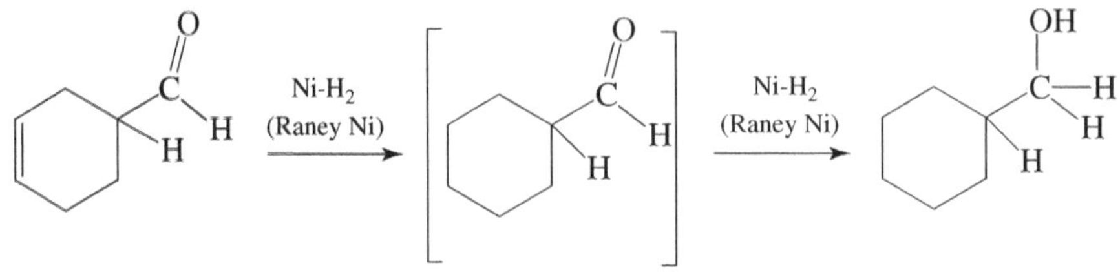

2.6.8.1.4 Reduction of Monosaccharides Similar to other aldehydes and ketones, monosaccharides (aldoses and ketoses) can be reduced to the corresponding polyalcohols (sugar alcohols) under catalytic hydrogenation using a nickel catalyst. Sugar alcohols are widely used in industry, primary as food additives and sugar substitutes. One of the examples is the reduction of glucose to glucitol known as sorbitol:

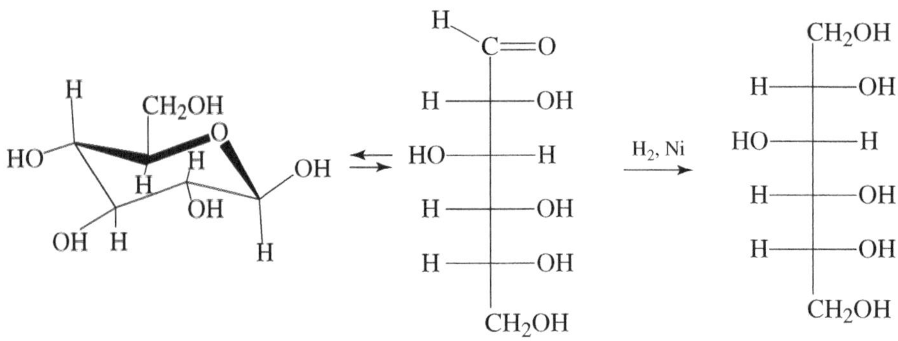

β-D-glucopyranose (β-D-glucose) — Open-chain aldehyde — D-glucitol (D-sorbitol), an alditol

2.6.8.1.5 Metal Hydride Reduction Hydrogen can act as a reducing agent not only in molecular form but also in the form of hydride ion H^-: sodium borohydride ($NaBH_4$) and lithium aluminum hydride ($LiAlH_4$) are common compounds used as a source of hydride ion, which is utilized to reduce the carbonyl group of an aldehyde or a ketone:

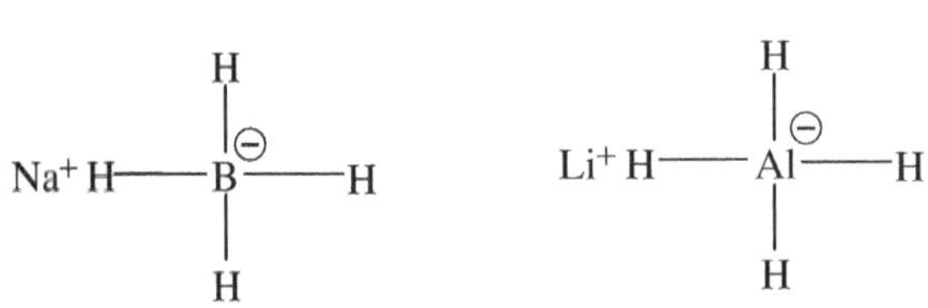

Sodium borohydride — Lithium–aluminum hydride

These compounds are very powerful reducing agents. Lithium–aluminum hydride is a special agent because of its ability to reduce not only the carbonyl groups of an aldehyde or a ketone but also those of carboxylic acids or their derivatives:

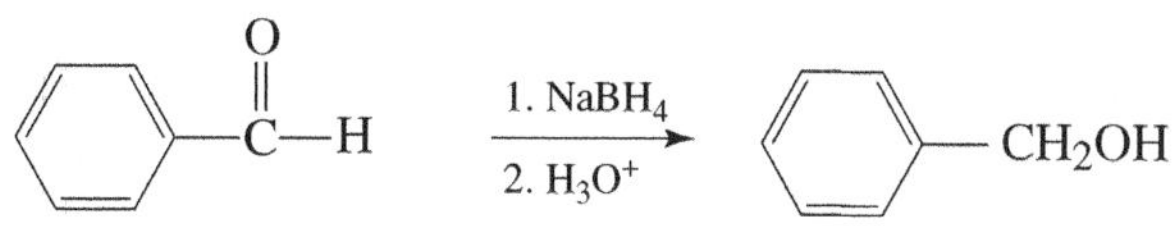

Benzaldehyde — Benzyl alcohol (primary alcohol)

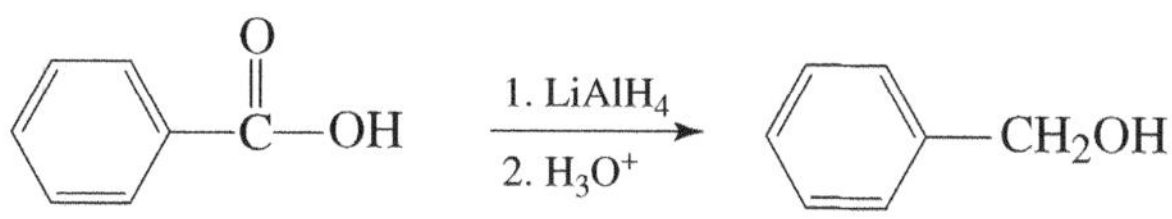

Carboxylic acid — Primary alcohol

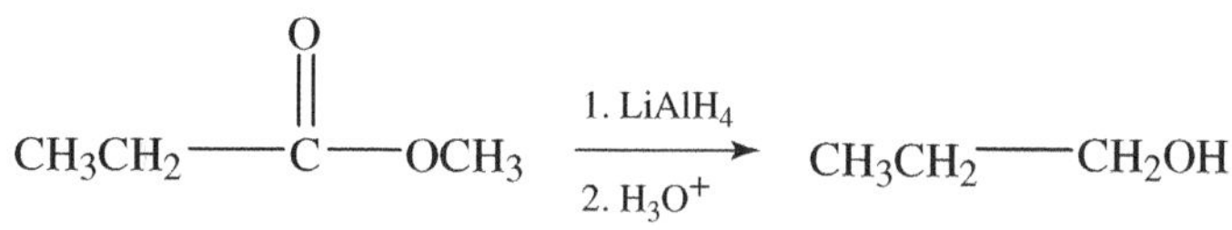

Ester — Propanol (primary alcohol)

2.6.8.1.6 Hydride Ion in Biological Reduction Reactions In living systems, reduction reactions are very important; they take place in cells. The biological reducing agents act in the same way as metal hydrides: they donate hydride ions. The most common source of hydride ions in living systems is reduced nicotinamide adenine dinucleotide:

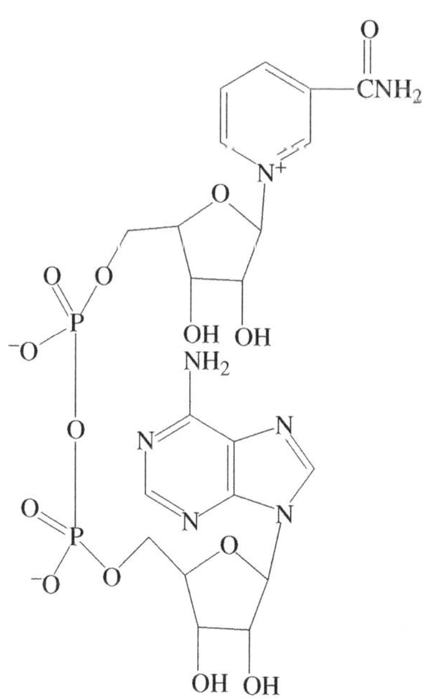

Nicotinamide adenine dinucleotide
NAD^+

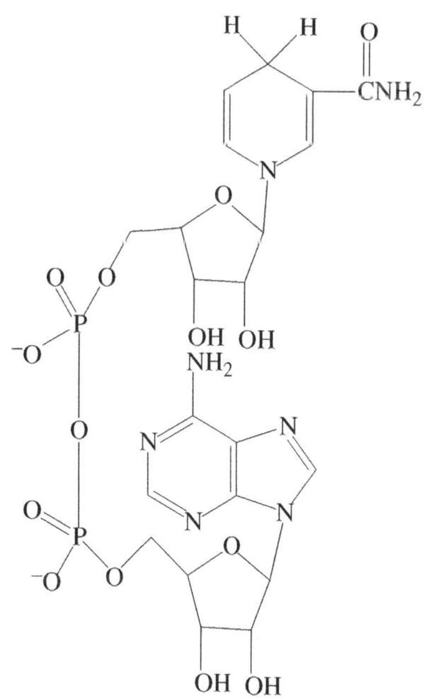

Reduced nicotinamide adenine dinucleotide
NADH

NADH reduces compounds such as aldehydes and ketones by donating a hydride ion of the six-member ring:

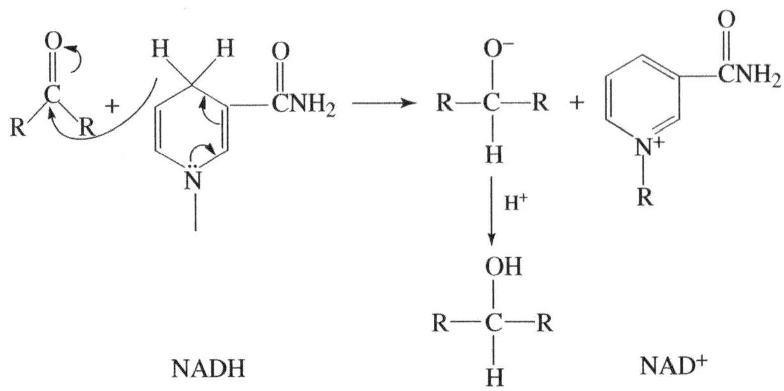

2.6.8.2 Hydrogen as a Catalyst Most organic reactions take place only in the presence of catalysts. The catalyst provides a more favorable pathway for the reaction. Acid catalysis is common for various organic reactions. An acid catalyst increases the rate of a reaction by donating a proton to a reactant. For example, the rate of hydrolysis of an ester, amides, acid chlorides, or anhydrides is markedly increased by the acid catalyst:

$$\underset{\text{Ester}}{CH_3C(=O){-}OCH_3} + H_2O \overset{H^+}{\rightleftharpoons} \underset{\text{Acid}}{CH_3C(=O){-}OH} + CH_3OH$$

$$\underset{\text{Amide}}{CH_3CH_2C(=O){-}NH_2} + H_2O \xrightarrow{H^+} \underset{\text{Acid}}{CH_3CH_2C(=O){-}OH} + NH_4^+$$

$$\underset{\text{Acid chloride}}{CH_3C(=O){-}Cl} + H_2O \xrightarrow{H^+} \underset{\text{Acid}}{CH_3C(=O){-}OH} + Cl^-$$

$$\underset{\text{Anhydride}}{CH_3{-}C(=O){-}O{-}C(=O){-}CH_3} + H_2O \xrightarrow{H^+} 2\ \underset{\text{Acid}}{CH_3C(=O){-}OH}$$

2.7 POLYMER CHEMISTRY

Polymers are large molecules, typically with molar masses between 10,000 and 500,000 g/mol. The name *polymer* originates from Greek, with "poly" meaning many, and "mer" standing for unit, together translated as a substance that is composed of many units. Polymers are typically formed from small molecules as building blocks, consequently called monomers ("mono": Greek for single). The bonds that connect the monomers to a polymer are covalent, which is one reason why salts, which are held together by ionic bonding, would not be considered polymers. An alternate term for polymer is the Latin-based *macromolecule*, which means large molecule and is often used in academia. Most polymers contain predominantly carbon and hydrogen atoms, which places them in the category of organic compounds, discussed previously.

Polymers exist in nature, for example, as cellulose in cotton or polypeptides in wool. Cellulose is a polymer of glucose. Glucose forms another type of common polymer as starch. The structures of these natural polymers are shown in Figure 2.28. While in cellulose the molecules of glucose are bonded by β-glycosidic bonding, in starch, those molecules are bonded by α-glycosidic bonding. As a consequence, starch can be digested by humans, while cellulose cannot be. The proteins in our body forming parts of our skin and hair and internal tissue are polymers of amino acids. DNA (deoxyribonucleic acid), the substance found in every form of life, which carries the genetic information of each organism, is a polymer. Nature-based polymers have been used as materials such as cloth throughout human civilization. These types of polymers have become more recently important as a resource for energy. For example, corn starch and hay are being considered as alternative fuel sources. They are also environmentally more friendly than their synthetic counterparts as they are more readily absorbed by the soil.

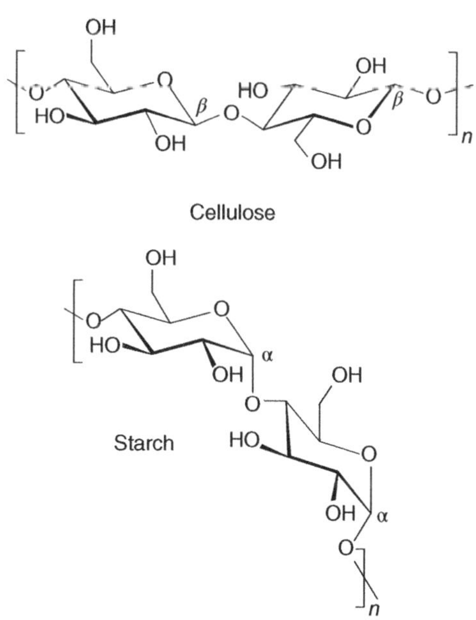

Figure 2.28 Two natural polymers of glucose: cellulose and starch (F13-020 from Olmsted). [*Source*: Olmsted and Williams (2006) [103]. Reproduced with permission of John Wiley & Sons.]

Natural polymers, such as cellulose, had been modified chemically during the second half of the 19th century, for example, with acetic acid to form cellulose acetate. However, the first truly synthetic polymers were made not earlier than 1907, when Leo N. Bakeland produced phenol/formaldehyde resin from the two monomers. This tough and brittle polymer, occasionally named Bakelite after its inventor, was used for the production of telephones, the housing of radios, and turntable records. It is still an important polymer in the electronics industry, for example, in the production of photoresists to produce microchips, and is used for imitation wood in furniture and automobiles.

Synthetic polymers represent human-made materials, which have become very important in our modern lives. At home and at work, we are surrounded by synthetic polymers and have gotten used to them. Polymers are used as consumer goods, such as apparel, soda bottles, food containers, plastic bags, credit cards, and CD-disks. More parts in a modern car are made of polymers than metal, including even gears and the car body. Polymers are used in industry for many products, including cable insulation, glazing, adhesives, paints, and heat shields. Certain synthetic polymers are applied in fuel cells as proton exchange membranes. Other types of polymers are electroconductive and are suitable to store electricity. Because some polymers are important for the conversion of hydrogen fuel cells, we will discuss them in more detail at the end of this chapter, after an overview of the basics of synthetic polymers.

2.7.1 Structural Formulas and Names of Polymers

There are several ways to name polymers and to write the structures of polymers. Since polymers are very large molecules, it would not be efficient to show the whole molecule. For most polymers, the monomers link up to form linear chains (see Section 2.7.2). Occasionally, it is useful to show a segment of the polymer chain. Alternative representations are given for the case of the simplest polymer, polyethylene, in Figure 2.29. The ball-and-stick

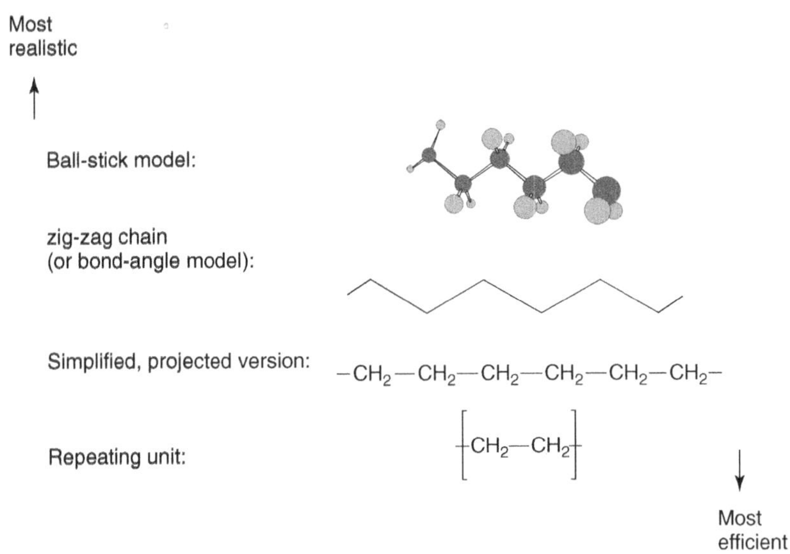

Figure 2.29 Four representations of polyethylene.

model would be the most realistic among those four representations; however, writing the repeating unit is usually the most efficient way to convey information about a polymer.

In Table 2.9, the formulas and names of some simple polymers are given. It is most common to use source names, that is, to use the name of the monomer from which the polymer was made from, and to place the prefix "poly-" in front of the source name. For example, poly(ethylene) is made from the monomer ethylene (structure explained in Section 2.6). Brackets around the source name are permitted but often omitted. A few polymers have unconventional source names, in which the prefix "poly" is not used but instead a word is added, for example, in "phenol/formaldehyde resin" or certain types of rubber, such as "isobutyl rubber" (alternatively called polyisobutylene).

Since most polymers are "organic," we can also use IUPAC names for them. In these more systematic names, we start again with "poly," but name each structural segment within the repeating unit. The IUPAC name of polyethylene would be simply poly(methylene), since the smallest structural segment, which repeats itself is a methylene group. However, IUPAC names of polymers are not used commonly in the industry or by consumers and are more relevant for scientists and technologists. They appear often in reference books on polymers.

Another way to designate polymers is to use acronyms or abbreviations based on their source names, for example, pp for polypropylene. Many polymers have also commercial or trade names, which in some cases are preferred over their common source names, for example, in case of Nylon or Teflon.

2.7.2 The Overall Chain Structure in Polymers

In Figure 2.30, the main types of overall structures or topologies of polymers are shown. As mentioned earlier, many polymers have basically the structure of chains. Though they have a zigzag structure along the main chain, they are generally called linear. The simplest example of a linear polymer is high-density polyethylene (HDPE, shown in Table 2.9). If any atoms (except those of hydrogen) or groups of atoms are connected to the main chain, they are called pendant atoms or groups. For example, chlorine forms pendant atoms in polyvinylchloride and phenyl groups represent pendent groups in polystyrene (Table 2.9).

Polymers can also be branched, which means that the repeating units of the polymer are also present in side chains, which are connected to the main chain. An example of a branched polymer is low-density polyethylene (LDPE). The branching point is formed by a tertiary carbon atom (see Section 2.6). Branched polymers usually are softer than their linear counterparts and are more easily processed. Similar to the linear polymers, they are thermoplastic, that is, they can be melted and reshaped. Such polymers can be easily recycled (see major recycled polymers in Table 2.10).

The third group of polymers that can be distinguished by their overall architecture are crosslinked (or network) polymers. In these polymers, all chains are connected, which restricts the polymer from being melted repeatedly and being reshaped. Therefore, these types of polymers are also called thermosets (i.e., you can heat them, but their shape is already set). Cross-linking of polymers can be caused by their formation mechanism. For example, phenol and formaldehyde form a tightly crosslinked polymer, with only a few carbon atoms between the network crosslinked. The resulting polymer is rigid and brittle.

In contrast, most types of synthetic rubber and natural rubber consist of linear chains when they are initially formed or produced. However, humans deliberately add sulfur or other cross-linking agents to these polymers and heat them to cross-link the chains. In this

TABLE 2.9 The formulas and names of some simple polymers

Monomer		Polymer				
Structure	Name	Structure	Source name	Acronym (recycle number)[a]	IUPAC name	Example of commercial name(s)
$CH_2{=}CH_2$	Ethylene	$-[CH_2-CH_2]_n-$	High density poly(ethylene) or low density poly(ethylene)	HDPE (2) or LDPE (4)	Poly(methylene)	Marlex Dowlex
$CH_2{=}CH(CH_3)$	Propylene	$-[CH_2-CH(CH_3)]_n-$	Poly(propylene)	PP (5)	Poly(1-methylethylene)	Hostalen
$CH_2{=}CH(Cl)$	Vinylchloride	$-[CH_2-CH(Cl)]_n-$	Poly(vinylchloride)	PVC or V (3)	Poly(1-chloroethylene)	Geon
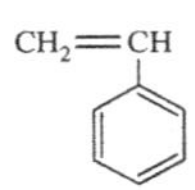	Styrene	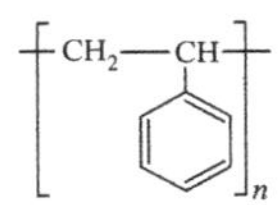	Poly(styrene)	PS (6)	Poly(1-phenylethylene)	Styron

[a]The polymer with the recycle number 1 is poly(ethylene terephthalate), which is more complex and will be discussed later.

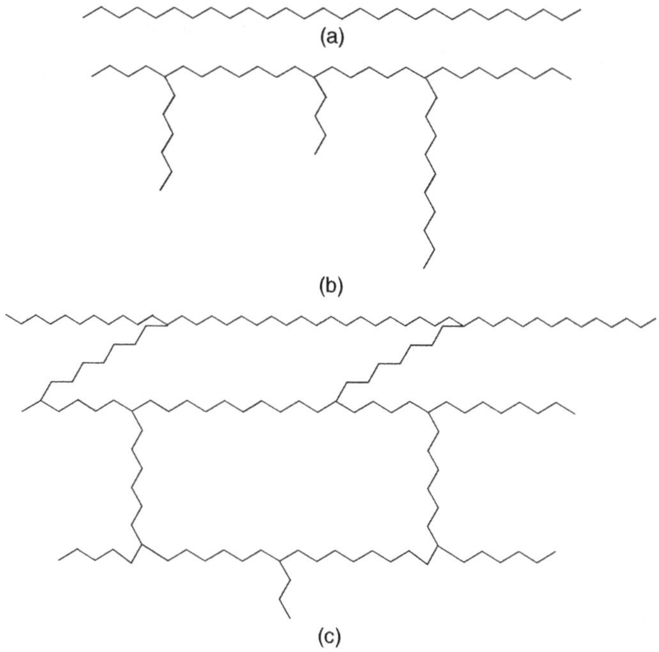

Figure 2.30 Polymers can form three types of overall structures: (a) linear, (b) branched, and (c) cross-linked.

TABLE 2.10 Overview of major chain-addition polymers

Polymer's (source) name	Repeat unit	Major applications (frequently used trade names)
Low-density polyethylene	$-CH_2-CH_2-$ (branched)	Soft bags, toys
High-density polyethylene	$-CH_2-CH_2-$	Containers, thin grocery bags
Derived from monomers as monosubstituted ethylene		
Polypropylene	$-CH_2-CH(CH_3)-$	Ropes, carpeting, food containers
Polyvinylchloride	$-CH_2-CH(Cl)-$	Water pipes, garden hoses
Polystyrene	$-CH_2-CH(C_6H_5)-$	Styrofoam cups, insulation
Polyacrylonitrile	$-CH_2-CH(C\equiv N)-$	Apparel (Orlon)
Polyvinylacetate	$-CH_2-CH(O-C(=O)-CH_3)-$	Paints, adhesives

(*continued overleaf*)

TABLE 2.10 (***Continued***)

Polymer's (source) name	Repeat unit	Major applications (frequently used trade names)
Polymethacrylic acid	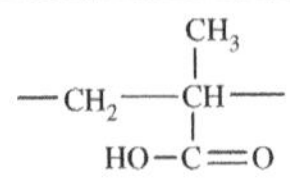	In copolymers as ionomer
Derived from monomers as 1,1-disubstituted ethylene		
Polyvinylidene dichloride	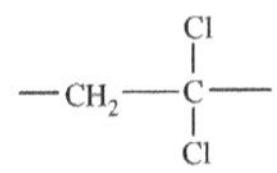	Food wrapping (Saran)
Polyvinylidene difluoride	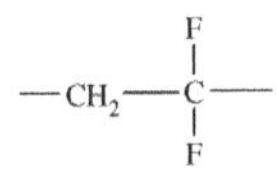	Piezoelectric, hydrophones, loudspeakers
Polymethyl methacrylate	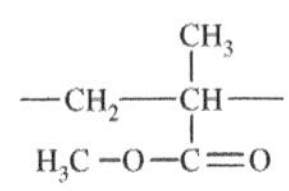	Windows, furniture (Lucite, Pexiglass)
Derived from monomers as tetrasubstituted ethylene		
Polytetrafluoroethylene (Teflon)	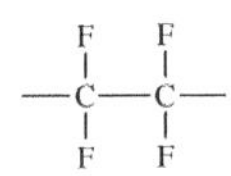	Coating for nonstick pans, electrical parts

manner, the rubber does not remain sticky and soft but becomes stretchable and snaps back to its original shape when released. This method of cross-linking is called "rubber vulcanization" (after Vulcan, the Roman god of fire) and was discovered by Charles Goodyear in 1835. There are usually 100–500 carbon atoms between the cross-links in such vulcanized or cured rubber, making their cross-linking density much lower than that in the rigid thermosets.

2.7.3 Mechanical Properties of Polymers

Polymers can be categorized according to their mechanical behavior into four major groups: (a) fibers, (b) rigid plastics, (c) regular (flexible) plastics, and (d) elastomers (i.e., vulcanized rubber). In Figure 2.31, the stress/strain diagram for polymers is shown. Stress is here equivalent to force or strength, and strain means the same as elongation or stretching. While fibers withstand the highest stress at the lowest strain, elastomers show the lowest stress at the highest strain. Ropes made out of fibers based on polyamide (Nylon) can be used in rescue operations to pull humans out of a dangerous situation or even to support hanging bridges. As we all know, (vulcanized) rubber indeed stretches the most with relatively small effort or force.

Regular and rigid plastics fall in between these two extremes. Thermoplastic polymers tend to "yield" a little before being pulled further and breaking apart. Yielding means that, when the polymer sample is pulled quickly and at high force, softening occurs in the sample because segments of the polymer start to flow even below its melting temperature. Good examples for regular, flexible thermoplastics are polyethylene and polypropylene, which have glass transition temperatures (defined in Section 2.7.8) below their typical usage temperatures, that is, room temperature or slightly higher temperatures. In contrast, polymers

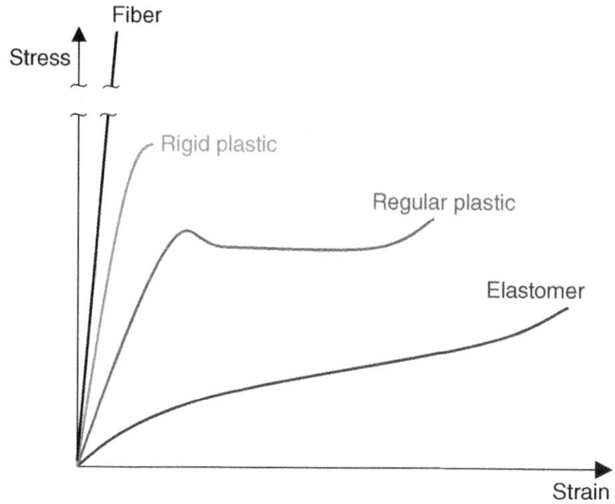

Figure 2.31 Stress/strain diagram for polymers.

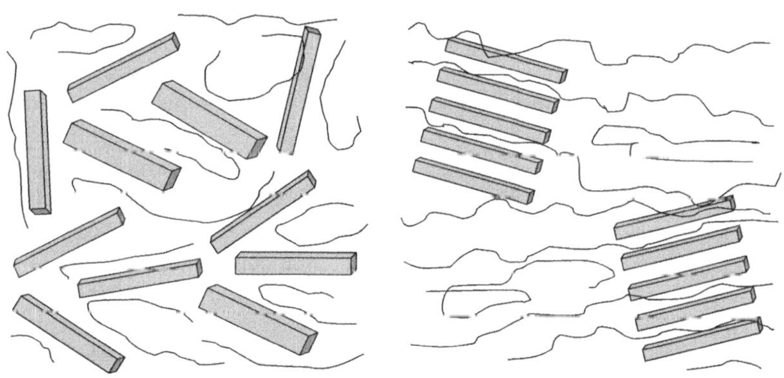

Figure 2.32 The exfoliated (left) and intercalated (right) forms of clay/polymer composites. The gray sheets represent the clay, while the chains represent the polymer. [*Source*: Miri (2006) [51]. Reproduced with permission of Pearson.]

such as polystyrene and polyvinylchloride are rigid plastics with higher stiffness and brittleness. That is why transparent (nonfoamed) polystyrene cups break when you squeeze them a little. In contrast, cups or bottles made of polyethylene are much more flexible.

Some engineering plastics, such as polyethersulfone (repeat unit: $–[–O–C_6H_4–SO_2–C_4H_6–]–$), have higher tensile strengths than polyolefins. By adding fillers, such as carbon or clay to the polymers, these composites can be made even stronger, reaching at least twice the normal strength of plastics. In Figure 2.32, it is shown that the separation of the sheet-like clay and better distribution of it in the Nylon matrix, called "exfoliated," leads to better reinforcement of the composite than the less uniform distribution with the clay maintaining its stacked formation, called "intercalated." Regular Nylon has a tensile strength of 85 MPa, whereas addition of about 10% clay results in a tensile strength of 170 MPa.

2.7.4 Synthesis of Polymers

The properties of a polymer can be controlled by the way polymers are made, that is, polymer properties can be tailored by choosing the right reaction conditions. The synthesis of polymers is also termed polymerization. We distinguish between two major groups by which polymers are formed: *chain-addition polymerization* and *step-growth polymerization*. Polycondensation represents part of the latter group in which small molecules such as water are formed as the by-product to the polymer. An older term for the latter group is *condensation polymerization*.

2.7.4.1 Chain-Addition Polymers In chain-addition polymerization, one monomer at a time is added to a growing polymer chain. Monomers that polymerize by chain-addition polymerization must have at least one double bond or, less frequently, a triple bond. Single-bonded hydrocarbons would not be reactive enough to polymerize. Most commercial polymers are chain-addition polymers, with polyethylene being the most produced polymer. In 2005, about 32 billion pounds of all types of polyethylene were produced in the United States. Several important chain-addition polymers are listed in Table 2.10. This group also includes most synthetic rubber, among which SBR (styrene/butadiene rubber) has the highest production volume.

To polymerize a monomer by chain-addition polymerization, an initiator is required. Four types of initiators can be potentially used: radical, cationic, anionic, or coordinative. Radical initiators are formed from compounds that relatively easily split a bond to produce radical electrons (see the following paragraphs). Coordinative initiators are metalorganic catalysts or catalyst combinations, such as $TiCl_4$ and $Al(C_2H_5)_3$. These form coordination bonds between the metal of the catalyst and the double-bonded carbon atoms. Cationic initiators can be produced by adding a strong Lewis acid, such as aluminum chloride, to a monomer, leading to the formation of a positively charged carbon, or "carbocation," in the monomer. Anionic initiators are metalorganic compounds in which a metal such as lithium or sodium is bonded to a carbon of an alkyl or aryl group, causing it to become negatively charged, a so-called carbanion. For the polymerization of ethylene, either radical or coordinative polymerization works well, the former leading to LDPE, whereas the latter leads to HDPE. In contrast, styrene works well with any of the four initiators because its aromatic phenyl group can stabilize any type of adjacent carbon.

Based on the electronic interactions, certain initiators are more suitable for a given monomer. An overview of some preferred initiator/monomer combinations is given in Table 2.11.

The mechanism for a chain-addition polymerization is given in Figure 2.33. As an example, the initiator could be a radical initiator, a peroxide, and vinylchloride could be a

TABLE 2.11 Some preferred monomer/initiator combinations for chain-addition polymerizations

Monomer	Preferred initiator	Example of precursor of initiator	Polymer
Ethylene	Radical	Benzoylperoxide	LDPE
Ethylene	Coordinative	$TiCl_4/Al(C_2H_5)_3$	HDPE
Propylene	Coordinative	γ-$TiCl_3/Al(C_2H_5)_2Cl$	PP
Isobutene	Cationic	BF_3	Polyisobutene
1,3-Butadiene	Anionic	Li-C_4H_9	Polybutadiene
Styrene	All	Many	Polystyrene

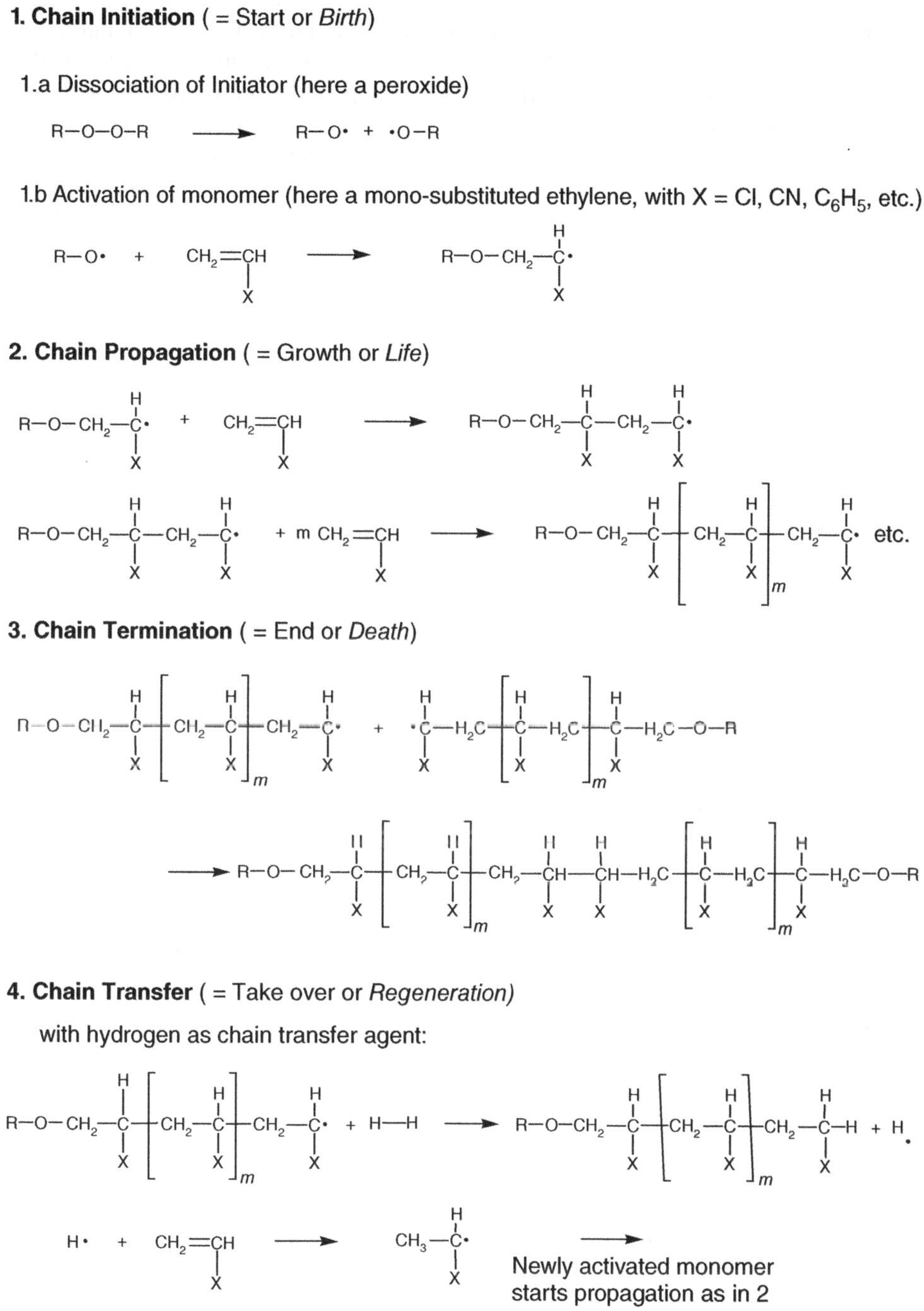

Figure 2.33 Mechanism of chain-addition polymerization.

monomer. The polymerization could be performed with ethylene, styrene, or acrylonitrile. In principle, all other chain-addition reactions follow this mechanism. The Latin terms *initiation*, *propagation*, and *termination* are used to describe the different stages of the polymerization. Their English counterparts and corresponding terms from biology are actually also used by polymer scientists (e.g., a "living polymerization" leads to very long polymer chains and deactivation of initiator or growing polymer chains result in "dead polymer"). During the initiation phase, the initiator must split into radicals, which typically results in two radical molecules per initiator molecule. Then the radical will react or activate one monomer molecule. During propagation, which is the main phase of the polymer formation, the activated monomer molecule can start off a chain reaction with the other molecules, whereby the growing polymer chain continues to carry a radical electron at its active end. This mechanism is based on the assumption that the length of the growing polymer chain does not have a significant effect on its reactivity. Experimental evidence shows that this assumption is valid.

Once a polymer has formed, it becomes probable that the radical electrons at each of the two growing chains meet and react to form a bond. This implies an end, or the termination of the polymerization, because the active radical electrons now cannot add new monomers. The molecular weight of the chain doubles in this termination by the combination of two radical electrons. (There are other ways the two chains can terminate, for example, by disproportionation resulting in a transfer of a hydrogen from one growing chain to the other without combination, not shown in Figure 2.33.)

There is also frequently another reaction observed in chain polymerization named chain transfer, in which the radical leaves the growing chain and starts a new chain. The net result of the presence of chain transfer agents is that the molecular weight is kept low, while the polymer yield keeps increasing. Frequently, hydrogen is used as a chain transfer reagent to decrease the molecular weight of a polymer.

2.7.4.2 Step-Growth Polymers

Step-growth polymerization occurs with monomers that have at least two functional groups at their ends, even without the presence of a double or triple bond. In contrast to a chain-addition polymer, a step-growth polymer always contains a heteroatom, that is, another atom besides carbon, in its main chain. Step-growth polymerizations do not require an initiator. In principle, heating of a monomer suffices for polymerization; however, catalysts may be added to increase the polymerization rate. In Table 2.12, an overview of important step-growth polymers is given, including their structures and applications.

In Figure 2.34, the formation of important step-growth polymers from their monomers is presented. In the simplest case, a monomer has the form X–Y, that is, it is then bifunctional with two different end-groups and polymerizes (e.g., Nylon 6, shown in Table 2.12). Alternatively, two different types of difunctional monomers, X–X and Y–Y, can react (e.g., forming Nylon 6,10, shown in Table 2.12).

In step-growth polymerization, monomers initially form larger segments, generally called multimers, for example, dimers formed from two monomers or trimers formed from three monomers. Then these multimers can combine in single steps to even larger multimers, which eventually form high-molecular-weight polymer. As two monomers react in step-growth polymerization, typically a few atoms at their ends combine to a small

1. For homopolymers from bifunctional monomers

AX—YB + AX—YB → (−AB) AX—Y—X—YB

Bifunctional monomer → A dimer

AX—Y—X—YB + AX—YB → (−AB) AX—Y—X—Y—X—YB

A trimer

AX—Y—X—YB + AX—Y—X—YB → (−AB) AX—Y—X—Y—X—Y—X—YB

A tetramer

and so on

2. For Copolymers from difunctional monomers

AX—XA + BY—YB → (−AB) AX—X—Y—YB

Two difunctional monomers → A dimer

AX—X—Y—YB + AX—XA → (−AB) AX—X—Y—Y—X—XA

A trimer

AX—X—Y—YB + AX—X—Y—YB → (−AB) AX—X—Y—Y—X—X—Y—YB

A tetramer

and so on

Figure 2.34 Mechanism of step-growth polymerizations for bifunctional and difunctional monomers.

Examples

Nylon 6 using caprolactam after ring opening due to heating, X = NH, Y = CO, A = H, B = OH
Nylon 6 using 6-aminocaproic acid, X = NH, Y = CO, A = H, B = OH
Nylon 11 using 11-aminoundecanoic acid, X = NH, Y = CO, A = H, B = OH

Examples

For PETE using monomers ethylene glycol and terephthalic acid, X = O, Y = CO, A = H, B = OH
For Nylon 6,6 using monomers hexamethylene diamine and adipic acid, X = NH, Y = CO, A = H, B = OH
For Nylon 6,6 using monomers hexamethylene diamine and adipoyl chloride, X = NH, Y = CO, A = H, B = Cl

molecule, for example, water, while the remaining larger portion of the monomers form a linkage. The small molecule then ends up as a by-product of the polymer. Therefore, step-growth polymerizations are also referred to in the older nomenclature as condensation polymerizations. However, some step-growth polymers, such as polyurethane (see

TABLE 2.12 Important step-growth polymers

Monomer 1 Common name	Monomer 2 (if any) Common name	Repeat unit Common source name (trade name)	Major applications (specific recycle no., if any)
Polyester			
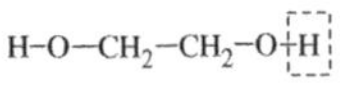 Ethylene glycol	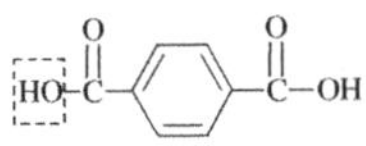 Terephtalic acid	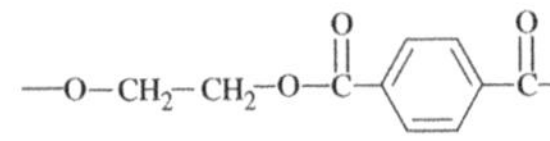 Polyethylene terephtalate (Dacron)	Soda bottles, apparel, car body parts
Polyamides			
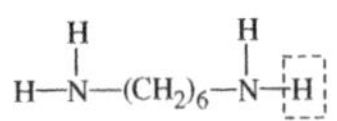 Hexamethylenediamine	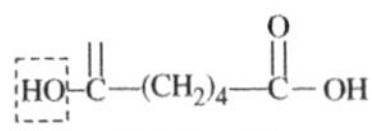 Adipic acid	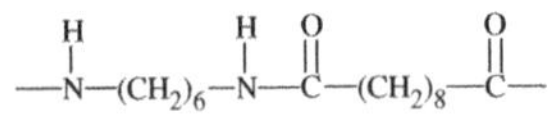 Polyhexamethyleneadipamide (Nylon 6,6)	Textiles, apparel, brush bristles, machine parts, ropes, tire cord
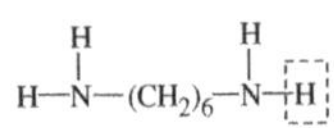 Hexamethylenediamine	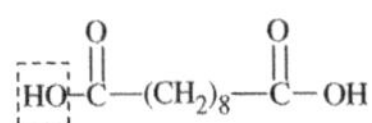 Sebacic acid	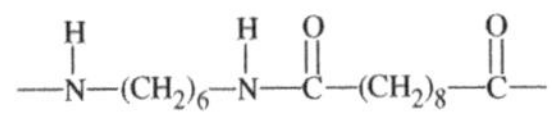 Polyhexamethylenesebacamide (Nylon 6,10)	Similar to Nylon, 6,6 (less produced)
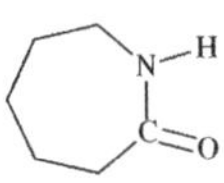 Caprolactam	None	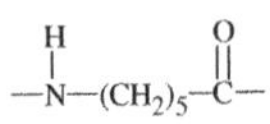 Polycaprolactam	Similar to Nylon 6,6 (lower melting than Nylon 6,6)
Other classes			
$H_2C{=}O$ Formaldehyde	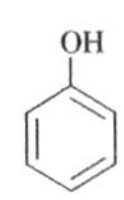 Phenol	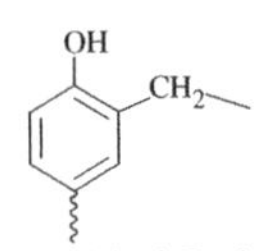 Phenol/Formaldehyde Resin (Bakelite)	Electronic parts, photoresists, imitation wood
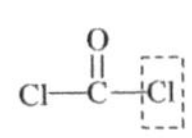 Phosgene	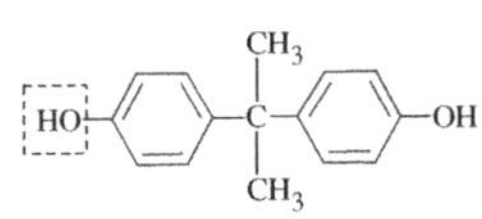 Bis-phenol-A	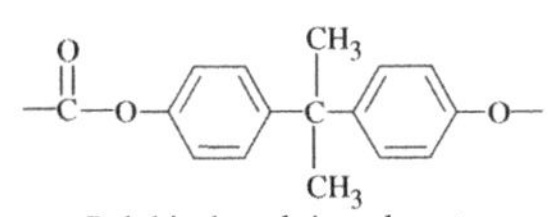 Polybisphenol-A-carbonate	Windows, eyeglasses, CDs
$HO{-}(CH_2)_6{-}OH$ Hexamethylenediol	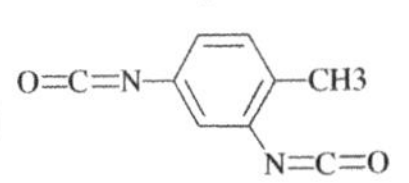 2,4-Toluene diisocyanate	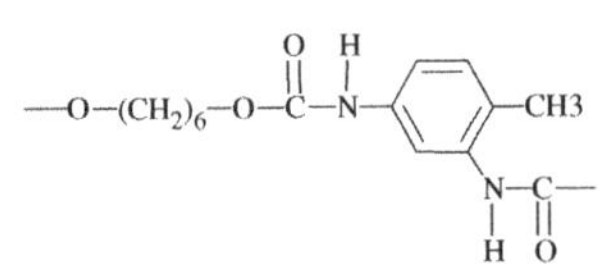 A polyurethane	Insulating foam, cushions, footwear

Table 2.12), conserve all the atoms of their monomers and their formation is distinguished as a polyaddition.

The numbering of the members of the Nylon group, the polyamides, is straightforward. If the polymer was made from only one monomer, such as caprolactam, take the number of carbon atoms in it and add it to the name, for example, Nylon 6. In the case of polyhexamethylene adipamide, there are two monomers, each containing six carbon atoms; thus, it is called Nylon 6,6 (or Nylon 66). Polyhexamethylene sebacamide is called Nylon 6,10 because the second monomer sebacic acid contains 10 carbon atoms.

Example

What should be the name of the polymer formed from the polymerization of 11-aminoundecanoic acid, an amino acid with this formula: $H_2N–(CH_2)_{10}–COOH$?

Answer: Nylon-11.

Incidentally, one can produce a polymer with the same repeat unit as polycaprolactam from this amino acid: $H_2N–(CH_2)_5–COOH$. It would have a different source name: poly(6-aminocaproic acid); however, its IUPAC name would be identical, since the latter is only based on the structure of the repeat unit (IUPAC name: polyiminocaproyl).

A special group of polyamides are polyaramides, their best-known representative being Kevlar, which is used for bullet-proof vests and fire-retardant clothing.

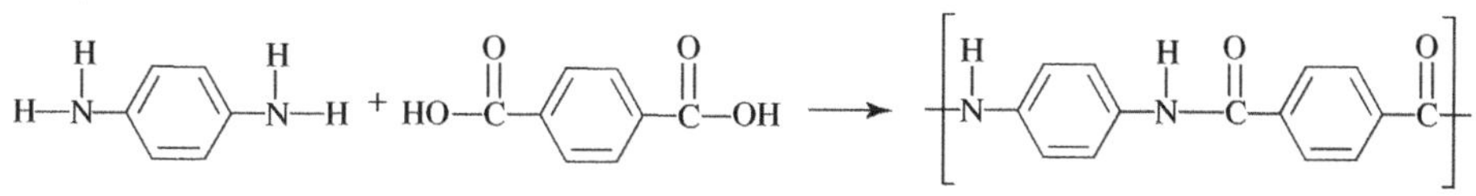

1,4-Diaminobenzene Terephthalic acid Kevlar (a polyaramide)

Another remarkable step-growth polymer is polysiloxane, a polymer that carries only silicon and oxygen in its main chain. It is used for many medicinal applications, such as skin replacement, contact lenses, and implants. It is also industrially important as a sealant.

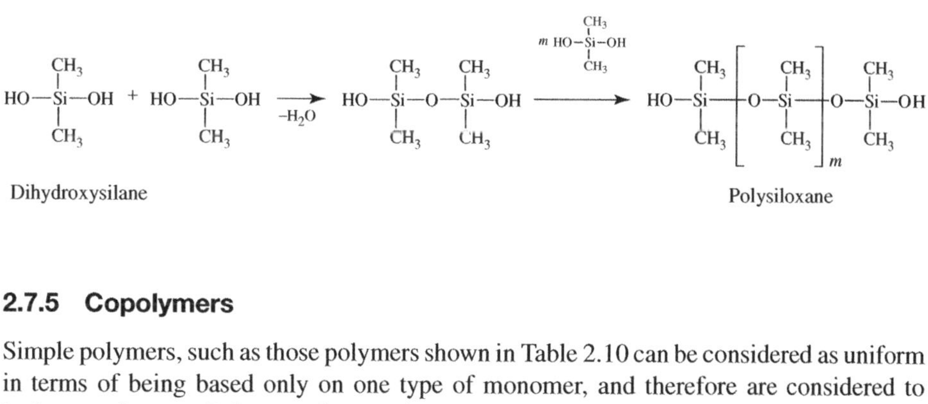

Dihydroxysilane Polysiloxane

2.7.5 Copolymers

Simple polymers, such as those polymers shown in Table 2.10 can be considered as uniform in terms of being based only on one type of monomer, and therefore are considered to be homopolymers. Polymers that are formed from two or more types of monomers are referred to as copolymers. Copolymers have two important base properties: the copolymer composition and the monomer sequence distribution, which really means the distribution of the different types of repeating units for a copolymer.

Representative sequences of three copolymers with different copolymer compositions:

Contents of B	
10%	AAABAAAAAA
20%	ABAAAAABAA
50%	AABBBABAAB

Representative sequences of three copolymers with same composition (e.g., 50%), but different monomer sequence distributions:

Type of monomer	Sequence distribution
Alternating	ABABABABAB
Random	ABBABAAABB
Block	AAAAABBBBB
Graft	AAAAAAAAAA
	B B
	B B
	B B
	B B
	B B

Nylon 6,6, for example, could be considered a case of a perfectly alternating copolymer. It is relatively easy to make linear step-growth copolymers from two monomers because the alternation of the monomers is a consequence of the mechanism of the step-growth polymerization. This is not the case for chain-addition polymerization, for which type of monomer sequence distribution depends on the monomers, which are copolymerized and other polymerization conditions. If one of the monomer's repeat unit is present in the copolymer to a lesser amount (e.g., 1–30 mol%), that monomer is called the "comonomer." For example, linear low-density polyethylene (LLDPE) can be obtained if 1–5 mol% repeat units from 1-hexene are present in the copolymer besides ethylene. This copolymer is made by adding 1-hexene as comonomer to the same process as for the polymerization of HDPE using a catalyst as initiator. The remaining four carbons of the comonomer 1-hexene, which do not polymerize into the main chain, form a butyl-branch that leads to the lower density, much like in LDPE, in which the branches are produced due to the radical initiator. A schematic of this process is shown here:

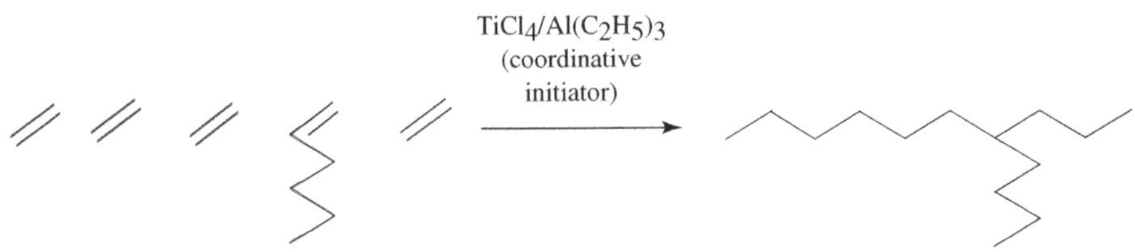

Ethylene 1-Hexene Ethylene/1-hexene copolymer (an LLDPE)

Most synthetic rubbers, which are copolymers, such as EPR (ethylene/propylene rubber), have close to random distribution of the monomers. SBR (styrene/butadiene rubber) has typically random distribution, but is also produced as block-type copolymer and graft copolymer.

2.7.6 The Molecular Weight and Molecular Weight Distribution of Polymers

As mentioned, polymers stand apart from "normal" chemical substances because of their large molecular size and molecular weight. If we are given the molecular weight of a specific type of polymer, we can determine how many repeating units a typical chain in that polymer sample contains. Instead of number of repeating units, the term degree of polymerization can also be used as a synonym. Notice that for polymers it is common to use the term *molecular weight*, although strictly speaking, according to definitions in general chemistry and based in the units of g/mol, it should be called molar mass.

$$\overline{M} = \overline{X} \times M_o \tag{2.126}$$

where $\overline{M}$ is the average molecular weight the of polymer (in g/mol), $\overline{X}$ is the degree of polymerization or number of repeating units in the typical polymer chain, and M_o represents the molar mass of the repeat unit (in g/mol).

Example

How many repeating units does a typical chain of a sample of polypropylene contain, which has a molar mass of 280,000 g/mol? We can compare the polymer to a chain of beads. If we know the weight of the chain and the weight of a single bead, we can determine the number of beads in the chain. For our problem, we first need to determine the molar mass of the polymer's repeat unit (shown in Table 2.10). Then, we sum up all atoms' weights as given in the Periodic Table of Elements and round to values without decimals. For polypropylene with its repeating unit, –(–CH_2–$CH(CH_3)$–)–, we obtain $M_o = (3 \times 12\ \text{g/atom}) + (6 \times 1\ \text{g/atom}) = 42\ \text{g/mol}$. Then, the number of repeating units must be $X = \overline{M}/M_o = 280,000\,\text{g/mol}/42\,\text{g/mol} = 6,667$ repeating units. (If a sample of polyethylene had the same molecular weight, it would contain 10,000 repeating units. Verify this by calculating the molar mass of the repeat unit of polyethylene.)

Most polymers are relatively strong enough for their main applications at average molecular weights in the tens of thousands, as, for example, Nylon fiber or polyethylene plastic bags. Generally, a polymer, which has a higher molecular weight, is more robust in its mechanical properties, that is, it will be stiffer. Polyethylene with molecular weights exceeding 1,000,000 g/mol is used for parts in artificial hips, so they can withstand longer the constant stress on the hips, as the person who received the implant moves around. Such UHMPE (ultra high molecular weight polyethylene) also can be used as artificial ice for skating parlors, saving on cooling costs for natural ice.

A polymer is different from a small molecule in its size and its molecular weight, but also in that its molecular weight is not uniform. For a typical synthetic polymer, one can define only an average molecular weight. However, many chains will be lower than the average, while some chains will exceed the average molecular weight. The relative amount of chains that differ from the average can be different between polymer samples. For example, two polymer samples may have the same molecular weight average of 40,000 g/mol. One sample may contain 20% of chains that are smaller than the average and 15% that are larger, while the other sample only may contain 2% shorter chains and 1% larger chains. The macroscopic properties of these two polymer samples will be different, with the latter usually being stiffer than the former.

Therefore, another important property of a polymer is its molecular weight distribution (MWD), or the so-called polydispersity. It can be defined using the PDI (polydispersity index):

$$\text{PDI} = \overline{M}_\text{w}/\overline{M}_\text{n} \tag{2.127}$$

Hereby $\overline{M}_\text{w}$ is the weight average molar weight and $\overline{M}_\text{n}$ is the number average molecular weight of a polymer. The latter average is the simple arithmetic average of the molecular weights of all polymer chains in a sample, and we used it in our earlier calculation using equation 2.126. It is important to know that, typically, a polymer with a smaller PDI or narrower molecular weight distribution will be stiffer and resists higher impact. For example, given a certain molecular weight average a polymer with a narrower molecular weight distribution has a higher impact resistance, whereas a polymer with a broader

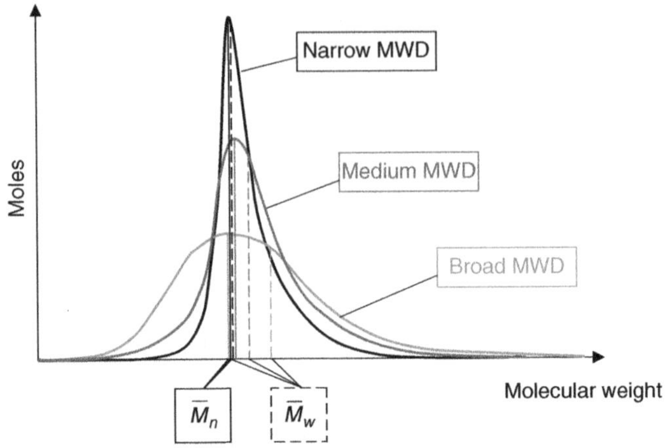

Figure 2.35 The molecular weight distribution (MWD) of three polymer samples.

distribution is easier to process. Figure 2.35 shows the molecular weight distributions of three polymer samples.

If for simplicity a polymer would consist of only two major (monodisperse) portions with different molecular weights and these would be 5 mmol with a molecular weight of 10,000 g/mol and 1 mmol with a molecular weight of 100,000 g/mol, then

$$\overline{M}_\mathrm{n} = \frac{(5 \times 10^{-3}\mathrm{mol} \times 10{,}000\ \mathrm{g/mol}) + (1 \times 10^{-3}\ \mathrm{mol} \times 100{,}000\ \mathrm{g/mol})}{5 \times 10^{-3}\mathrm{mol} + 1 \times 10^{-3}\mathrm{mol}}$$
$$= 25{,}000\ \mathrm{g/mol}$$

The weight average molecular weight can be found by summing the weight fractions for each molecular weight, which is identified in a polymer sample. If, as in the already mentioned example for the calculation of $\overline{M}_\mathrm{n}$ with only two major portions, 33.3 wt% of the sample has a molecular weight of 10,000 g/mol and 66.7 wt% a molecular weight of 100,000 g/mol, then

$$\overline{M}_\mathrm{w} = (0.333 \times 10{,}000\ \mathrm{g/mol}) + (0.667 \times 100{,}000\ \ \mathrm{g/mol}) = 70{,}030\,\mathrm{g/mol}$$

The polydispersity index in this case is PDI $= \overline{M}_\mathrm{w}/\overline{M}_\mathrm{n} = 70{,}300\ \mathrm{g/mol}/25{,}000\ \mathrm{g/mol} =$ 2.8.

For more realistic cases in which several monodisperse fractions for a polymer have been determined (instead of just two as in the simple case described earlier), we can use these two formulas for $\overline{M}_\mathrm{n}$ and $\overline{M}_\mathrm{w}$:

$$\overline{M}_\mathrm{n} = \frac{\sum N_i \times M_i}{\sum N_i} \quad (2.7.2) \text{ and } \overline{M}_\mathrm{w} = \frac{\sum N_i \times M_i^2}{\sum N_i \times M_i} \qquad (2.128)$$

where N_i means the number of moles with a certain molecular weight, M_i.

We already applied the equation to calculate $\overline{M}_\mathrm{n}$ in the simpler example above (there N_i was expressed in millimoles). You may recalculate $\overline{M}_\mathrm{w}$ using the two values for N_i and M_i given in equation 2.128 to see that you will get a very similar result for $\overline{M}_\mathrm{w}$ as when the weight fractions (33.3% and 66.7%) were used.

Other molecular weight averages, such as a viscosity molecular average and a z-average molecular weight, are occasionally useful; however, we will not go more in detail on the definitions of these molecular weight averages.

2.7.7 Polymer Crystallinity

Polymers are different from typical solids in that they are typically softer. Most inorganic solids are salts, in which ions are attracted to each other. Pure elements form solids as well, such as metals as the majority of all elements, or solid nonmetals, such as sulfur. These solids, which are nonpolymeric, often are highly ordered or crystalline materials. However, polymers are only partially crystalline. They also contain a portion that is noncrystalline, or *amorphous*. Polymers are soft due to the amorphous portions they contain, and we use them for applications, in which this is to our advantage, for example, for the interior parts of a car or toys. Important macroscopic properties such as stiffness and strength depend also on the degree of crystallinity of a polymer: polymers with higher portions of crystallinity are usually stiffer and stronger.

In Figure 2.36, the fringed micelle model for polymers is shown, in which regions of highly crystalline polymer and other regions for amorphous polymer are presented. When cross-polarized light is shined through a polymer film, regions can be observed in which Maltese crosses appear in circles and in between these circles we typically can observe regions that appear darker. Those circular regions represent, for the most part, the crystalline regions in the polymer, while the remaining portions make up the amorphous portion of the polymer. Such images show cross sections of the polymer, and the crystalline regions form spheres and are called spherulites. Under other conditions, the crystalline portions in a polymer can form other ordered, rod like structures (called axialites).

The degree of crystallinity of a polymer sample can be indirectly determined from its density, if two additional constants for the specific type of polymers are available: the polymer's density if it would be 100% crystalline and its density at which it would be not

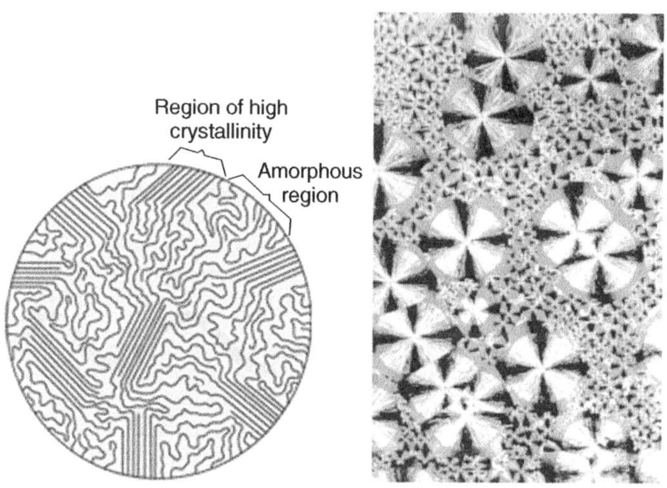

Figure 2.36 Fringed micelle model representing ordered, crystalline and unordered, amorphous regions in a polymer. [*Source*: (Left image) Haydem (1965) [106]. Reproduced with permission of John Wiley & Sons. (Right image) Painter and Coleman (2004) [107]. Reproduced with permission of DEStech Publications, Inc.]

crystalline at all or 100% amorphous. As an equation we can write

$$\%\text{Crystallinity} = \frac{\rho_{\text{Sample}} - \rho_{100\%\text{amorphous}}}{\rho_{100\%\text{crystalline}} - \rho_{100\%\text{amorphous}}} \times 100 \tag{2.129}$$

where ρ represents the density.

For example, let us determine the degree crystallinity, as a percentage, of a sample of polyethylene terephthalate, the density of which has been determined as 1.420 g/mL. The density of 100% crystalline PETE is 1.515 g/mol, and that of the 100% amorphous PETE is 1.335 g/mL. Then,

$$\%\text{Crystallinity} = \frac{1.420 - 1.335}{1.515 - 1.335} \times 100$$
$$= 47.2\%$$

2.7.8 Thermal Behavior of Polymers: Glass and Melting Transition Temperature

Another way of measuring the degree of crystallinity of a solid is by determining its melting behavior, that is, to heat the polymer sample. In contrast to most monomeric materials, polymers do not have a sharp melting point, and some have no melting point at all. To understand how a polymer sample behaves as the temperature changes we need to cool it to sufficiently low temperatures. For HDPE, we need to lower the sample's temperature to about −150°C. At that cold temperature, the polyethylene will be quite brittle and in a glasslike or glassy state. Indeed, if we were to hit a piece of the sufficiently cooled polyethylene with a hammer, it would shatter into pieces as glass would. On the microscopic level, the atoms in the polyethylene chains only perform vibrational motions as in any solid (recall that there is always some motion in matter above 0 K).

When the HDPE sample is heated to about −120°C, it starts to slowly soften and it becomes flexible as we know it in our typical everyday applications. This transition temperature from the glassy to the flexible state is called the *glass transition temperature*, and carries the symbol, T_g. Above the T_g, the shorter segments in the polymer chain can move, looping around the main polymer axis. We therefore call this motion that occurs at the microscopic level *segmental motion*. Figure 2.37 illustrates this type of motion.

As we heat the polyethylene further to about 130°C, we obtain a melt in the form of a viscous liquid, the temperature corresponding to the *melting temperature*, T_m, of the polymer. In contrast to monomolecular compounds, polymers do not have a sharp melting point

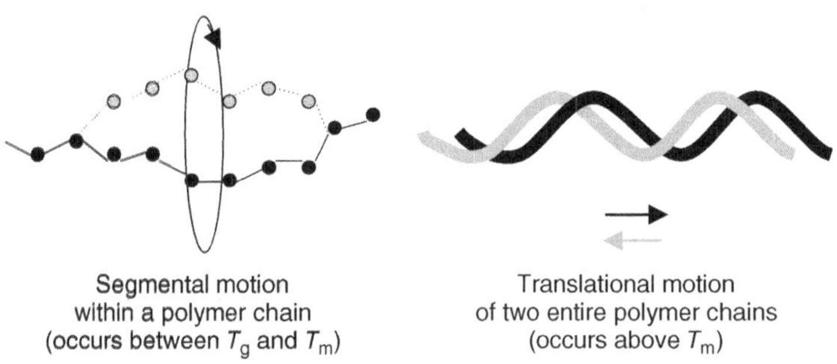

Figure 2.37 Illustration of microscopic behavior of polymer chains at the glass transition and melting temperature of the polymer. [*Source*: Miri (2006) [51]. Reproduced with permission of Pearson.]

because they are not uniform, and they exert a higher viscosity because the chains are partially intertwined (entangled). However, at the microscopic level, the *entire* polymer chains can freely glide by each other, the motion being called *translational motion*. Using the analogy of two intertwined but still separable snakes, this type of motion is also called "reptation." Above 400°C, the covalent bonds in the polymer chains of HDPE, similar to most hydrocarbon-based polymers, will break, and the polymer decomposes. In Figure 2.37, the state of the polymer in the melt is illustrated. Linear and branched polymers can have observable glass transition and melting temperatures; however, as we explained earlier, thermosets and crosslinked rubber cannot be melted. Due to the cross-links in these latter materials, the polymer chains cannot freely move or separate for melting to occur.

Both the glass transition temperature and the melting temperature are related to the ability of the polymer chains to move, in segments or as entire chains. This motion in turn depends on the crystallinity of a polymer. A higher-ordered or crystalline polymer is less flexible and has both a higher glass transition temperature and melting temperature. Therefore, one can use calorimetry measurements of polymers to determine the two transition temperatures, T_g and T_m, and the enthalpies (heat values) for melting and its reverse process, crystallization. This method is referred to as differential scanning calorimetry (DSC).

2.7.9 Influence of the Polymer's Microstructure on Its Crystallinity

The crystallinity of a polymer depends on the attraction between the polymer molecules, the so-called intermolecular forces between them. These forces in turn depend on two factors in decreasing order: the chemical makeup (the repeat unit's formula) and the overall structure of the polymer chains. Furthermore, the degree of crystallinity of a specific polymer does also depend on what temperatures and for how long it was exposed. We will discuss these temperature-related aspects later.

We now know that low-density polyethylene (LDPE) is branched compared to high-density polyethylene (HDPE), which is linear. The reason for the lower density of LDPE is that the branches act like spikes, causing a disorder in the alignments of the main polymer chains, thereby leading to a lower crystallinity. In contrast, the linear polymer chains in HDPE can neatly align to form a closer packing. Consequently, LDPE has a melting point that is typically 10–20°C lower than that of HDPE. The polymers used as rubber typically have very low crystallinity if only to obtain better elasticity. Tightly cross-linked thermosets, such as phenol/formaldehyde resin, are brittle; however, also in their case, crystallinity does not come to bear.

The chemical makeup of the repeat unit has a very important influence on the crystallinity of a polymer. For example, the attractive, intermolecular forces between chains of polypropylene are stronger than between polyethylene chains, due to the stronger dispersion forces between methyl groups than those between hydrogen atoms, respectively. For the same reason, in polystyrene, which has large phenyl groups (each contain six carbon atoms), the attractive forces between the polymer chains are even higher. Polyamides (Nylon) have also relatively high transition temperatures because they contain functional groups in their repeat unit that allow hydrogen bonding between polymer chains, as shown in Figure 2.38.

2.7.9.1 *Tacticity* All polymers formed from monosubstituted and some disubstituted ethylene can form three major structures with different degrees of order and crystallinity. They exert a phenomenon generally called *tacticity*. We will use as an example polypropylene, in which, as we have seen earlier, one hydrogen of ethylene has been substituted with a methyl group. In Figure 2.39, these structures are shown. Because the composition

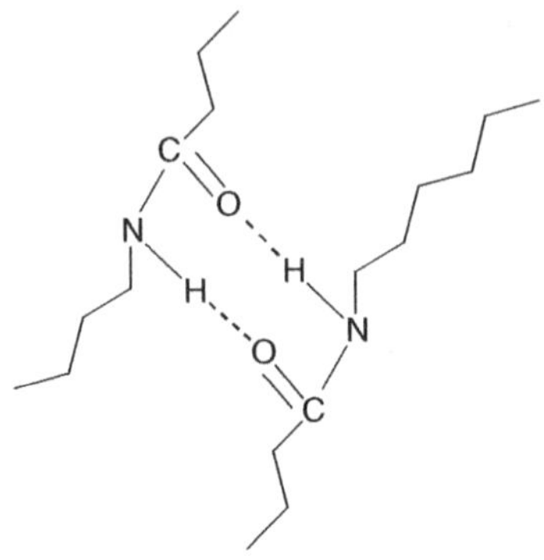

Figure 2.38 Hydrogen bonding (dotted) between two polymer chains in a polyamide, that is, nylon.

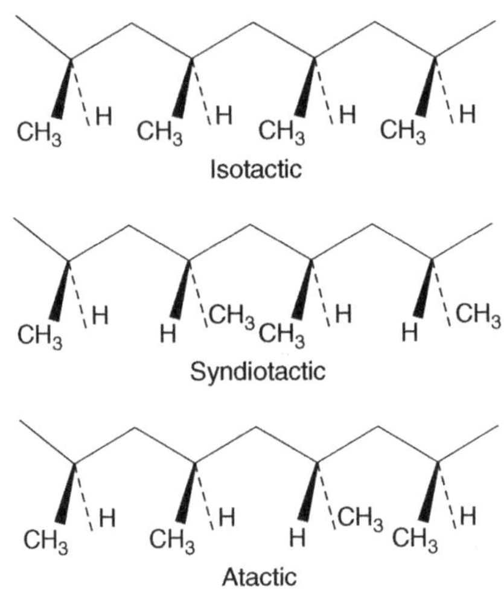

Figure 2.39 The three types of tacticities, which can occur in polymers of monosubstituted ethylene, such as propylene shown here.

of atoms and the bonds in the main chain are still the same, we call these three forms of polypropylene stereoisomers. In the figure, a more realistic unprojected representation of these polymer chains, we can see the single bonds with the tetrahedron angles and the possibility that at the methyl group at every second carbon in the polymer chain can point *toward* us from the paper plane or *away from* us into the paper plane. If the methyl groups all point in one and the same (Greek: iso) direction, that is, as shown toward us, that polypropylene is called *isotactic*. If every other methyl group along the polypropylene chain points in the opposite direction, the polymer is called *syndiotactic*. Finally, if there is no order in the orientation of the methyl groups at all, there is no tacticity and the polymer is called *atactic*.

Table 2.13 shows the glass transition and melting temperatures of several of the above-discussed polymers. The glass transition temperature of a polymer can be lowered to make it less brittle by adding a softener or "plasticizer." Typically, hydrocarbons are added to hydrocarbon-based polymers, for example, dioctyl phthalate (DOP) is added to PVC. If 10% DOP is added, the T_g of the PVC drops by 21°C (from 81 to 60°C).

TABLE 2.13 The glass transition and melting temperatures of several of the above-discussed polymers [55]

Polymer	T_g (°C)	T_m (°C)
HDPE	−120	135
LDPE	−125 to 130	110 to 120
PP, isotactic	−10	165
Syndiotactic	15	155
Atactic	20	None
PS	95	195
Nylon 6	47	220

PP = Polypropylene, PS = Polystyrene.

2.7.10 Environmental and Energy-Related Issues with Polymers

A major issue with our increasing use of polymers is the waste. If consumer products such as food containers and plastic bags are disposed of in the environment, they could take several hundred years to disintegrate. We will discuss the different methods to resolve the problem with the waste disposal of polymers, with the most preferable being discussed first. The waste issue and other environmental issues with polymers, such as those related to their production, should also be seen in context with our problem with energy resources. The production of polymers not only requires energy but also adds value to the raw materials, that is, in essence, the oil the polymers were made from. Many of the following options are also considered as the so-called waste-to-energy conversions.

2.7.10.1 Source Reduction The simplest way to avoid waste of polymers is to use less of them. Many consumer products are now much thinner or smaller than they were several decades ago. For example, supermarket bags are now made of HDPE, rather than LDPE, because the HDPE has a higher tensile strength and a smaller thickness of it is necessary to carry the same weight of groceries.

2.7.10.2 Reuse Another environmentally beneficial method for dealing with the polymer waste issue is to reuse consumer products and commercial items. Instead of throwing plastic dishes, cups, and "silverware" out after a picnic, we should clean them and reuse them for another event. Some plastic milk bottles for doorstep delivery are now cleaned and reused. A few companies engage in reusing ink cartridges.

2.7.10.3 Recycling Recycling is different from reuse in that the material is broken down. We have recycled plastic since 1988, when the Society of Plastic Industry (SPI) established a universal resin identification code, as shown in Figure 2.40. The six recyclable polymers are all commodity thermoplastics, and can be easily melted and reshaped. For example, PETE is known to be used for soda bottles, and is also recycled to be used in apparel, such as fleece, or to produce parts of car bodies, for example, doors. An issue with plastic recycling is finding an efficient method of sorting plastic into the six groups and some resistance to the use of recycled materials in industry, some due to slight deterioration effecting physical properties. Another problem is that many polymers that do not belong in the specific six groups fall under group 7 and cannot be recycled. Though these polymers

Figure 2.40 Universal resin identification code.

are thermoplastic and melt, different types of polymers are not compatible and do not lead to uniform and useful blends. Currently, less than 20% of plastics are recycled.

2.7.10.4 Depolymerization and Biodegradation The polymers, which are formed by step-growth polymerization, have the potential be depolymerized at moderate temperatures, which is in essence the reverse reaction of the polymerization from their monomers. An example of a depolymerization is

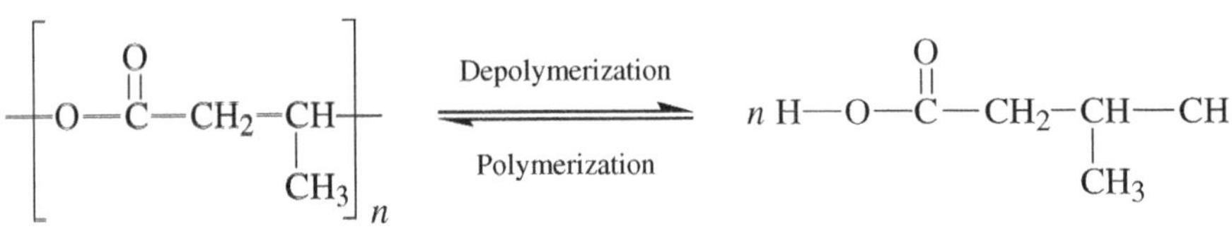

Poly-3-hydroxybutyrate (PHB)

The depolymerization of PHB is relatively fast under composting conditions, that is, bacteria in the soil can open up the ester linkages. PHB therefore is an example of a biodegradable and compostable polymer and does not release any harmful products into the environment. An example of a step-growth polymer, whose biodegradation would result in toxic products is polyurethane based on 2,4-toluene diisocyanate. Biodegradation can proceed in the presence of oxygen, that is, it can be aerobic, or in the absence of oxygen, that is, anaerobic. Composting is more typically performed as aerobic biodegradation in which microorganisms feed on the polymer and mineralize it, that is, convert the organic material to inorganic compounds. Genetic engineering has been applied to increase the efficiency of these bacteria. More recently, several companies have carried out research on finding better methods of producing PHB.

Polylactic acid or polylactide, –[–C(CH_3)–COO–]–, also is compostable though at a slower pace. Advantageously, polylactates can be produced from renewable crops, such as cornstarch, as starting material. Polylactates are used to produce various consumer goods, such as plastic forks and knives or food packaging. The depolymerization of polylactate is also applied in medical surgeries as sutures, which dissolve and are absorbed in the human body by the time a patient's wounds have healed.

PETE can also be depolymerized. However, it requires more vigorous chemical conditions to undergo separation into the original monomers and such a process cannot occur naturally. For example, X-ray films made of PETE used in hospitals are depolymerized and subsequently repolymerized to produce new virgin polymer with better target properties than recycled PETE.

There are also several ways to make existing commodity polymers partially biodegradable. A relatively simple method, which is typically employed during polymer processing, is to blend a solid filler into the polymer. Starch can be used for this purpose, which also can result in oxygen and water permeability desired for packaging to keep certain foods, such as fruits and vegetables, from spoiling. However, in the soil only the starch will disintegrate, while the polyethylene portion would stay unchanged. A method to make polyethylene waste biodegradable is to distribute small amounts of iron or cobalt compounds within a batch, which will accelerate the degradation process in soil (termed "oxobiodegradable" polyethylene).

Compostability is certainly a highly sustainable solution to the waste issues caused by polymers, since it causes the least harm to the environment. The major advantage of depolymerization is that monomers can be reclaimed and typically have a high value compared to the raw materials they were made from. One disadvantage is that depolymerization cannot be applied at moderate conditions to many polymers, including all chain-addition polymers. A major concern with biodegradable polymers is that one has to design the product so that the intended degradation does not interfere with the desired performance of the product during its expected time of usage.

2.7.10.5 Pyrolysis At temperatures close to 500°C, polymers can undergo pyrolysis, that is, there is sufficient heat to break some bonds in the polymer's main chain. However, one does not obtain necessarily the monomers as products as was the case in the depolymerization. For example, in the case of HDPE, the aromatic benzene and its two derivatives called toluene and xylene (dimethyl-benzene) are energetically more favored than the original monomer. Pyrolysis seemed to be most beneficial for the waste disposal of vulcanized or cured rubber, such as tires, since the previous waste disposal methods cannot be used. The disadvantages are that pyrolysis requires relatively high energy costs and results in products with a relatively low value.

2.7.10.6 Incineration The goal of the incineration is to heat the polymer waste so high that a complete burning leads to mostly carbon dioxide and water. This can be only achieved at temperatures close to 1000°C or higher. The high fuel value of plastics makes them attractive for incineration; for example, burning polystyrene produces about two and a half times the energy than burning wood. However, many plastics contain heteroatoms (atoms other than hydrogen and carbon atoms), which lead to the formation of toxic gases such as hydrogen chloride or oxides of nitrogen. Therefore, the emissions of incineration plants need to be controlled. In addition, most polymer waste contains some toxic heavy metals, for example, cadmium and mercury, which were added during its processing, for example, as color pigments, or ended up with the polymer because of its application. These hazardous metals will remain as residue in the ashes of the incineration process and require special disposal. An additional drawback associated with burning polymers is that it adds to the CO_2 emissions and therefore is another contributing factor to global warming. Incineration is considered by some as a renewable energy source; however, one has to consider that synthetic polymers are made mostly from oil, which is nonrenewable.

2.7.10.7 Landfill Currently, most polymers end up in landfills, where they are exposed to mostly anaerobic conditions. Plastic makes up about 20% of the waste in landfills. It takes typical plastic at least several centuries to disintegrate in a landfill, which makes this the least desirable method of treatment of polymer waste. The relative inertness, that is, low reactivity of polymers in the environment, has been considered a blessing by some, since these materials would release less harmful poisons into the environment. However, over the long run, the amount of plastic will keep growing and with it the size of our landfills.

2.7.11 Electroconductive Polymers

While most polymers we discussed so far are electric insulators, electroconductive polymers are semiconductors, some of which almost being as conductive as metals. These types of polymers have delocalized electrons, typically conjugated double bonds along their main polymer chains, that is, one double bond alternates with a single bond, including aromatic groups. This structure allows electrons or positive charges (holes) to move along the chains. In Figure 2.41, the repeating units of several electroconductive polymers are given. *trans*-Polyacetylene is slightly more conductive than *cis*-polyacetylene. The morphology of the polymer must be suitable, and most conductive samples are prepared as films, for example, by spin coating, a process by which a polymer solution is poured on a quickly rotating disk thereby spreading out due to the centrifugal forces.

The conductivity occurs rarely over the whole chain but over shorter segments called *polarons and solitons*. There is also some conduction occurring between separate polymer chains. To reach higher electroconductivities and reduce the band gap, polymers need to be doped. Electroconductive polymers can be doped with an electron donor (n-dopant), for example, iodine, or an electron acceptor (p-dopant), for example, Li^+ pr As^{5+}. The former (n-dopant) is less preferred because exposure to the oxygen in the air would lead to reversal and inefficiency of that process. When polyacetylene is doped with AsF_5, its conductivity is measured to be 10^5 S/cm, reaching that of copper. To increase the solubility of some of these conductive polymers, alkyl groups can be added as substituents. These side groups also change the morphology of the polymers and can lead to variations of conductivity or wavelengths in color for light-emitting diode applications (PLED, polymer light-emitting diode).

2.7.11.1 Applications of Polymers in Batteries and Capacitors

Modern batteries used, for example, in cell phones and laptops are based on lithium polymer. Instead of using a salt as an electrolyte, a polymer can be used in these cases, such as polyvinylidene fluoride or polyacrylonitrile (see Figure 2.10). In lithium polymer batteries, the electrodes

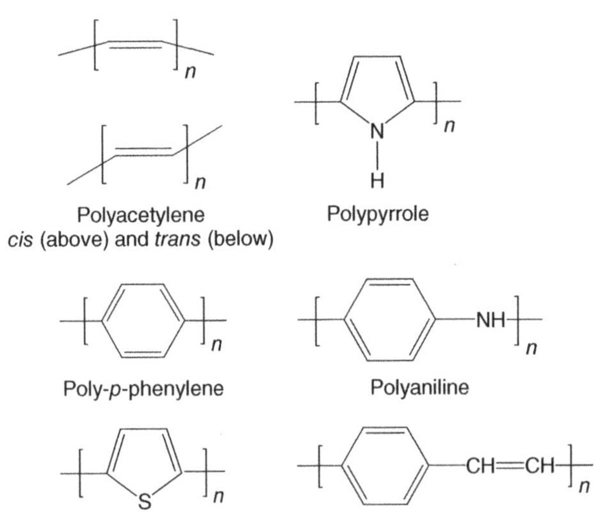

Figure 2.41 Examples of electroconductive polymers.

can be simply laminated. There are several advantages with these polymer-based batteries. First they are lighter than their metal-encased counterparts. They can be shaped in any form, simplifying the design of products, and they last longer in terms of recharging cycles. Hybrid polymer batteries are based on the polymer acting as a long-term capacitor. In contrast to conventional capacitors, they can release the charge more slowly, thereby acting as a battery. Polypyrrole has been used for this kind of application. More recently, it has been shown that polyacetylene coated with a metal such as titanium was excellent for hydrogen storage.

2.7.12 Ionomers and the Protoype of a Fuel Cell Membrane: Nafion

Ionomers are copolymers that contain up to 15 mol% of an ionic comonomer. The ionic groups are often used to cause a higher attraction between the polymer chains and function as reversible cross-links that increase the mechanical properties of the polymer. For example, the copolymer of ethylene with methacrylic acid (see repeat unit of polymethacrylic acid in Table 2.10) is used as coating for cut-proof golf balls (trade name SURLYN).

For the membranes in hydrogen fuel cells, Nafion is frequently applied: Nafion. Nafion is a copolymer of Teflon units (see Table 2.10) and sulfonated perfluoro (alkyl vinyl ether) as a comonomer. It is produced by chain-addition polymerization with radical initiators. The monomers from which Nafion is formed and its repeat units are shown as follows.

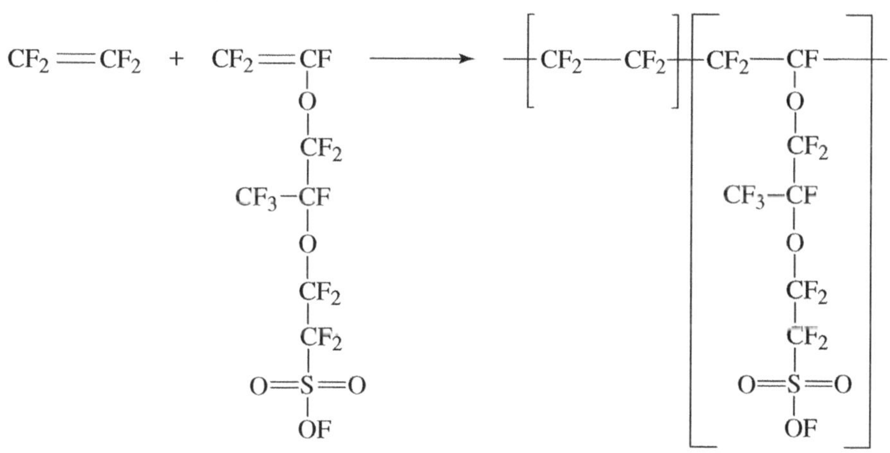

Nafion

To activate the Nafion as a proton exchange the free acid is obtained in the following manner:

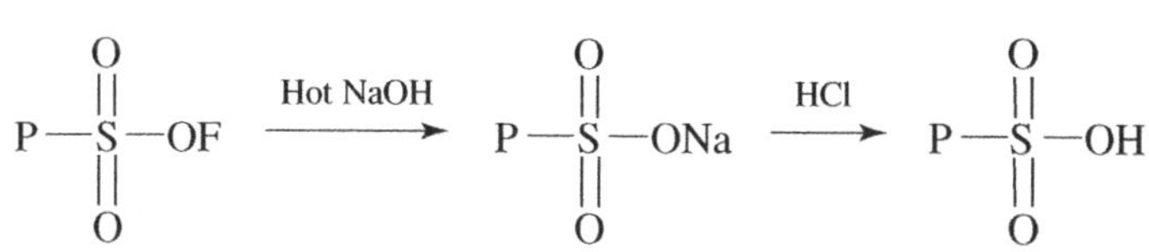

Nafion
(P = polymeryl)

Nafion is preferably used as PEM (proton exchange membrane) in hydrogen fuel cells because it is only permeable to cations, including protons, but impermeable for anions or electrons. In the hydrogen fuel cell, it allows the protons to move to the oxygen and form water. The protons can move along the polymer chain by "hopping" from one acid group to the next. It is a relatively stable polymer that can be used at temperatures close to 200°C. The molecular weight of Nafion is not easy to determine due to its insolubility, and is estimated to range between 100,000 and 1,000,000 g/mol. The major drawback of Nafion is that it is relatively expensive. As a less expensive alternative, porous polyethylene (Solupor) has been introduced. Besides its application as PEM and membranes in other electrochemical cells, such as the chloralkali process, Nafion is also used as a super acid in organic synthesis in sensors and in actuators.

2.8 PHOTOCHEMISTRY

2.8.1 Introduction

The earth receives a vast amount of energy from the sun in the form of solar radiation, which provides more energy in 1 h to the earth than all of the energy consumed by humans in an entire year. If just 0.2% of the solar radiation that falls on the United States were converted to usable energy, it would meet the total energy demand of the United States. The goal of this section is to describe the fundamental interactions of solar radiation with "noncarbon-containing (carbon-neutral)" materials to convert the energy of sunlight into hydrogen fuel.

2.8.2 Electromagnetic Spectrum

Light possesses not only wave-like properties, that cause diffraction, but also particulate properties. Light may be considered to be made up of distinct "packets" of energy, called quanta or photons. A single packet is termed a quantum or photon and has energy (E) proportional to the frequency (ν) of the radiation

$$E = h\nu$$

where the proportionality constant, h, is called Planck's constant and is equal to 6.626×10^{-34} J s. Since the speed of light, c (2.998×10^{8} m/s), is equal to the product of the frequency and wavelength (λ) of the light, the energy is inversely proportional to the wavelength.

$$E = hc/\lambda \text{ since } c = \nu\lambda$$

Wavelengths are usually reported in units of nanometers (1 nm = 10^{-9} m), angstroms (1 Å = 10^{-10} m), or microns (1 μm = 10^{-6} m). From the above equation, shorter wavelengths of radiation have more energy than longer wavelengths and the energy of a photon is also proportional to $1/\lambda$ called the wave number.

The electromagnetic spectrum of radiation consists of a range of wavelengths, wave numbers, frequencies, and energies called radio frequency, microwave, far infrared, near infrared, visible, ultraviolet, vacuum ultraviolet, X-rays, and γ-rays (Table 2.14) that have a variety of chemical and physical effects on materials.

TABLE 2.14 Energies of electromagnetic radiation

Description	Wavelength Range	Wave Number cm^{-1}	Frequency (Hz)	Energy (kJ/mol)	Energy eV
	3×10^3 m	3.33×10^{-6}	10^5	3.98×10^{-8}	4.12×10^{-10}
Radio frequency					
	0.30 m	0.0333	10^9	3.98×10^{-4}	4.12×10^{-6}
Microwave					
	0.0006 m (600 μm)	16.6	4.98×10^{11}	0.191	2.07×10^{-3}
Far infrared					
	30 μm	333	10^{13}	3.98	0.0412
Near infrared					
	0.8 μm (800 μm)	1.25×10^4	3.75×10^{14}	149.8	1.55
Visible					
	400 nm	2.5×10^4	7.5×10^{14}	299.2	3.10
Ultraviolet					
	150 nm	6.06×10^4	19.98×10^{14}	795	8.25
Vacuum ultraviolet					
	5 nm	2×10^6	6×10^{16}	2.39×10^6	247.8
X-rays and γ-rays					
	10^{-4} nm	10^{11}	3×10^{21}	1.19×10^9	1.24×10^7

2.8.3 Laws of Photochemistry

The first law of photochemistry states that only the radiation that is absorbed by the molecule can produce changes in the molecule. Radiation transmitted through the material does not cause a photochemical change.

The second law says that one particle is excited for each quantum of radiation absorbed. Many photochemical processes involving intense sources, such as lasers, have more than one quantum absorbed by a single molecule. These multiphoton processes do not violate the second law since they involve excitation to successively higher energy levels of the molecule, each step requiring a single photon.

2.8.4 Absorption of Radiation

The Beer–Lambert law describes the absorption of a monochromatic beam of light in terms of the dependence of absorbance, A, on concentration, c (mol/l), pathlength, b (cm), through which the beam has passed, and the constant of proportionality, "a" (1/(M cm)), that is known as the molar extinction coefficient or molar absorptivity:

$$A = abc$$

The absorbance is defined in terms of the incident (I_o) and transmitted (I) intensity:

$$A = \log_{10}(I_o/I)$$

The exponential form of the law is often employed:

$$I_o/I = 10^{abc} \; or \; I/I_o = 10^{-abc}$$

The natural logarithm, *e*, is also used to express the law:

$$I/I_o = e^{-\sigma n l}$$

where *n* is the concentration in molecules/cm^3, *l* is the pathlength in cm, and σ is the photoabsorption cross section in units of cm^2/molecule.

2.8.5 Photoabsorption Spectrum, Threshold Wavelength, and Quantum Yield for Photodissociation

A plot of the molar absorptivity, "*a*," or photoabsorption cross section, σ, on the *y*-axis versus wavelength, λ, on the *x*-axis is called the photoabsorption spectrum of the molecule. Experimentally, UV/visible and infrared spectrophotometers provide the UV/visible and IR photoabsorption spectra for a molecule.

Figure 2.42 shows the measured photoabsorption spectrum of water vapor. Water vapor is colorless and is therefore transparent to the visible wavelengths; however, it begins to absorb in the UV region around 190 nm (1900 Å). The absorption cross sections for water vapor near this long wavelength absorption edge are listed in Table 2.15. Continuous absorption extends into the shorter vacuum UV (VUV) wavelength region down to 145 nm with a maximum at around 165 nm (Figure 2.42). Between 145 and 69 nm, the absorption spectrum consists of diffuse bands.

The photodissociation threshold is the wavelength of the photon that just has sufficient energy to break the chemical bond of interest. For the photodissociation of gaseous water,

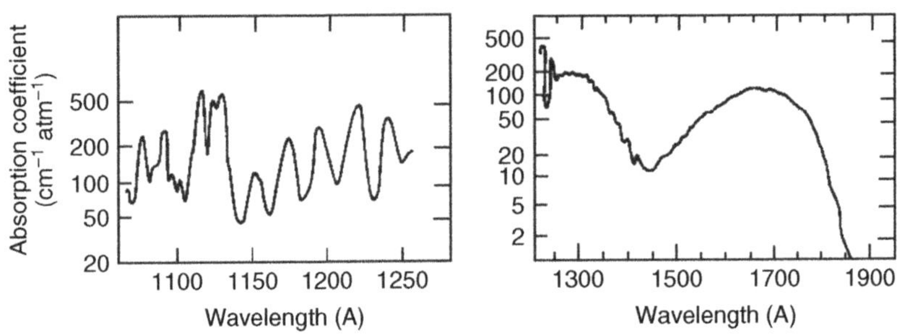

Figure 2.42 The photoabsorption spectrum of water vapor.

TABLE 2.15 Photoabsorption cross sections for water vapor

λ (nm)	$10^{20}\sigma$ (cm^2/molecule)
189.3	0.7
187.5	1.6
186.0	3.1
185.0	5.5
182.5	23.0
180.0	78.1
177.5	185.4
175.5	262.8

the photon should break the HO–H bond, which is often one step in the process known as water splitting.

$$HO - H(g) \rightarrow H(g) + OH(g)$$

$$hc/\lambda = \Delta_f H^\circ(H) + \Delta_f H^\circ(OH) - \Delta_f H^\circ(H_2O)$$

The heats of formation for H(g), OH(g), and $H_2O(g)$ are 217.965, 38.95, and −241.818 kJ/mol at 298 K, which results in a HO–H bond energy of 498.745 kJ/mol that corresponds to a wavelength of 240 nm. Therefore, if water vapor absorbs a photon of wavelength 240 nm or shorter, there will be sufficient energy to break the HO–H bond and potentially produce $H_{2(g)}$ by the following series of reactions.

$$H_2O + h\nu \rightarrow H + OH\lambda \leq 240\,\text{nm}$$

$$H + H_2O \rightarrow H_2 + OH$$

$$OH + OH \rightarrow O_2 + H_2$$

$$\overline{2H_2O + h\nu \rightarrow 2H_2 + O_2}$$

Since water vapor does not absorb photons with wavelengths $190 < \lambda \leq 240$ nm (Figure 2.42), solar radiation having wavelengths < 190 nm are needed to form hydrogen by the above sequence of chemical reactions. Using other sources of radiation between 175 and 190 nm, the above photochemical reaction has been shown to produce one H and one OH for each photon absorbed. Thus, the photochemical reaction is said to have a quantum yield of unity for the wavelengths where each absorbed photon breaks one bond. In general, the quantum yield is defined as the number of molecules of reactant consumed per photon absorbed. At shorter wavelengths between 105 and 145 nm, H_2 and O are also formed as primary products with a reported quantum yield of 0.11.

2.8.6 Solar Radiation

One way of reporting the solar flux of radiation is in power per unit area per unit interval of wavelength (W/m^2 nm). The solar flux from the sun at the earth's surface is dependent on a number of factors such as the composition and concentration of gases, particulate matter and clouds in the earth's atmosphere through which the radiation passes, as well as the angle of the sun, time of day, latitude, and season. Figure 2.43 shows an example of the solar flux for different wavelengths outside the atmosphere and at sea level for a solar zenith angle of 0° which is for overhead, noonday sun. The solar flux outside the atmosphere approximates blackbody radiation at 6000 K. The hatched areas in Figure 2.43 illustrate the attenuation of the radiation by photoabsorption of some naturally occurring components O_3, O_2, H_2O, and CO_2. Very little of the UV radiation required to photodissociate water vapor reaches the earth's surface primarily due to UV absorption by ozone. Table 2.16 shows representative integrated solar flux data (W/m^2) at the earth's surface as a function of wavelength interval and solar zenith angle for mid-latitude during summer.

The altitudes above the earth for maximum light absorption by various atomic and molecular species as a function of wavelength for an overhead sun are shown in Figure 2.44. Therefore, if hydrogen is to be made directly from the photodissociation of water, it will have to be carried out at high altitudes where there is a presence of wavelengths <190 nm.

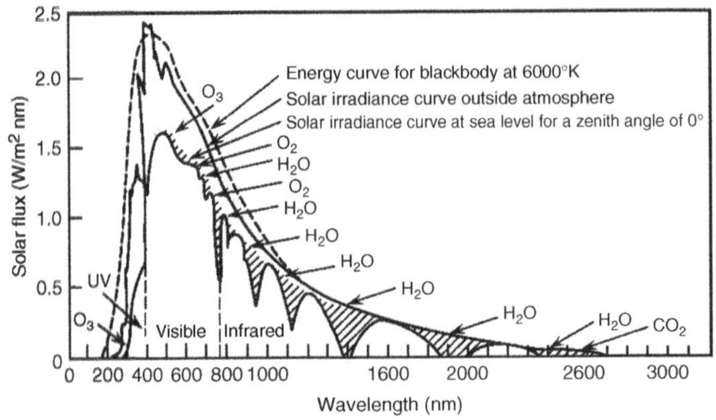

Figure 2.43 Solar flux outside the atmosphere and at sea level, respectively. The emission of a blackbody at 6000 K is also shown for comparison. The species responsible for light absorption in the various regions (O_3, H_2O, etc.) are also shown.

TABLE 2.16 Estimates of integrated solar flux values at the earth's surface as a function of wavelength interval and solar zenith angle within specific wavelength intervals[a]

Wavelength interval (nm)	Solar zenith angle (°)[b]			
	20	40	60	80
	Solar fluxes (W/m^2)			
280–300	0.0	0.0	0.0	0.0
300–320	3.0	2.0	0.8	0.1
320–340	14.5	10.8	5.6	1.1
340–360	16.8	12.8	7.0	1.6
360–380	17.4	13.4	7.4	1.7
380–400	21.5	16.6	9.4	2.2
400–420	26.1	20.3	11.6	2.7
420–440	24.7	19.3	11.1	2.6
440–460	33.5	26.3	15.3	3.6
460–480	33.6	26.4	15.5	3.7
480–500	34.1	26.8	15.9	3.7
500–520	32.7	25.8	15.3	3.6
520–540	32.1	25.4	15.0	3.5
540–560	31.6	24.9	14.8	3.4
560–580	30.5	23.9	14.0	3.0
580–600	28.4	22.2	12.9	2.7
600–620	29.6	23.4	13.9	3.1
620–640	28.5	22.6	13.5	3.0
640–660	27.6	21.8	13.2	3.2
660–680	27.3	21.8	13.7	3.4
680–700	24.1	19.1	11.5	2.8

[a]The authors are grateful to Dave Rutan for his help in assisting with calculations using the Coupled Ocean Atmosphere Radiation Transfer (COART) model (Jin, Z, Charlock, TP, Rutledge, K, Stamnes, K, Wang, Y. *Appl. Optics* 2006;45:7443), which may be found at http://www-cave.larc.nasa.gov/jin/rtset.html.

[b]The zenith angle is the angle between the direction of the sun and the vertical. Therefore, an overhead noonday sun has a zenith angle of 0°, while sunset and sunrise are approximately 90°. The calculations were done for mid-latitude during summer, Modtran rural aerosol layer, and fraction of light incident on the land surface that is reflected (surface albedo) of 0.2.

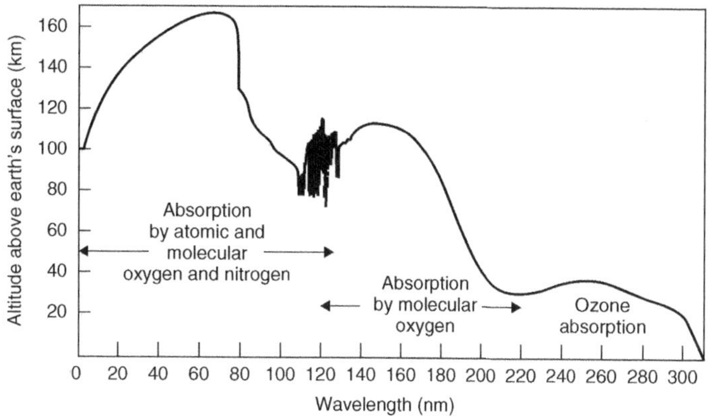

Figure 2.44 Atmospheric regions of maximum light absorption of solar radiation in the atmosphere by various atomic and molecular species as a function of altitude and wavelength with the sun overhead.

Water in the upper atmosphere is mainly photodissociated with vacuum UV radiation emitted from excited hydrogen atoms, called the "Lyman alpha line," at 121.6 nm.

2.8.7 Solar Photochemical Sensitizers and Semiconductors

Since pure water does not absorb photons at sea level that can directly convert water to hydrogen, a photochemical process involving a sensitizer, a molecule that absorbs sunlight and transfers some of that energy to water, may be necessary to generate hydrogen. Figure 2.45 shows a schematic for photochemical splitting of water where S is the photochemical sensitizer that initiates the oxidation and reduction reactions, which would generate hydrogen.

Solar radiation may also be absorbed by a semiconductor catalyst in water, which has an energy level diagram as illustrated in Figure 2.46 that displays an energy band gap, E_g, between a valence band and a conduction band. At 0 K, the conduction of a semiconductor is zero because all of the energy states in the valence band are filled and all of the states in the conduction band are vacant. As the temperature increases and/or photons are absorbed, electrons are excited from the valence band to the conduction band where they become mobile. Both the electrons in the conduction band and the vacancy of electrons, called holes, in the valence band contribute to electrical conductivity. Some examples of the energy band gaps for semiconductors are given in Table 2.17.

The addition of donor impurities to the semiconductor may substantially increase the concentration of electrons in the conduction band since the donor energy level lies in the band gap slightly below the conduction band and thermal energies and/or photons are sufficient to promote the electrons into the conduction band. Acceptor impurities may also be added, which have a band gap that is slightly above the valence band. The acceptor atom produces a hole in the valence band since it accepts an electron.

Semiconductors, in contact to liquid electrolytes, have fixed energies where the charge carriers enter the solution as determined by the chemistry of the semiconductor/electrolyte interface, the nature of the surface and the electrolyte composition. For spontaneous photochemical water splitting, the oxygen and hydrogen reactions (see Section 2.5) must lie

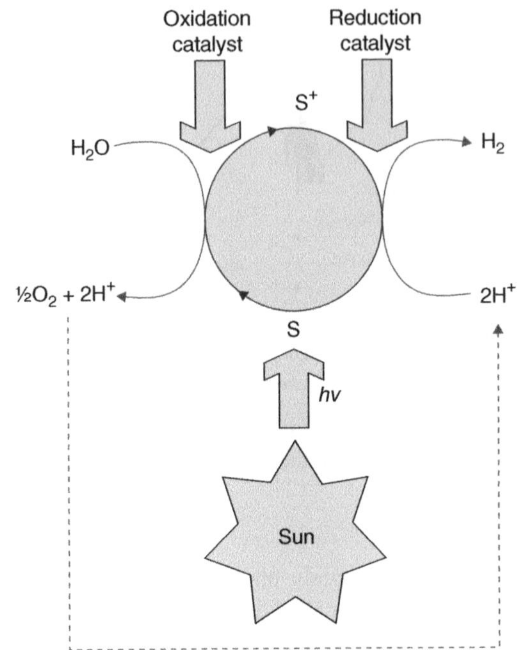

Figure 2.45 Minimal scheme for photochemical splitting of water with a photochemical sensitizer S.

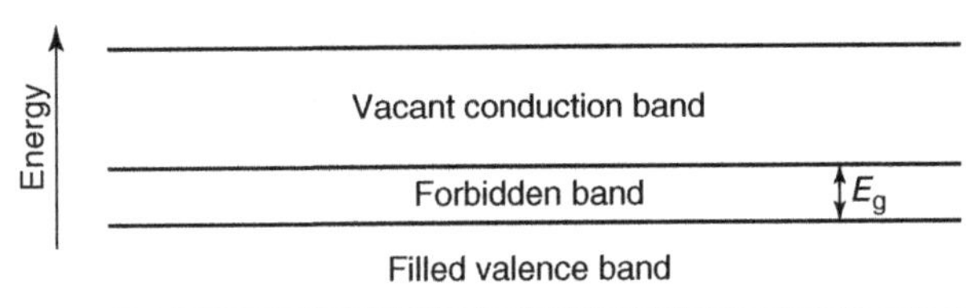

Figure 2.46 Energy band scheme for intrinsic conductivity in a semiconductor.

TABLE 2.17 Semiconductor energy gaps between the valence and conduction bands

	E_g (eV)			E_g (eV)	
Crystal	0 K	300 K	Crystal	0 K	300 K
Diamond	5.4				
Si	1.17	1.11	PbS	0.286	0.34–0.37
Ge	0.744	0.66	PbSe	0.165	0.27
αSn	0.00	0.00	PbTe	0.190	0.29
InSb	0.23	0.17	CdS	2.582	2.42
InAs	0.43	0.36	CdSe	1.840	1.74
InP	1.42	1.27	CdTe	1.607	1.44
GaP	2.32	2.25	ZnO	3.436	3.2
GaAs	1.52	1.43	ZnS	3.91	3.6
AlSb	1.65	1.6	AgCl	—	3.2
SiC(hex)	3.0	—	AgI	—	2.8
Te	0.33	—	Cu_2O	2.172	—
ZnSb	0.56	0.56	TiO_2	3.03	—

between the valence and conduction band edges, which for efficient solar light absorption must be less than 2.0 eV, and this is almost never the case. Active research is being done to develop materials that absorb visible radiation and have a high photocatalytic activity for producing hydrogen from water.

One of the difficulties with photosensitization is that the absorption of a single photon causes the transfer of only one electron and, therefore, a catalyst is needed to induce more electrons because a hydrogen molecule requires two electrons, one for each proton, and oxygen produces four electrons to complete the redox reaction. Few direct photochemical systems have been able to reach an efficiency in excess of 10% and the method has a number of technical challenges, such as the photosensitizer and catalyst must not photodegrade, and hydrogen and oxygen need to be separated.

Photocatalytic decomposition of H_2S is being considered as an alternative to water splitting since H_2S is readily produced as a waste product in a number of industrial processes and requires less energy than water to generate hydrogen. The polysulfide and elemental sulfur that are formed may be converted back to H_2S by a hydrothermal process.

2.8.8 Introduction to Production of Hydrogen Using Solar Radiation

Although it is difficult to produce molecular hydrogen directly from the photodissociation of water or with a sensitizer or catalyst using solar radiation at sea level, the fundamental principles described earlier are essential for a number of alternate pathways whereby hydrogen fuel has been synthesized from water with sunlight.

2.8.9 Solar Thermochemical Production of Hydrogen

As shown in Figure 2.47, a substantial amount of solar radiation is absorbed by liquid water in the infrared region of the electromagnetic spectrum, which causes heating. Solar thermochemical processes are based on the use of concentrated solar radiation as the energy source for high temperature heat to drive an endothermic chemical transformation. Solar energy can be concentrated with parabolic reflectors to generate temperatures over 2000 K to synthesize hydrogen from water via thermochemical cycles. Large-scale collection systems to concentrate solar energy are described in terms of their mean flux concentration ratio $\check{C}$, as defined by, $\check{C} = Q_{\text{solar}}/(IA)$, where Q_{solar} is the solar power received per targeted area, A, and normal beam isolation, I. $\check{C}$ is often expressed in the units of "suns" when normalized to $I = 1$ kW/m^2 and may have values as high as 10,000 suns. These concentrating systems have been proven to be technically feasible in large-scale (MW) pilot and commercial plants aimed at producing electricity where a fluid (typically air, water, synthetic oil, helium, sodium, or molten salt) is solar-heated. The high temperature generated has been demonstrated to produce hydrogen gas such as through the following two-step thermochemical cycle using metal oxide redox reactions.

$$\begin{aligned}
\text{(solar)}: &\quad M_xO_y \rightarrow xM + y/2O_2 \\
\text{(nonsolar)}: &\quad xM + yH_2O \rightarrow M_xO_y + yH_2 \\
\hline
&\quad yH_2O \rightarrow yH_2 + y/2O_2
\end{aligned}$$

Since H_2 and O_2 are formed in different steps, the need for a high temperature gas separation is thereby eliminated.

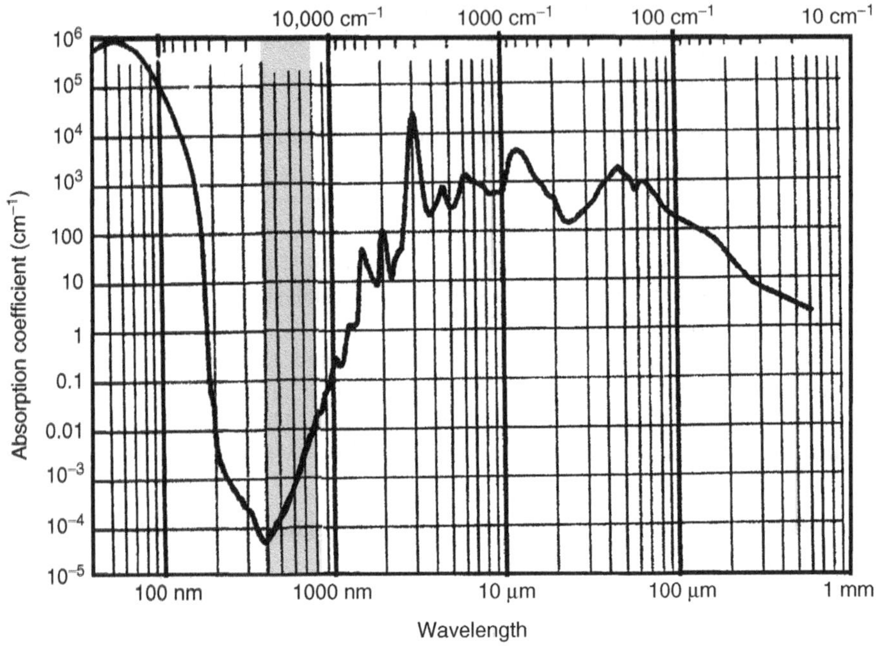

Figure 2.47 The absorption spectrum of liquid water. [*Source*: Reproduced with kind permission from Professor Martin Chaplin.]

2.8.10 Solar Photoelectrochemical Production of Hydrogen

Efficient hydrogen production by photoelectrochemical water splitting from sunlight has been desired for a long time. Photoelectrolysis work by Fujishima and Honda in 1972 synthesized hydrogen as a clean and renewable energy source. When the semiconductor TiO_2 was mounted on an indium plate and exposed to sunlight while connected to a platinum electrode, electrons (e^-) and holes (p^+) were formed, which resulted in the formation of hydrogen at the positive electrode.

$$TiO_2(s)(+2h\nu) \rightarrow TiO_2(s) + 2e^- + 2p^+$$

$$2p^+ + H_2O(l) \rightarrow 1/2O_2(g) + 2H^+$$

$$2H^+ + 2H_2O(l) \rightarrow 2H_3O^+(l)$$

$$2e^- + 2H_3O^+(l) \rightarrow H_2(g) + 2H_2O(l)$$

$$\overline{H_2O(l)(+2h\nu) \rightarrow H_2(g) + 1/2O_2(g)}$$

A German–Saudi partnership conducted a decade of intense research from 1986 to 1996 in the field of solar hydrogen production (HYSOLAR) concentrating on photovoltaics. The program erected and operated photovoltaic–electrolytic test facilities in the power ranges of 2, 10, and 350 kW. Harvesting energy from sunlight using organic-based materials in photovoltaics offers a long-term potential for large-scale power generation.

A recent mini review of the literature surveyed state-of-the-art electrocatalysts and photocatalysts for use in the area of solar-driven water splitting by improving the photoactivity of photoanodes and -cathodes through the utilization of nanostructured materials, which would help meet the requirements for an ideal semiconductor: (a) proper positions of the conduction and valence bands to straddle the hydrogen and oxygen evolution reaction potentials; (b) relatively small band gaps for visible light absorption; (c) high stability under sunlight irradiation in acidic/basic aqueous solutions; (d) efficient charge transfer; and (e) low overpotentials for the oxidation/reduction of water. They recommended that binary or ternary alloys might comprise an excellent additional approach for the exploration of new materials, which would meet all the requirements for solar-mediated water splitting [83].

One limiting factor for the efficiency of water splitting is the recombination of the excited electrons and positive charges (holes) formed by the absorption of visible light. A nanoparticle-based photocatalyst system was designed to keep the charges separated and achieved 100% photon-to-hydrogen production efficiency. A light-harvesting CdS quantum rod was used with a quantum dot of CdSe near one end and a Pt tip on the other. When exposed to visible light in water, the CdS rod absorbs photons and releases electrons. The rod then transfers the electrons to the Pt tip where photons are reduced to hydrogen leaving behind the positively charges holes in the CdSe dot. By adjusting the pH and adding isopropyl alcohol to the water, charge recombination was prevented and the record efficiency was achieved [84].

The future use of inexpensive, earth-abundant materials as catalysts for electrochemical and photoelectrochemical water splitting is readily being explored [85].

2.8.11 Solar Photobiological Production of Hydrogen

Solar-driven biological processes to produce hydrogen consist of the following:

1. direct biophotolysis in which a photosynthetic apparatus captures light and the recovered energy is used to couple water splitting to the generation of a low-potential reducing agent that reduces a hydrogenase enzyme (Figure 2.48);
2. indirect biophotolysis that separates the oxygen generation by solar radiation from the hydrogen evolution step (Figure 2.49); and
3. photofermentation involving photosynthetic bacteria (Figure 2.50).

More details of these solar methods to generate hydrogen will be provided in later sections of this book.

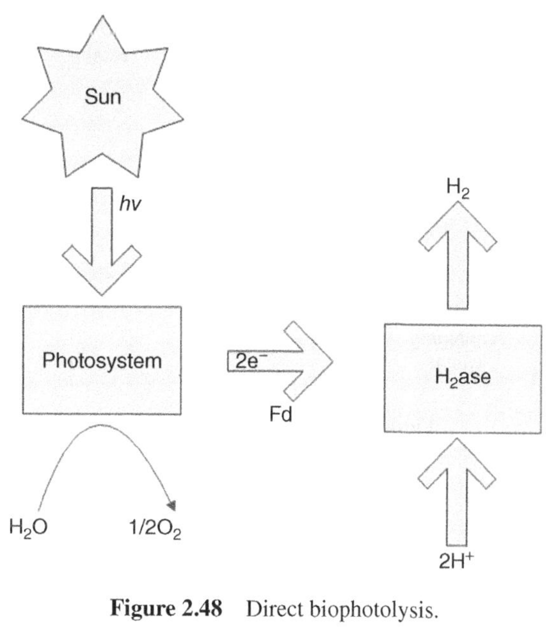

Figure 2.48 Direct biophotolysis.

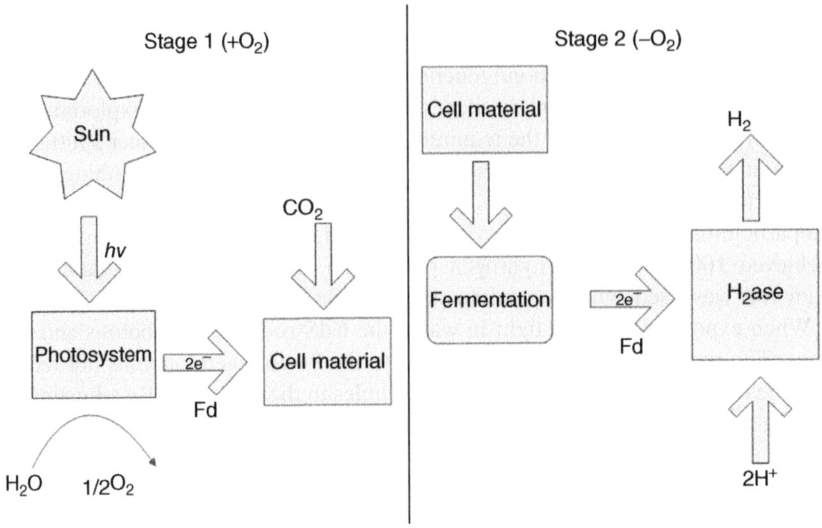

Figure 2.49 Indirect biophotolysis.

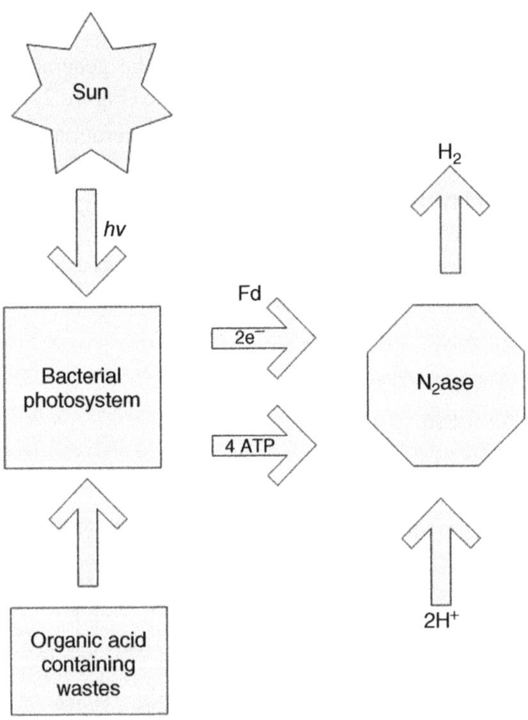

Figure 2.50 Photofermentation.

2.8.11.1 Hydrogen Production by Molecular Solar Photocatalysis Active research is being pursued to store solar energy in the chemical bonds of hydrogen using solar photocatalysis to achieve water splitting. Catalysts acting directly on water require charge-separation features employing some type of membrane so that the protons created on the anodic side of the cell are transported to the cathodic side. Alternative methods also include oxidative cleavage of X–H(X = C, N) bonds, as in organometallic chemistry, and the water gas shift reaction to convert CO to CO_2.

2.9 PLASMA CHEMISTRY

2.9.1 Introduction

Irving Langmuir and Levy Tonks, in the 1920s, were the first to use the term "plasma" when the oscillating glow of an electric discharge in gases reminded them of the yellowish liquid in which blood cells are transported. However, there is no real connection between the gaseous plasma state and the liquid blood plasma.

2.9.2 The Fourth State of Matter

According to the "Big Bang" theory, 10–20 billions of years ago our entire universe, which was very small, erupted at a very high temperature so that all matter was in the form of plasma. The expansion created the stars and the earth's sun. Cooling occurred during the expansion of the universe producing the other states of matter: gas, liquid and, finally, the solid state. Although the plasma was really the first state of matter, the plasma is often referred to as the fourth state of matter consisting of an ionized gas of electrons, ions, and neutral charged particles. The matter in the universe is often described as being over 99% plasma since it includes stellar interiors and atmospheres, gaseous nebulae, and interstellar hydrogen. In the upper altitudes of the earth's atmosphere, often called the ionosphere, the plasma state is present in the Van Allen radiation belts and the solar wind. On earth, the plasma state does not often naturally occur but is observed in lightening, the glow of the Aurora Borealis, ionization in rocket exhaust, emission from a conducting gas in a neon sign or fluorescent tube, and in plasma TV screens.

To form plasma on earth, a sufficient amount of energy must be placed within the neutral gas by subjecting the molecules, for example, to UV radiation, X-rays, electric discharge, or intense heat. The energy must be able to remove electrons from the gaseous molecules and atoms to create positive nuclei. The gas is said to be ionized and the plasma consists of moving negatively charged electrons and positively charged ions. This ionization process is counteracted by recombination where the electrostatic forces of positive ions attract the electrons that are close enough to recombine into a neutral atom or molecule. Plasmas may be completely ionized, in which case the molecules are completely separated into ions and electrons, or partially ionized with only a fraction of the molecules ionized and the remainder electrically neutral molecules. The percentage of ionization may vary from small values for glow discharges to very high values as in the case of nuclear reactors.

2.9.3 Criteria for Plasma

The plasma has nearly equal concentrations of electrons and positive ions and is said to be a quasineutral (almost equal) system, which exhibits collective behavior. The concentration of electrons and ions is usually called the number density of the plasma that is the number of electrically charged particles per unit volume (cm^3). The collective behavior may be envisioned by considering a positive ion in the plasma. A group of negatively charged electrons are attracted around the ion by the attractive coulomb electrostatic force. The electrons closest to the ion form a shield so that the other electrons will experience a smaller coulomb force than without the shielding. The force of attraction between the positive ion and electrons only extends to a finite distance called the Debye radius or Debye length. The Debye length, λ_D, is a function of temperature and number density. For a higher temperature, there is more energy in the motion of the particles and the shielding becomes less effective and the Debye length increases. Thus, the first criterion for plasma is that the dimensions of the system must be much larger than the Debye length. The collective behavior of the ions and electrons in the plasma cause them to move similarly, which is termed the phenomenon of ambipolar diffusion.

As the number density of the plasma increases, the shielding becomes more effective and the Debye length decreases. A small amount of ionization does not mean there is plasma. Therefore, a second condition for the plasma is that there exists a large enough number density of charged particles to create shielding.

2.9.4 Plasma Regions

Figure 2.51 shows a wide variety and range of plasmas that differ by temperature and number density. The plasma regions have well-defined boundaries between which no plasma is known to exist. Plasma regions associated with (a) the sun to produce solar energy and (b) industrial plasma processing will be discussed with regard to hydrogen technology.

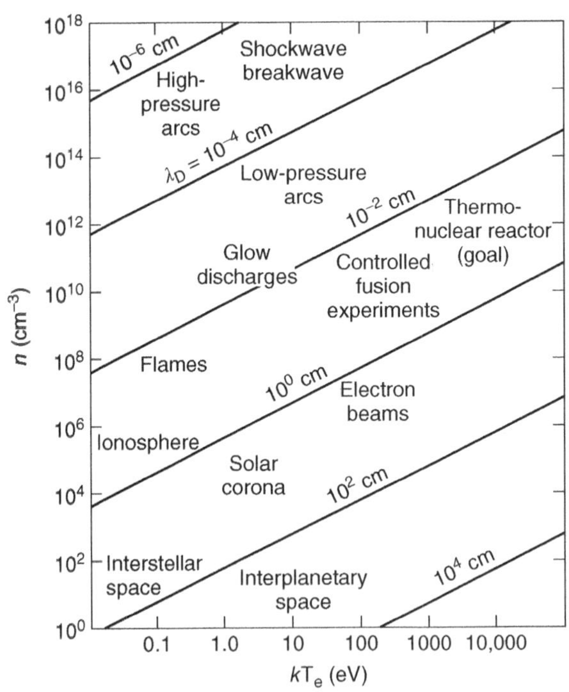

Figure 2.51 Typical plasmas characterized by their number density, electron energy, and Debye length.

2.9.5 Sun as a Plasma Generator

All life on earth depends on plasma energy since the sun is a plasma energy generator. The sun is constantly converting four hydrogen nuclei into one helium nucleus through a process called thermonuclear fusion, where the light-weight hydrogen nuclei are fused into heavier helium nucleus. During the solar fusion process, the helium nucleus is 0.7% lighter than the four hydrogen nuclei that were originally present. This loss in weight is converted into energy, some of which is emitted as solar radiation. The proton–proton reaction mechanism shown as follows produces most of the energy in the sun where H^1, H^2, He^3, and He^4, ν and γ represent the hydrogen nucleus (proton), deuterium nucleus (deuteron), helium-3 and -4 nuclei, neutrino and gamma ray, respectively.

$$H^1 + H^1 \rightarrow H^2 + e + \nu$$

$$H^2 + H^1 \rightarrow He^3 + \gamma$$

$$He^3 + He^3 \rightarrow He^4 + 2H^1$$

The previous section on photochemistry described how radiation generated from the plasma in the sun is useful in the production of hydrogen on earth via solar photothermochemical, photoelectrochemical, photobiological, and photocatalytic processes. More details of these processes will be presented in Section 4.5 on renewable energy as sources of hydrogen.

2.9.6 Industrial Plasma Processing Related to Hydrogen Technology

To sustain plasma in the laboratory, a gas at a reduced pressure is often sealed in a tube containing two metal electrodes: an anode and a cathode. A direct current (DC) is applied with a battery so that the anode is connected to the positive battery terminal and the cathode to the negative terminal. When the switch is turned on, an electric discharge takes place in the sealed tube changing the poor electrically conducting gas into conducting plasma resulting in a DC current flowing through the plasma. Figure 2.52 shows a few distinct electric discharges that are categorized by the values for the current and potential difference (V) between the anode and the cathode. The corona discharge has very low currents and very high voltages and is frequently observed around electric power wires. The glow discharge has currents less than 1 A and voltages around a few hundred volts while the arc discharge is characterized as having currents between 1 and 100,000 A and a low voltage of about 10 V. The two regions usually of greatest interest to plasma chemistry in industry are the glow discharges and arcs. Another difference between glow discharges and arcs is that glow discharges are said to be nonequilibrium plasmas because at their low number densities there is a lack of equilibrium between the electron temperature, T_e, and the gaseous ion temperature, T_g, as shown in Figure 2.53. The gas temperature is near ambient conditions while the energetic electrons interact with molecular bonds promoting chemical reactions. In arc discharges, also known as thermal plasmas, there exists an equilibrium between the electron and gas temperature. The very high temperature in arc plasmas ($>5 \times 10^3$ K) makes them useful for processing many materials.

The electric discharge may be driven by a high-frequency, alternating current (AC) power supply often operating at radio frequency (RF), for example, 13.56 MHz (megahertz), or microwave (MW), 2.45 GHz (gigahertz) frequencies. In such systems, there

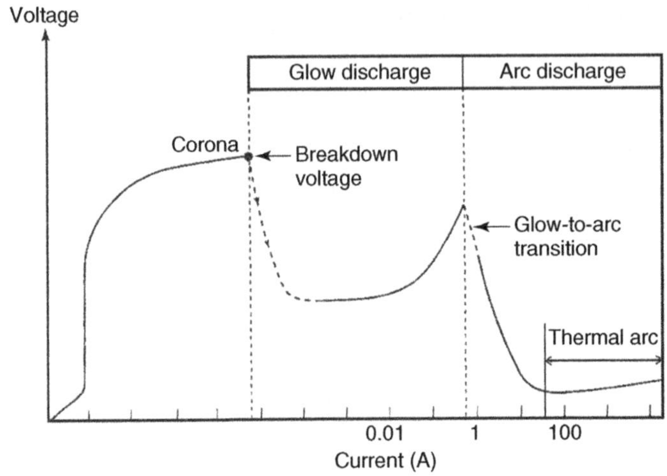

Figure 2.52 Voltage–current characteristics of a DC electric discharge *E*. [*Source*: Eliezer and Eliezer (2001) [86]. Reproduced with permission of Taylor & Francis.]

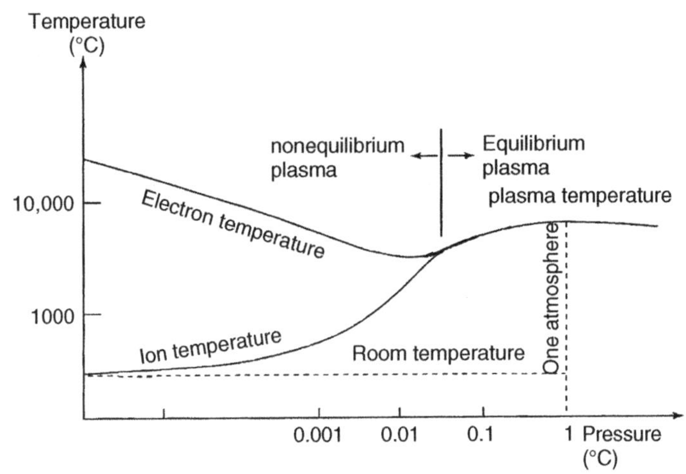

Figure 2.53 Temperature and pressure domain for equilibrium and nonequilibrium plasmas for DC discharges E. [*Source*: Eliezer and Eliezer (2001) [86]. Reproduced with permission of Taylor & Francis.]

is no real anode or cathode since the negative charge accumulated during one half-cycle is neutralized by the positive charge during the next half-cycle. The greater mobility of electrons helps achieve almost continuous energetic ions while these high frequencies are used.

When an electric discharge is applied to the electrodes, energetic electrons are formed that collide with the molecules of the gas resulting in dissociation, ionization, recombination, and deactivation as illustrated in Figure 2.54. The figure also shows that the components of the plasma may modify a substrate, which is placed on an electrode, by processes that are defined as sputtering, implantation, deposition, and etching. In addition, a wide range of excited states are formed in the plasma, which may be involved in energy transfer, chemical reactions, ionization, and radiation processes as described in Table 2.18 for excited hydrogen atoms (H*).

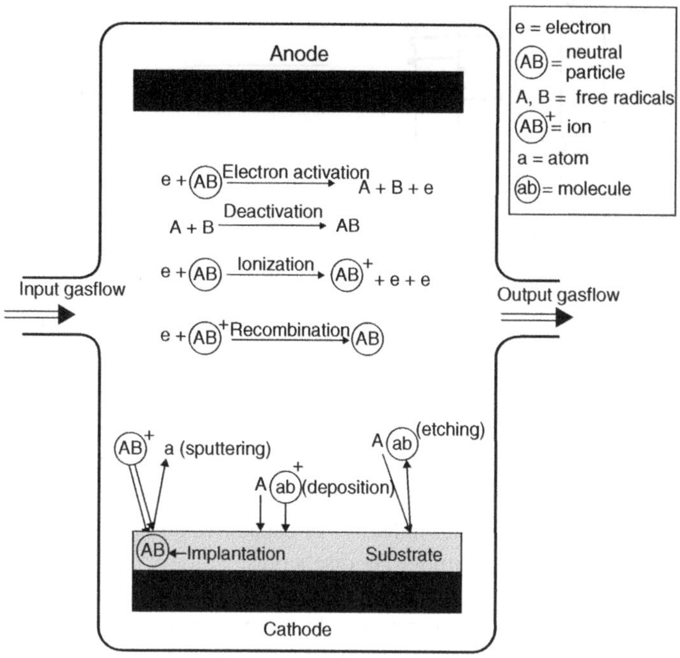

Figure 2.54 Reactions taking place in a plasma processing reactor. [*Source*: Eliezer and Eliezer (2001) [86]. Reproduced with permission of Taylor & Francis.]

TABLE 2.18 Physical and chemical processes involving excited hydrogen atoms (H*)

Energy transfer	
Resonant transfer	$H^* + H \rightarrow H + H^*$
Electronic excitation	$H^* + M \rightarrow H + M^*$
Energy pooling	$H^* + M^* \rightarrow H^{**} + M$
Vibration, rotation, or translation excitation	$H^* + M \rightarrow H + M^+$
Chemical reaction	
Dissociation	$H^* + M \rightarrow H + E + F$
Abstraction or fragmentation	$H^* + M \rightarrow A + B$
Addition or insertion	$H^* + M \rightarrow HM$
Ionization	
Penning ionization	$H^* + M \rightarrow H + M^+ + e^-$
Dissociative ionization	$H^* + M \rightarrow HE^+ + F + e^-$
Associative ionization	$H^* + M \rightarrow HM^+ + e^-$
Collisional ionization	$H^* + M \rightarrow H^+ + M + e^-$ or $H^+ + M$
Radiation	
Spontaneous emission	$H^* \rightarrow H + hv$
Collision-induced emission	$H^* + M \rightarrow H + M + hv$
Stimulated emission	$H^* + hv \rightarrow H + 2hv$

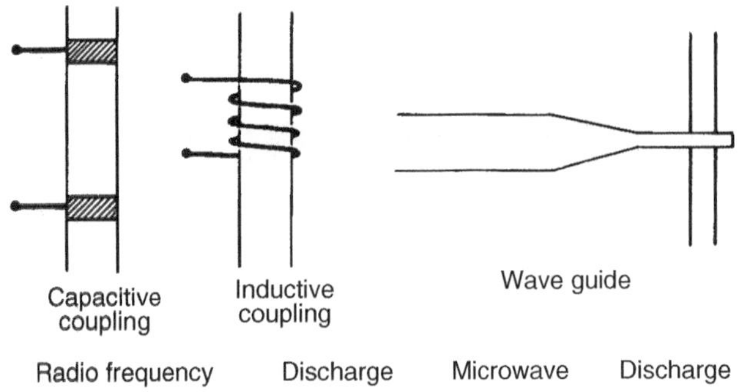

Figure 2.55 Some methods for exciting high-frequency discharges. [*Source*: Bell (1967) [94]. Reproduced with permission of MIT Press.]

RF and MW plasmas are of special interest since they may be induced without the presence of electrodes that may react with the gaseous plasma or introduce impurities into the plasma. Figure 2.55 illustrates several methods for coupling RF and MW power into a discharge. For radio frequencies, either inductive or capacitive coupling are often employed, while a waveguide is used for exciting MW discharges.

Many external operating controls may be varied to change the plasma including gas flow rates, gas composition, pressure in the reactor, electric power and frequency, magnitude and direction of applied magnetic fields that are used to confine or guide the plasma, substrate temperature, and reactor geometry. The external parameters affect the internal features of the reactor (electron and ion number densities and fluxes, electron and ion temperatures, concentration of neutral and excited states, flux of photons) that are important for industrial applications.

Following are a few examples of plasma processing associated with hydrogen technology.

2.9.7 Plasma Reforming of Fuels into Hydrogen-Rich Gases

Because the plasma is a highly energetic state of matter characterized by high temperature and high degree of ionization, chemical reactions involved in the reforming or partial oxidation of fuels are greatly accelerated. Some of the relevant chemical reactions for steam reforming of the fuel methane are presented as follows:

$$CH_4 + H_2O \rightarrow 3H_2 + CO \quad \text{(Steam reforming equation)}$$

$$\underline{CO + H_2O \rightarrow CO_2 + H_2} \quad \text{(Water gas shift reaction)}$$

$$CH_4 + 2H_2O \rightarrow 4H_2 + CO_2 \quad \text{(Composite steam reforming reaction)}$$

Carbon deposition or sooting is often observed in a plasma reactor. Three of the important carbon deposition reactions are as follows:

$$CH_4 \rightarrow 2H_2 + C$$

$$2CO \rightarrow C + CO_2$$

$$H_2 + CO \rightarrow C + H_2O$$

The following three chemical reactions show the oxidation of methane with increasing amounts of the reactant oxygen. The first reaction indicates the partial oxidation of methane to produce hydrogen gas.

$$CH_4 + 1/2O_2 \rightarrow CO + 2H_2$$

$$CH_4 + 3/2O_2 \rightarrow CO + 2H_2O$$

$$CH_4 + 2O_2 \rightarrow CO_2 + 2H_2O$$

The mechanism for the oxidation of methane in air using a plasma technique is quite complex since many reactive neutral species, charged particles and excited states are present. For example, in the methane conversion by air microwave plasma, over 50 elementary steps have been written in the model with each step having its own chemical kinetics. Hydrogen molecules are formed in many of the reaction steps.

Plasma-enhanced partial oxidation/reforming with O_2/air/water and injection of a variety of fuels including biofuels (corn, canola and soybean oils, ethanol), natural gas, diesel fuel, CH_4, methanol, propane, and isooctane (a representative of gasoline) have been studied under both homogeneous gas phase (noncatalytic) and heterogeneous (catalytic) conditions. Glow discharge of a gaseous ethanol–water mixture resulted in 70% hydrogen production at an energy cost of about 0.6 Wh/L of hydrogen produced. Unreacted ethanol and water may be condensed out and thus this technique appears easier to develop than the methane steam reforming that generates substantial CO, which requires further treatment such as the water gas shift reaction. Liquid ethanol–water mixtures are being investigated using diaphragm discharge systems. The plasma treatment produces hydrogen-rich gases for application with fuel cells, stationary electric power production, and vehicular propulsion combustion engine systems. An onboard compact plasma reformer is useful in processing hydrocarbon fuel into hydrogen-rich gas for a high-efficient fuel cell as described in Chapter 5. Recycling a small portion of electric energy to power the plasma reactor is an acceptable compromise especially in light of the energy-rich, hydrogen-rich gas produced from the fuels and the substantially reduced emissions of NO_x, CO_x, sulfur-containing compounds, soot, and hydrocarbons.

2.9.8 Plasma Deposition of Metals

The sputtering process shown in Figure 2.54 is often initiated when Ar^+ ions formed in an Ar plasma are accelerated with an electric potential toward a metal target. The kinetic energy of the Ar^+ ions sputter metal atoms from the target into the gas phase, which may then be deposited as a thin film onto a selected substrate. For hydrogen technology, sputtered metal atoms have been used to (a) catalyze low temperature (25°C) 1,3-butadiene hydrogenation, (b) enhance high-temperature partial oxidation of hydrocarbons for syngas production, (c) improve the properties of fuel cell electrodes and proton exchange membranes, such as Nafion, and (d) produce membranes for the separation of hydrogen.

Plasma techniques have been used to develop solid oxide fuel cells for the conversion of chemical energy in hydrogen to electric energy (Chapter 5). In the fabrication of the cell, the anode, electrolyte, and cathode layers are consecutively deposited onto a metallic substrate by a multistep vacuum plasma spray process. DC plasma spray technology applies powders for the deposition of ceramic layers while RF spray methods have employed liquid precursors for the deposition of porous solid cathode layers.

2.9.9 Plasma Water Splitting

Hydrogen production (0.33 mol% in products) from H_2O splitting has been achieved at atmospheric pressure using reactors consisting of plasma and catalyst integrated technologies based on dielectric barrier discharges and AC plasmas.

In addition, the plasma has been used as a component in other methods for water splitting. Higher photocatalytic water-splitting efficiencies are obtained with plasma preparation of Ni-loaded Ta_2O_5 and ZrO_2 semiconductors. In hydrogen generation employing a calcium–bromine thermochemical water-splitting cycle, a plasma chemical stage for the recovery of HBr as H_2 and Br_2 has a lower demand for energy than a steam-electrolysis system.

2.9.10 Plasma Modification of Proton Exchange Membranes

Plasma treatment is one of the most versatile methods for modifying surfaces and improving the properties of polymer membranes. The action of the plasma may cause polymer etching, deposition of polymer layers, and/or change the surface by enhancing cross-linking and introducing different chemical functional groups. The etching process may result in roughening of the surface and an increase in the pore diameters of the polymer while the deposition of a polymer layer may help plug the pores. Altering the chemical functional groups can help control the hydrophilic (water-liking) and adhesion properties of the membrane surface. Thin films of plasma polymerized materials often have a high degree of cross-linking and are pinhole-free even for films of only a few hundreds of nanometers in thickness. The thin film technologies of plasma processing, such as plasma polymerization and sputtering, are therefore excellent techniques for fabricating the membrane electrode assemblies of miniaturized fuel cells.

Some plasma modification studies of Nafion and perfluorosulfonic acid membranes have included argon plasma treatment, and plasma polymerization of arylene and hexane/hydrogen mixtures to produce hydrocarbon coatings. Nafion solution has been impregnated into porous polypropylene membranes that were plasma treated with fluorine containing Freon 116 gas. Pore-filling electrolyte membranes consisting of a porous poly(tetrafluoroethylene) (PTFE) substrate has been prepared by plasma grafting with poly(acrylic acid). Ion beam-assisted plasma polymerization of PTFE with various sulfur components (SO_2, CF_3SO_3H, or $ClSO_3H$) to form sulfonic acid groups produced proton conductive thin films. Plasma polymerization has also been used to synthesize electrolyte membranes from the monomers: tetrafluoroethylene, 1,3-butadiene, styrene, halogenated hydrocarbons, and hexafluoropropylene to generate the polymeric backbone of an ion-conductive membrane and vinylphosphonic, trifluoromethane sulfonic (CF_3SO_3H), and triflic acids to incorporate the acidic functional groups that promote the conductivity of protons. Waterproof treatment by plasma polymerization to increase the hydrophobic nature of the substrate has been applied to gas flow channels, and to porous current collector carbon paper using hexafluoropropylene monomer prior to binding to Nafion in polymer electrolyte fuel cells. Plasma-enhanced chemical vapor deposition of p-doped SiO_2 films as a proton exchange membrane in micro fuel cells has yielded more than one order of magnitude improvement in power density compared to undoped SiO_2 membranes.

2.9.11 Plasma Treatment of Municipal Solid Waste

Millions of tons of municipal solid waste are generated annually in the world, which is usually deposited in land-fill sites. A plasma process provides gasification of materials to produce hydrogen and carbon monoxide-rich synthesis gas that can be used after purification for electricity and heat generation.

The environmental impacts of the gasification and plasma processes have better environmental performance than conventional waste treatment technologies and incineration plants.

PROBLEMS

Section 2.5

1. Electrochemistry deals with oxidation–reduction reactions.
   What is oxidation? Write a typical oxidation reaction. What is reduction? Write a typical reduction reaction.
2. Use the oxidation number principle to identify what is being oxidized and reduced.
3. What is a half reaction? Write a half cell representation for zinc metal in contact with zinc sulfate.
4. If I combine the following two half cells, what is the expected voltage from the cell? Cu/Cu^{2+} and Pb/Pb^{2+}
5. Since single electrode potentials cannot be measured, how are the potentials of several different half cells obtained?
6. What are reference electrodes? Name two reference electrodes.

MULTIPLE CHOICE QUESTIONS

THERMODYNAMIC QUESTIONS

1. The equation of state for hydrogen as an ideal gas is
   (a) $PT = nRV$
   (b) $nPV = RT$
   (c) $PT = nRV$
   (d) $PV = nRT$
2. Work is NOT:
   (a) $W = mgh$
   (b) $W = -P\Delta V$
   (c) A quantity that flows across a boundary because of a difference in temperature.
   (d) Force moved through a distance.
3. Which of the following is NOT a statement of the First Law of Thermodynamics?
   (a) The energy of the universe is a constant.
   (b) A change in energy is dependent on the pathway during the change in state.
   (c) Energy cannot be created or destroyed.
   (d) Conservation of energy.

4. The change in Gibbs free energy is NOT:
   (a) $\Delta G = \Delta H - T\Delta S$ at constant temperature.
   (b) Greater than zero for a spontaneous change.
   (c) Related to the tendency for a reaction to take place spontaneously.
   (d) Related to the number of moles of electrons transferred in an electrochemical cell.
5. When "n" is the number of moles of electrons transferred, the Nernst equation defines the electromotive force at 298 K as
   (a) $E = H + nPV$
   (b) $H = H^\circ - (0.059/n)\log Q_p$
   (c) $E = E^\circ - (0.059/n)\log Q_p$
   (d) $E = E^\circ + (0.059/n)\log Q_p$

PHOTOCHEMISTRY QUESTIONS

1. The energy of a photon
   (a) varies as the speed of light changes
   (b) is different in the four phases of matter
   (c) is proportional to the frequency
   (d) is a constant in the plasma state
2. Infrared radiation typically causes
   (a) ionization
   (b) vibrational excitation
   (c) formation of a plasma
   (d) promotion of an electron to the conduction band
3. Which part of the electromagnetic spectrum has the least amount of energy?
   (a) X-rays
   (b) visible radiation
   (c) radio frequency radiation
   (d) vacuum UV radiation
4. Photochemistry results from the following:
   (a) transmission of energetic photons through a material
   (b) having many photons from a laser passing through the sample
   (c) exciting many molecules with one photon
   (d) absorption of one photon
5. The absorption of radiation is NOT proportional to
   (a) the pathlength of the sample
   (b) the concentration of the sample
   (c) the wavelength of radiation
   (d) the molar extinction coefficient of the material

PLASMA CHEMISTRY QUESTIONS

1. The fourth state of matter is
   (a) plasma
   (b) solid
   (c) liquid
   (d) gas
2. The process where an energetic ion collides with a solid surface and knocks material into the gas phase is called
   (a) sputtering
   (b) recombination
   (c) ionization
   (d) plasma activation
3. The plasma that has the lowest number density is
   (a) electron beam
   (b) flame
   (c) thermonuclear reactor
   (d) interplanetary space
4. Which of the following mechanisms represents Penning ionization?
   (a) $H^* + M \rightarrow H + M^+ + e^-$
   (b) $H^* + M \rightarrow HE^+ + F + e^-$
   (c) $H^* + M \rightarrow HM^+ + e^-$
   (d) $H^* + M \rightarrow H^+ + M + e^-$ or $H^+ + M^-$
5. Which of the following mechanisms represents energy pooling?
   (a) $H^* + H \rightarrow H + H^*$
   (b) $H^* + M \rightarrow H + M^*$
   (c) $H^* + M^* \rightarrow H^{**} + M$
   (d) $H^* + M \rightarrow H + M^\dagger$
6. The direct current (DC) discharge characterized by low current and very high voltages is
   (a) arc
   (b) microwave
   (c) corona
   (d) glow
7. The plasma is used in which of the following hydrogen technology:
   (a) reforming of fuels
   (b) water splitting
   (c) modification of fuel cell membranes
   (d) treatment of municipal waste
   (e) all of the above

CHEMICAL EQUILIBRIUM

1. Consider the reaction: $2CO(g) + O_2(g) \rightleftharpoons 2CO_2(g)$

   The equilibrium expression for this reaction is

   (a) $\frac{[CO_2]}{[CO]+[O_2]}$

   (b) $\frac{[CO_2]2}{[CO]^2[O_2]}$

   (c) $\frac{[CO][O_2]}{[CO_2]}$

   (d) $\frac{2[CO_2]}{2[CO][O_2]}$

   (e) $\frac{[CO]^2[O_2]}{[CO_2]^2}$

2. The position of the equilibrium for a system where $K = 3.7 \times 10^{-14}$ can be described as being favored to ______; the concentration of products is relatively ______.

   (a) the right; large
   (b) the right; small
   (c) the left; large
   (d) the left; small
   (e) neither direction; large

3. Which change to this reaction system would cause the equilibrium to shift to the right?

$$N_2(g) + 3H_2(g) \rightleftharpoons 2NH_3(g) + \text{heat}$$

   (a) addition of a catalyst
   (b) addition of $NH_3(g)$
   (c) removal of $H_2(g)$
   (d) heating the system
   (e) lowering the temperature

4. Consider the reactions:

   (a) $2CO_2(g) \rightarrow 2CO(g) + O_2(g)$
   (b) $N_2(g) + O_2(g) \rightarrow 2NO(g)$
   (c) $Si(s) + 2Cl_2(g) \rightarrow SiCl_4(g)$

   When the equilibria are disturbed by increasing the pressure, does the concentration of reaction products increase, decrease, or remain the same?

   Answer:

   (a) Decreases
   (b) Remains the same
   (c) Increases

5. Consider the reaction:

$$H_2(g) + I_2(g) \rightarrow 2HI(g) \quad \Delta H = -2.2\,\text{kcal/mol}$$

   Is this reaction exothermic or endothermic?

Exothermic

Write the equilibrium expression for the reaction.

$$K_c = \frac{[HI]^2}{[H_2][I_2]}$$

Will the equilibrium concentration of HI increase or decrease when:

(a) I_2 is added? Will increase

(b) H_2 is removed? Will decrease

(c) A catalyst is added? Doesn't change the equilibrium position

(d) The temperature is increased? Will decrease

6. Calculate the equilibrium constant for this reaction: $3H_2(g) + N_2(g) \leftrightarrows 2NH_3(g)$ if in equilibrium mixture of gases at 500°C, the partial pressure of H_2 is 0.928 atm, the partial pressure of N_2 is 0.432 atm and that of NH_3 is 2.24×10^{-3} atm. Calculate the equilibrium constant for this reaction.

   $K_p = 1.45 \times 10^{-5}$

ACID-BASE CHEMISTRY

1. Which of the following would you expect to be a Bronsted–Lowry acid?
   (a) Ca^{2+}
   (b) H_2CrO_4
   (c) NH_3
   (d) NO_2^-
2. Which of the following would you expect to be a Bronsted–Lowry base?
   (a) HNO_2
   (b) NH_4^+
   (c) Ni^{2+}
   (d) PO_4^{3-}
3. Some weak acids and pK_a values are listed below. Which is the strongest acid?
   (a) Hydrocyanic acid, $pK_a = 9.31$
   (b) Ascorbic acid, $pK_a = 4.10$
   (c) Formic acid, $pK_a = 3.74$
   (d) Propanoic acid, $pK_a = 4.89$
   (e) Acetic acid, $pK_a = 4.74$
4. Hydrogen cyanide, HCN, is a weak acid. Which equation **best** represents its aqueous chemistry?
   (a) $HCN + H_2O \rightleftharpoons CN^- + H_3O^+$
   (b) $HCN + H_2O \rightleftharpoons H_2CN^+ + OH^-$
   (c) $HCN \rightleftharpoons H^+ + CN^-$
   (d) $HCN \rightleftharpoons H^- + CN^+$
   (e) $H_2O \rightleftharpoons H^+ + OH^-$
5. $CH_3NH_2 + HCl \rightleftharpoons CH_3NH_3^+ + Cl^-$

   A conjugate acid–base pair in the reaction shown is _______ and _______.

(a) CH_3NH_2 and HCl
(b) CH_3NH_2 and Cl^-
(c) $CH_3NH_3^+$ and Cl^-
(d) HCl and Cl^-
(e) HCl and H_3O^+

6. What is the pH of a solution in which $[H_3O^+] = 4.1 \times 10^{-2}$ M?
   (a) 12.61
   (b) 1.39
   (c) 1.10
   (d) 12.90
   (e) 4.10
7. Which of the following will not result in a buffer solution?
   (a) 100 mL 1.0 M HCN + 100 mL 1.0 M NaCN
   (b) 100 mL 1.0 M HNO_3 + 50 mL 1.0 M KNO_3
   (c) 50 mL 1.0 M HCO_2H + 25 mL 1.0 M KOH
   (d) 50 mL 1.0 M CH_3CO_2Na + 25 mL HCl
8. Write the formula for the conjugate base of the following: (a) H_2O, (b) CO_3^{2-}, and (c) $H_2PO_4^{1-}$
   (a) OH^-
   (b) Since CO_3^{2-} does not have an H, it cannot be an acid
   (c) HPO_4^{2-}
9. Write the formula for the conjugate acid of the following: (a) H_2O, (b) CO_3^{2-}, (c) NH_3, and (d) $H_2PO_4^{1-}$
   (a) H_3O^+
   (b) HCO_3^-
   (c) NH_4^+
   (d) H_3PO_4
10. Write the equations for the (a) HBr and (b) HSO_4^{-1} reacting with water and acting as a monoprotic acid. Label the conjugate acid and base.

(a)

HBr	$+ H_2O \Leftrightarrow$	Br^{-1}	$+ H_3O^{+1}$
Acid	Base	Conjugate base	Conjugate acid

(b)

HSO_4^{-1}	$+ H_2O \Leftrightarrow$	SO_4^{-2}	$+ H_3O^{+1}$
Acid	Base	Conjugate base	Conjugate acid

11. Write the equations for the (a) I^- and (b) CO_3^{2-} reacting with water and acting as a monoprotic-accepting base. Label the conjugate acid and base.

(a)

I^-	$+ H_2O \Leftrightarrow$	HI	$+ OH^-$
Base	Acid	Conjugate acid	Conjugate base

(b)

CO_3^{2-}	$+ H_2O \Leftrightarrow$	HCO_3^-	$+ OH^-$
Base	Acid	Conjugate acid	Conjugate base

12. Predict the position of equilibrium for this acid–base reaction: $HCN + OH^- \rightleftharpoons CN^- + H_2O$
    Favored to the right

KINETICS

1. Which of the following does not have an influence on the rate of a chemical reaction?
   (a) temperature
   (b) catalyst
   (c) surface area
   (d) enthalpy
   (e) concentration
2. For the combustion of propane: $C_3H_8 + 5\ O_2$ --> $3\ CO_2 + 4\ H_2O$, determine the rate of the formation of carbon dioxide, if the rate of formation of the water vapor is 0.00258 M/s.
   (a) 0.00344
   (b) 0.00194
   (c) 0.00258
   (d) 0.00430
   (e) 0.00323
3. A reactant has a reaction order of 3. How will the rate change if the concentration of that reactant is doubled?
   (a) the rate will increase by 2
   (b) the rare will increase by 3
   (c) the rate will increase by 6
   (d) the rate will increase by 8
   (e) the rate will increase by 9
4. For the decomposition of hydrogen peroxide (a first-order reaction) determine the concentration, in M, of H_2O_2 after 14 h, if the original concentration was 0.750 M. Use 10 h as the half-life for the decomposition of H_2O_2 for your calculations.
   (a) 0.122
   (b) 0.284
   (c) 0.375
   (d) 0.505
   (e) 0.813
5. If changed, which of the following can cause a change in the activation energy of a chemical reaction?
   (a) catalyst
   (b) concentration
   (c) entropy
   (d) temperature
   (e) time

6. Which of the following statements is incorrect?
   (a) For a given reaction at constant temperature, the rate of a reaction depends on its rate constant.
   (b) The slowest elementary step is rate determining.
   (c) Most chemical reactions are bimolecular.
   (d) For an elementary reaction one can determine its rate law from the balanced reaction.
   (e) A rate constant is dependent of temperature.

ORGANIC CHEMISTRY

1. Which of the following is an organic compound?
   (a) H_2O
   (b) NaOH
   (c) HNO_3
   (d) $CaCl_2$
   (e) CH_4
2. Which of the following compounds contains a triple carbon–carbon bond?
   (a) Benzene
   (b) Propane
   (c) Ethene
   (d) Ethyne
   (e) Cyclohexane
3. Which of the following represents a hydrocarbon with 4 carbon atoms?
   (a) Butane
   (b) Ethane
   (c) Methane
   (d) Pentane
   (e) Propane
4. Which of the following compounds is not an alkane?
   (a) C_2H_6
   (b) C_4H_{10}
   (c) CH_4
   (d) C_3H_6
   (e) C_5H_{12}
5. To which group does the following compound belong? $H_5C_2\text{-}O\text{-}C_2H_5$
   (a) alcohol
   (b) aldehyde
   (c) ether
   (d) carboxylic acid
   (e) ester

6. Which is currently the main source for the production of hydrogen?
   (a) ocean water
   (b) metal hydrides
   (c) natural gas
   (d) cellulose
   (e) ammonia

POLYMERS

1. Which of the following is not a chain-addition polymer?
   (a) Polyethylene
   (b) Polypropylene
   (c) Polyvinyl chloride
   (d) Polystyrene
   (e) Polyethylene terephthalate
2. Which does a typical polymer NOT have?
   (a) high molecular weight
   (b) crystalline and amorphous portions
   (c) glass transition temperature
   (d) boiling point
   (e) molecular weight distribution
3. A sample of polystyrene has an average molecular weight of 155,000 g/mol. Determine its degree of polymerization (= average number of repeat units per chain)?
   (a) 1,476
   (b) 1,490
   (c) 1,505
   (d) 3,690
   (e) 5,536
4. Which method of polymer waste disposal is used currently the most, but is not sustainable?
   (a) Composting
   (b) Incineration
   (c) Landfill
   (d) Pyrolysis
   (e) Recycling
5. The density of completely (i.e. 100%) crystalline polyethylene is 1.000 g/ml and the density of completely amorphous polyethylene is 0.850 g/ml. Determine the degree of crystallinity, in %, of a sample of LDPE, which has a density of 0.930 g/ml.
   (a) 42.4
   (b) 44.7
   (c) 46.6

(d) 53.3
(e) 58.1

5. Which is not true for Nafion, used as proton exchange membrane in fuel cells?
   (a) It is a copolymer.
   (b) In its activated form it contains acid groups.
   (c) It is relatively inexpensive.
   (d) It is also used as actuator.
   (e) It is similar to Teflon.

BIBLIOGRAPHY

Section 2.2

1. Kauffman, G.B. (1988) The Brønsted–Lowry acid–base concept. *Journal of Chemical Education*, **65**, 28–31.
2. Kolob, D. (1978) Acids and bases. *Journal of Chemical Education*, **55**, 459–464.
3. Adcock, J.L. (2001) Teaching Brønsted–Lowry acid–base theory in a direct comprehensive way. *Journal of Chemical Education*, **78**, 1495–1496.
4. Fainzilberg, V.F. and Karp, S. (1994) Chemical equilibrium in the general chemistry course. *Journal of Chemical Education*, **71**, 769–770.
5. Van Driel, J.H., de Vos, W., and Verloop, N. (1999) Introducing dynamic equilibrium as an explanatory model. *Journal of Chemical Education*, **76**, 559–561.
6. Treptow, R. (1980) Le Chatelier's principle. *Journal of Chemical Education*, **57** (6), 417.
7. Howard, R.A. (1999) The fizz keeper, a case study in chemical education, equilibrium, and kinetics. *Journal of Chemical Education*, **76**, 208–209.
8. De Lange, A.M. and Pogieter, J.H. (1991) Acid and base dissociation constants of water and its associated ions. *Journal of Chemical Education*, **68**, 304–305.
9. Ault, A. (1999) Do pH in your head. *Journal of Chemical Education*, **76**, 936–938.
10. Silverstain, T.P. (2000) Weak vs strong acids and bases: the football analogy. *Journal of Chemical Education*, **77**, 849–850.
11. Brown, T.L., LeMay, H.E. Jr., Bursten, B.E., and Burdge, J.R. (2003) *Chemistry. The Central Science*, 9th edn, Prentice-Hall, Inc., NJ.
12. Kotz, J.C., Treichel, P.M., and Weaver, G.C. (2006) *Chemistry & Chemical Reactivity*, 6th edn, Thomson Brooks/Cole, CA.
13. Miri, M.J. (2006) *Aspects of Chemical Reactions and Chemistry of Materials*, 3rd edn, Pearson Custom Publishing, MA.
14. McMurry, J. and Castelion, M.E. (2007) *Fundamentals of General, Organic, and Biological Chemistry*, 5th edn, Prentice-Hall, Inc., NJ.

Section 2.3

15. Silbey, R.J., Alberty, R.A., and Bawendi, M..G. (2005) *Physical Chemistry*, Wiley, Hoboken, NJ.

16. Goswami, D.Y., Mirabel, S.T., Goel, N., and Ingley, H.A. (2003) *A review of hydrogen production technologies*. Int. Conf. Fuel Cell Science, Engineering and Technology, Rochester, NY; pp. 61–74.
17. Hess, G. (2006) Incentives boost coal gasification. *Chem. & Eng. News*, January 16. p. 22.

Section 2.4

18. Steinfeld, J.I., Francisco, J.S., and Hase, W.L. (1999) *Chemical Kinetics and Dynamics*, 2nd edn, Prentice Hall, Upple Saddle River, NJ.
19. Miri, M.J. (2006) *Aspects of Chemical Reactions and Chemistry of Materials*, 3rd edn, Pearson Custom Publishing, Massachusetts.
20. Moore, J.W. and Pearson, R.G. (1981) *Kinetics and Mechanism*, 3rd edn, Wiley, New York.
21. Pilling, M.J. and Seakins, P.W. (1995) *Reaction Kinetics*, University Press, Oxford.
22. Dawson, B.E. (1973) *Kinetics and Mechanisms of Reactions*, Methuen Educational Ltd, London.
23. Espenson, J.H. (1995) *Chemical Kinetics and Reaction Mechanism*, 2nd edn, McGraw-Hill, New York.
24. Laidler, K.J. (1987) *Chemical Kinetics*, 3rd edn, Harper and Row, New York.
25. Entelis, S.G. and Tiger, R.P. (1976) *Reaction Kinetics in the Liquid Phase*, Wiley, New York.
26. Nicholas, J. (1976) *Chemical Kinetics*, Wiley, New York.
27. Eyring, H. and Eyring, E.M. (1963) *Modern Chemical Kinetics*, Reinhold, New York.
28. Chorkendorff, I. and Niemantsverdriet, J.W. (2003) *Concepts of Modern Catalysis and Kinetics*, Wiley, New York.

Section 2.6

29. Morrison, R.T. and Boyd, R. (1992) *Organic Chemistry*, 6th edn, Prentice Hall, Englewood, NJ.
30. Solomon, T.W.G. and Fryhle, C.B. (2003) *Organic Chemistry*, 8th edn, Wiley.
31. McMurry, J. (2000) *Organic Chemistry*, 5th edn, Brooks/Cole, Pacific Grove.
32. Carey, F.A. (2008) *Organic Chemistry*, 7th edn, McGraw-Hill, New York.
33. Hanson, J.R. (2002) *Functional Group Chemistry*, Wiley, New York.
34. Cahn, R.S. (1979) *Introduction to Chemical Nomenclature*, 5th edn, Butterworths, Boston.
35. Bruice, P.Y. (2004) *Organic Chemistry*, 4th edn, Prentice-Hall, NJ.
36. Vollhardt, K.P.C. and Schore, N.E. (2003) *Organic Chemistry. Structure and Function*, 4th edn, W. H. Freeman and Company, New York.
37. Miri, M.J. (2006) *Aspects of Chemical Reactions and Chemistry of Materials*, 3rd edn, Pearson Custom Publishing, Boston, MA.
38. Klein, D.R. (2006) *Organic Chemistry 2 as a Second Language*, John Willey & Sons, Inc., New York.
39. McMurry, J. and Castelion, M.E. (2003) *Fundamentals of General, Organic, and Biological Chemistry*, 4th edn, Prentice-Hall, Inc., Englewood, NJ.
40. Doyle, E. (1997) Trans fatty acids. *Journal of Chemical Education*, **74**, 1030–1032.
41. Seymour, R.B. (1989) Alkanes: abundant, pervasive, important, and essential. *Journal of Chemical Education*, **66**, 59–63.

Section 2.7

42. Flory, P.J. (1953) *Principles of Polymer Chemistry*, Cornell University Press, Ithaca, NY.
43. Campbell, I.M. (2000) *Introduction to Synthetic Polymers*, Oxford University Press, Oxford, UK.
44. Painter, P.C. and Coleman, M.M. (1997) *Fundamentals of Polymer Science*, 2nd edn, CRC Press, Lancaster.
45. Odian, G. (2004) *Principles of Polymer Science*, 4th edn, Wiley Interscience, Hoboken, NJ.
46. Sperling, L.H. (2001) *Physical Polymer Science*, 3rd edn, Wiley, New York.
47. Allcok, H.R. and Lampe, F.W. (1990) *Contemporary Polymer Chemistry*, 2nd edn, Prentice Hall, Englewood Cliffs, NJ.
48. Morawetz, H. (1985) *Polymers – The Origins of a Science*, Wiley, New York.
49. Billmeyer, F.W. (1984) *Textbook of Polymer Science*, 3rd edn, Wiley-Interscience, New York.
50. Elias, H.G. (2005/2006) *Macromolecules*, Vols 1 and 2, Wiley, New York.
51. Miri, M. (2006) *Aspects of Chemical Reactions and Chemistry of Materials*, 3rd edn, Pearson/Prentice Hall, Boston.
52. Tonelli, A.E. (2001) *Polymers from the Inside Out*, Wiley-Interscience, New York.
53. Strobl, G. (1996) *The Physics of Polymers*, Springer-Verlag, Berlin.
54. Bassett, D.C. (1981) *Principles of Polymer Morphology*, Cambridge University Press, Cambridge.
55. Polymer Processing, Scott BE, ed. Data on polymers. Available at: http://www.polymerprocessing.com/polymers/PA6.html (accessed 14 June 2017)
56. Physics web, Top hydrogen storing polymer revealed; August 24, 2006. Available at: http://physicsweb.org/articles/news/10/8/15/1 (accessed 14 June 2017)
57. Mauritz, K.A. and Moore, R.B. (2004) State of understanding Nafion. *Chemical Reviews*, **104**, 4535–4585.

Section 2.8

58. Alberty, R.A. (1983) *Physical Chemistry*, 6th edn, John Wiley & Sons, New York, p. 433.
59. Silbey, R.J., Alberty, R.A., and Bawendi, M.G. (2005) *Physical Chemistry*, John Wiley & Sons, Inc., NJ, p. 870.
60. Watanabe, K. and Zelikoff, M.J. (1953) Absorption coefficients of water vapor in the vacuum ultraviolet. *Journal of the Optical Society of America*, **43**, 753.
61. NIST chemistry webBook. NIST standard reference database, No. 69, March 2003. Available at http://webbook.nist.gov/chemistry.
62. Finlayson-Pitts, B.J. and Pitts, J.N. Jr. (1986) *Atmospheric Chemistry: Fundamentals and Experimental Techniques*, John Wiley & Sons, New York, p. 96.
63. Howard, J.N., King, J.I.F., and Gast, P.R. (1960) Thermal radiation, in *Handbook of Geophysics* (eds C.F. Campen Jr., A.E. Cole, T.P. Condron *et al.*), Macmillan, New York.
64. Finlayson-Pitts, B.J. and Pitts, J.N. (2000) *Chemistry of the Upper and Lower Atmosphere*, Academic Press, New York.
65. Lewis, N.S. and Nocera, D.G. (2006) Powering the planet: chemical challenges in solar energy utilization. *Proceedings of the National Academy of Sciences*, **103**, 15729–15735.
66. Esswein, A.J. and Nocera, D.G. (2007) Hydrogen production by molecular photocatalysis. *Chemical Reviews*, **107**, 4022–4047.

67. Friedman, H. (1960) The Sun's Ionizing Radiations, in *Physics of the Upper Atmosphere* (ed. J.A. Ratcliffe), Academic Press, New York, pp. 133–218.

68. Bolton, J.R. (1996) Solar photoproduction of hydrogen: a review. *Solar Energy*, **57**, 37.

69. Bolton, J.R. (1996) "Solar photoproduction of hydrogen." Int. Energy Agency Technical Report; IEA/H2/TR-96.

70. Kittel, C. (1986) *Introduction to Solid State Physics*, 6th edn, John Wiley & Sons, New York.

71. Omar, M.A. (2000) *Elementary Solid State Physics*, Addison-Wesley, Delhi, India.

72. Khaselev, O. and Turner, J.A. (1998) A monolithic photovoltaic-photoelectrochemical device for hydrogen production via water splitting. *Science*, **280** (5362), 425–427.

73. Materials Research Society Symposium; Nov. 28–Dec. 2, 2005; Boston, MA.
Maggard, P., Porob, D., and Luo, J. (2005) Multicomponent metal-oxide photocatalysts for hydrogen generation from solar energy. Materials Research Society Symposium; Nov. 28–Dec. 2, 2005; Boston, MA.
Van de Krol, R., Lloyd, D.A., Damen, M.R., Enache, C.S., and Schoonman, J. (2005) Synthesis and characterization of thin film $InVO_4$ photocatalysts. Materials Research Society Symposium; Nov. 28–Dec. 2, 2005; Boston, MA.
Yu, Z.G., Pryor, C.E., Lau, W.H., Berding, M.A., MacQueen, D.B. (2005) Core–shell nanorods for efficient photoelectrochemical hydrogen production. Materials Research Society Symposium; Nov. 28–Dec. 2, 2005; Boston, MA.
Kim, H.-J. and Misture, S.T. (2005) Structure and stability of protonated aurivillius ceramics for photocatalysis applications. Materials Research Society Symposium; Nov. 28–Dec. 2, 2005; Boston, MA.
Fujii, K., Ono, M., Ito, T., Ohkawa, K. (2005) Characteristics of H_2 gas generation using GaN photoelectrolysis. Materials Research Society Symposium; Nov. 28–Dec. 2, 2005; Boston, MA.

74. Goswami, D.Y., Mirabel, S.T., Goel, N., and Ingley, H.A. (2003) A review of hydrogen production technologies. Int. Conf. Fuel Cell Science, Engineering and Technology, Rochester, NY; 2003. pp. 61–74.

75. Materials Research Society Symposium; 2005 Nov. 28- Dec. 2; Boston, MA.
Tohji, K., Arai, T., Sato, Y., Jeyadevan, B., Hongfei, L., and Yamasaki, N. (2005) Hydrogen production using sulfur circulation. Materials Research Society Symposium; Nov. 28–Dec. 2, 2005; Boston, MA.
Senda, S., Arai, T., Sato, Y., Shinoda, K., Jeyadevan, B., and Tohji, K. (2005) Influence of Cu on the photocatalytic activity of ZnS nanoparticles. Materials Research Society Symposium; Nov. 28–Dec. 2, 2005; Boston, MA.
Matsumoto, H., Arai, T., Sato, Y., Jeyadevan, B., Tohji, K. (2005) Hydrogen generation from hydrogen sulfide using a two-compartment photoelectrochemical cell with the stratified CdS nanoparticles. Materials Research Society Symposium; Nov. 28–Dec. 2, 2005; Boston, MA.
Ishiyama, T., Arai, T., Sato, Y., Shinoda, K., Jeyadevan, B., Tohji, K. (2005) The role of film morphology on the photocatalytic efficiency of CdS film synthesized by CBD method. Materials Research Society Symposium; Nov. 28–Dec. 2, 2005; Boston, MA.
Hirayama, M., Arai, T., Sato, Y., Shinoda, K., Jeyadevan, B., Tohji, K. (2005) Synthesis of $Zn_xCd_{1-x}S$ photocatalyst thin film by dip-coating method and its photoreactivity. Materials Research Society Symposium; Nov. 28–Dec. 2, 2005; Boston, MA.

76. Steinfeld, A. (2005) Solar thermochemical production of hydrogen – a review. *Solar Energy*, **78**, 603–615.

77. Bard, A.J., Whitesides, G.M., Zhare, R.N., and McLafferty, F.W. (1995) Holy Grails of chemistry. *Accounts of Chemical Research*, **28**, 91.

78. Penner, S.S. (2006) Steps toward the hydrogen economy. *Energy*, **31**, 33–43.

79. Murphy, A.B., Barnes, P.R.F., Randeniya, L.K. *et al.* (2006) Efficiency of solar water splitting using semiconductor electrodes. *International Journal of Hydrogen Energy*, **31** (14), 1999–2017.
80. Abaoud, H. and Steeb, H. (1998) The German-Saudi HYSOLAR program. *International Journal of Hydrogen Energy*, **23** (6), 445–449.
81. Shaheen, S.E., Ginley, D.S., and Jabbour, G.E. (2005) Organic-based photovoltaics: toward low-cost power generation. *MRS Bulletin*, **30**, 10–15.
82. Hallenbeck, P.C. and Benemann, J.R. (2002) Biological hydrogen production; fundamentals and limiting processes. *International Journal of Hydrogen Energy*, **27**, 1185–1193.
83. Chen, S., Thind, S.S., and Chen, A. (2016) Nanostructured materials for water splitting – state of the art and future needs: a mini-review. *Electrochemistry Communications*, **63**, 10–17.
84. Kalisman, P., Nakibli, Y., and Amirav, L. (2016) Perfect photon-to-hydrogen conversion efficiency. *Nano Letters*. **16** (3), 1776–1781.
85. Roger, I., Shipman, M.A., and Symes, M.D. (2017) Earth-abundant catalysts for electrochemical and photoelectrochemical water splitting. *Nature Reviews Chemistry*, **1**(003): 1–13.

Section 2.9

86. Eliezer, S. and Eliezer, Y. (2001) *The Fourth State of Matter: An Introduction to Plasma Science*, 2nd edn, Institute of Physics (IOP) Publishing, Philadelphia.
87. Vukanovic, V. (1995) *Science and Faith*, Light and Life Publishing, Minneapolis.
88. Bova, B. (1971) *The Fourth State of Matter: Plasma Dynamics and Tomorrow's Technology*, St. Martin's Press, New York.
89. Chen, F.F. (1974) *Introduction to Plasma Physics*, Plenum Press, New York.
90. Venugopalan, M. (1971) The plasma state, in *Reactive Plasma Conditions* (ed. M. Venugopalan), Interscience Publisher, New York, pp. 1–34.
91. Chapman, B. (1980) *Glow Discharge Processes: Sputtering and Plasma Etching*, John Wiley & Sons, New York.
92. Boenig, H.V. (1982) *Plasma Science and Technology*, Cornell University Press, Ithaca, NY.
93. Hollahan, J.R. and Bell, A.T. (1974) *Techniques and Applications of Plasma Chemistry*, John Wiley & Sons, New York.
94. Bell, A. (1967) The physical characteristics of electric discharges, in *The Applications of Plasma to Chemical Processing* (eds R.F. Baddour and R.S. Timmons), The MIT Press, Cambridge, MA.
95. J. Mostaghimi, T.W. Coyle, V.A. Pershin, and H.R. Salimi Jazi (2005) *Proceedings 17th International Symposium on Plasma Chemistry*; 2005 Aug 7–12; Toronto, Canada.
96. Khot, A., Lu, F., Debies, T., and Takacs, G.A. (2012) Surface activation of perfluorosulfonic acid membrane used in fuel cells with vacuum UV photo-oxidation. *Journal of Adhesion Science and Technology*, **27**, 309–323.
97. Wang, H., Song, L.-J., Li, X.-H., and Yue, L.-M. (2015) Hydrogen production from partial oxidation of methane by dielectric barrier discharge plasma reforming. *Acta Physico-Chimica Sinica*, **31** (7), 1406–1412.
98. Valincius, V., Grigaitiene, V., Valatkevicius, P., and Tamosiunas, A. (2013) The simultaneous synthesis of hydrogen-rich gas and oxidation of fine metal particles in water vapour plasma. *Problems of Atomic Science and Technology, Series: Plasma Physics*, **19**, 160–163.
99. Maccato, C., Barreca, D., Carraro, G. *et al.* (2013) Photoassisted H_2 production by metal oxide nanomaterials fabricated through CVD-based approaches. *Surface & Coatings Technology*, **230**, 219–227.

100. Zhang, H., Zhu, F., Li, X., and Cen, K. (2016) Rotating gliding arc assisted water splitting in atmospheric nitrogen. *Plasma Chemistry and Plasma Processing*, **36**, 813–834.

101. Evangelisti, S., Tagliaferri, C., Clift, R. *et al.* (2015) Life cycle assessment of conventional and two-stage advanced energy-from-waste technologies for municipal solid waste treatment. *Journal of Cleaner Production.*, **100**, 212–223.

102. Evangelisti, S., Tagliaferri, C., Clift, R. *et al.* (2015) Integrated gasification and plasma cleaning for waste treatment: a life cycle perspective. *Waste Management*, **43**, 485–496.

103. Olmsted, J. and Williams, G.M. (2006) *Chemistry—The Molecular Science*, 4th edn, John Wiley & Sons, Inc., p. 688.

104. Malone, L.J. (2004) *Basic Concepts of Chemistry*, 7th edn, John Wiley & Sons, Inc., New York.

105. Brady, J.E. (2004) *Chemistry: Matter and Its Changes*, 4th edn, John Wiley & Sons, New York.

106. Haydem, H.W., Moffatt, W.G., and Wulff, J. (1965) *The Structure and Properties of Materials, Vol. III, Mechanical Behavior*, John Wiley & Sons.

107. Painter, P. and Coleman, M. (2004) *Painter and Coleman on Polymers*, DEStech Publications, Inc., Lancaster, PA.

108. Bard, A.J. and Faulkner, L.R. (2001) *Electrochemical Methods*, Wiley, New York.

CHAPTER 3

Hydrogen Production

In this chapter, the recent developments in hydrogen production, the processes involved, and the relative efficiencies of different methods are discussed at a basic level that may shape the future technology. A survey of hydrogen gas produced by several different methods is presented in Chapter 4.

3.1 ELECTROLYSIS

The electrolytic method has several distinct advantages over the other methods of producing hydrogen. This method is a simple decomposition of water into hydrogen and oxygen as shown by equation 3.1,

$$H_2O \rightarrow H_2 + {}^1\!/_2\, O_2 \tag{3.1}$$

For splitting water by reaction 3.1, electrical energy is required. This then poses the question of the source for this electricity; is the method adopted suitable for achieving a cleaner energy environment? The electricity used today is predominantly obtained from nonrenewable energies. If we use this method to decompose water, then we are contributing to pollution and greenhouse gas effect. On the other hand, using renewable energies, this problem could be overcome. This method has the advantage of using the available infrastructure. Let us examine the water splitting by electrolysis of water. Three types of electrolyzers have been developed. Figure 3.1 shows the electrolytic cell with a membrane. This type of electrolyzer is called polymer electrolyte membrane electrolyzer. The diagram shows that at the cathode, hydrogen ion picks up electrons to produce hydrogen gas. In the second type of electrolyzer, a liquid alkaline solution made of sodium or potassium hydroxide as the electrolyte is used. This type of electrolyzer is called alkaline electrolyzer. This has been commercially available for decades. A third type of electrolyzer uses solid ceramic material as the electrolyte and operates at 700–800°C. Hydrogen produced through electrolysis is expected to result in no greenhouse gas emissions depending on the source of electricity. Another advantage of electrolytic method is the high efficiency of hydrogen production [1–6] making it suitable for commercial production.

Introduction to Hydrogen Technology, Second Edition.
K.S.V. Santhanam, Roman J. Press, Massoud J. Miri, Alla V. Bailey, and Gerald A. Takacs

Book Companion site : www.wiley.com/go/santhanam/hydrogentech_2e

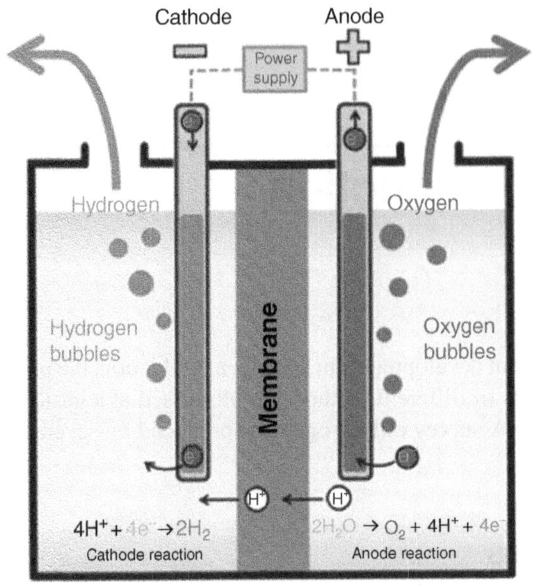

Figure 3.1 Single electrolyzer. [*Source*: Taken from http://energy.gov/eere/fuelcells/hydrogen-production-electrolysis.]

Sample Exercise

How is an electrolyzer generating hydrogen and why should a membrane be used in the electrolyzer?

Water is present as H^+ ions and OH^- ions in the electrolyzer. When a direct current (DC) is passed through the electrolyzer, the following reactions occur. At the anode, water is oxidized by the following reaction:

$$2H_2O \rightarrow O_2 + 4H^+ + 4e^-$$

At the cathode, hydrogen ion is reduced to hydrogen gas by the following reaction:

$$2H^+ + 2e^- \rightarrow H_2$$

Membrane serves the purpose of separating the anode and cathode compartments of the electrolytic cell such that the gases from anode and cathode are isolated. It prevents reversal reactions to occur in the electrolytic cell.

Calculate how much volume of hydrogen gas is generated by passing 2 A current through the electrolytic cell for 1 h. Assume the NTP conditions.

For this purpose, we use Faraday's law of electrolysis. Assuming a current efficiency of 100% (only hydrogen ion reduction is occurring at the cathode), passage of 1 F of electricity (1 F = 96,494 C/mol) will bring out

Moles of $H_2 = \{Q/2F\}$, where Q is the quantity of electricity $[Q = I\,(\text{A}) \times t]$

1 mol of a gas occupies at NTP (normal temperature and pressure) 22.4 L or 22,400 mL.

Volume of Hydrogen = [Moles of $H_2 \times 22.4$ L/1 mol H_2]

In the problem, $I = 2$ A; $t = 1$; $h = 3600$ s

$Q = 2\,A \times 3600\,s$
Moles of $H_2 = [2 \times 3600/2 \times 96{,}494] = 0.0373$ mol of H_2
Volume of $H_2 = 0.0373 \times 22.4\,L = 0.83\,L$

Efficiency: In a practical experiment if the electrolysis gave 0.50 L of hydrogen, the efficiency of the electrolytic process is

$$\eta = [0.50\,L/0.83\,L] \times 100 = 60.24\%$$

The reason for not obtaining the expected volume of hydrogen is caused by several factors such as (a) electrolysis not going 100% for the conversion of hydrogen ion to hydrogen, (b) part of the current is used for heating the electrolytic cell, (c) the side reactions in the electron transfer processes, and (d) measurement errors.

3.2 THERMOLYSIS (THERMAL REACTIONS INVOLVING SOLAR ENERGY)

Water splitting is carried out at high temperatures in the range 500–2000°C using solar energy, nuclear power reactions, or chemical catalyst. This method of producing hydrogen may be called low or no green gas emissions. A typical solar concentrator that provides the energy for decomposition of water is given in Figure 3.2. In the diagram, solar energy is concentrated using heliostats onto solar receiver. In method (a), the solar rays are projected by flat reflectors directly to the solar receiver, while in method (b), the solar rays are collected using a parabolic dish concentrator.

A maximum dish temperature of 1125°C has been reported by Sandia Laboratories [7]. The thermochemical decomposition of water efficiency is discussed in the later section. Thermochemical water splitting using chemical catalysts have been reported. The

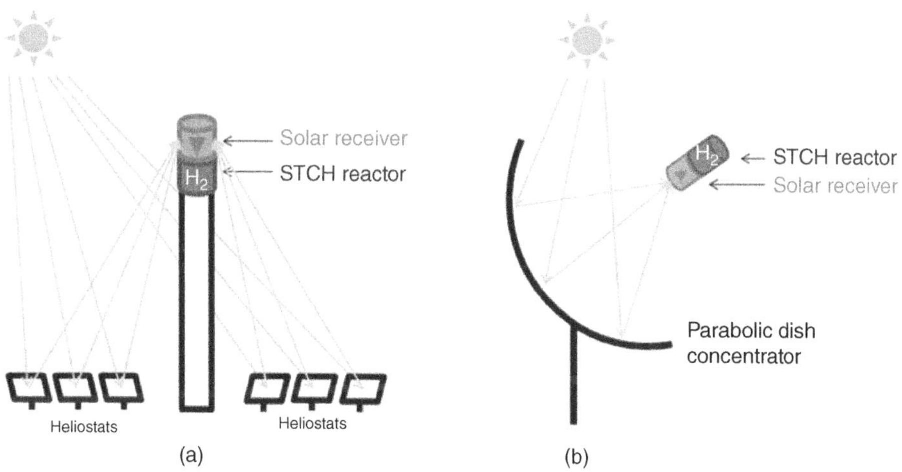

Figure 3.2 Solar method of decomposition of water splitting. (a) Central receiver/reactor tower with heliostats. (b) Modular dish-mounted receiver/reactor. [*Source*: Taken from https://energy.gov/eere/fuelcells/hydrogen-production-thermochemical-water-splitting]

decomposition is carried out using either cerium oxide or copper chloride-hybrid catalyst. The following reactions occur in the reactor for producing hydrogen.

In method (a) [see Figure 3.2a], CeO_2 is used as a catalyst in STCH (solar thermal chemical) reactor, and when it reaches a temperature of 2000°C, the following reaction occurs in the reactor:

Step 1:

$$2CeO_2 = Ce_2O_3 + {}^1\!/\!{}_2\,O_2 \tag{3.2}$$

After the above conversion, water is fed to the reactor at 400°C when reaction (3.3) takes place.

Step 2:

$$Ce_2O_3 + H_2O = 2CeO_2 + H_2 \tag{3.3}$$

Note that Ce_2O_3 is produced from CeO_2 using concentrated sunlight at 2000°C as shown in Figure 3.3. This method is considered as a two-step cycle for producing hydrogen gas. In method (b) [see Figure 3.3b], copper chloride (CuCl) is used as a catalyst in the STCH reactor when thermolysis followed by electrolysis proceeds. The reaction that produces hydrogen gas is given by reaction 3.4.

$$2CuCl + 2HCl = 2CuCl_2 + H_2 \tag{3.4}$$

Figure 3.3b shows the thermochemical methods employing chemical catalysts.

Sample Exercises

What is the purpose of heliostats in the thermochemical method of producing hydrogen?

Heliostat is a device that consists of a plane mirror that reflects the sun's radiation to a predetermined target (solar receiver). It takes into account the movement of the sun to

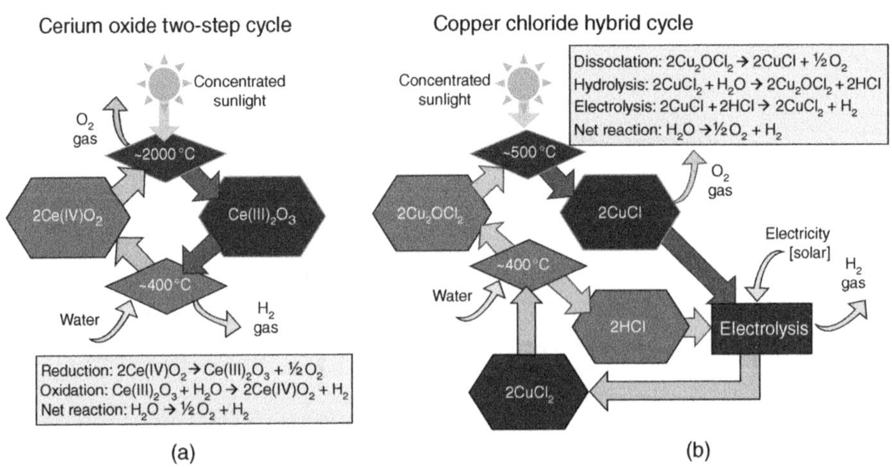

Figure 3.3 Thermochemical method of producing hydrogen using chemical catalysts. [*Source*: Taken from https://energy.gov/eere/fuelcells/hydrogen-production-thermochemical-water-splitting]

optimize maximum reflectivity. The mirror is kept perpendicular to the bisector of the angle between the two directions: the sun and the target. It was invented by William Gravesande. In Figure 3.2a, heliostats reflect the radiation to a target named as solar receiver. STCH reactor is solar thermochemical reactor. The reactions shown in Figure 3.3 occur in the reactors.

What is the alternative method of concentrating solar energy for the thermochemical method?

The solar radiation is reflected into a target (solar receiver) by parabolic dish concentrator. The thermochemical reactor is surrounded by the solar receiver. Figure 3.2b describes this methodology.

In the thermoelectrochemical method, how is CuCl produced for electrolysis?

The starting material is aqueous $CuCl_2$ (cupric chloride) in the STCH reactor. At 400°C, this decomposes to HCl by the following reaction:

$$2CuCl_2 + H_2O = Cu_2OCl_2 + 2HCl$$

HCl electrolysis produces H_2 and $CuCl_2$ by the following reactions:

$$HCl \rightarrow H^+ + Cl^-$$

and at the cathode,

$$2H^+ + 2e^- = H_2$$

and at the anode,

$$Cl^- = Cl^{\cdot} + e^-$$

$$CuCl + Cl^{\cdot} = CuCl_2$$

Thus, the starting material is regenerated. This is a cyclic process.

Problem: Write the chemical reactions that produce hydrogen gas in the cerium oxide method?

It is a two-step process. Aqueous solution of CeO_2 undergoes thermolysis at 2000°C in the presence of sunlight. It is converted to Ce_2O_3 and O_2 by the following reaction:

$$2CeO_2 = Ce_2O_3 + 1/2\,O_2$$

$$\text{At } 400°C, Ce_2O_3 + H_2O = 2CeO_2 + H_2$$

It is again a cyclic process as CeO_2 is regenerated. CeO_2 is considered as a catalyst in the thermochemical method.

3.3 PHOTOVOLTAIC ELECTROLYSIS

This method of producing hydrogen uses electrochemical splitting of water using solar energy. This technology is currently economically not cost competitive. However, marked improvement has recently been reported using two polymer electrolyte membrane electrolyzers with triple junction solar cell as shown in Figure 3.4. The highest solar to hydrogen efficiency of 30% has been reported. The solar cell used in Figure 3.4 is a triple junction

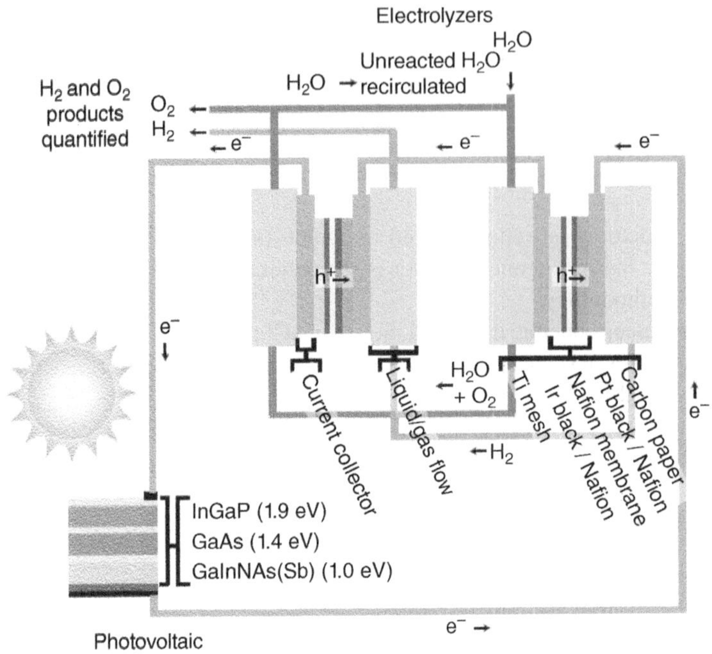

Figure 3.4 Photovoltaic splitting of water to hydrogen. [*Source*: Jia *et al.* 2016 [8]. https://www.ncbi.nlm.nih.gov/pmc/articles/PMC5095559/. Licensed under CC BY 4.0.]

solar cell with an area of $0.316\,cm^2$. The photovoltaic cells use InGaP, GaAs, and GaInNAsSb in the following configuration:

$$InGaP/GaAs/GaInNAsSb \tag{3.5}$$

In equation 3.5, / indicates a boundary between the two semiconductors. The output of the triple junction solar cell is connected to two polymer electrolyte membrane electrolyzers containing Nafion membranes coated with $0.5\,mg/cm^2$ platinum black as cathode and $2\,mg/cm^2$ iridium black as catalyst as anode. The two electrolyzers shown in Figure 3.4 are connected in series in the above photovoltaic cell with water circulating in one of the electrolyzers. The output of the first electrolyzer (water and oxygen) is flown into the anode compartment of the second electrolyzer. Similarly, hydrogen gas from the cathode compartment is flown into the second electrolyzer. The overall reactions are

$$\text{Solar cell} + \text{sunlight}(h\nu) = \text{DC(direct current) output} \tag{3.6}$$

$$\text{DC output} + \text{Electrolysis of } H_2O = H_2 + O_2 \tag{3.7}$$

In Figure 3.4, the flow of electrolytes, oxidant (h^+) and reductant (e^-) are shown for depicting

$$2H^+ + e^- = H_2 \tag{3.8}$$

and

$$2H_2O + 4h^+ = O_2 + 4H^+ \tag{3.9}$$

This method is claimed to have 30% efficiency over 48-h period of performance. Note in the above cell that GaInNAsSb and InGaP are the anodes where oxidation reaction occurs producing O_2 gas. GaAs is the cathode where reduction of hydrogen ion produces H_2 gas.

3.4 PLASMA ARC DECOMPOSITION

This method is used for fossil fuels for generating hydrogen from natural gas. This method decomposes the natural gas into hydrogen and carbon.

$$CH_4 \rightarrow C + 2H_2(g) \tag{3.10}$$

This method is claimed to produce no carbon dioxide.

Sample Exercise

How many liters of methane gas are required to produce 1 gal of hydrogen at NTP?

$$1\,\text{gal} = 3.7854118\,\text{L}$$
$$1\,\text{mol}\,CH_4 = 2\,\text{mol of}\;H_2$$
$$\text{At NTP, }1\,\text{mol} = 22.4\,\text{L}$$
$$22.4\,\text{L}\;CH_4 = 2 \times 22.4\,\text{L}\;H_2$$
$$CH_4 = [3.7854118\,\text{L}/2 \times 22.4\,\text{L}] \times 22.4\,\text{L}\;CH_4 = 1.8927\,\text{L}$$

3.5 THERMOCHEMICAL PROCESS (THERMAL DECOMPOSITIONS BY PROCESSES OTHER THAN SOLAR ENERGY)

Water decomposition into hydrogen is carried out through cyclic chemical reactions. The redox pairs used for this purpose are innumerable [9a]; the redox couples Fe_3O_4/FeO and ZnO/Zn are suitable for two-step splitting of water. The two-step process for water decomposition is represented as

$$XO \rightarrow X + {}^1\!/_2\,O_2$$
$$X + H_2O \rightarrow XO + H_2$$

XO is metal oxide. When hydrides are used for thermochemical process, the following steps are involved:

$$XH_2 \rightarrow X + H_2$$
$$X + H_2O \rightarrow XH_2 + {}^1\!/_2\,O_2$$

where XH_2 represents metal hydrides. When hydroxide is used for thermochemical process,

$$2XOH \rightarrow 2X + H_2O + {}^1\!/_2\,O_2$$
$$2X + 2H_2O \rightarrow 2XOH + H_2$$

TABLE 3.1 Hydrogen production by thermochemical process

Notation	Runs of cyclic reaction	Average of H_2 production/ 10^{-4} g g^{-1} material	Ferrite conversion
Undoped	6	5.7	33
MnO_4	6	3.9	26
MnO_7	6	6.7	41
Mn_{10}	4	4.7	28
MgO_2	4	6	40
MgO_6	4	5.1	29
CoO_2MnO_5	6	7	42
CoO_4MnO_1	6	4.3	30
CoO_4MnO_3	6	5.9	44
CoO_4MnO_5	6	4.6	34
CoO_6MnO_2	6	6.5	42
CoO_4	6	5.6	39
CoO_7	6	8	46
Co_{10}	6	8.7	51
Ni_{10}	6	11	69
Fe_3YSZ	7	4.5	26
Fe_8YSZ	6	4.7	31
$Fe_{10}YSZ$	6	4.6	31

Uses oxide ZrO_2 or Yittria stabilized zirconia (YSZ) supported ferrites. Ferrite conversion in the table represents fraction of Fe^{3+} converted to Fe^{2+} in the thermal process.

Source: Kodama (2007) (9a). Reproduced with permission of American Chemical Society.

where XOH is metal hydroxide. X is a metal in all the abovementioned cases. Thermochemical process of producing hydrogen is an engineering challenge and is actively being perused in combination with solar energy (Table 3.1).

3.6 PHOTOCATALYSIS

A material that catalyzes a chemical reaction that is initiated by photons is termed as photocatalysis. A number of semiconductors have been reported to catalyze the decomposition of water to hydrogen. Figure 3.5 shows a model by which hydrogen is produced.

A semiconductor absorbs the solar radiation (light) and excites the electrons from valence band (VB) of the semiconductor to conduction band (CB). A metallic coating (Co-cat) brings out the conversion of water to hydrogen. On the semiconductor surface, water is oxidized to oxygen by the hole (h^+). The abovementioned method of producing hydrogen and oxygen on a coated semiconductor is described as short-circuited electrochemical cell [9b].

Some examples of the photocatalysts are TiO_2, CdS, $CuAlGaO_4$, and 3d transition-metal-doped $In_{1-x}M_xTaO_4$ (M = Sc, Ti, V, Cr, Mn, Fe, Co, Ni, Cu, Zn). Table 3.2 gives the amount of hydrogen produced with UV and visible light irradiation. Among other semiconductors used in photocatalysis for hydrogen production are $CuAlGaO_4$. Using a pyrex reactor with a quartz window to permit UV rays to pass through and 250 mL 0.5 M KOH solution and 0.5 g photocatalyst ($CuAlGaO_4$) decomposed H_2S gas flowing into KOH

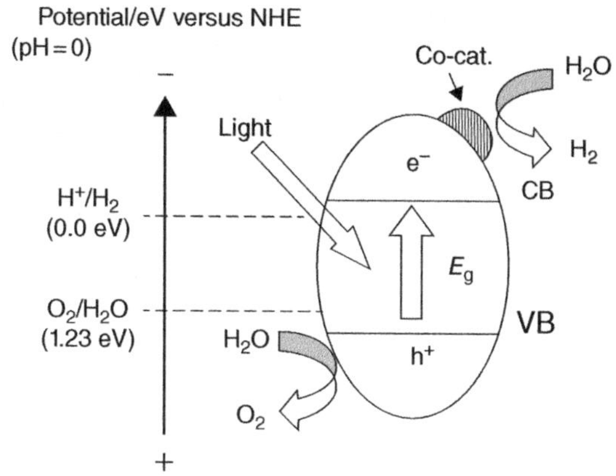

Figure 3.5 Photocatalysis of water decomposition. [*Source*: Ye and Zou 2005 [9b]. Reproduced with permission of Elsevier.]

TABLE 3.2 Photocatalytic production of hydrogen

	[a]Surface area $(m^2/g)^c$	UV irradiation (μmol/h)	[b]Visible irradiation $(\mu mol/h)^c$
$InTaO_4$	1.35	270	1.2
$In_{0.8}Sc_{0.2}TaO_4$	1.2	250	None
$In_{0.8}Ti_{0.2}TaO_4$	0.82	70	None
$In_{0.8}V_{0.2}TaO_4$	0.93	110	< 0.1
$In_{0.8}Cr_{0.2}TaO_4$	0.97	95	< 0.1
$In_{0.8}Mn_{0.2}TaO_4$	0.95	310	0.5
$In_{0.8}Fe_{0.2}TaO_4$	0.82	80	None
$In_{0.8}Co_{0.2}TaO_4$	0.93	370	0.6
$In_{0.8}Ni_{0.2}TaO_4$	0.97	470	3.1
$In_{0.8}Cu_{0.2}TaO_4$	0.81	100	None
$In_{0.8}Zn_{0.2}TaO_4$	0.81	220	None

[a]*Light source*: 400 W high-pressure Hg lamp. Sample amount: 1.0 g powder.
[b]*Light source*: 300 W Xe arc lamp. A 420-mm cutoff filter was applied. Sample amount: 0.5 g.
[c]Cocatalyst: 0.1 wt%Pt.
Ye and Zou (2005) (9b). Reproduced with permission of Elsevier.

solution at 2.5 mL/min for 1 h under constant agitation produced hydrogen gas when the solution was irradiated with a xenon lamp (450 W) [10].

Table 3.3 gives the amount of hydrogen produced and the apparent quantum yield (AQY) with different semiconductors. Figure 3.6 shows a cartoon of a photocatalytic generation of hydrogen.

The effect of catalyst in hydrogen generation is demonstrated in Figure 3.7. The photocatalysts are labeled as TS_2 (made of TiO_2-SnO_2 (Ti:Sn = 98:2)), TS_5 (made of TiO_2-SnO_2 (Ti:Sn = 95:5)), and TS_{10} = TiO_2-SnO_2 (Ti:Sn = 90:10). Pd-TS_2, Pd-TS_5, and Pd-TS_{10} refer to palladium coated materials of TS_2, TS_5, and TS_{10}. Figure 3.7 demonstrates that Pd coated

TABLE 3.3 Photocatalytic evolution of hydrogen from H_2S

Catalyst	Band gap (eV)	H_2 (μmol/h)	AQY (%)
$CuAlGaO_4$	1.87	3214	10.8
NiO–$CuAlGaO_4$	1.87	3616	12.7
$CuAlGaO_4$ (used)	1.87	3169	10.7
NiO–$CuAlGaO_4$ (used)	1.87	3571	12.6

Catalyst, 0.5 g; 250 mL (0.5 M aqua. KOH); H_2S (2.5 mL/min); 450 W Xe lamp (Oriel), cutoff filter ($\lambda \geq 420$ nm).
Source: Biswas *et al.* 2011 [10a]. Reproduced with permission of Elsevier.

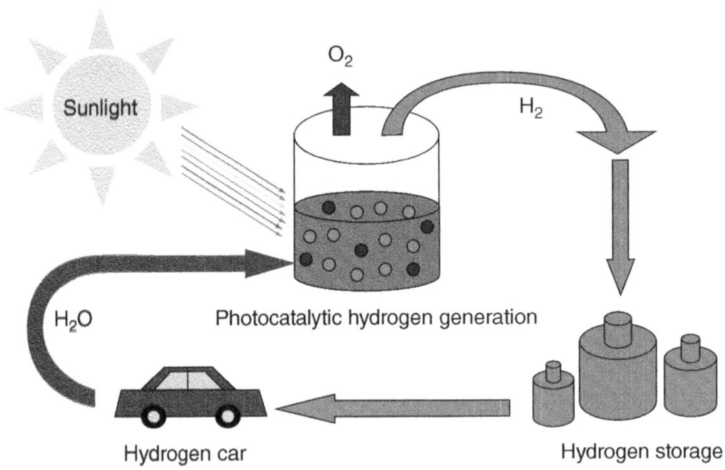

Figure 3.6 Photocatalytic hydrogen generation. [*Source*: Chen *et al.*, (2010) [17] Reproduced with permission of American Chemical Society.]

catalysts gives higher yields over the uncoated catalysts. In eight hours irradiation, hydrogen yield of 500 μmol/g of the catalyst can be obtained as a function of irradiation time. Figure 3.7 demonstrates that the hydrogen yield is higher with palladium-coated semiconductors [12].

Recently, Zn-doped $K_2La_2Ti_3O_{10}$ has been found to have a high photocatalytic activity for hydrogen production under exposure to visible light. Another photocatalytic material using visible light is Cr^{3+}-doped $Bi_4Ti_3O_{12}$ for splitting water to hydrogen gas.

Photocatalysts made of sulfur compounds show high activity for hydrogen production. Hydrogen is observed as bubbles in the solution during the photodecomposition of water. The photocatalysts are Pt-coated cadmium sulfide, solid solution consisting of $CuInS_2$, $AgInS_2$, and ZnS and Ru/$Cu_{0.25}Ag_{0.25}In_{0.5}ZnS_2$. Figure 3.8 shows the hydrogen bubble evolution from a solution containing potassium sulfide and sodium sulfide under solar radiation (AM-1.5).

Hydrogen is evolved at 8 L/m^2 h with this catalyst. The sulfide solutions can utilize visible light of wavelengths up to 700 nm [14]. Carbon-nanotube-coated materials have been examined for photocatalytic activity; examples of such materials show high hydrogen gas production over graphite-coated silica (GS).

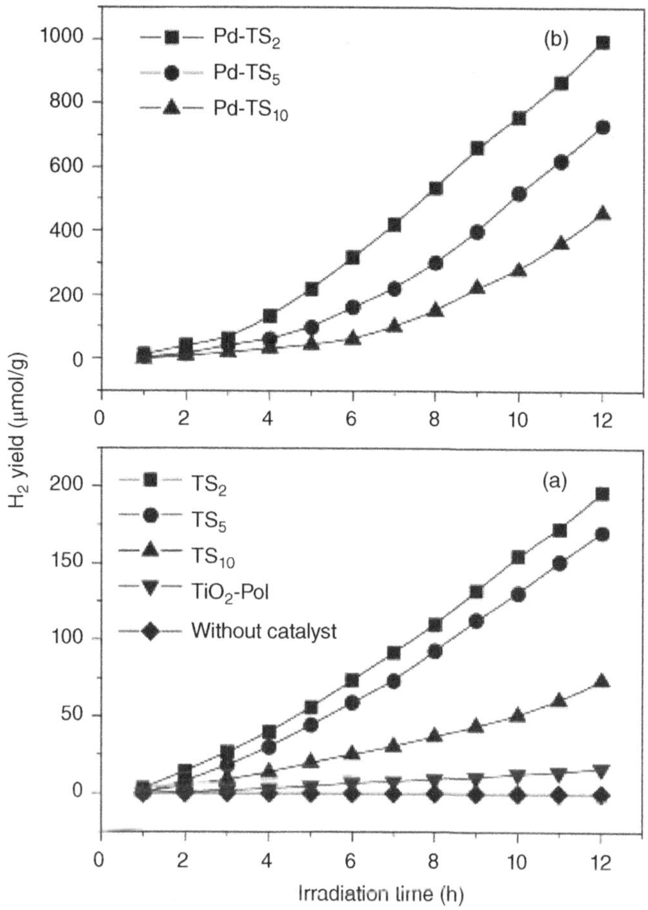

Figure 3.7 Hydrogen generation under different photocatalysts. $TS_2 = TiO_2–SnO_2$ (Ti:Sn = 98:2), $TS_5 = TiO_2–SnO_2$ (Ti:Sn = 95:5), and $TS_{10} = TiO_2–SnO_2$ (Ti:Sn = 90:10). Ratios are by atomic weight. Pd–TS_2, Pd–TS_5, and Pd–TS_{10} refer to palladium-coated photocatalysts. [*Source*: Sasikala *et al.* 2009 [12]. Reproduced with permission of Elsevier.]

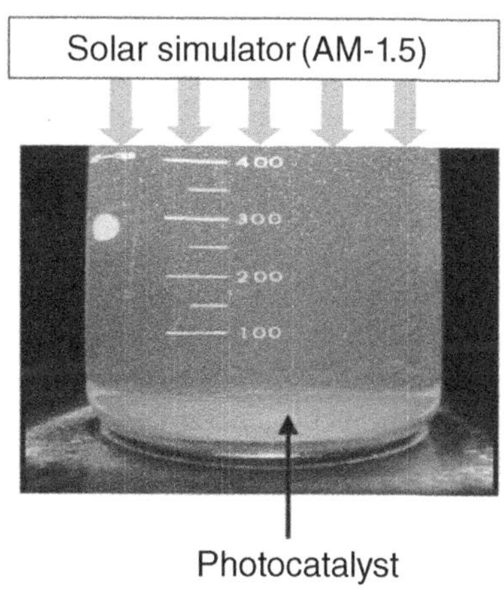

Figure 3.8 Hydrogen production using a photocatalyst, $Ru/Cu_{0.25}Ag_{0.25}In_{0.5}$ ZnS_2. The aqueous solution is a mixture K_2SO_3 and Na_2S. Notice the bubbles produced using a solar simulator (AM-1.5). [*Source*: Reproduced with kind permission of International Union of Pure and Applied Chemistry (IUPAC) [14].]

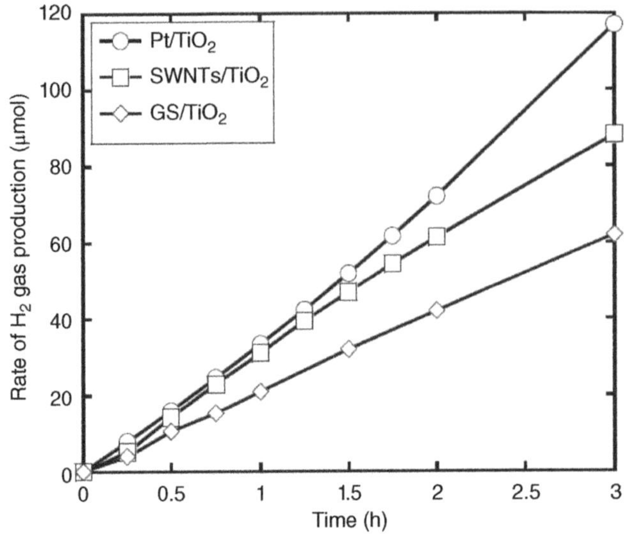

Figure 3.9 Single-walled carbon-nanotube-coated TiO_2 photocatalyst for hydrogen evolution. Pt/TiO_2, platinum-coated titanium dioxide; SWNT/TiO_2, single-walled carbon-nanotube-coated TiO_2; and GS/TiO_2, graphite-silica-coated TiO_2. [*Source*: Data from International Journal of Chemical, Molecular, Nuclear, Materials and Metallurgical Engineering, 8, No:1, 2014, Reproduced with permission World Academy of Science, Engineering and Technology.]

The performance of single-walled carbon-nanotube-coated titanium dioxide (TiO_2) is nearing the noble-metal-coated platinum-coated TiO_2 photocatalyst, as shown in Figure 3.9.

3.7 BIOMASS CONVERSION

This method involves the use of biomass and water to decompose water to hydrogen gas by the action of microbes or bacteria. If light (photons) are also involved in the decomposition, it is called biophotolysis. The total annual production of biomass is beyond 100 billion tonnes of carbon/year, which corresponds to approximately 150×10^3 tera watt-hours (TWh) required for the current world energy.

Annual world biomass energy equivalent = 16.7–33.4 TWh.

Annual world energy consumption = 17.7 TW h. On average, biomass production is 1.4 times larger than world energy consumption.

The biomass conversion is carried out by a number of methods. They are (a) chemical, (b) biochemical, (c) electrochemical, and (d) thermal methods of conversion. The biomass has been used to generate electricity either directly or through fuel cells. The use of biomass as a fuel generally results in air pollution by generating CO, CO_2, NO_x, and volatile organic compounds. Black carbon from biofuels is considered as the second largest pollutant and has been a contributor to global warming.

Biomass method of producing fuel is projected to be expensive by 2017 [16]. It would be nearly two times costlier than natural gas and considered to be little more expensive

Figure 3.10 Pyrolytic plant for biomass conversion. https://energy.gov/eere/fuelcells/hydrogen-production-biomass-gasification.

than nuclear power. However, it would compete with solar power. The environmental information agency shows that the US government will plan to implement a 25% renewable energy standard by 2025; it projects 598 million tons of biomass would be available. Figure 3.10 shows a pyrolysis plant for biomass for hydrogen production.

The problems associated with biomass are (a) inefficiencies of the conversion process, (b) high water content of biomass feed stock, and (c) complex multicomponents.

3.8 GASIFICATION

In this method, coal is converted to syn gas; also fossil fuels are converted to carbon dioxide and hydrogen gas. Coal is a complex and valuable substance that can be transformed to several different products. Gasification of coal can produce power, liquid fuels, and chemicals including hydrogen. By reacting coal with oxygen and steam under high pressures and temperatures, a mixture having a composition of a mixture of CO and hydrogen can be produced. This reaction is generally represented as

$$CH_{0.8} + O_2 + H_2O \rightarrow CO + CO_2 + H_2 + \text{other products}$$

Hydrogen is removed by a separation system. Figure 3.11 shows the four methods of gasification.

Countercurrent fixed-bed gasifier uses a fixed bed of carbonaceous fuel (biomass) through which steam, oxygen, and/or air flows in countercurrent direction. The ash is removed in the dry condition. Thermal efficiency is high as the temperature in the gas exit

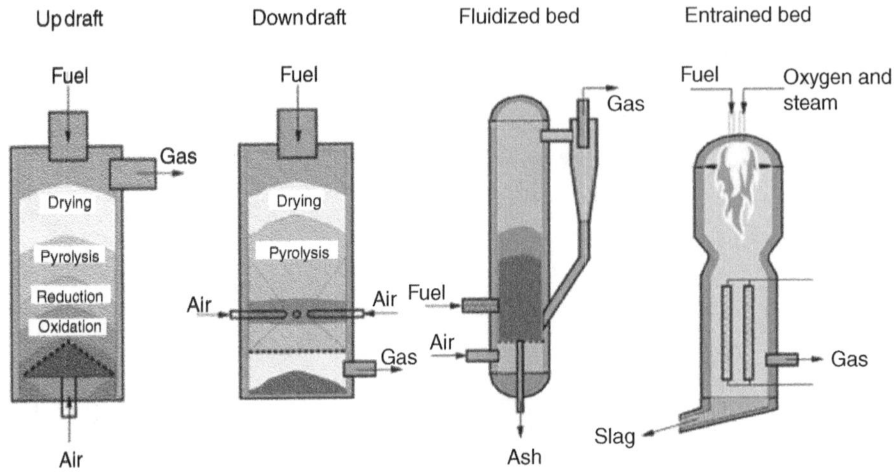

Figure 3.11 Four gasification methods. [*Source*: https://en.wikipedia.org/wiki/Gasification. Used under CC BY SA 3.0 https://creativecommons.org/licenses/by-sa/3.0/.]

is kept at minimum. Tar and methane production is significant. Rice hulls used as biomass in the gasifier at 1000°C produce hydrogen and carbon monoxide. Cocurrent fixed-bed gasifier flows in concurrent configuration as shown in Figure 3.11. Using external heaters or other heating sources, the upper part of the body is heated. Tar levels are much lower in this method. Fluidized bed reactor uses oxygen and steam or air mixed with the fuel. The ash falls out as dry and is removed. In comparison to the fixed bed, the fuel throughput is higher. The conversion efficiency is low. High levels of corrosive ash are formed in this method. Entrained flow gasifier uses a dry pulverized solid, atomized liquid fuel for gasification with oxygen in a cocurrent flow. Very fine particles are formed.

If the gasification reaction is generating greenhouse gases, why is it being considered? The United States of America has abundant domestic coal resource. By relying on this method, the country can reduce the dependence on imported petroleum. Besides it can create jobs for the domestic industry. It also provides environmental benefits when integrated with advanced technologies in coal gasification for power generation, carbon capture, and storage. The integration facilitates capture SO_2, NO_x, Hg, and greenhouse gases and other pollutants.

3.9 HIGH-TEMPERATURE ELECTROLYSIS

This method uses a combination of electrical and thermal energy for decomposition of water to hydrogen gas. Electrolysis of water is carried out at high temperature. Figure 3.12 shows a typical arrangement for high-temperature electrolysis. It is more efficient than room temperature electrolysis. The temperature at which electrolysis is carried out is about 100–800°C. If the temperature of 2500°C is reached, then thermal decomposition of water will be dominant and reduce the need for electrical input to decompose water. By carrying the electrolysis at this temperature, the electrical input necessary to decompose water to hydrogen is reduced and becomes cheaper.

$$\text{Heat energy necessary} = 141.86 \times 10^6\ \text{J/kg}\,\text{H}_2$$

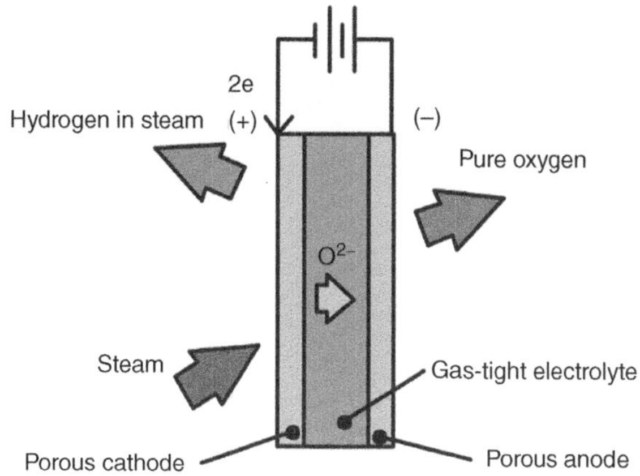

Figure 3.12 High-temperature electrolysis using porous electrodes. Electrolyte Yittria stabilized Zirconia. [*Source*: https://en.wikipedia.org/wiki/High-temperature_electrolysis. Used under CC BY SA 3.0 https://creativecommons.org/licenses/by-sa/3.0/.]

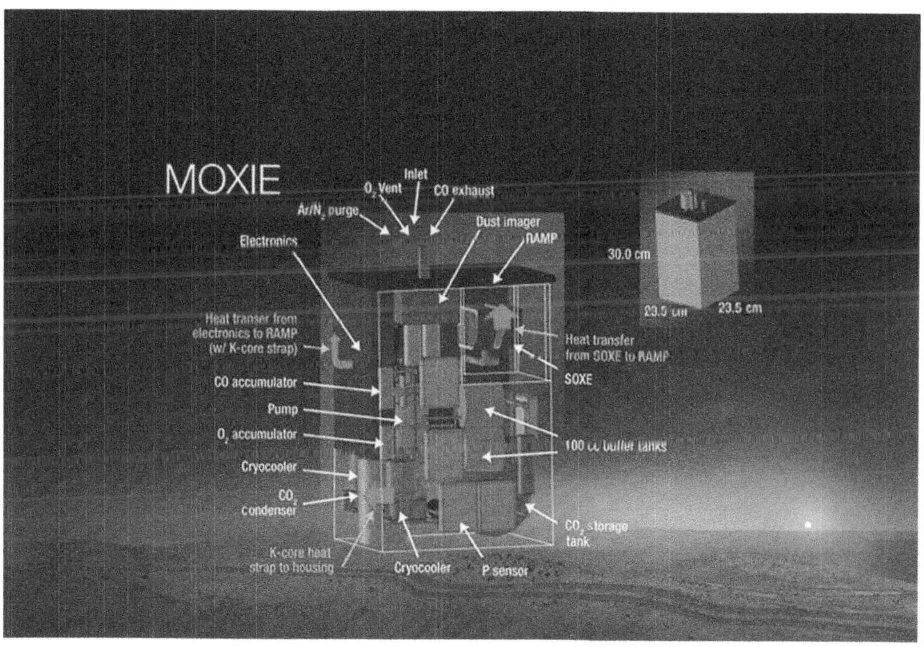

Figure 3.13 MOXIE for converting carbon dioxide to oxygen by high-temperature fuel cell.

High-temperature electrolyzer has been modified to produce oxygen. This method is to convert

$$2CO_2 \rightarrow 2CO + O_2$$

utilizing high-temperature solid oxide fuel cell. The planned mission to Mars in 2020 utilizes this method as Mars atmosphere is made up of 95% CO_2. This technology will allow it to be converted to oxygen. The rover named MOXIE (Figure 3.13) is expected to produce

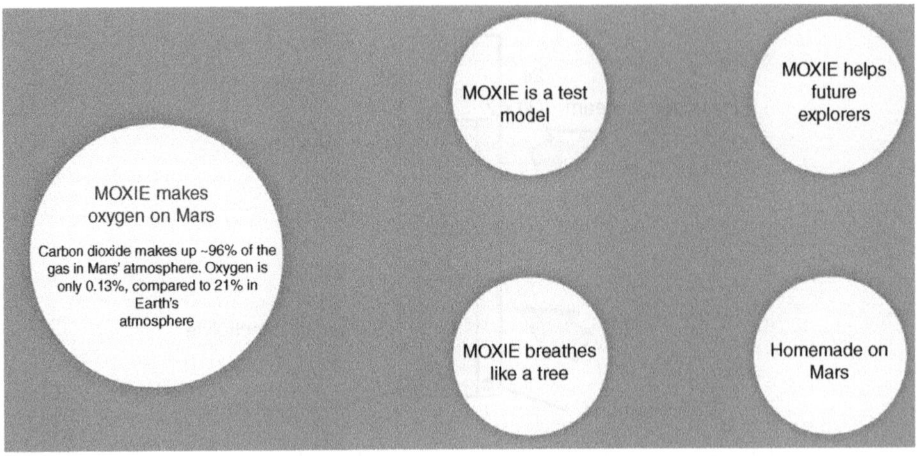

Figure 3.14 MOXIE and its functions. [*Source*: http://mars.nasa.gov/mars2020/mission/instruments/moxie/. Reproduced with kind permission of NASA.]

22 g of oxygen per hour with 99.6% purity. This exploration will result in five ways MOXIE functions, as given in Figure 3.14.

3.10 MISCELLANEOUS METHODS

A variety of methods used for the generation of hydrogen are listed as follows:

a) Dark fermentation (biological reactions producing hydrogen gas in the absence of radiation)
b) Fossil fuels reforming
c) Photofermentation
d) Artificial photosynthesis (chemical methods of producing photosynthesis and hydrogen).

3.11 COMPARATIVE EFFICIENCIES

Figure 3.15 shows the efficiencies of the different methods of producing hydrogen gas. Note that this figure contains methods discussed in this chapter as well as some others not discussed here. The discussion in this chapter has been confined to the methods that are scalable to an industrial level. The other methods listed in Figure 3.15 caption are based on laboratory-scale research results.

Energy efficiency is defined as

$$\eta = [m\text{LH}\,W_{H_2}]/E_{in} \tag{3.11}$$

where m is mass flow rate of hydrogen, LH is the lower heating value of hydrogen (121 MJ/kg), W_{H_2} is the weight of hydrogen, and E_{in} is the energy input. Chemical

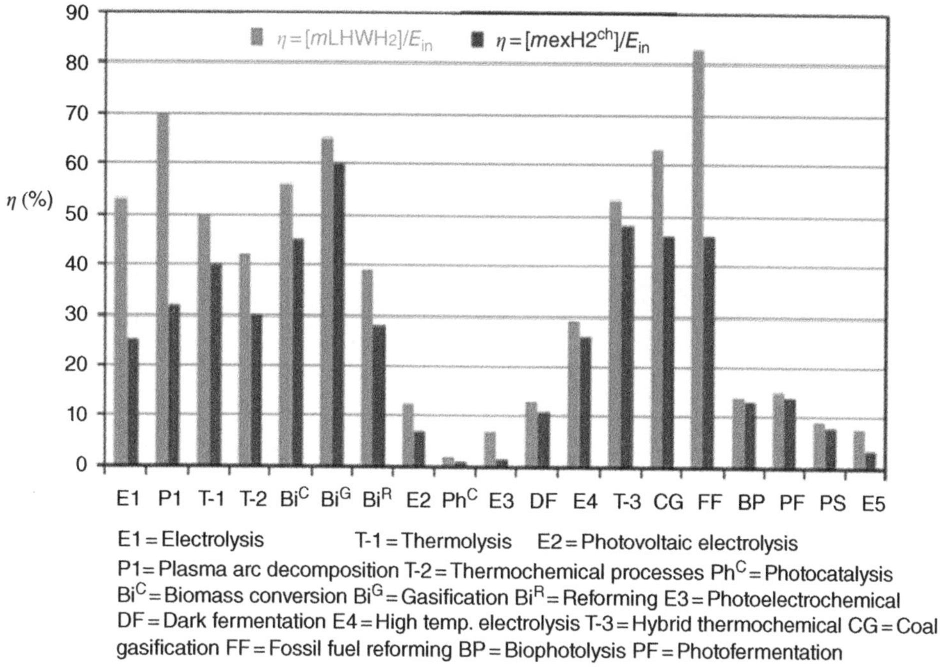

Figure 3.15 Hydrogen gas production efficiencies. [*Source*: Taken from Int. J. Hydrogen Energy,40, 11095 (2015) [1] Reproduced with permission of Elsevier.]

efficiency is defined as

$$\eta_{ch} = [m\,ex_{H_2}]/E_{in} \tag{3.12}$$

where ex_{H_2} is the chemical exergy of hydrogen.

From Figure 3.15, it is to be inferred that fossil fuel (FF) method is having an efficiency of greater than 80%. Plasma arc and hybrid thermochemical methods also have comparable efficiencies. However, these methods produce gases that are contributing to greenhouse gas effect. In comparison, the electrolytic method (E1) is having 52% efficiency that produces no greenhouse gas effect gases. Hence, currently, there is focus on improving the efficiency of the electrolytic method. In this context it is appropriate to define three other efficiencies for maintaining our environment free from greenhouse gas effect. We define environment polluting factor (EPF)

$$\text{EPF} = \frac{\text{kg of } CO_2 \text{ produced}}{\text{kg of fuel used}} \tag{3.13}$$

Sample Exercise

In FF method of producing hydrogen gas, the reactions are

$$CH_4 + H_2O = CO + 3H_2 \tag{3.14}$$

$$CO + H_2O = CO_2 + H_2 \tag{3.15}$$

The overall reaction is

$$CH_4 + 2H_2O = CO_2 + 4H_2 \tag{3.16}$$

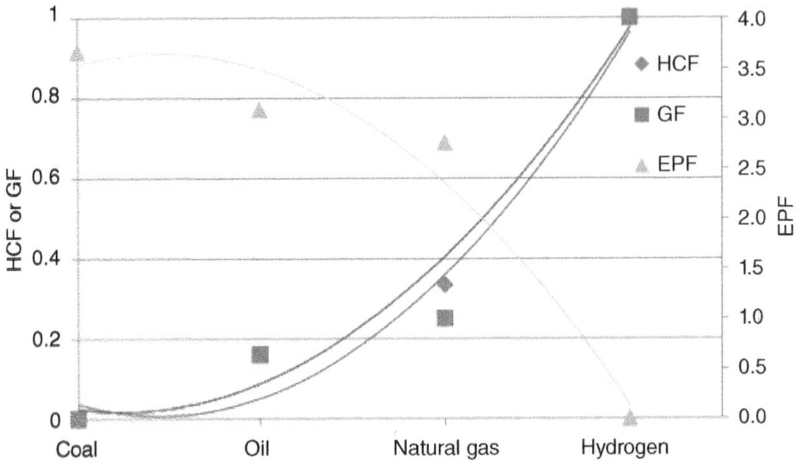

Figure 3.16 Correlation between hydrogen content of the fuel and environmental pollution factor. [*Source*: Taken from Int. J. Hydrogen Energy, 40, 11094 (2015) [1] Reproduced with permission of Elsevier.]

If 5 L of methane gas is used in the FF steam reformer at a pressure of 508 psi and 815°C, calculate the EPF. 1 psi = 0.068046 atm; $R = 82.057$ atm/K mol. Molar mass of $CO_2 = 44$ g/mol and molar mass of $CH_4 = 16$ g/mol.

$$[508\,\text{psi} \times 0.068046\,\text{atm}/1\,\text{psi}] \times 5000\,\text{cc} = n \times 82.057\,\text{cc atm/K mol} \times 1088\,\text{K}$$

$$\text{No. moles of CH}_4 = 1.9355$$

$$1\text{mol CH}_4 = 1\,\text{mol CO}_2$$

$$\text{Moles of CO}_2 = 1.9355$$

$$\text{Grams of CO}_2 = 1.9355\,\text{mol} \times 44\,\text{g CO}_2/1\,\text{mol CO}_2 = 85.162\,\text{g} \times 1\,\text{kg}/1000\,\text{g}$$
$$= 0.08516\,\text{kg}$$

$$\text{Grams of fuel used} = 1.9355\,\text{mol} \times 16\,\text{g}/1\,\text{mol CH}_4 = 30.968\,\text{g} \times 1\,\text{kg}/1000\,\text{g}$$
$$= 0.03096\,\text{kg}$$

$$\text{EPF} = \frac{0.08516\,\text{kg}}{0.03096\,\text{kg}} = 2.746$$

EPF for other methods can be similarly calculated. For example, in E_1 in Figure 3.16, the EPF value is zero as the process produces no carbon dioxide. The reaction is simple decomposition of water to hydrogen and oxygen.

Two other important factors that are considered in hydrogen production are greenization factor (GF) and hydrogen content factor (HCF). These are defined as follows:

$$\text{GF} = [\text{EPF}_{max} - \text{EPF}]/\text{EPF}_{max} \quad (3.17)$$

and

$$\text{HCF} = [\text{kg H}_2 \text{ in fuel}]/\text{kg fuel} \quad (3.18)$$

TABLE 3.4 Hydrogen production—important benefits and challenges

Methods	Benefits	Challenges
1. Fossil fuel reforming	Most viable approach Lowest current cost Existing infrastructure	High operational maintenance cost Reliability
2. Thermochemical method	Robust, low-cost material Ease of manufacture Sustainable Recycled chemicals	Reliability Efficient heat transfer
3. Coal and biomass gasification	Low-cost syn gas production Abundant and cheap	High cost, tolerance for impurities, efficiency
4. Photoelectrochemical method	Low operational cost Clean and sustainable	Low system efficiency Effective photocatalytic material, low-cost materials
5. Electrolysis	No pollution with renewable energy sources Existing infrastructure Interfacial with fuel cells	Low system efficiency High capital costs Durable materials
6. Biofuel reforming	Viability and existing infrastructure	High capital costs, hydrogen yield, and efficiency
7. Biological method	Clean and sustainable Self-sustaining	Efficient microorganisms for sustainable production, new organisms

Source: Dincer, I. and Acar, A.C. (2015) [1] Reproduced with permission of Elsevier.

where EPF_{max} is the environmental impact factor of coal as it produces maximum amount of greenhouse gas; it is taken as 3.6. Figure 3.16 shows the relationship observed between hydrogen content of the fuel and GF. This diagram shows decreasing polluting factor from coal to hydrogen used as fuel.

In Table 3.4, the important benefits and challenges are given in seven selected methods for hydrogen production. While FF method is a viable approach, it involves high cost and high EPF. The photoelectrochemical method offers significant advantages over the other methods in terms of lower cost.

PROBLEMS

1. Explain how the hydrogen production efficiency is calculated in the electrolytic and thermochemical methods.
2. In a practical experiment if the electrolysis gave 1.8 L of hydrogen upon electrolysis by passing a current of 5 A for 1 h, calculate the efficiency of hydrogen production.
3. What is the difference between photovoltaic and photocatalytic methods of generating hydrogen gas?
4. Explain the reasons for Pd-coated TS giving higher hydrogen yield as compared with TiO_2-coated semiconductors?

5. If 50 L of methane gas is used in the FF steam reformer at a pressure of 600 psi and 815°C, calculate the EPF. 1 psi = 0.068046 atm; $R = 82.057$ atm/K mol. Molar mass of $CO_2 = 44$ g/mol; molar mass of $CH_4 = 16$ g/mol.
6. At what temperature the conversion of CeO_2 to Ce_2O_3 occurs in the thermochemical method for producing hydrogen gas. How much thermal energy is used in the reaction?
7. At what temperature water is decomposed to hydrogen gas in the thermochemical method by Ce_2O_3. Write the chemical conversion reaction.
8. How is HCl produced in the electrolytic cell in the $CuCl_2$ catalyzed decomposition of water to hydrogen gas?
9. Write the mechanism by which hydrogen gas is produced in photocatalysis.
10. With $CuAlGaO_4$ having a band gap of 1.8 eV and an absolute quantum efficiency of 10.8% for producing hydrogen gas, estimate the number of hours required to produce 1 gal of hydrogen gas. Use the data in Table 3.3.
11. In Figure 3.16, natural gas used as fuel gives an EPF of about 2.7. In the same figure, hydrogen gas gives an EPF of zero. What is the reason for one giving a high value and the other a low value?
12. What are the benefits and challenges in using hydrogen as fuel?

REFERENCES

1. Dincer, I. and Acar, A.C. (2015) *International Journal of Hydrogen Energy*, **40**, 1094.
2. Santhanam, K.S.V. (2015) *ECS Transactions*, **66** (30), 23–34.
3. Marcelo, D. and Dell'Era, A. (2008) *International Journal of Hydrogen Energy*, **33**, 3041.
4. Herring, J.S., O'Brien, J., Stoots, C.M. *et al.* (2007) *International Journal of Hydrogen Energy*, **32**, 440.
5. Munoz, L.d., Bergel, A., Feron, D., and Basseguy, R. (2010) *International Journal of Hydrogen Energy*, **35**, 8561.
6. Rosen, M.A. and Scott, D.S. (1998) *International Journal of Hydrogen Energy*, **23**, 653459.
7. http://energy.gov/sites/prod/files/2014/03/f9/solar_thermo_h2.pdf (accessed 26 May 2017).
8. Jia, J., Seitz, L.C., Benck, J.D. *et al.* (2016) *Nature Communications*, 7.
9. (a) Kodama, T. and Gokan, N. (2007) *Chemical Reviews*, **107**, 4059;
   (b) Ye, J. and Zou, Z. (2005) *Journal of Physics and Chemistry of Solids*, **66**, 266.
10. (a) Biswas, S.K., Baeg, J., Kale, B.B. *et al.* (2011) *Catalysis Communications*, **12**, 651;
    (b) Esswein, A.J. and Nocera, D.G. (2007) *Chemistry Review*, **107**, 4022–4047.
11. Sasikalaa, R., Chen, A.X., Shen, S. *et al.* (2010) *Chemistry Review*, **110**, 6503–6570.
12. Sasikalaa, R., Shirolea, A., Sudarsana, V. *et al.* (2009) *International Journal of Hydrogen Energy*, **34**, 3621.
13. Esswein, A.J. and Nocera, D.G. (2007) *Chemistry Review*, **107**, 4022–4047.
14. Kudo, A. and IUPAC (2007) *Pure and Applied Chemistry*, **79**, 1917–1927.
15. Ahmmad, B., Kanomata, K., and Hirose, F. (2014) *International Journal of Chemical, Molecular, Nuclear, Materials and Metallurgical Engineering*, **8** (1), 24–29, World Academy of Science, Engineering and Technology.
16. U.S. Energy Information Administration (2010). *Annual Energy Outlook 2010* (PDF). Report no. DOE/EIA-0383(2010), Washington, DC.
17. Chen, X., Shen, S., Guo, L., and Mao, S.S. (2010) *Chemistry Review*, **110**, 6503.

CHAPTER 4

Hydrogen Properties

4.1 OCCURRENCE OF HYDROGEN, PROPERTIES, AND USE

4.1.1 General Characteristic and Physical Properties of Hydrogen

Hydrogen (Latin: *hydrogenium*, from Greek: *hydro*: water and *genes*: forming) is a chemical element with the atomic number 1 in the periodic table. It was discovered and recognized as an element by English chemist and physicist Henry Cavendish in 1766. However, hydrogen was observed and collected as a unique gas long before by Robert Boyle in 1671 when he dissolved iron in diluted hydrochloric acid.

Hydrogen is the simplest of all elements; the hydrogen atom is made up of a nucleus with a positive charge and only one electron. Electron configuration is $1s^1$ (one electron per shell). Most of the mass of a hydrogen atom is concentrated in its nucleus: the proton is more than 1800 times more massive than an electron; mass of neutrons is almost the same as mass of protons. The radius of the electron orbit is approximately 100,000 times larger than the radius of the nucleus. Since the radius of the electron orbit defines the size of the atom, a hydrogen atom actually consists mainly of empty space.

Hydrogen is the only element that has different names for its isotopes. There are three major hydrogen isotopes:

1. Protium (${}^{1}{}_{1}H$); mass 1; the nucleus consists of a single proton; found in nature in more than 99.985% (Figure 4.1).
2. Deuterium (${}^{2}{}_{1}H$, often written D_2); mass 2; the stable, heavy isotope of hydrogen; the nucleus contains both a proton and a neutron; found in nature in approximately 0.015%; was discovered in 1931 by American chemist Harold Urey.
3. Tritium (${}^{3}{}_{1}H$, the symbol T is sometimes used); mass 3; the nucleus consists of two neutrons and one proton; appears in small quantities in nature but can be artificially produced by various nuclear reactors; it is unstable and radioactive, has a half-life of 12.32 years. Tritium occurs in nature due to interaction of cosmic rays and atmospheric gases. Radioactive water T_2O can be formed in the atmosphere when tritium reacts with oxygen, and a form of slightly radioactive rain can enter the earth's oceans and lakes; however, the short half-life prevents the input of essential hazardous radioactivity.

Introduction to Hydrogen Technology, Second Edition.
K.S.V. Santhanam, Roman J. Press, Massoud J. Miri, Alla V. Bailey, and Gerald A. Takacs

Book Companion site : www.wiley.com/go/santhanam/hydrogentech_2e

Figure 4.1 Atomic structure of protium.

With a single electron orbiting a nucleus in its atomic arrangement, an individual hydrogen atom is highly reactive and exists in nature only in a form of hydrogen molecules H_2. The energy concentrated in a single hydrogen molecule is lower than energy containing in two separate H atoms and is equal to 436 kJ/mol. In addition, in a molecule of hydrogen, each proton has a field associated with a spin. There are two kinds of hydrogen molecules: "orthohydrogen," in which both protons have the same spin, and "parahydrogen," in which the protons have opposite spins. At room temperature, over 75% of hydrogen exists as orthohydrogen and 25% as parahydrogen; however, at very low temperatures, orthohydrogen becomes unstable and converts to more stable parahydrogen, releasing heat, which can complicate some low-temperature hydrogen technological processes. Since the two forms of hydrogen differ in energy, their physical properties also differ.

Hydrogen is a colorless, odorless, tasteless, nonirritating, and nonpoisonous gas at room temperature and under normal pressure. It is also the lightest element, approximately 1/15th as heavy as air; it rises and dissipates quickly. Hydrogen has high capacity for adsorption; it is readily attached to some substances. It is highly soluble in water, alcohol, and ether.

Under extreme pressure, for example, at the center of gas giant planets, hydrogen becomes a metal; in this case, the molecules lose their identity. The opposite situation is observed under extremely low pressure, close to vacuum: only individual atoms exist because there is no way for them to combine.

When cooled to its boiling point, 252.7°C, hydrogen becomes a transparent, odorless liquid. Liquid hydrogen is 1/14th as heavy as water, not corrosive, and particularly reactive. Hydrogen expands approximately 840 times when converted from liquid to gas. Tables 4.1 and 4.2 list some of the atomic and physical properties of hydrogen:

TABLE 4.1 Atomic properties of hydrogen

Oxidation states	Atomic radius	Atomic radius (calc.)	Covalent radius	Van der Waals radius	Ionization energy	Electron affinity	Electronegativity
Amphoteric oxide	pm	pm (Bohr radius)	pm	pm	kJ/mol	kJ/mol	Pauling scale
1,–1	25.0	53.0	37.0	120.0	1st: 100	72.8	2.2

TABLE 4.2 Physical properties of hydrogen

Property	Value
Molecular weight	
Molecular weight	2.016 g/mol
Solid phase	
Melting point	−259°C
Latent heat of fusion (1.013 bar, at triple point)	58.158 kJ/kg
Liquid phase	
Liquid density (1.013 bar at boiling point)	70.973 kg/m^3
Liquid/gas equivalent (1.013 bar and 15°C)	844 vol/vol
Boiling point (1.013 bar)	−252.8°C
Latent heat of vaporization (1.013 bar at boiling point)	454.3 kJ/kg
Critical point	
Critical temperature	−240°C
Critical pressure	12.98 bar
Critical density	30.09 kg/m^3
Triple point	
Triple point temperature	−259.3°C
Triple point pressure	0.072 bar
Gaseous phase	
Gas density (1.013 bar at boiling point)	1.312 kg/m^3
Gas density (1.013 bar and 15°C)	0.085 kg/m^3
Compressibility factor (Z) (1.013 bar and 15°C)	1.001
Specific gravity (air = 1) (1.013 bar and 21°C)	0.0696
Specific volume (1.013 bar and 21°C (70°F))	11.986 m^3/kg
Heat capacity at constant pressure (C_p) (1 bar and 25°C)	0.029 kJ/mol K
Heat capacity at constant volume (C_v) (1 bar and 25°C)	0.021 kJ/mol K
Ratio of specific heats (Gamma:C_p/C_v) (1 bar and 25°C)	1.384259
Viscosity (1.013 bar and 15°C)	0.0000865 Poise
Thermal conductivity (1.013 bar and 0°C) :	168.35 mW/m K
Others	
Solubility in water (1.013 bar and 0°C)	0.0214 vol/vol
Concentration in air	0.00005 vol%
Autoignition temperature	560°C
Diffusion	1.697 m^2/h
Flame temperature	2318°C
Flammable range	4–74% by vol in air
Heat of combustion by mass	28,670 kcal/kg
Ignition energy	0.02 mJ
Volumetric energy density	57.8 kcal/kg mo
Heat of fusion (H_2)	0.117 kJ/mol

4.1.2 Occurrence of Hydrogen

Hydrogen is the most abundant element in the universe and makes up about 90% of the atoms or 75% of the mass of the universe. Hydrogen is a major constituent in the plasma state of all visible matter in stars and galaxies, including the earth's sun. "Every chemical element heavier than hydrogen was manufactured and released into space by stars.

Our sun, like all normal stars, is powered by nuclear fusion—hydrogen nuclei combine with each other in nuclear reactions to build helium atoms and then larger ones. The temperature (15,000,000°C; 27,000,000°F) and pressure (340 billion times earth's air pressure at sea level) are so intense within the core of the sun that nuclear reactions take place. These high energy reactions create the light and heat that pour from the sun and most other stars into space. So, stars spend their lives changing hydrogen and helium to heavier elements. The planets are made mostly of matter assembled in a star (or stars) that disappeared billions of years ago. We are made of that stardust which is initially made from hydrogen!

Hydrogen is a major component of the planet Jupiter and scientists suggest that, in that planet, hydrogen is converted into metallic hydrogen because of the huge pressure in the planet's interior.

Hydrogen accounts for up 1% of earth's total mass. However, hydrogen is not commonly found in the pure form on our planet; it occurs in the free state only in volcanic gases and some natural gases. Only trace amounts of free hydrogen are found for two reasons: being the lightest element, hydrogen escapes the earth's gravity, and being extremely reactive, hydrogen forms numerous chemical compounds.

As a component of compounds, hydrogen is the 10th most abundant element on earth. These compounds include water, minerals, and hydrocarbons (compounds made of hydrogen and carbon) such as petroleum and natural gas. Water, the compound of hydrogen and oxygen (H_2O) is the most common source of hydrogen. Water covers about two-thirds of our planet. Air also contains water vapor: about 6% in humid places and about 0.1% in a desert. Water is absolutely essential to life and it is present in all living organisms. Almost 70% of our body is water. Hydrogen is also an important part of all organic matter. This includes vegetable, animal, and fossil matter. Hydrogen is in the molecules in food that provide energy: fats, proteins, and carbohydrates. Hydrogen is a major part of biomass and can take up to 14% of total mass. Hydrogen is in DNA, the molecules that code our genetic information.

4.1.3 Chemical Properties of Hydrogen

Hydrogen forms compounds with all the known elements except the noble gases. Hydrogen has an electronegativity of 2.2, so it acts either as a more nonmetallic element forming compounds with metals such as NaH or CaH_2 (they are ionic salts called hydrides, where hydrogen exists as H^- ions), or it acts as a more metallic element forming covalent bonds with nonmetals such as S, N, and halogens. Among these hydrogen compounds there are the strong acids such as HCl, H_2SO_4, and HNO_3. In this case, the H^+ ion is baring nucleus and so has a strong tendency to pull electrons to itself.

Hydrogen also forms numerous compounds with carbon, called organic compounds. Organic compounds that include carbon are the hydrocarbon fuels—methane (CH_4), ethane (C_2H_6), propane (C_3H_8), and butane (C_4H_{10})—and alcohols such as methanol (CH_3OH) and ethanol (C_2H_5OH). Among organic compounds are acids, amines, and numerous compounds that contain various heteroatoms such as sulfur, halogens, and phosphorus.

How can such a high ability of hydrogen to combine with other elements and such a large variety of chemical compounds and reactions be explained? The explanation can be found in terms of the unique $1s^1$ electron configuration of hydrogen atom. Since hydrogen has the half-filled valence shell, this circumstance allows hydrogen the possibility to gain one electron to form the complete shell of the helium configuration $1s^2$ and be H^-, the hydride ion, or lose one electron to be the proton H^+. Moreover, to complete the half-filled shell hydrogen can share electrons with other atoms, forming covalent bonds.

4.1.3.1 Hydrogen Bonding One of the very important properties of hydrogen is the possibility to provide very strong intermolecular force of attraction called hydrogen bonds. Hydrogen bonding occurs between molecules containing hydrogen directly bonded to a small, highly electronegative atom such as oxygen, nitrogen, or fluorine. In such extremely polar bonds, the hydrogen has partial positive charge (δ^+) and the other atom (O, N, or F) has partial negative charge (δ^-). When molecules are close together, their positive and negative regions are attracted to the oppositely charged regions of nearby molecules (Figure 4.2). The δ^+ H atom is attracted to a lone pair of electrons on F, O, or N atom other than the atom to which it is covalently bonded. This is possible because the hydrogen atom has no inner-shell electrons to act as a shield around its nucleus; in addition, it has a small size so it can be approached closely.

The molecule that contains a hydrogen-bonding δ^+ H atom is often referred to as the hydrogen-bond donor; the δ^- atom to which it is attracted is called the hydrogen-bond acceptor. We depict the hydrogen bond using the dashed line. Thus, hydrogen bonding is a special case of very strong dipole–dipole interactions; it influences the properties of substances in the same way as dipole–dipole interaction but to a larger degree. Typical hydrogen bond energies are in the range of 15–20 kJ/mol, which is four to five times greater than the energies of other dipole–dipole interactions.

Hydrogen bonding is responsible for the unusually high melting and boiling points of compounds such as water, alcohols, and ammonia compared with other compounds of similar molecular weight and molecular geometry.

The hydrogen bond has only 5% or so of the strength of a covalent bond. However, when many hydrogen bonds can form between two molecules, the resulting union can be sufficiently strong. As a result, in many ways, hydrogen bonding is responsible for life on earth. For example, multiple hydrogen bonds determine the shape of biologically important molecules such as proteins and nucleic acids

- hold the two strands of DNA double helix together,
- help enzymes bind to their substrate, and
- help antibodies bind to their antigen.

4.1.3.2 Chemical Reactions of Hydrogen At normal temperature hydrogen is not a very reactive substance, unless an appropriate catalyzer has activated it. At high temperature, it is very reactive; diatomic molecular hydrogen dissociates into free atoms. In the atomic state, hydrogen is a powerful reductive agent, even at ambient temperature.

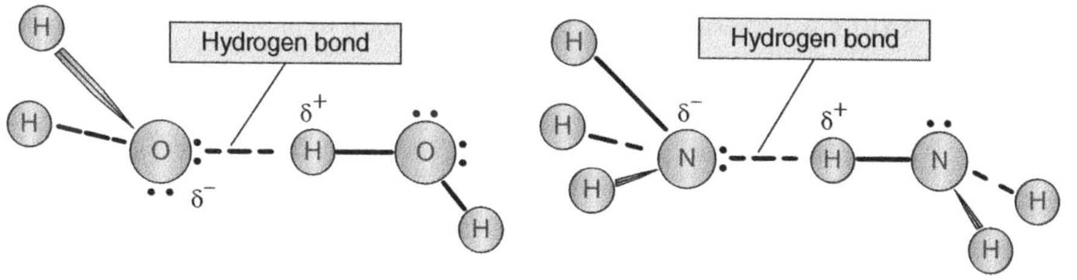

Figure 4.2 Hydrogen bonding (indicated by dashed lines) in water, (H_2O); and ammonia (NH_3).

Some important reactions of hydrogen are as follows:

1. Direct reaction of hydrogen gas with the highly reactive metals of groups IA and IIA of the periodic table leads to the formation of salts known as metal hydrides:

$$H_2(g) + 2Na(l) \rightarrow 2NaH(s)$$

$$H_2(g) + Ca(l) \rightarrow CaH_2(s)$$

2. Hydrogen is able to reduce metal oxides of many metals, such as silver, copper, lead, bismuth, and mercury, to free metals. For example, the reaction is used industrially to produce tungsten:

$$3H_2(g) + WO_3(s) \rightarrow W(s) + 3H_2O(g)$$

3. Halogen halides are formed when hydrogen reacts with the halogens:

$$H_2(g) + Cl_2(g) \rightarrow 2HCl(g)$$

4. The catalytic reduction of nitrogen by hydrogen is used industrially for the production of ammonia, an important source of fertilizers. This is the basis of the Haber process. The process runs at temperatures of 450–500°C and pressures of 35,000–40,000 kPa:

$$3H_2(g) + N_2(g) \rightleftarrows 2NH_3(g)$$

5. Atomic hydrogen reacts with unsaturated hydrocarbons, alkenes and alkynes, to form saturated alkanes. An example is reaction with ethylene:

$$H_2(g) + H_2C = CH_2(g) \rightarrow H_3C - CH_3(g)$$

6. Hydrogen reacts with pure oxygen to form water:

$$H_2(g) + O_2(g) \rightarrow 2H_2O(g)$$

This reaction is extraordinarily slow at ambient temperature; however, it goes explosively if it is accelerated by a catalyst such as platinum or a single electric spark. The ability to burn in air is the most important property of hydrogen.

Hydrogen has the highest combustion energy release per unit of weight of any commonly occurring material. It is such a powerful fuel that it is used for engines of the space shuttle. Table 4.3 details the energy content for 1 kg of hydrogen in the reaction with oxygen to form water.

TABLE 4.3 Energy content for 1 kg of hydrogen in the reaction with oxygen to form water

Higher heating value	39.3 kWh	141,600 kJ	33,800 kCal
Lower heating value	33.2 kWh	119,600 kJ	28,560 kCal

4.1.4 Health Effects of Hydrogen

There are some hazards and risks associated with hydrogen:

- Hydrogen is dangerous because it is extremely flammable if mixed with air or oxygen; heating may cause violent combustion or explosion. Hydrogen burns in air with a pale blue, almost invisible flame that makes a hydrogen fire difficult to see.
- It reacts violently with halogens and strong oxidants, causing fire and explosion. Metal catalysts, such as platinum and nickel, enhance these reactions.
- Hydrogen is not toxic; it does not cause mutagenicity.
- It can be absorbed into the body by inhalation and cause oxygen deficiency by denying the body access to oxygen. When breathing hydrogen, one may experience headaches, dizziness, drowsiness, unconsciousness, nausea, vomiting, and depression; high concentrations of hydrogen can even cause death.

Since hydrogen has no color or odor, there is no warning when toxic concentrations are present. Hydrogen concentration should be under control and measured with suitable gas detector.

4.1.5 Use of Hydrogen

There are numerous consumers of hydrogen; most important are listed as follows:

1. *The food and beverage industry*. A huge amount of hydrogen gas is used in the catalytic hydrogenation of unsaturated vegetable oils (fatty acids) to obtain solid fat. For example, margarine is produced when two-thirds of the double bonds present in vegetable oil are hydrogenated.
2. *Ammonia synthesis*. The Haber ammonia process is the most important use of hydrogen.
3. *Manufacture of chemical compounds*. Hydrogen is used in large quantities as a raw material in the chemical synthesis of methanol, hydrogen peroxide, hydrochloric acid, polymers, and solvents.
4. *The pharmaceutical industry*. Hydrogen is used to manufacture vitamins and other pharmaceutical products.
5. *Glass, cement, and lime*. In combination with nitrogen, hydrogen is used as a reductive agent in the float glass process, for example, to prevent oxidation of the large tin bath. As a component of oxyhydrogen flame, hydrogen is also used for heat treatment of glass preforms.
6. *The metal industry*. Hydrogen is mixed with inert gases to obtain a reductive atmosphere for heat treating steel.
7. *Oil and gas*. The use of hydrogen is extending quickly in fuel refinement: desulfurization of fuel-oil and gasoline and breaking down by hydrogen (hydrocracking) to convert heavy and unsaturated compounds to lighter and more stable compounds.

8. *Welding, cutting, and coating*. Hydrogen is used for heat treatment of various metals; it is often used in annealing stainless steel alloys, magnetic steel alloys, sintering, and copper brazing.
9. *Electronics*. Hydrogen diluted in nitrogen is used as carrier gas to eliminate oxygen in high-temperature semiconductor processes, in the manufacturing of semiconducting layers in integrated circuits, especially for silicon deposition or crystal growing.
10. *Laboratories and analysis*. Hydrogen is used as carrier gas in gas chromatography and various analytical instrument applications as a fuel component of combustion gases for flame ionization and flame photometric detectors.
11. *Space and aeronautics*. Huge quantities of hydrogen are consumed in the liquid state as rocket fuels and as a rocket propellant propelled by nuclear energy.
12. *Automotive and transportation*. Hydrogen can be burned in internal combustion engines being a carbon-free energy source.
13. *Miscellaneous*.
    - Hydrogen gas is used for filling balloons since it is much lighter than air.
    - Hydrogen is used for proton–proton reactions and carbon–nitrogen cycle.
    - Liquid hydrogen is used in the study of superconductivity because its melting point is close to absolute zero.
    - Deuterium is used as a moderator to slow down neutrons and as a tracer.
    - Tritium is used in the production of hydrogen (fusion) bomb.
    - Tritium is also used in making luminous paints and as a tracer.

As we can see Figure 4.3, the largest consumers of hydrogen today are the production of ammonia and methanol, and oil refining. However, much has been said about hydrogen being the "fuel of future" due to its abundance and nature. Considering the physical and chemical properties of hydrogen, we can notice that the most important properties affecting its use as a fuel are as follows:

- Hydrogen combines with oxygen to form water, releasing heat and energy.
- It has a high energy content per weight (almost three times as much as gasoline).
- Hydrogen is highly flammable: a small amount of energy is needed to ignite it; hydrogen is the most flammable of all the known substances.
- It has a broad flammable range: it can burn when there is 4–74% of hydrogen in the air by volume.
- The combustion of hydrogen makes no environmental impact since it does not produce carbon dioxide, acid rain, or any emissions.
- Modern technology has made it possible to transport large quantities of hydrogen in liquefied form at temperature close to absolute zero. Hydrogen can also be absorbed into metal hydrides to be transported.
- Hydrogen can be produced from renewable resources such as reforming ethanol or electrolysis of water.

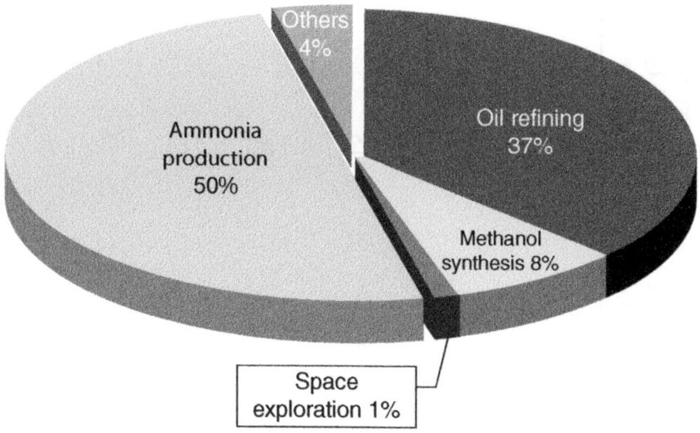

Figure 4.3 The largest consumers of hydrogen today.

4.2 HYDROGEN AS AN ENERGY CARRIER

Similar to electricity, hydrogen is not a source of energy rather it is an energy carrier that is produced from naturally occurring sources. An energy carrier is a substance that moves energy in a usable form from the place of production to the place of utilization. Electricity is a classical example of an energy carrier that shifts the energy condensed in fossil, nuclear, and other sources from the production location to industrial and private consumers. It is much more convenient to utilize energy in a transformed electrical form rather than in the form of the original source. Unlike electricity, large quantities of hydrogen can be stored for the future use or moved to the place of consumption.

4.2.1 Comparison to Other Fuels

To evaluate the energy content of commonly used fuels, we can mention that one gallon of gasoline energy is equivalent to 380 SCF H_2 (10 Nm^3) or 113 SCF (3 Nm^3) of natural gas or 8.3 lb of coal.

The comparison of hydrogen properties to other fuels raises major concerns regarding its storage and transportation. Hydrogen's massive volume for a given amount of energy far exceeds that of gasoline, natural gas, and any other possible fuel with the same energy content. Table 4.4 shows this comparison for the major physical parameters. The specific energy values of compressed gases and gasoline are presented in Table 4.5.

As can be seen from Table 4.5, hydrogen, in comparison per unit volume with other fuel sources, performs unsatisfactorily. For instance, hydrogen at 1000 psi pressure delivers only 2.5% of the energy delivered by gasoline. At 5000 psi pipeline pressure, hydrogen would carry about 114,285 BTU/ft^3, which is about 5 times higher than at 1000 psi and 10 times higher than at 10,000 psi. Utilization of such high pressure in pipelines along with requirements for high hydrogen volume will require larger pipe diameters. In addition, we have to take into account the supplemental engineering and equipment costs plus complexity related to pipelines installation in urban areas.

TABLE 4.4 Comparative fuels properties

Substance	Density	Main constituent (%/wt.)	Boiling point (°C)	Latent heat of vaporization (kJ//kg)	Specific heating value (MJ/kg)	Ignition temperature (°C)	Lower ignition limit % by gas volume in air	Upper ignition limit % by gas volume in air
Regular gasoline	0.715 … 0.765 kg/liter	86 C 14 H	25 … 215	380 … 500	42.7	~300	~0.6	~8.0
Diesel fuel	0.81 0.815 … 0.855 kg/liter	86 C 13 H	180 … 360	~250	42.5	~250	~0.6	~6.6°
Crude oil	0.70 … 1.0 kg/liter	80 … 83 C 10 … 14 H	25 … 360	222 … 352	339.8 … 46.1	~220	~0.6	~6.5
Ethanol (C_2H_5OH)	0.79 kg/liter	52 C 13 H 35 O	78	904	26.8	420	3.5	15
Methanol (CH_3OH)	0.79 kg/L	38 12 H 50 O	65	1110	19.7	450	5.5	26
Natural gas	~0.83 kg/m^3	76 C 24 H	−162		47.7	—	—	—
Hydrogen (H_2)	0.09 kg/m^3	100 H	−253		120	560	4	77
Methane (CH_4)	0.72 kg/m^3	75 C 25 H	−162		50	650	5	15
Propane (C_3H_8)	2 kg/m^3	82 C 18 H	−43		46.3	470	1.9	9.5

TABLE 4.5 Specific energy value of compressed gases and gasoline

Fuel	BTU/ft^3 at STP	BTU/ft^3 1,000 psi	BTU/ft^3 5,000 psi	BTU/ft^3 10,000 psi
Hydrogen	320	22,857	~114,300	~230,000
Methane	985	70,357		~680,000
Propane (gas)	2,450	~175,000		
Propane (noncompressible liquid)	782,000	782,000	782,000	782,000
Gasoline (noncompressible liquid)	894,740	894,740	894,740	894,740

4.3 HYDROGEN STORAGE

If hydrogen has to be used on a large-scale basis, the storage of hydrogen becomes a crucial issue for mobility and transport application. Fuel cell technology would be practical if hydrogen could be stored in a safe, efficient, compact, and economic manner.

The difficulties arise because, although hydrogen has one of the highest specific energies (energy per kilogram), its density is very low: it has one of the lowest energy densities (energy per cubic meter). For example, the volume of 90 g of hydrogen is 1 m^3 (Figure 4.4). For this reason, existing chemical and petroleum industries mostly use hydrogen close to its point of manufacture, and hydrogen is always transported in a compressed or condensed state as a gas or liquid. Evidently, the small hydrogen distribution infrastructure that exists today cannot support the distribution or scaling needs of the hydrogen economy.

The question is whether the infrastructure of pipelines of the United States that are used to transport gasoline, natural gas, or propane can be used for transport of compressed hydrogen, and what alternative ways could be found. Comparative data for other fuels (see Table 4.5) shipped by pipeline shows that hydrogen performs very inadequately, possessing the lowest volumetric efficiency while requiring the most thermal and pressure conditioning. Even at 5000 psi pressure, a hydrogen pipeline will carry only 12.8% of the equivalent gasoline line.

It has been suggested that, rather than pipelining the gas across the country, hydrogen-fueling stations should be developed at the production place. Hydrogen could be stored as a compressed gas in pressure tanks in underground cavities and as a liquid in superinsulated tanks. Besides compressed gas and liquid hydrogen, there are some alternative states of

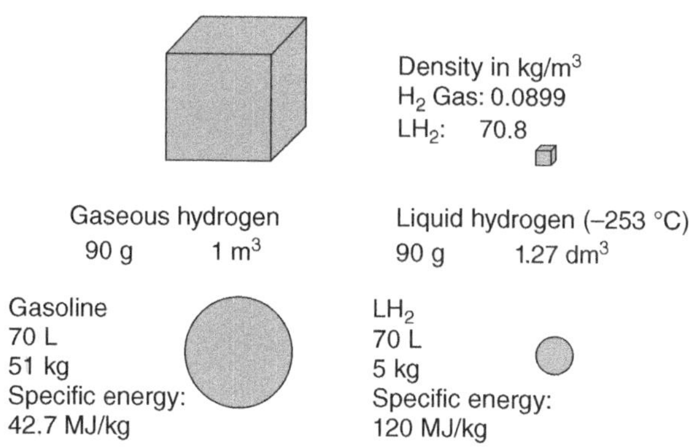

Figure 4.4 Hydrogen/gasoline properties.

hydrogen in which it can be stored, as we will discuss in the following paragraphs. Storing hydrogen at the production plant or fueling station shows the potential to store other kinds of energy—solar, wind, biological, nuclear, and fossil fuel—for later use. The local production plant would require further distribution of hydrogen by tanker trucks. The fueling station would be able to supply hydrogen for direct use for a vehicle in the form of hydrogen contained in tanks that will be placed onboard a vehicle (e.g., an automobile or bus).

Hydrogen-powered vehicles require a driving range of greater than 300 miles in order to meet customer requirements and compete effectively with other automotive technologies. Achieving such a driving distance requires that approximately 4–5 kg of hydrogen must be stored onboard the fuel cell–powered passenger vehicle. Storage of this quantity of hydrogen within a vehicle's weight, volume, and system faces cost constraints in parallel with major scientific and engineering challenges. Development of a lightweight hydrogen storage system is essential for onboard automotive applications.

There are basically three storage options for hydrogen that we will consider:

- Hydrogen may be compressed and stored in a pressure tank.
- Hydrogen may be cooled to a liquid state and kept cold in a properly insulated tank.
- Hydrogen may be stored chemically in a solid compound.

4.3.1 Storage of Hydrogen as a Compressed Gas

Storing hydrogen under pressure has been done successfully for many years. This method is broadly accepted for transportation of the small amounts of gas. Cylinders (tanks) in a wide range of sizes are supplied today to thousands of industrial and research establishments. It may be used, for example, for storing the hydrogen that is a product from electrolyzers. Hydrogen light-weight polymer tanks are shown in Figure 4.5. The tanks filled with compressed hydrogen can be held in the vehicle, typically on the roof space of buses or in trunks of the passengers' cars.

A high-pressure compressed gas system consists of a cylindrical tank, a pressure regulator that reduces the pressure of the compressed hydrogen to a lower value for delivery of the hydrogen to the propulsion system (fuel cell or ICE), gas flow control

Figure 4.5 Hydrogen light-weight polymer tanks.

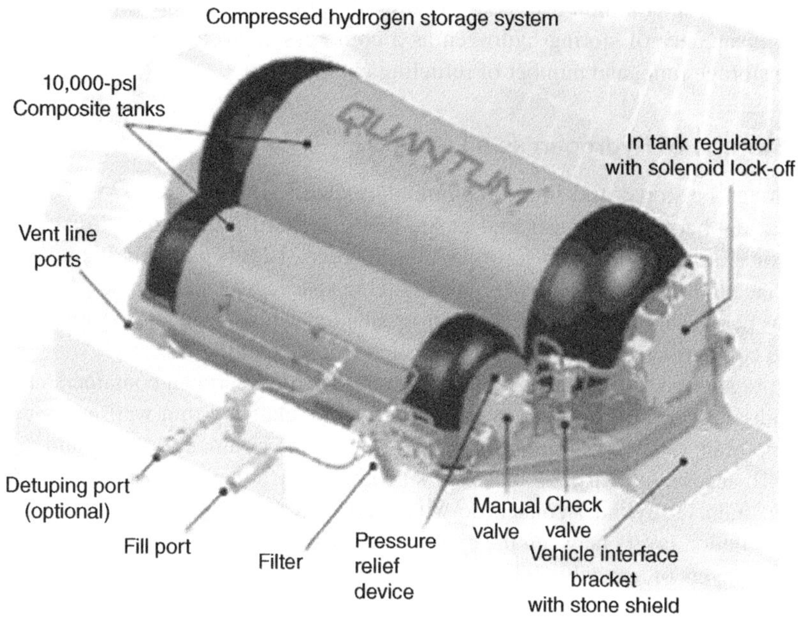

Figure 4.6 Typical high-pressure hydrogen storage system. [*Source:* Roadmap on Manufacturing R&D for the Hydrogen Economy. Based on the Results of the Workshop on Manufacturing R&D for the Hydrogen Economy Washington, D.C. July 13–14, 2005 and http://www.1.eere.energy.gov/hydrogenandfuelcells/storage/hydrogen_storage.html.]

valves, tubing, mounting brackets, and some environmental protection. A representation of such a high-pressure hydrogen storage system is shown in Figure 4.6.

For remote and transportation applications, the storage system should consist of containers that are manufactured at low cost and with minimal weight. The material that the pressure vessel is made of is one of key problems. Hydrogen is a very small molecule, has a high velocity, and is capable of diffusing into materials that are impermeable to others gases. Penetrated hydrogen can react with carbon, which is a component of steel, and, as a result, buildup of internal blistering causes a decrease in strength of the walls (called hydrogen-induced cracking). To solve this problem, certain chromium-rich steels and alloys have been found. Composite-reinforced plastic materials are also used for larger tanks. Most high-pressure hydrogen storage systems in use today operate at 350 bar (5000 psi); some tanks made of advanced carbon composites are available that can operate at 700 bar (10,000 psi) at temperatures ranging from −30° to +50°C. The hydrogen content is typically of 45 m^3.

The three main types of tanks are as follows:

- Steel
- Aluminum core encased with composites based on fiberglass, carbon, or aramide fiber
- Plastic core encased with composites based on fiberglass, carbon, or aramide fiber.

In the composites, the fiber is impregnated with a resin to form a continuous matrix. There are also safety problems associated with storing hydrogen at high pressure; hydrogen, like all fuels, must be carefully handled. However, there is no more danger than any other flammable gases in common use today. Taking all things into account, it is obvious that the main advantages of storing hydrogen as a compressed gas are simplicity, practically indefinite storage time, and number of refueling cycles.

4.3.2 Storage of Hydrogen as a Liquid

Hydrogen can be stored as a liquid (commonly called LH_2) at 20 K (−253°C) and pressures of a few bars in superinsulated tanks. LH_2 has been produced and distributed for various kinds of industrial needs for more than 70 years. LH_2 is particularly interesting for long-distance transportation purposes and as fuel in spacecraft and airplanes. For example, NASA has been using hydrogen in space programs for several decades. Figure 4.7 demonstrates the comparison data for LH_2 and gasoline storage parameters.

The storage of LH_2 requires special equipment. Liquid hydrogen containers are generally cylindrical, and are currently manufactured with relatively thin wall metal alloys. It is very important to keep heat input from the external environment to the liquid as low as possible. To accomplish this, the tanks are surrounded by a thermal insulation barrier that is a larger diameter cylinder concentric with the liquid container, with high vacuum and thermal insulation layers between the cylinder walls. The multilayer insulation consists of a number of layers of a metallic foil with a thin layer of glass wool between each foil layer. A typical liquid hydrogen container is shown in Figure 4.8.

The design and dimensions of an LH_2 storage system ultimately depend on its destination, whether in an airplane, a bus, a truck, or a passenger car. As an example, following are some characteristics of a Messer liquid hydrogen storage system for an LH_2 vehicle:

Hydrogen content	5 kg
Temperature	20 K
Pressure_{max}	0.6 MPa
Gross volume	130 liter
Net volume (liquid hydrogen)	71 liter
Tank mass	50 kg
Auxiliary mass	40 kg
Coolant flow rate	100 liter/h

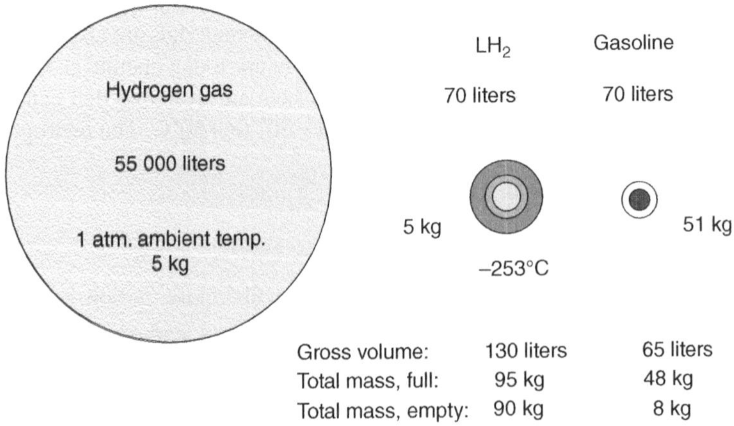

Figure 4.7 Comparison data for LH_2 and gasoline storage.

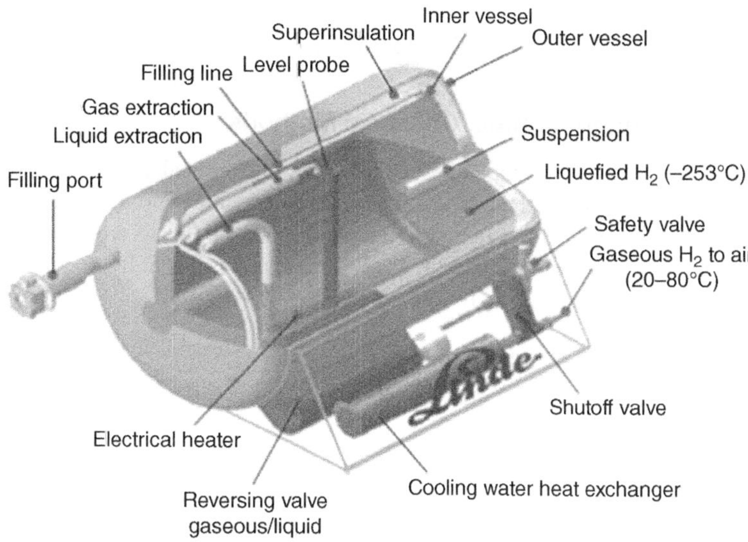

Figure 4.8 Liquid hydrogen storage system. [*Source*: From http://www.1.eere.energy.gov/hydrogenandfuelcells/storage/hydrogen_storage.html.]

To supply vehicles with LH_2, a suitable fuel station is necessary. This station needs to be connected to an LH_2 storage vessel and must have the proper control and dispensers to transfer the fuel from the storage vessel to the vehicle tank. For an LH_2 vehicle tank system with fuel capacity of about 100 L, the feeling process takes less than 2 min.

There are some problems with LH_2 storage:

1. The cooling process requires a great deal of energy equal to 30–40% of that in the fuel needed, so it is important to develop new efficient cooling processes that would cut the energy use. Currently, LH_2 is the most expensive hydrogen state of aggregation.
2. Tank manufacturing is expensive.
3. Boil off gas that results in the following:
   - Extra measures required to avoid H_2 release to the surrounding.
   - Reduction of tank energy during parking (4% per day; a full tank can be empty in less than 4 weeks). The car has to be towed to a filling station; refueling the warm tank takes longer time and causes energy loss.
4. Infrastructure requires a large investment to achieve safe and convenient refueling. The transfer of LH_2 requires special lines that have to be adequately insulated; insulation methods are similar to those used in the storage system.
5. Filling and withdrawing requires additional safety cautions: when the LH_2 tank is being filled, and when fuel is being withdrawn, it is important that air not be allowed into the system; otherwise, an explosive mixture could form. The tank should be purged with nitrogen before filling.

The comparison of the LH_2 storage system with the compressed hydrogen storage system in terms of the weight of hydrogen relative to carrier weight shows the benefit of LH_2 use. For example, a trailer for LH_2 with a total weight 40 tons can have a hydrogen load of 3370 kg, whereas a trailer for compressed hydrogen with the same total weight, 40 tons, can have a hydrogen load of only 530 kg at 20 MPa.

4.3.3 Solid Hydrogen Storage, Chemical Methods

Hydrogen can also be transported, and especially stored, chemically. Chemical-based and solid-state hydrogen storage methods use materials that retain hydrogen, which can subsequently be released by heating or via a chemical reaction. There are more than 1000 chemical compounds that can hold, for their mass, large quantities of hydrogen. To be useful, these compounds must pass the following tests:

- They have to give up their hydrogen very easily.
- The manufacturing process must be simple and use little energy so as not to be expensive.
- They must be safe to handle.

In other words, to reach this goal a number of vital objectives have to be achieved, and these include a high wt% and volume density of stored hydrogen (particularly for mobile applications), attractive absorption (or adsorption) and desorption kinetics at convenient temperatures and pressures, and cheap and readily available material. Long-term resistance to poisoning by trace impurities in the hydrogen streams is another very critical hurdle. Some of the perspective compounds are listed in Table 4.6.

4.3.3.1 Reversible Metal Hydride Hydrogen Store Some metals, such as titanium, iron, manganese, nickel, and chromium, and their mixtures (alloys) can react with hydrogen to form metal hydrides. This property of hydrogen on Pd, for example, has been used in the construction of hydrogen sensors; hydrogen absorption by Pd is very efficient and stable. Some of the compounds are also used in such products as nickel-metal hydride batteries, which can be found in cell phones and computers. For large-scale storage, other metals and alloys have been investigated. Due to their high volumetric density, metal hydrides are attractive for the storage of hydrogen in fuel cell–driven cars.

The reaction of hydrogen absorption by metal or metallic alloy is reversible and based on the general equation:

$$AB_z + xH_2 \rightarrow AB_zH_{2x} + \Delta Q$$

where A and B are metals (AB_z is an alloy); ΔQ is the heat released upon absorption of hydrogen. The heat ΔQ is usually characterized by the enthalpy ΔH, determined from the pressure–composition–temperature isotherms. To the right, the reaction is slightly exothermic. To release the hydrogen, a small amount of heat must be supplied and the reaction becomes endothermic.

The reaction of hydrogen absorption is usually conducted on an alloy placed into a pressurized vessel and takes a few minutes. The vessel containing the metal hydride is kept sealed, and is connected to the hydrogen-consuming equipment. Fueling might typically take place at a refueling station: the hydrogen would be pumped into the car much the same as gasoline, filling the storage material in several minutes. In the car, the system can be warmed slightly to increase the rate of hydrogen supply, using, for example, warm water. Typical system architecture of metal hydride–hydrogen storage with high temperature metal hydrates is shown in Figure 4.9.

TABLE 4.6 Hydrogen storage material

Name	Formula	Percent hydrogen	Specific gravity	Vol.(L) to store 1 kg H_2	Notes
Simple hydrides					
Liquid H_2	H_2	100	0.07	14	Cold, −252°C
Lithium hydride	LiH	12.68	0.82	6.5	Caustic
Beryllium hydride	BeH_2	18.28	0.67	8.2	Very toxic
Diborane	B_2H_6	21.86	0.417	11	Toxic
Liquid methane	CH_4	25.13	0.415	9.6	Cold −175°C
Ammonia	NH_3	17.76	0.817	6.7	Toxic, 100 ppm
Water	H_2O	11.19	1.0	8.9	
Sodium hydride	NaH	4.3	0.92	25.9	Caustic, but cheap
Calcium hydride	CaH_2	5.0	1.9	11	
Aluminum hydride	AlH_3	10.8	1.3	7.1	
Silane	SiH_4	12.55	0.68	12	Toxic 0.1 ppm
Potassium hydride	KH	2.51	1.47	27.1	Caustic
Titanium hydride	TiH_2	4.40	3.9	5.8	
Complex hydrides					
Lithium borohydride	$LiBH_4$	18.51	0.666	8.1	Mild toxicity
Aluminum borohydride	$Al(BH_4)_3$	16.91	0.545	11	Mild toxicity
Lithium aluminum hydride	$LiAlH_4$	10.62	0.917	10	
Hydrazine	N_2H_4	12.58	1.011	7.8	Toxic 10 ppm
Hydrogen absorbers					
Palladium hydride	$PdH_{0.7}$	0.471	10.78	20	
Titanium iron hydride	$TiFeH_2$	1.87	5.47	9.8	

Source: Larmine (2000) [34]. Reproduced with permission of John Wiley & Sons.

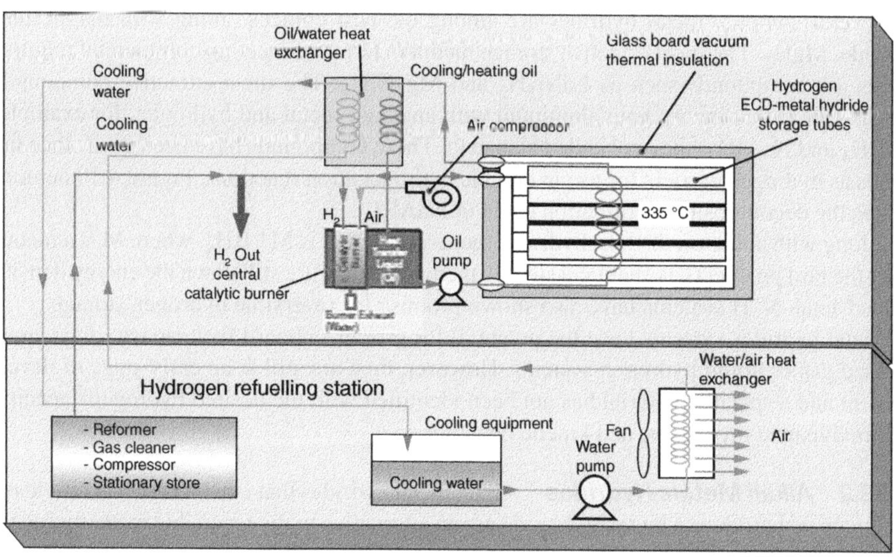

Figure 4.9 Metal hydride–hydrogen storage.

TABLE 4.7 Key properties of metal hydrides suitable for gas-phase applications

Metal hydrates parameters for hydrogen storage					
Hydrate phase	Max. H_2 capacity (wt%)	Reversible H_2 capacity (wt%)	Desorption pressure (bar)	T (°K)	Enthalpy (kJ/mol H_2)
MgH_2	7.66	<7.0	$\sim 10^{-6}$	552	74.5
VH_2	3.81	1.9	2.1	285	40.1
$MgNiH_4$	3.59	3.3	$\sim 10^{-5}$	528	64.5
$TiFeH_2$	1.89	1.5	4.1	265	28.1
$ZrNiH_3$	1.96	1.1	$\sim 5 \times 10^{-6}$	573	68.6
$TiMn_{1.4}V_{0.62}H_{3.4}$	2.15	1.1	3.6	268	28.6
$ZrM_{n2}H_{3.6}$	1.77	0.9	0.001	440	53.2
$LaNi_5H_{6.5}$	1.49	1.28	1.8	285	30.8
$LaNi_{4.8}Sn_{0.2}H_{6.0}$	1.40	1.24	0.5	312	32.8

Key properties of metal hydrides suitable for gas-phase applications are listed in Table 4.7. Typically, metal A is an early transition metal, rare-earth metal, or Mg and forms stable binary hydrides. The second metal in an alloy, B (e.g., Ni, Co, Cr, Fe, Mn, or Al) does not form stable hydrides, although it may help dissociate the H_2 molecule during the sorption. Table 4.7 describes representative hydrides from the five metal hydride families: A, A_2B, AB, AB_2, and AB_5.

The key hydrogen-storage properties, total hydrogen capacity and reversible portion, are usually presented as weight percentage of hydrogen. Other important parameters for metal hydrides to be considered as fuel storage are the pressure and temperature required to provide hydrogen gas. It is generally accepted that hydrides should provide hydrogen gas at 1–10 bar and over the temperature range from 270 K to 360–600 K, depending on the availability of a source of heat. The research goal is to develop the hydrate combination that will be operational at atmospheric or about atmospheric pressure and room temperature.

Several complex metal hydrides are among the best choices, along with magnesium hydride, MgH_2. The most effective storage media in terms of meeting commercial requirements are compounds such as $LaNi_5H_7$ and Mg_2NiH_4. The most attractive compounds are also those that incorporate aluminum with an alkali metal and hydrogen, for example, $LiAlH_4$ and $NaAlH_4$ (they are called alanates). These compounds have been used since the 1950s as hydrogen carriers for use in organic hydrogenation reactions. Figure 4.10 demonstrates the decomposition–formation cycle of $NaAlH_4$.

Along with alanates, the borohydrides (general formula is $M^+ BH_4$, where M is a metal) offer the best prospects as chemical storage media, but they are still down the energy density scale. Metal–N–H systems have also shown promise for reversible hydrogen storage.

Metal hydrides systems have the potential for greater onboard fuel capacity than compressed gas or liquid hydrogen systems. However, they are still at an early stage of development and a specific material has not been identified with the desired hydrogen capacity, thermodynamic properties, and kinetic behavior.

4.3.3.2 *Alkali Metals Hydrides*

Alkali metal hydrides that react with water to release hydrogen and produce a metal hydroxide are an alternative to the reversible metal hydrides. Some of these are shown in Table 4.6. As an example, there is a system using calcium hydride (the method was described in 1990), for which the reaction is

$$CaH_2 + 2H_2O \rightarrow Ca(OH)_2 + 2H_2$$

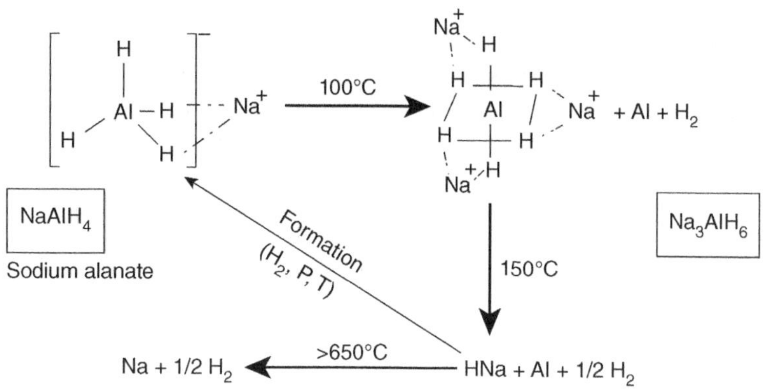

Figure 4.10 Decomposition–formation cycle of $NaAlH_4$.

The method that is used commercially (Powerballs) is based on the reaction of sodium hydride with water:

$$NaH + H_2O \rightarrow NaOH + H_2$$

Sodium hydride is supplied in the form of polyethylene-coated spheres of about 3 cm diameter. These spheres named Powerballs are stored under water and when necessary cut in half to allow sodium hydride to react with water and produce hydrogen; the cutting process is controlled with microprocessors to provide hydrogen when it is needed.

Compared to the methods we have considered so far, this is a very simple way of producing hydrogen with high energy density. Sodium hydride is not expensive. It shows serious promise for NaOH usage in fuel cells for automotive applications because of its high hydrogen storage potential relative to other methods. The major problem is the need to dispose of a corrosive mixture of hydroxide and water.

4.3.3.3 Carbon Nanostructures It has been shown that carbon-based nanomaterials (carbon nanotubes, carbon fibers), allowing a wide range of pore size, shape, and distribution, can be the promising contenders for hydrogen storage. Molecular engineering of the pore structure and surface chemistry of these carbons have opened up this new potential application. The possibility for hydrogen to be absorbed in nanotubes at room temperature, with only moderate overpressures and desertion upon small heating conditions, has initiated intensive investigation into nanotube technology. According to the U.S. Department of Energy, a carbon material needs to store 6.5% of its own weight in hydrogen to make fuel cells practical in cars. Researchers at MIT claim to have produced nanotube clusters with the ability to store 4.2% of their own weight in hydrogen. In 1997, NREL, National Renewable Energy Laboratory researchers demonstrated that single-walled nanotubes (SWNTs) are capable of storing hydrogen in the 5–10 wt% range. More recent work at NREL has shown that SWNTs can absorb up to 8 wt% of hydrogen when catalytic metal species are present. Some researcher reported that materials such as single-walled carbon nanotubes demonstrate hydrogen adsorption rates of approximately 5–10 wt% and higher.

Carbon nanotubes are typically produced in bundles that are lightweight and have a high density of small, uniform, cylindrical pores (individual nanotubes). SWNTs are formed from a single graphite layer; multiwalled nanotubes (MWNTs) consist of multiple concentric graphite layers. Properties of carbon nanotubes include a high surface area ranging from 50 to over 2000 m^2/g, and an ability to conduct electricity (metallic or semiconducting).

Mechanically, they are as strong as steel, light as aluminum, and elastic as plastic. They can have a length to diameter ratio of more than 100 million to 1. The diameters of SWNTs vary from 0.671 to 3 nm, whereas MWNTs usually have diameters of 30–50 nm. Carbon fibers differ from nanotubes in their inner structure: carbon fibers consist of graphite platelets stacked together, with interlayer spacing of 0.3355 nm. Figures 4.11 and 4.12 demonstrate the SWNT's and MWNT's laser-generated structures.

There are various methods of production of carbon nanotubes:

- Electric discharge method for SWNT and MWNT production—carbon nanotubes are produced from vapors of carbon containing a small amount of catalysts. The process takes place inside a stainless steel chamber filled with helium gas at low pressure (500 torr).
- Pulsed laser vaporation or laser ablation—the process of removing material by irradiating it with a laser beam for SWNT production. Carbon nanotubes are produced by pulsed laser vaporation of carbon containing metal catalysts.
- High-pressure CO Conversion (HiPCO)—synthesis of single-wall carbon nanotubes.
- Chemical vapor deposition (CVD)—catalytic decomposition of hydrocarbon gases is used for SWNTs and MWNT production in vacuum or at atmospheric pressure. CVD, thermal CVD, and microwave plasma–enhanced CVD is used. The method allows control of diameters and lengths of nanotubes with changing, for example, CH_4/H_2 gas ratio and growing time.

The hydrogen interaction with carbonaceous materials is based on van der Waals attractive forces (physisorption), or the overlap of the highest occupied molecular orbitals of

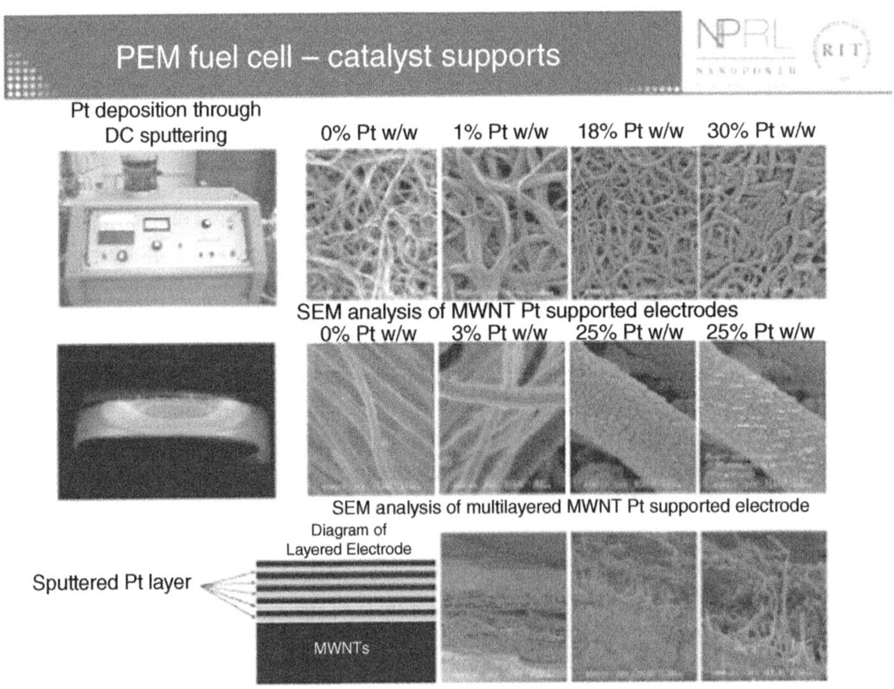

Figure 4.11 Multiwalled nanotubes structure. [*Source*: Reproduced with kind permission of RIT Nanopower Lab.]

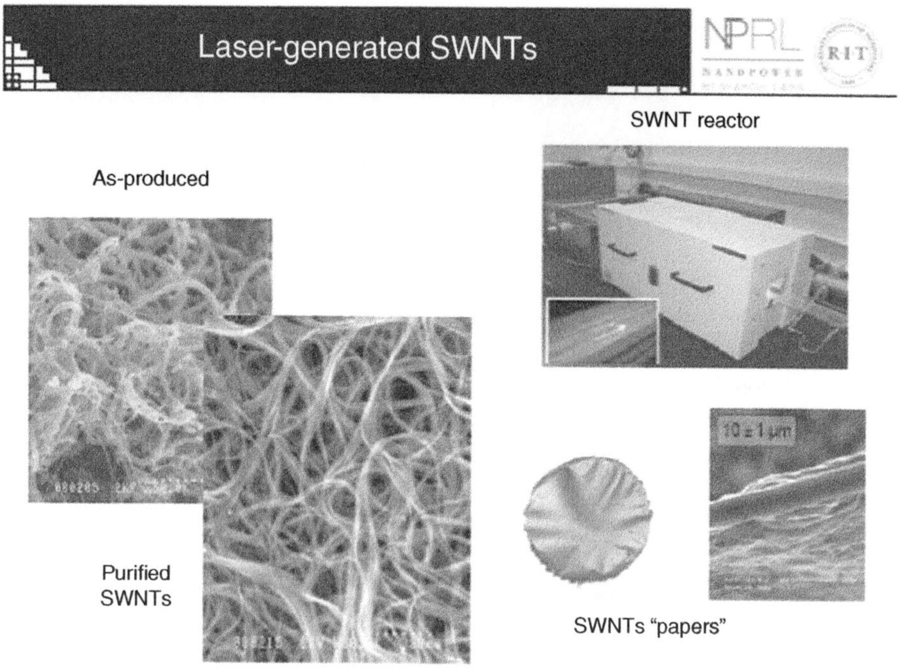

Figure 4.12 Single-wall nanotubes structure. [*Source*: Reproduced with kind permission of RIT Nanopower Lab.]

carbon with the hydrogen electron, overcoming the activation-energy barrier for hydrogen dissociation (chemisorptions). The crucial question is this: Can nanotubes store and release practical amounts of hydrogen under reasonable conditions? Experiments on hydrogen absorption and desorption in carbon nanostructures were carried out at many research facilities. Experiments at MIT showed a hydrogen storage capacity of 4.2 wt% was achieved at room temperature under pressure of about 10 MPa. Under ambient pressure, 78.3% of the absorbed hydrogen may be released at room temperature. About 20% of the absorbed hydrogen remained in the sample after desorption at room temperature and required some heating.

Nanostructured carbonaceous materials offer much promise for hydrogen storage and must therefore be investigated further (Figure 4.13).

4.3.3.4 New Technologies Currently, several approaches for storing hydrogen are being pursued, which have found that the sorbent material, capable of reversible uptake and release, offers key advantages if suitable gravimetric and volumetric uptake can be achieved. It has been determined that, while hydrogen–carbon interaction is too weak, the metal–hydrogen interaction is too strong for room temperature reversible storage. There are some novel ways to overcome this difficulty by forming artificial structures.

Promising new classes of sorbing materials are as follows:

- Nanotubes "activated" by the alloy particles incorporated into the carbon nanotubes. As a result, they absorb more hydrogen than would be in the absence of the metal. It is believed that metal, for example, titanium, may assist hydrogen uptake by a catalytic effect.
- Nanoporous metal-organic framework. These are materials consisting of metal clusters linked by organic linkers. This class of extended solids possesses extremely

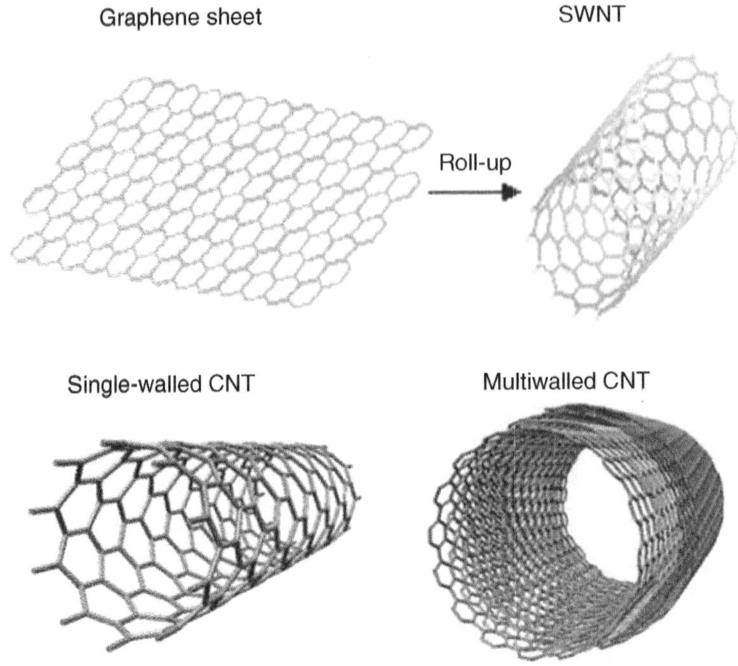

Figure 4.13 SWNT and MWNT are allotropes of carbon.

high surface areas with tunable pore sizes and has been suggested as a potential hydrogen storage medium. An example is ZnO $(BDC)_3$, where BDC is 1,4-benzenedicarboxylate. The hydrogen uptake of this compound at 30 K approached 10 wt%.

- Nanoscrolls. These are new pure carbon nanostructures formed by graphite sheets rolled into spirals. They have a great advantage over conventional nanotubes, in part due to their great superficial area and free internal volume.

While progress is being made, an ideal technology for hydrogen storage is not yet available.

MULTIPLE CHOICE QUESTIONS

Section 4.1

1. A neutral hydrogen atom has how many valence electrons?
   (a) 2
   (b) 0
   (c) 1
   (d) 3
2. The statement that best describes the formation of a chemical bond by hydrogen:
   (a) Electrons are transferred from hydrogen to a metal, and the resulting charged particles form a crystalline network.

(b) Electrons are transferred from a metal to hydrogen, and the resulting charged particles form a crystalline network.
(c) Electrons are shared between hydrogen and another atom and discrete molecules are formed.
(d) All of the above

3. What physical state hydrogen can exist?
(a) Gas
(b) Liquid
(c) Plasma
(d) Metallic
(e) All of the above

4. What is unique about hydrogen as an element?
(a) It is all around us.
(b) Has only one proton and one electron in the atomic structure
(c) Only exist in a gaseous state

5. In what form mostly does hydrogen exist on earth?
(a) Diatomic molecules
(b) Individual atoms
(c) Gas
(d) A component of compounds

Section 4.2

1. Why is liquid hydrogen used to power space rockets?
(a) Produces very high temperature when oxidized by liquid oxygen
(b) Hydrogen has the highest combustion energy release per unit of weight
(c) Easy to transport in a rocket fuel tank
(d) Only one type of material that can burn in a space

2. What is the approximate specific energy ratio for liquid hydrogen and gasoline at normal conditions?
(a) 12.3
(b) 8.6
(c) 2.8
(d) 1.3

Section 4.3

1. Which of the hydrogen storage types have highest gravimetric density (H_2 mass to the mass of storage)?
(a) Liquid hydrogen storage
(b) Compressed hydrogen gas storage
(c) Alkali metal–hydride storage
(d) Reversible metal hydride hydrogen storage

2. What type of energy is required to store and release hydrogen in "Reversible metal hydride storage?
   (a) Microwave
   (b) Heat
   (c) Electrical
   (d) Radiation
3. What advanced type of material is used for compressed hydrogen storage at high pressures?
   (a) Steel cylinder
   (b) Aluminum core encapsulated in a carbon fiber casing
   (c) Thick-wall plastic cylinder
   (d) Titanium tube welded at the ends
4. What are the main advantages of storing compressed hydrogen?
   (a) Simplicity
   (b) Practically indefinite storage time
   (c) Large number of refueling cycles
   (d) All of the above
5. What is a major disadvantage of storing liquid hydrogen (LH2) for a long period of time?
   (a) Very low temperature
   (b) Complex double-walled storage tank
   (c) Loss of hydrogen due to the "boil-off" (LH2 boils when stored and gas must be vented)
   (d) Complicated delivery and refueling the tanks
6. What type of reaction becomes when hydrogen is released in reversible metal hydride hydrogen storage?
   (a) Exothermic
   (b) Endothermic
   (c) Chemical
   (d) None of the above
7. What type of reaction is used commercially (Powerball) in alkali metal hydride storage?
   (a) Reaction of sodium hydride with water
   (b) Reaction of sodium with acid
   (c) Reaction of metal hydride with sodium
   (d) Reaction of sodium hydride with alcohol
8. What percentage of hydrogen need to be stored in carbon-based nanomaterials (carbon nanotubes, carbon fibers) to become practical for automotive applications?
   (a) 1–2% of its own weight
   (b) 50–60% of its own weight
   (c) 34% of its own weight
   (d) 6.5% of its own weight

9. What is a main criterion had to be used in selection of metal hydrides for storage?
   (a) They have to give up their hydrogen very easily.
   (b) The manufacturing process must be simple and use little energy so be not expensive.
   (c) They must be safe to handle.
   (d) All of the above

BIBLIOGRAPHY

Section 4.1

1. Rigden, J.S. (2002) *Hydrogen: The Essential Element*, Harvard University Press, Cambridge, MA.
2. Newton, D.E. (1994) *The Chemical Elements*, Franklin Watts, New York.
3. Stwertka, A.A. (2002) *Guide to the Elements*, Oxford University Press, New York.
4. Krebs, R.E. (1998) *The History and Use of Our Earth's Chemical Elements: A Reference Guide*, Greenwood Press, Westport, CT.
5. Brown, T.L., LeMay, H.E. Jr., Bursten, B.E., and Burdge, J.R. (2003) *Chemistry. The Central Science*, 9th edn, Prentice-Hall, Inc, NJ.
6. Kotz, J.C., Treichel, P.M., and Weaver, G.C. (2006) *Chemistry & Chemical Reactivity*, 6th edn, Thomson Brooks/Cole, Belmont, CA.

Section 4.2

7. Cammack, R., Frey, M., and Robson, R. (2001) *Hydrogen as Fuel: Learning from Nature*, Taylor and Francis, London.
8. John Wilson, J., Burgh, G. *The Hydrogen Report*. TMG/The Management Group. Winsor, Canada.

Section 4.3

9. Larmine, J. and Dicks, A. (2002) *Fuel Cell Systems Explained*, John Wiley & Sons., Chichester, England.
10. Schlapbach, L. and Editor, G. (2002) Hydrogen as a fuel and its storage for mobility and transport. *MRS Bulletin*, **27** (9), 675–676.
11. Dagani, R. (2002) Tempest in tiny tube. *Chemical Engineering News*, January 14, 2002. 4.
12. Zuttel, A. and Orimo, C. (2002) Hydrogen in nanostructured, carbon-related, and metallic materials. *MRS Bulletin*, **27** (9), 705–710.
13. Bowman, R.C. and Fultz, B. Jr. (2002) Metallic hydrides I: hydrogen storage and other gas-phase applications. *MRS Bulletin*, **27** (9), 688–691.
14. A.C. Dillon, K.E.H. Gilbert, P.A. Parilla, J.L. Alleman, G.L. Hornyak, K.M. Jones, and M.J. Heben (2002) Hydrogen Storage in Carbon Single-Wall Nanotubes, National Renewable Energy Laboratory Golden, CO 80401–3393 Proceedings of the 2002 U.S. DOE Hydrogen Program Review NREL/CP-61032405.

15. A.C. Dillon, P.A. Parilla, K.E.H. Gilbert, J.L. Alleman, T. Gennett, and M.J. Heben (2003) Hydrogen Storage in Carbon Nanotubes. National Renewable Energy Laboratory, Rochester Institute of Technology 2003 DOE HFCIT Program Review Meeting DOE Office of Energy Efficiency and Renewable Energy DOE Office of Science.
16. http://www.whitehouse.gov/infocus/energy/ (accessed October 2007).
17. http://en.wikipedia.org/wiki/Hydride (accessed September 2007).
18. http://www.nrel.gov/hydrogen/proj_storage.html. http://www.eoearth.org/article/Hydrogen_storage (Accessed February 2007).
19. Cammack, R., Frey, M., and Robson, R. (2001) *Hydrogenase Fuel: Learning from Nature*, Taylor and Francis, London.
20. Hydrogen as a Future Energy Carrier by A. Züttel, A. Borgschulte, L. Schlapbach, Wiley, March 2008
21. http://www.afdc.energy.gov/fuels/fuel_properties.php (Accessed 10 June 2017)
22. Hydrogen Compared with Other Fuels https://h2tools.org/bestpractices/h2properties (accessed 10 June 2017).
23. Larmine, J. and Dicks, A. (2003) *Fuel Cell Systems Explained*, John Wiley & Sons.
24. Zuttel, A. and Orimo, C. (2002) Hydrogen in nanostructured, carbon-related, and metallic materials. *MRS Bulletin*, **27** (9), 705–710.
25. Bowman, R.C. and Fultz, B. Jr. (2002) Metallic hydrides I: hydrogen storage and other gas-phase applications. *MRS Bulletin*, **27** (9), 688–691.
26. Hydrogen storage in carbon nanotube based materials: Solid state reversible hydrogen storage by D. Silambarasan, V. Vasu, K. Iyakutti LAP LAMBERT Academic Publishing, (accessed 20 December 2016).
27. Hydrogen Storage in Carbon Single-Wall nanotubes – NREL. www.nrel.gov/docs/fy02osti/32405b28.pdf 2002
28. Adsorption of Hydrogen Molecules on Carbon Nanotubes http://pubs.acs.org/doi/abs/10.1021/acs.jpca.5b12574 (accessed 28 July 2016)
29. Schlapbach, L. (2002) Hydrogen as a fuel and its storage for mobility and transport. *MRS Bulletin*, **27** (9), 675–676.
30. DOE Carbon-based Hydrogen Storage Center of Excellence by *NREL* Team, Rochester Institute of *Technology*, Rice University, May 2006; www.nrel.gov/docs/fy06osti/41726.pdf (accessed 10 June 2017).
31. http://en.wikipedia.org/wiki/Hydride (accessed 10 June 2017).
32. http://www.nrel.gov/hydrogen/proj_storage.html (accessed 10 June 2017).
33. https://energy.gov/eere/fuelcells/hydrogen_storage (accessed 10 June 2017).
34. Larmine, J. and Dicks, A. (2000) *Fuel Cell Systems Explained*. John Wiley & Sons, page 213.

CHAPTER 5

Hydrogen Infrastructure and Technology

Under a hydrogen infrastructure we assume the basic facilities, services, and installations for

a) hydrogen production,
b) storage,
c) transportation,
d) distribution, and
e) application

that operate on proven safety codes and standards.

Figure 5.1 shows the hydrogen produced by different methods such as from coal, gas, and biomass being stored and distributed for variety of usages. The cost projections by years from 2001 to 2023 are given graphically in the figure. The hydrogen storage methods vary with the technology to be adopted, starting from compressed gas to cryogenic storage. The distribution system has a built-in hydrogen safety for transportation purposes. The use of hydrogen to drive buses, ships, and so on has started to produce cleaner environment. This chapter discusses these aspects of hydrogen infrastructure.

Hydrogen technology is based on the availability of hydrogen gas for commercial usage. Although hydrogen is not available in abundance in free form, it is present in a combined form in the universe. About 70% of the world is filled with water as oceans, lakes, and wells. In Chapter 3, the current advancements in the field of hydrogen production and their efficiencies have been discussed besides the methodology adopted in calculating the hydrogen production efficiencies and the environmental impact factors caused by those methods. However, apart from those methods, hydrogen is produced for several hundreds of years by chemical methods and has also been used for technological purposes. A review of those conventional methods is discussed in this chapter first before discussing the hydrogen-based technology.

5.1 PRODUCTION OF HYDROGEN

Despite its unique abundance in the universe, hydrogen is surprisingly hard to produce in large amounts. Since only trace amount of free hydrogen can be found on our planet, hydrogen must be obtained from its compounds.

Introduction to Hydrogen Technology, Second Edition.
K.S.V. Santhanam, Roman J. Press, Massoud J. Miri, Alla V. Bailey, and Gerald A. Takacs

Book Companion site : www.wiley.com/go/santhanam/hydrogentech_2e

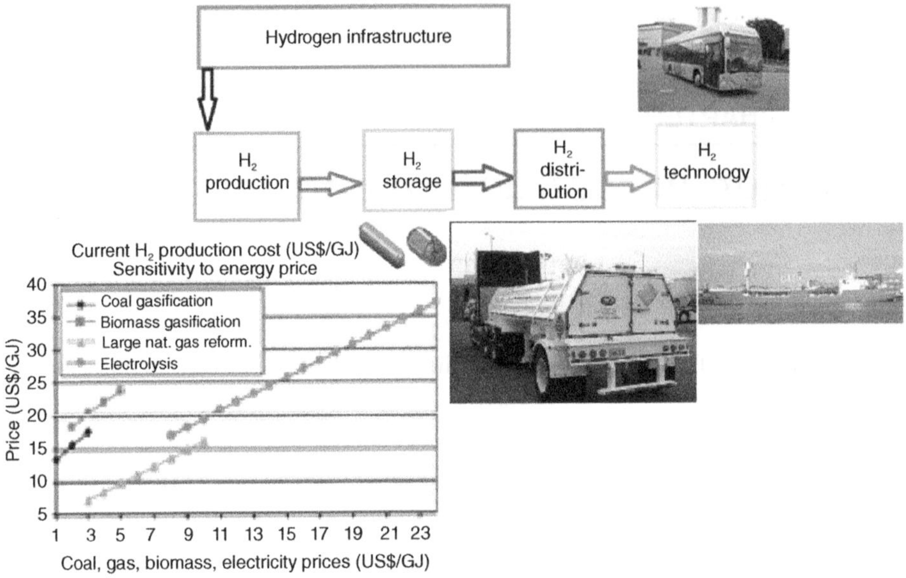

Figure 5.1 Hydrogen infrastructure.

Here is a short review of hydrogen production methods including laboratory and industrial methods and hydrogen production using renewable energy.

5.1.1 Laboratory Methods

It is easy to make small amounts of hydrogen gas; there are several very convenient methods:

1. Reaction of metal hydride (e.g., calcium or barium hydrides) with water:

$$CaH_2(s) + 2H_2O(l) \rightarrow Ca(OH)_2(s) + 2H_2(g)$$

$$BaH_2(s) + 2H_2O(l) \rightarrow Ba(OH)_2(aq) + 2H_2(g)$$

2. Reaction of certain metals, such as zinc, aluminum, or iron filings, with dilute sulfuric or hydrochloric acids:

$$Fe(s) + H_2SO_4(aq) \rightarrow FeSO_4(aq) + H_2(g)$$

$$2Al(s) + 6HCl(aq) \rightarrow 2AlCl_3(aq) + 3H_2(g)$$

3. Reaction of a strong base in an aqueous solution with aluminum:

$$2Al(s) + 2OH^-(aq) + H_2O(l) \rightarrow Al_2O_3(s) + 2H_2(g)$$

Hydrogen can be collected by displacement of air because hydrogen gas is about nine times lighter than air (Figure 5.2).

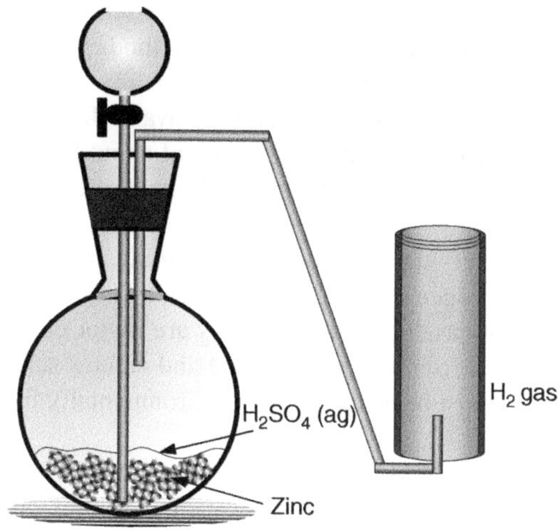

Figure 5.2 Hydrogen collection by displacement of air.

5.1.2 Industrial Methods

There are many industrial methods for the production of hydrogen as a gas or a liquid. The methods depend on local factors such as the quantity required and the raw materials on hand. The following describes some of the most common ways of producing hydrogen. Some of these are well-proven commercial techniques, while others are technologies under development.

5.1.2.1 Production of Hydrogen Based on Fossil Raw Materials Fossil fuels are hydrocarbon-containing natural resources such as coal, oil, and natural gas. They are also known as mineral fuels. The utilization of fossil fuels has enabled large-scale industrial development, and there is a strong and worldwide energy dependency on them. In modern times, this dependency can be seen as a major source of regional and global conflict.

5.1.2.1.1 Gasification of Coal and Steam Reforming of Natural Gas The most widely used method is heating coke, natural gas, methane, or other light hydrocarbons (ethane or propane) with steam in the presence of a catalyst:

$$\mathrm{C(coke)} + \mathrm{H_2O} \xrightarrow{(1100-1300^\circ\mathrm{C})} \mathrm{CO} + \mathrm{H_2}$$

$$\mathrm{CH_4} + \mathrm{H_2O} \xrightarrow{(700-925^\circ\mathrm{C})} \mathrm{CO} + 3\mathrm{H_2}$$

The process is endothermic, and the heat of reaction is supplied by the combustion of fossil fuels.

In both the cases, additional hydrogen can be produced by passing the CO and steam over iron oxide or cobalt oxide at high temperature (400°C); the reaction is called water–gas shift reaction:

$$\mathrm{CO} + \mathrm{H_2O} \rightarrow \mathrm{CO_2} + \mathrm{H_2}$$

Finally, in both the cases, a mixture of H_2 and CO_2 is sent to a gas purifier where the hydrogen is separated from CO_2 via one of the many methods (pressure swing absorption, wet scrubbing, or membrane separation).

Two coal gasification processes are commercially available; one is operated at atmospheric pressure, and another is operated at a pressure of about 5.5 MPa. Inorganic materials, remaining after gasification of coal, are removed as a molten slag at the bottom of the reactor.

Steam reforming is currently the cheapest way of producing hydrogen; it accounts for about half of the world's hydrogen production. However, this way would negatively impact the environment with the by-products: CO and CO_2 are major compounds among them. In addition, sulfur is released from the raw material and creates sulfur and nitrogen compounds. These compounds must be handled in an environmentally friendly way.

5.1.2.1.2 Partial Oxidation of Hydrocarbons This process refers to the conversion of heavy hydrocarbons (they are called feedstock) such as natural gas, naphtha, petroleum coke, or coal into a mixture of H_2, CO, and CO_2 using superheated steam and reduced amounts of oxygen. The process requires external energy, which can be obtained through the combustion of the feedstock itself. To increase the H_2 content, the mixture of H_2, CO, and CO_2 is subjected to the water–gas shift reaction. Final products are H_2 and CO_2. The overall efficiency of the process is about 50%.

5.1.2.1.3 Thermal Decomposition of Methane or Thermal Cracking of Methane This is a high-temperature (>700°C), endothermic process by which methane is decomposed into carbon and hydrogen:

$$CH_4 \rightarrow C(s) + 2H_2$$

The thermal decomposition occurs in a furnace where a methane–air flame heats the oven to about 1400°C. At this point, the flame is shut off and the methane decomposes into carbon black and hydrogen until the temperature falls to about 800°C. The carbon black is a by-product, which is filtered from hydrogen and can be used as a fuel, a filler in automobile tire production, or a reducing material in metallurgic industries. The main benefit of this process compared to the steam reforming of natural gas is the drastic reduction of CO_2 emissions. This process has the potential to be more cost-effective than the steam reforming of natural gas.

5.1.2.1.4 Petroleum Refining Operations Hydrogen is recovered from various refinery and chemical streams, which typically purge gas, tail gas, and fuel gas. For example, hydrogen can be obtained as an important by-product when hexane (the hydrocarbon of low octane rating) is converted to benzene (the hydrocarbon of higher octane rating):

$$C_6H_{14} \rightarrow C_6H_6 + 4H_2$$

As we see, all conventional technologies of production of hydrogen based on fossil raw materials and used by industry have numerous health and environmental impacts, including air and water pollution, ozone depletion, and global warming. Scientists are constantly researching new methods for hydrogen production that do not negatively impact the environment.

5.1.2.2 Ammonia Dissociation This process is based on the breaking up of ammonia, NH_3, into its simpler components, namely hydrogen and nitrogen. The process happens at about 1000°C with the aid of a catalyst. A resulting product, a mix of 75% hydrogen and 25% mononuclear nitrogen (N rather than N_2), can be directly used as a protective atmosphere for applications in the metal industry, for example, such as brazing or bright annealing.

5.1.2.3 Electrolysis of Water British scientist Sir William Robert Grove carried out the first experiments on electrolysis in 1839. He used electricity to split water into hydrogen and oxygen. He also carried out a process reverse to electrolysis, which is the reaction of oxygen and hydrogen, to generate electricity.

Electrolysis of water is a process that transforms water into its elemental parts through the use of an electric current. During this process, hydronium ions capture electrons from the cathode, producing hydrogen gas. Water molecules lose electrons to the anode, producing oxygen gas:

$$\text{Oxidation at anode} : 6H_2O(l) \rightarrow O_2(g) + 4H_3O^+(aq) + 4e^-$$

$$\text{Reduction at cathode} : 4H_3O^+(aq) + 4e^- \rightarrow 2H_2(g) + 4H_2O(l)$$

$$\text{Overall} : 2H_2O(l) \rightarrow 2H_2(g) + O_2(g)$$

There are different types of electrolyzers, depending on what electrolyte is used in electrolysis cells:

- *Alkaline electrolyzers*. In alkaline electrolyzers, a liquid electrolyte used is typically a 25% potassium hydroxide solution.
- *Polymer electrolyte membrane electrolyzers*. This type of electrolyzer utilizes polymer membranes as electrolytes.
- *Solid oxide electrolyzer*. The cell uses the ceramic materials both as an electrolyte and a separator membrane to isolate hydrogen and oxygen.
- *Seawater electrolyzer*. Seawater, the most abundant source of water in nature, is the ideal medium for hydrogen production through electrolysis. However, since it contains high sodium chloride (NaCl), almost all anode materials generate toxic chlorine gas.

5.1.3 Hydrogen Production Using Renewable Energy

Sharply rising energy prices have stimulated great interest in alternative, cheaper means of energy production from renewable sources. Because currently no commercial hydrogen production technique has been developed for geothermal or ocean renewable energy, we will concentrate on the use of wind, hydropower, solar, and biomass for hydrogen production or energy storage (Figure 5.3).

5.1.3.1 Renewable Wind Energy Storage in the form of Hydrogen The United States has many areas with abundant winds, particularly in the Midwest. Recognizing the wind as a resource is a crucial step in our national energy policy, and incentives such as the federal production tax credit and net metering provisions are available in

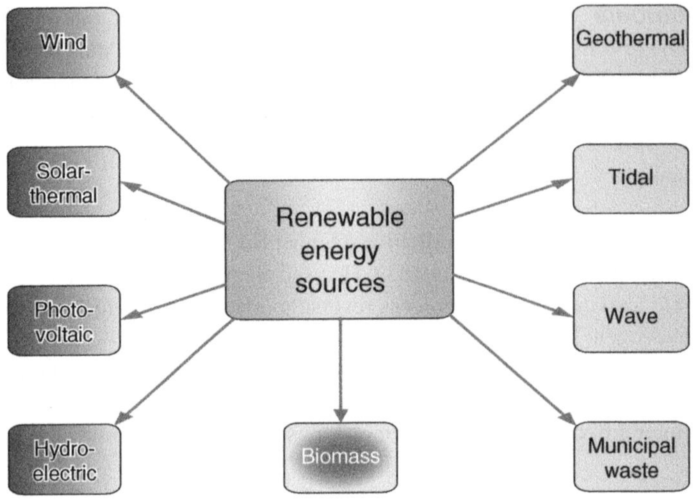

Figure 5.3 Renewable energy sources.

some areas. As an example, the New York State Energy Research and Development Authority (NYSERDA) is actively supporting small wind installations under 100 kW for agriculture, municipal, and commercial sectors in windy sites in New York. NYSERDA potential project development resulting from these activities is expected to be no less than 60 MW and, more likely, will exceed 100 MW.

Any means of energy production impacts the environment in some way, and wind energy is no different.

It is known that electric power has a tremendous weakness; it must always be used when it is produced. The rate of electricity production is based on supply and demand balance, and therefore it changes throughout the day and night. For this reason, an effective use of wind energy is not fully efficient due to its unpredictability. Decoupling the production and consumption of electricity produced by renewable sources can make its usage more competitive and viable.

Although electricity cannot be economically stored directly, it can be stored in other forms, such as an electric battery pack, flywheels, or in pneumatic or hydraulic storages. All these methods consume a lot of energy for conversion, and subsequent distribution is commonly associated with risks to the environment.

However, storing the produced energy in the form of hydrogen by means of a dedicated electrolysis system and the PEM fuel cell technology can be an economically attractive solution for wind energy storage. Currently, the electric output from a wind generator is connected to an electric grid and generates electricity mainly when wind is available, and especially during the night. By storing the power from renewable sources during off-peak periods and releasing it at peak times, coinciding with periods of peak consumer demand, energy storage can transform this spontaneous power into schedulable, high-value products.

To do this, the electric output from a wind turbine feeds the PEMFC electrolyzer. Hydrogen and oxygen from an electrolyzer are stored separately in pressure tanks or other storage media, such as metal hydrides or medium pressure gaseous storage devices and mixers for injecting those gases in the stream of natural gas, which feed the diesel generator or microturbine to produce electricity and heat.

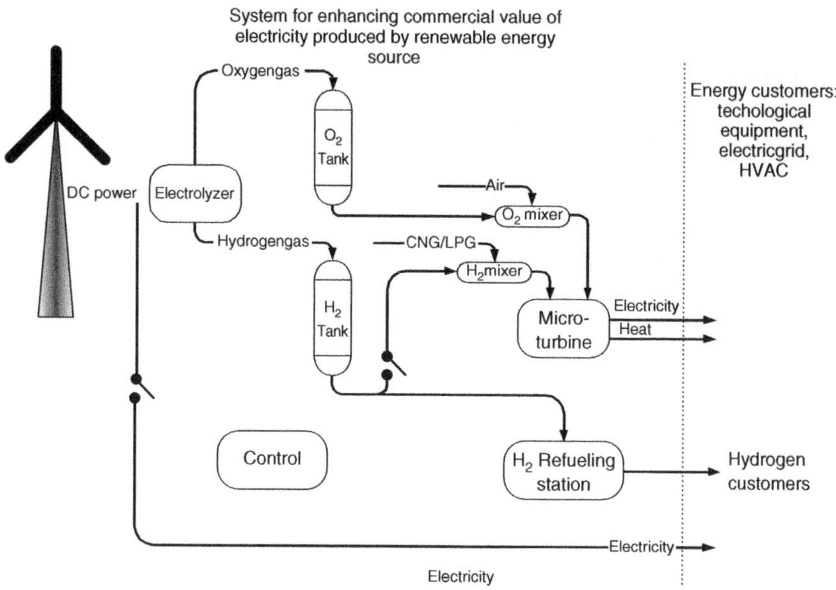

Figure 5.4 Wind power conversion into commercial-grade hydrogen and oxygen.

By-products of electrolysis are a limited amount of heat, which may be utilized for internal building use, together with heat generated as a result of combustion in the engine or microturbine. The heat also can be used for a variety of other heating applications, including greenhouse heating, heating pathways, technological processes, and supplementing boiler combustion. An additional economic efficiency lies in the area of dual use of stored energy: one is wind energy conversion into additional fuel for combustion generators and the other is refueling of hydrogen-powered vehicles (Figure 5.4).

5.1.3.2 Hydropower Energy Conversion to Hydrogen The total US hydropower capacity, including pumped storage facilities, is about 95,000 MW. Hydropower is currently the largest source of renewable power, generating nearly 10% of the electricity used in the United States. Electricity from hydropower is one of most promising sources of hydrogen production. Unfortunately, the biggest sources of hydropower are already in use. However, small water resources such as rivers and creeks can be used as distributed power/hydrogen production. It is worth mentioning that undesirable environmental effects, such as fish injury from passage through turbines, require development of new types of equipment for hydropower conversion into electricity.

Hydropower is an old and established source of electricity in New York State, dating back to the 1880s. In New York, hydropower accounts for 19% of electricity generation (1998). Though very little new hydropower has come on line in the state, the potential for more capacity exists by retrofitting existing facilities or adding turbines at developed hydropower sites. The following example illustrates this possibility.

Niagara Falls American Plant production:

- 2,400,000 kW × 24 = 57,600,000 kWh NY Power authority information

Assume that we use 25% off-peak power for H_2 production:

- 14,400,000 kWh

Average H_2 consumption per vehicle:

- 20 lb H_2 per fill ~10.7 gal of gas equivalent for BMW 745H
- Assume that the total efficiency of H_2 production equals 50%, and the useful energy output is 7,200,000 kWh (electrolyzer efficiency, pumping, storage, and distribution losses included).

In hydrogen equivalents, it is equal to 400,000 lb H_2 per day

- Or in number of fills: 400,000/20 = 20,000
- In number of miles: 3.6 million miles (180 miles per fill).

Assuming that average car goes 36 miles/day:

- Fill-up 100,000 vehicles/day.

New York has 8.83 million "standard series" vehicles (NY Department of Motor Vehicles). Niagara County has 130,000 "standard series" vehicles.

Conclusion: about 75% of the entire Niagara County can run on H_2.

5.1.4 Solar Direct Conversion to Hydrogen

5.1.4.1 Photoelectrochemical Water Splitting Multijunction semiconductor devices developed by the photovoltaic industry are being used by the Tandem Cell Company to build modular arrays of semiconductor cells that are scalable to any application's size. In the photoelectrochemical process, only solar energy is used to split water into its constituent parts, and the purity of produced hydrogen is currently 99.99%. Research in solar energy has developed equipment and methods to split water directly into pure hydrogen fuel and oxygen without external power requirements. Instead of first converting sunlight to electricity by the photovoltaic cells and then using electrolyzer to produce hydrogen from water, it is possible to combine these two steps into one using sunlight to directly split water into hydrogen and oxygen. Technology known as photoelectrochemical (PEC) light collecting systems uses solar energy to break water molecules into hydrogen and oxygen. The National Renewable Energy Lab (NREL) PEC system produces direct electricity conversion electricity from sunlight with a solar-to-hydrogen conversion efficiency of 12.4% using captured light.

Many other organizations, including GE Global Research, the University of California at Santa Barbara, are pursuing research on photoelectrochemical hydrogen production. Caltech University professor of chemistry Nathan Lewis said that integrated systems that convert solar energy photoelectrochemically are more efficient than splitting water through the more extensively researched electrolysis technique. In his opinion, the nanotech-based metal oxide photoelectrochemical materials could lower the cost of hydrogen production "somewhere between a factor of 4 and 10." "Visible light has enough energy to split water," according to John Turner, from the NREL who is working on developing materials for photoelectrochemically producing hydrogen. The serious limitation is susceptibility to corrosion of the nanomaterials that are immersed in water. Research is in progress to find more efficient, lower-cost components and systems that are durable and stable against corrosion in water surroundings.

5.1.4.2 Solar Thermal Methane Splitting Highly concentrated sunlight can be used to produce the high temperatures required to split methane into hydrogen and carbon. Concentrated solar energy can also be used to generate temperatures of several hundreds to over 2000°F at which thermochemical reaction cycles can be used to produce hydrogen. Such high-temperature, high-flux, and solar-driven thermochemical processes offer a novel approach for the environmentally benign production of hydrogen. Very high reaction rates at these elevated temperatures give rise to very fast reaction rates that enhance the production rates significantly and more than compensate for the intermittent nature of the solar resource.

5.1.4.3 Photobiological Water Splitting Certain photosynthetic microbes produce hydrogen from water in their metabolic activities using light energy. Photobiological technology holds great promise, but because oxygen is produced along with the hydrogen, the technology must overcome the limitation of oxygen sensitivity of the hydrogen-evolving enzyme systems. Researchers are addressing this issue by screening for naturally occurring organisms that are more tolerant of oxygen, and by creating new genetic forms of the organisms that can sustain hydrogen production in the presence of oxygen. A new system is also being developed that uses a metabolic switch (sulfur deprivation) to cycle algal cells between a photosynthetic growth phase and a hydrogen production phase.

5.1.4.4 Advanced Water Splitting In 2016, the energy department announced the financial support for Hydrogen Advanced Water splitting Materials Consortium (HydroGEN). This consortium will utilize the capabilities of the national laboratories to accelerate the development of commercially viable pathways for hydrogen production from renewable energy sources. HydroGEN focuses on advanced water splitting materials for the photoelectrochemical, solar thermochemical, and low- and high-temperature electrolytic water splitting technologies.

5.1.5 Hydrogen Production from Nuclear Energy

Nuclear power again continues to attract attention due to a desire for independence from foreign oil-producing sources, very limited greenhouse gases generation, and political security. During off-peak hours, nuclear plants generate more electricity than is requested by grid consumers, and hence electricity is at its cheapest; this excess electricity can be used to produce hydrogen.

The current approach for producing hydrogen from nuclear energy employs off-peak-generated electricity by exploiting the three methods for producing hydrogen: electrolysis, high-temperature steam electrolysis, and thermochemical water splitting cycles. The hydrogen produced by light water nuclear reactors (LWRs), unlike hydrogen produced via coal or steam methane reforming, does not require any cleaning for chemical impurity. In the United States today, there are 103 LWRs situated on 64 sites in 31 different states, which account for approximately 20% of US electricity needs. Utilization of off-peak electricity output represents a good energy reserve for domestic production of hydrogen. It is also worth mentioning that additional opportunities exist in combining electricity generation, hydrogen production, and ocean water desalination.

5.1.6 Hydrogen from Biomass

Biomass is organic substance made from plants and animals. Biomass contains stored energy from the sun. Plants absorb the sun's energy in a process called photosynthesis. The chemical energy in plants gets passed on to animals and people that eat them. Biomass is a renewable energy source because we can always grow more trees and crops, and waste will always exist. Some examples of biomass fuels are wood, crops, manure, and some garbage. When burned, the chemical energy in biomass is released as heat. Wood waste or garbage can be burned to produce electricity and to heat industries and homes.

Biomass resources (Figure 5.5) include agricultural residues such as corn stover, wheat, and rice straw; forest residues such as trees, stumps, and leaves; energy crops such as tall grasses and fast-growing trees; municipal solid waste (MSW); and sewage. In the future, biomass resources may be replenished through the cultivation of energy crops, such as fast-growing trees and grasses, called biomass feedstocks. Biomass fuels provide about 3% of the energy used in the United States. The most common form of biomass is wood. For thousands of years, people have burned wood for heating and cooking. Today only about 20% of the wood burned in the United States is used for heating and cooking, the rest is used by industries.

Unlike other renewable energy sources, biomass can be converted directly into liquid fuels for transportation needs. The two most common biofuels are ethanol and biodiesel. These fuels are usually blended with the petroleum fuels, gasoline and diesel fuel, but they can also be used on their own. Ethanol, an alcohol, is made by fermenting any biomass high in carbohydrates, such as corn, through a process similar to brewing beer. Biodiesel is made using vegetable oils, animal fats, algae, or even recycled cooking greases. These days, many specially equipped vehicles can run on E85, a fuel that is 85% ethanol and 15% gasoline.

There are a number of ways to convert biomass to hydrogen, including all of the methods of conversion of fossil fuels. Biomass can be burned directly to produce steam for electricity production or manufacturing processes or to heat production structures.

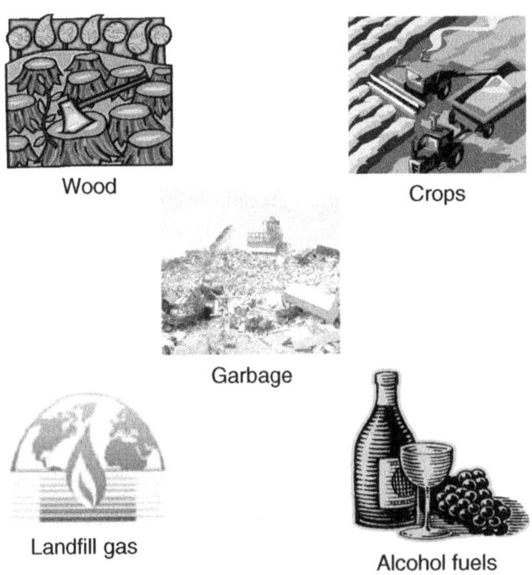

Figure 5.5 Types of biomass for conversion to fuel.

The gas also can be produced from biomass for electricity generation. Gasification systems use high temperatures to convert biomass into a gas. The gas consists mainly of CH_4, H_2, and CO. Steam is then introduced to reform CH_4 to H_2 and CO. CO is then put through the shift process to attain a higher level of H_2. The by-product from this process is CO_2, but it does not increase the CO_2 concentration in the atmosphere and therefore it is considered "neutral" with respect to greenhouse gas.

Another process is pyrolysis of biomass where heat is used to chemically convert biomass into a fuel oil that can be burned like petroleum. This oil can be converted into hydrogen and CO_2 by reforming. The bio-oil can be stored and reformed to hydrogen as needed. Where there is no infrastructure for natural gas, producing hydrogen from biomass may be economically more competitive than hydrogen production from natural gas.

5.1.6.1 Municipal Solid Waste Utilization Another source of biomass is our garbage, also called MSW. Municipal waste and biomass are a biodegradable fraction of products such as waste and residues from agriculture, the food industry, lawn clippings and leaves, forestry and related industries, as well as the biodegradable fraction of industrial and municipal waste. Materials that are made out of glass, plastic, and metals are not biomass because they are made out of nonrenewable materials. MSW can be a source of energy by either burning MSW in waste-to-energy plants, or by capturing biogas. In waste-to-energy plants, trash is burned to produce steam that can be used either to heat buildings or to generate electricity (Figures 5.6 and 5.7). In landfills, biomass rots and releases landfill gas (LFG) that can be used as a fuel source. Some dairy farmers collect biogas from tanks called "digesters," where they put all of the muck and manure from their barns.

MSW and biomass are the largest producers of anthropogenic methane emissions in the world. Its minimization by controllable anaerobic decomposition has become a proven technology today. LFG that is generated from solid municipal waste and biomass is available around any large city [1–3] and, in comparison to solar and wind, is more predictable in time of delivery. It is formed as a by-product of decomposition and is

Figure 5.6 Typical landfill gas power generation family.

Figure 5.7 Typical landfill view.

composed of approximately 50% methane, 45% carbon dioxide, 3% nitrogen, 1% oxygen, and 1% nonmethane organic vapors. LFG contains approximately 50% of the heating value of compressed natural gas (CNG). There are three different mechanisms for LFG production:

- Bacterial decomposition that occurs when organic waste is broken down by bacteria naturally present in the waste and in the soil used to cover the landfill.
- Volatilization, based on organic compounds in a liquid or a solid form into a vapor. Certain chemicals disposed of in the landfill may lead to nonmethane organic compound (NMOC) creation.
- Chemical reactions, such as LFG, including NMOCs, can be created by the reactions of chemicals including chlorine bleach and ammonia that may be present in waste, resulting in harmful gas producing.

In contemporary high gas-producing sites, LFG is usually collected via vertical wells that are drilled deep into the waste mass (~1 per acre). A well-head is installed on each vertical well to control gas flow and provides a means to sample the gas. The wells are joined together by perforated plastic pipes, which are connected to a suction pump to extract the gas from the well field. Since the mid-1980s, over 350 LFG projects are currently on line in the United States. Some of the biggest corporations and institutions in North America, including General Motors, Rolls Royce, Georgia Pacific, Nestle, General Electric, Ford Motor Company, NASA, International Paper, and Lucent, as well as many academic institutions, are now successfully using LFG directly in boiler, furnace, and kiln applications.

Rochester Institute of Technology conducted the technical and economical feasibility of piping gas generated from the Mill Seat Landfill for use as a fuel source for a possible cogeneration thermal/electric plant. The project, which has been funded by NYSERDA, reviews the landfill's current gas generation potential and summarizes the capital investment and annual operating costs for and LFG utilization project, including a dedicated pipeline to

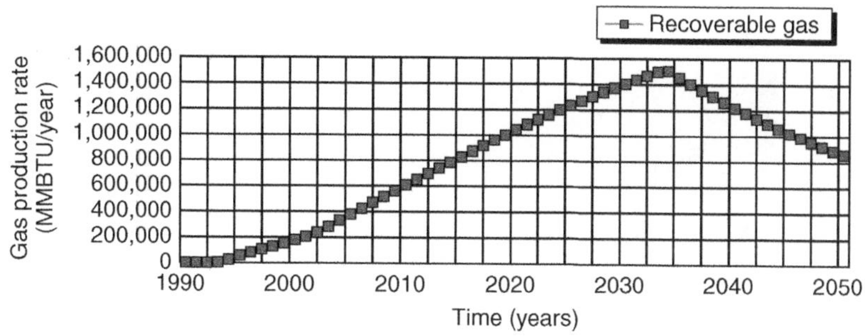

Figure 5.8 Typical landfill gas production. [Courtesy of Waste Management.]

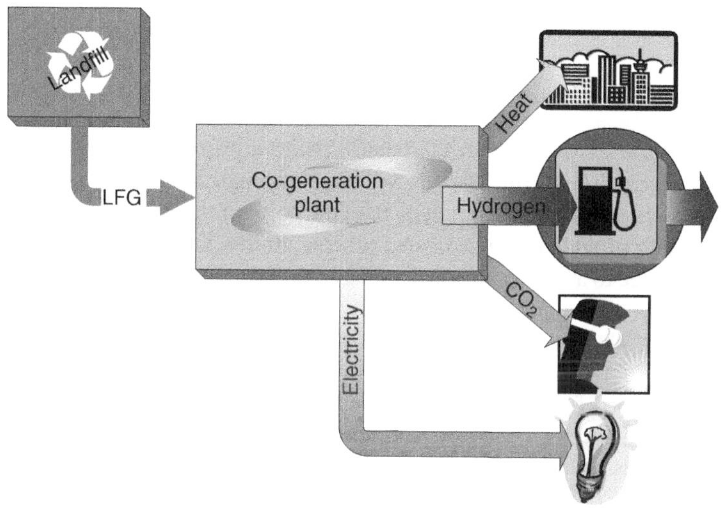

Figure 5.9 LFG cogeneration plant integrated with hydrogen production.

transport partially treated, high-quality, medium BTU LFG. Typically, the landfill will provide abundant renewable energy for decades. Figure 5.8 details the expected gas production life cycle at the Mill Seat Landfill.

In addition to power and heat cogeneration, there is an opportunity to develop a petroleum-free energy conversion system that is suitable for hydrogen production. The hydrogen supply will be constantly replenished by LFG, and hydrogen will be collected and stored in high-pressure storage tanks and then transported to the place of use (Figure 5.9).

The technical approach in this problem solving is to build a hydrogen production plant that will generate hydrogen using a dedicated LFG reformation process. Output from the plant will be twofold:

- Clean, compressed hydrogen stored at high pressure
- Compressed carbon dioxide for a range of potential industrial needs such as dry ice manufacturing, welding, soft drink carbonation, and CO_2 utilization for commercial greenhouses.

Figure 5.10 Hydrogen refueling station. [*Source*: Reproduced with permission of Fuel Cells today.]

The LFG will be extracted from the landfill using a primary compressor that will be driven from the local power grid. The gas will travel through an impurity removal filter and passed through a heat exchanger to reduce the dew point of the gas to approximately 35°F using chilled water provided by a dedicated process chiller. From there, LFG will pass the desulfurization stage to avoid catalyst poisoning in the hydrogen production block, and will be compressed and filtered again.

The production plant will be integrated with a hydrogen refueling station, where the plant's gaseous output will be used for long-term high-pressure hydrogen storage with fueling and dispenser ports for distribution (Figure 5.10).

The commonly used technology for the recovery of pure hydrogen from the hydrogen-rich gases is pressure swing adsorption (PSA). PSA is a cyclic process based on the capacity of certain materials, such as activated carbon and zeolites. Typically, PSA utilizes a dual adsorbent bed consisting of an initial layer of silica gel and a subsequent layer of activated carbon to adsorb and desorb particular gases as the gas pressure is raised and lowered. Vessels connected together containing adsorbent material undergo successive pressurization and depressurization cycles in order to produce a continuous stream of purified gas. The hydrogen product obtained using the PSA method meets high purity requirements (up to 99.999%). The particular features of PSA technology are low operating cost and operational simplicity.

The other method for hydrogen purification, which received substantial support from U.S. Department of Energy, is the use of a high-pressure gas separation membrane (HPGSM).

5.1.7 Biological Methods for Hydrogen Production

Biological processes for hydrogen production involve organic compounds and microorganisms. Low conversion efficiencies of biological systems can be compensated for by low energy requirements. Currently, the two ways of biological hydrogen production are as follows:

- Bacterial fermentation
- Biophotolysis.

5.1.7.1 Bacterial Fermentation

Fermentation of the bacteria is an anaerobic process that converts organic substances to H_2 and CO_2 without the need for sunlight and oxygen. Fermentation of materials such as starch, cellobiose, sucrose, and xylose can produce also other metabolites such as butyric acid, lactic acid, and acetic acid; water pollution creates a need to further treat fermented broth before disposal.

5.1.7.2 Biophotolysis

This is a process that uses microalgae-cyanobacteria and green algae to produce hydrogen in the presence of sunlight and water. Photosynthesis is the basis for almost all life on earth. The first step in photosynthesis involves splitting water into oxygen and hydrogen. It is the most common biochemical process on earth. Plain sunlight cannot directly break water down, but with the help of special pigments in microorganisms that engage in photosynthesis, the energy in sunlight can be utilized.

In theory, algae can produce hydrogen with an efficiency of up to 25%. The problem is that during this process, oxygen is also produced. The oxygen inhibits the hydrogen-producing enzyme hydrogenase, so only small amounts of hydrogen are actually produced.

5.2 HYDROGEN TRANSPORTATION, STORAGE, AND DISTRIBUTION

Under a hydrogen fuel infrastructure we assume amenities for installation and operation.

Rough calculation shows that feeding only 100,000 fuel cell vehicles would require 30×10^6 kg/year of liquid hydrogen. To produce, store, and transport such amount of hydrogen, the new infrastructure should be built. In "Blueprint for Hydrogen Fuel Infrastructure Development," J. Ohi suggests that the following production options may be considered:

- Off-site stream methane reforming of natural gas with tanker-truck delivery of liquid hydrogen to the refueling station with on-site storage of liquid and gaseous hydrogen
- On-site electrolysis with on-site storage of gaseous hydrogen
- On-site natural gas reforming with on-site storage of gaseous hydrogen.

Currently, hydrogen production is 48% from natural gas, 30% from oil, and 18% from coal; water electrolysis accounts for only 4%. The possibility exists that small sources of renewable energy should make hydrogen on-site without direct connections to the electrical grid or fossil fuel reformation. Future considerations may be given for producing hydrogen by excessive heat generated by nuclear reactors and coal liquefaction.

5.2.1 Logistics for Hydrogen Distribution

The primary roadblock for the initiation of a hydrogen economy is the development of the industry and transport systems needed for hydrogen production, storage, and distribution. These facilities and systems will take years for implementation and extensive monetary investments. However, the economic reality of nonrenewable fuel supplies is seen as making the transition economically viable in the future.

Cost estimation for producing hydrogen is not well established because it depends on variables such as production quality, chosen technology, and capital investment. The hydrogen physical properties are demanding new scientific and engineering solutions for storage and transportation. It is difficult to predict that in the next decade hydrogen will become an acceptable economical alternative for energy transportation over long distances. Advances in electrolysis and other hydrogen production technologies have not yet addressed the underlying cost issues.

5.2.2 Transportation, Storage, and Distribution

Hydrogen is stored and transported in a compressed or condensed condition as a gas or liquid. However, each step in transportation, storage, and distribution to the end users results in additional energy consumption and accordingly increasing cost penalties.

5.2.2.1 Hydrogen Liquefaction For hydrogen transportation in liquid form, industry uses vehicles equipped with an insulated large-scale Dewar or vacuum containers. Eliasson and Bossel point out that the cooling process for hydrogen is very energy consuming, with a Carnot efficiency of only 7%. Actual plant experience suggests that even at the most efficient, very large plant sizes, about 30% of the heating value of the gas, are required to liquefy hydrogen. The very low cryogenic temperature at which hydrogen is transported requires both pressurization and refrigeration to prevent the liquid from boil off. Due to ambient heating, hydrogen would require substantial rerefrigeration along any long pipeline, and energy losses for transportation may be comparable with energy than is delivered by the hydrogen itself.

5.2.2.2 Hydrogen Compression Due to its very low volumetric energy density, hydrogen must be compressed before it is transported by a piping infrastructure. The compressed hydrogen cannot utilize existing natural gas transportation systems because hydrogen accelerates the cracking of steel due to hydrogen embrittlement. In addition, the higher pressure requirements of hydrogen transportation through a piping infrastructure require more expensive materials, fabrication, valves for metering, higher leak-resistant joints, and greater maintenance costs. An analysis provided by Eliasson and Bossel has concluded that the energy to compress hydrogen from 1 to 800 bars will require about 10% of its energy content. They came to the conclusion that the distribution of hydrogen by trucks will require an increase in the number of trucks than are currently used to deliver gasoline.

5.2.2.3 Pipelining Hydrogen The ability of a fuel pipeline to transport energy is a function of the energy per unit volume of the fuel being delivered at the operating pressure of the pipeline. Direct utilization of the existing natural gas infrastructure for hydrogen delivery is not realistic due to engineering limitations created by hydrogen gas physical peculiarities. Land transmission system for hydrogen should be very specific in design to cover energy losses, material science, and safety. As natural gas infrastructure, the pipelines for hydrogen will include compressor stations to compensate for the transmission pressure losses. Utilizing 5000 psi pipeline pressure will partially improve the energy per unit volume ratio; however, increasing the pipeline's diameter entails additional engineering and fabrication costs and might not be suitable in populated areas.

There are other important issues related to hydrogen transportation. High-pressure hydrogen will easily leak through the smallest of holes and even straight through a lot of materials that would be impervious to natural gas. In addition, the hydrogen gas is capable of the metal embrittlement, especially in the mild steel used for pipeline construction. The transport of hydrogen gas over pipelines will also require different welding procedures, piping and equipment materials, gaskets, seals, and many other components widely used in the natural gas delivery to a user. In addition, use of a very high pressure for pipelining sufficient amounts of hydrogen will require different designs for valves, pumps, sensors, and safety devices to prevent additional failures and leaks due to the consequence of using high pressure.

Storage, transportation, and utilization of hydrogen require special precautions and subsequent regulations similar to the flammable gases handling procedures used for other applications.

5.2.3 Centralized Hydrogen Station

In this approach, the high-efficiency hydrogen generators would combine with a distribution system. This system would be similar to today's natural gas distribution system but would be modified to address a different set of operational challenges associated with hydrogen, such as diffusion through seals and embrittlement of pipe walls. Small end users, such as fuel cell vehicles, when not in use, may produce electricity as cogeneration units connected to local electrical distribution system.

5.2.4 Decentralized Hydrogen Station

An attractive option to simplify the pipelining of gas across country is to concentrate the hydrogen production on-site by electrolysis or on-site reformers. The current approach for a decentralized hydrogen station producing hydrogen gas from renewable energy is based on water electrolysis utilization. The typical system for a solar-powered water electrolyzing hydrogen station will consist of solar cells, power converter, water purifier, electrolyzer, piping, hydrogen purification components, compressor, pressure tanks, and a hydrogen dispenser.

To increase efficiency of hydrogen production, the solar-powered system may integrate with wind turbines and a reformer utilizing the natural gas. The excess electricity can be delivered to the grid and therefore the small, decentralized hydrogen system would become one of many cells that feed the electricity distributed generation system for the local users.

5.2.5 Refueling Stations Design

Vehicle refueling stations (Figure 5.10) mostly will be constructed as a combination of hydrogen production modules, storage with controls, and metered dispensers. This structure is very similar to the current gasoline or CNG refueling stations and is not complex. Most of the safety standards and regulations for CNG refueling stations could be applied to the hydrogen systems, with additional features for controlling distribution of high-pressure flammable gas.

Each on-site hydrogen production module would require specific electrical components for electricity transformation, AC–DC conversion, and the electrolysis process itself. In the planning of hydrogen refueling stations attention must be paid to the environmental challenges presented by oxygen disposal and substantial water resources consumption availability and storage. When the liquid form of high-pressure hydrogen is needed to fuel vehicles, a special type of dispenser is needed to maintain appropriate refueling rate and provide a safe environment. Some companies even experiment with using robotic arms to refuel vehicles.

5.3 HYDROGEN SAFETY

Any fuel by its very nature is volatile and, while hydrogen is well known as an element in the periodic table, its extensive use as an energy carrier needs to be further examined. For hydrogen to be considered a viable fuel and energy carrier on the commercial market, the actual and perceptual safety of the consumer must be fully addressed.

Hydrogen has, over the years, acquired an image of an immensely powerful and possibly threatening substance. There are still some people who can recall the experience of the Hindenburg in 1937, as well as the cold war threat of the hydrogen bomb (H-bomb). The nuclear fusion reactions involved in the H-bomb have, of course, nothing to do with the chemical reactions involved in the use of hydrogen as a fuel for internal combustion engines (ICE) and hydrogen fuel cells. Essentially, hydrogen has been used for years by industry with a commendable record for safety. In some ways, the properties of hydrogen make it safer than gasoline or other fuels.

5.3.1 Hydrogen Safety-Related Properties

Hydrogen is a colorless, odorless gas that is lighter than air. It is nontoxic, does not create fumes, does not contribute to groundwater pollution, and the only output created from a fuel cell using hydrogen are pure water and electricity. Other common fuels, such as oil and gasoline, are particularly toxic and poisonous to people and the environment. If an oil spill occurs, massive cleanup efforts are required with irreparable damage still being done by uncontrollable seepage in the environment. Hydrogen is the lightest element on earth. A hydrogen spill would result in the evaporation of the hydrogen, with water being the only by-product left behind. Though it is light in weight, when compared to other fuels, such as natural gas, hydrogen's energy-to-weight ratio is the highest of all known fuels.

Table 5.1 lists the characteristics of gasoline, methane, and hydrogen relating to ignition and explosion hazards.

Hydrogen's perceived explosive nature is of great concern. At a 4% concentration, hydrogen does become volatile in the atmosphere, creating the possibility of igniting. While that appears to be too low a level for safety concerns, when compared to gasoline, which becomes volatile at 1% concentration, the level of safety provided by hydrogen as a fuel is much higher than other commonly used fuel carriers.

TABLE 5.1 Fire hazard characteristics

Property	Gasoline	Methane	Hydrogen
Density (kg/m^3)	4.40	0.65	0.084
Diffusion coefficient in air (cm^2/s)	0.05	0.16	0.610
Specific heat at constant pressure (J/Gk)	1.20	2.22	14.89
Ignition limits in air (vol%)	1.0–7.6	5.3–15.0	4.0–75.0
Ignition energy in air (mJ)	0.24	0.29	0.02
Ignition temperature (°C)	228–471	540	585
Flame temperature in air (°C)	2197	1875	2045
Explosion energy (G TNT/kJ)	0.25	0.19	0.17
Flame emissivity 1%	34–43	25–33	17–25

Source: Reproduced with kind permission of T. Nejat Veziroglu.

In actuality, hydrogen properties, such as its small molecular size, diffusivity, and buoyancy, have many safety advantages over other commonly used fossil fuels. These properties allow it to quickly disperse into the atmosphere when spilled, making a combustible situation less likely. Other fossil fuels such as propane or gasoline, with their high densities and slow dispersal, allow the fuels to congregate near the ground, thereby increasing the risk of ignition and explosion. In order for a hydrogen fire to occur, there must be a concentration of hydrogen, an ignition source, and the right amount of oxidizer present at the same time.

At low concentration levels, the energy required to cause hydrogen to ignite is quite high, similar to that of gasoline and natural gas in their ranges of flammability, making hydrogen the more difficult of the fuels to ignite when in the lower flammability level.

Explosions cannot occur in hydrogen-only tanks or containers, oxidizers must be present in at least a 10% concentration of pure oxygen or 41% of air. Gasoline's potential for explosion requires a much lower concentration of an oxidizer, becoming volatile at between 1% and 3% of pure oxygen. The likelihood that hydrogen would explode in open air is also reduced because of its tendency to rise quickly. Gasoline fumes hover near the ground and pose a greater danger for explosion.

When hydrogen burns, it has a nearly invisible, pale blue flame. It has a very high flame speed and, with its rapid dissipation, a hydrogen fire will burn itself out quickly. There is also little heat radiation in a hydrogen fire so other materials near the flames are less likely to catch fire, reducing the likelihood that a fire would spread. A gasoline fire in a car lasts approximately 30 min; a hydrogen fire would last only a tenth of that time. The same is the case for an airplane, and hydrogen, in its liquid form, is considered by many to be a safer fuel for airplanes than the fuel used today.

5.3.2 Codes and Regulating Safety

Though it has been over 70 years since it crashed, the Hindenburg disaster in 1937 is still a black mark on hydrogen's reputation as a fuel. And while NASA recently proved that hydrogen was not the cause of the fire that brought the "Titanic of the Sky" crashing to the ground (it was the fabric covering the vessel that was highly flammable and caught fire with a spark of static electricity), the decades-long mystery and perception of danger surrounding hydrogen has left a deep distrust in the public mind. To ensure public safety and allow the commercialization of hydrogen as an energy carrier, new model building codes and other technical standards for hydrogen must be developed and accepted by local, state, and federal governments.

Well-known organizations in the United States such NASA, the National Hydrogen Association (NHA), the Society of Automotive Engineers (SAE), and organizations across the globe work together to develop agree-upon standards. Codes and standards for storage equipment, electrolyzers, fuel cell transportation, and infrastructure issues can help to overcome the barriers to commercialization and foster more understanding in the public arena. Standards are a compilation of technical guidelines, definitions, and instructions for manufacturers and their engineers. They are developed through a consensus process involving numerous experts in the field and provide a systematic and precise means for measuring and communicating product risk to the consumer, the public at large, and government officials. Standards are generally followed on a voluntary basis to ensure compatibility, consistency, and safety. Once developed, standards are usually incorporated into codes that, in turn, must be adopted by state and local jurisdictions to become legal and binding.

New standards and codes are being developed in preparation for the advance of hydrogen fuel systems by 13 US and two international organizations working with public and private sectors. While the federal government has an indirect role in the voluntary process through which codes and standards are developed in the United States, to expedite the development of the hydrogen infrastructure, the Department of Energy (DOE) is coordinating a collaborative national effort by government and industry to prepare, review, and disseminate model hydrogen codes and standards.

The DOE's Hydrogen, Fuel Cells and Infrastructures Technologies Program (HFCIT) web site posts the current status of all codes and standards related to hydrogen and fuel cells. Domestic codes and standards are listed on the site by application, geographic location, and organization. There is also a listing of international codes and standards that is organized by individual countries.

HFCIT has done more than just provide a site for codes and standards. It has provided funding for several projects to move the commercialization of hydrogen to the forefront. It has supported the American National Standards Institute in its development of the Hydrogen Codes and Standards Portal and the National Hydrogen Association's monthly posting of news and information in the Hydrogen and Fuel Cell Safety Report.

Another report prepared for the DOE states, "There is no significant evidence of difference between the frequency and severity of structurally damaging explosions due to natural gas and town gas." Town gas is a mixture of hydrogen and carbon monoxide that was used as a heating source in the United States prior to the development of the natural gas supplies and infrastructure and is still routinely used by industry. The DOE report found that, "There is no evidence of unusual safety risks that would preclude use of hydrogen as a motor vehicle fuel, and there is no indication that a properly designed hydrogen powered vehicle and hydrogen refueling infrastructure would pose any more risk than conventional motor vehicle fuels."

This report and other research from around the world leaves a positive indication that the commercialization of hydrogen as an energy carrier and fuel of the future is moving in the right direction and that safety concerns, while valid, are not insurmountable.

5.4 HYDROGEN TECHNOLOGY ASSESSMENT

5.4.1 Introduction

To develop the hydrogen economy, processes and systems must be developed to extract the hydrogen from its source, contain it, transport it, and deliver it to the point of use.

The hydrogen economy appears to have as many supporters as critics. It seems that the major differences between these opposing positions tends to be related to the effect that the displacement of the use of fossil fuels with hydrogen-based technologies will have on the environment along with the time line for the necessary technological development of related applications for general use. Proponents have the expectation that displacing fossil fuels with a hydrogen energy source will

- reduce dependency on fossil fuels;
- yield viable fuel replacements for ICEs used for transportation;
- improve air quality and radically reduce greenhouse gas emissions; and
- substantially reduce the rate of global warming.

Hydrogen's use in place of nonrenewable fuels would result in their displacement and, therefore, their conservation. In current fuel cell applications, hydrogen yields about twice the energy in its conversion into power as current technology, gasoline-fueled, ICEs. With engine modifications, hydrogen has been used directly in an internal combustion engine as shown by BMW, Ford, and other vehicle manufacturers.

Displacement of nonrenewable fossil fuels with hydrogen will naturally lead to reduction of greenhouse gas emissions because hydrogen contains no carbon molecule and therefore undergoes combustion without generating carbon dioxide or carbon monoxide emissions. The primary by-product of hydrogen combustion process is water. Without a carbon-based by-product being exhausted into the atmosphere, one of the precursors of global warming is reduced. Given these many advantages, hydrogen appears to represent an "ideal" fuel solution for transportation, electrical generation, home heating, and other similar applications.

Supporters also believe that there are no major impediments to the widespread production and distribution of hydrogen fuels. The timeline for development and mass production has begun with the current use of fuel cell technologies in small power and automotive applications, which is in line with the supporters' point of view. The limited rollout of fuel cell usage also supports the opponents' point of view. They state that significant and complicated issues remain to be resolved regarding the production, transportation, and utilization of hydrogen as a fuel. Since hydrogen is a gas at atmospheric pressure, it is easy to make the assumption that hydrogen can be easily transported from its place of production to the point of use. This may be true if the gas can be compressed in sufficient volumes to meet the operational needs of the application, but the compression process is energy intensive and high-pressure vessels are not readily accepted by the general public. Liquefying before transportation and distribution also presents problems with cost of processing, logistics, and acceptance.

Opponents point out that hydrogen is not easily distributed in the volumes necessary for widespread displacement of fossil fuels. All fuels require energy consumption during their production and distribution. The amount of energy required to make the equivalent of a gallon of gasoline with hydrogen raises essential doubts about economical aspects of the proposed hydrogen economy. The current technologies for hydrogen production and transportation appear to indicate that the energy required to produce hydrogen may exceed its energy contribution as a fuel.

5.4.2 The State of Fossil Fuel Reserves

The US total oil consumption in 2005 reached 20.7 million barrels per day (Wilson & Burgh). Approximately 60% of this oil is imported from different countries. To provide this supply without interruption, the United States must pay attention to world events that may influence the import of fuel.

Knowing that the amount of oil reserves is limited, geophysicist Marion King Hubbert created in 1956 a mathematical model describing correlations between oil reserves and production rates. According to the Hubbert peak theory, the rate of petroleum production from an individual oil well follows a bell-shaped curve (Figure 5.11). The model shows how to calculate the point of maximum production in advance based on discovery rates, production rates, and cumulative production. Early in the curve (prepeak), the production rate increases due to the new discoveries. The next stage is the peaking of oil production (Hubbert's peak), combined with increased consumption and following a time plateau that

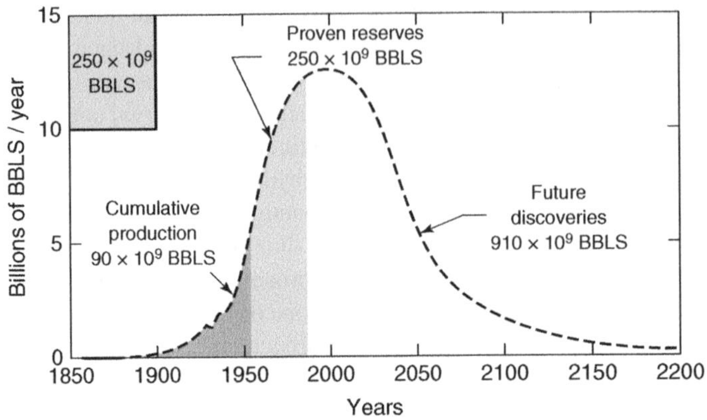

Figure 5.11 The Hubbert curve. [*Source*: http://en.wikipedia.org/wiki/Image:Hubbert-fig-20.png.]

has been observed for many geological regions. Late in the curve, (postpeak) oil production on earth declines due to resource depletion that consequently results in cost increases.

Based on the Hubbert curve prediction, researchers have different opinions on the particular timing for oil peak production because it is difficult to foresee the future oil resources' discovery, contents of available resources, and future consumption. As an example, Ben Ebenhack believes that the crisis point will hit in the year 2030 if the popular estimate of 2 trillion barrels of oil is accurate. It is a common opinion that we have nearly 25–50 years to develop replacements for petrochemical goods before energy supply issues pose overwhelming problems to energy and food supply for future generations (Figure 5.11).

These supply issues will be particularly troublesome for the US economy due to our consumption requirements. Great distances separate major US cities (as compared with Europe), and suburban spread out contributes to the inherent inefficiency of our transportation systems. US citizens consume huge quantities of energy for lifestyle needs, far exceeding that of citizens of other countries. Finding a truly independent replacement for oil and natural gas is thus already a critical issue for the United States.

5.4.3 Hydrogen Production Assessment

For mass production of hydrogen, the most developed commercial technologies are water electrolysis and thermochemical treatment of petrochemical sources. To satisfy the consumer demand in the United States, facilities have produced nearly 350 million gallons of gasoline a day per year. Since 1 kg of hydrogen contains approximately the same energy as 1 gal of gasoline, to replace 50% of our country's gasoline consumption will require 175,000 tons/day of hydrogen. Assuming that existing electrolysis technologies provide process efficiency of 75% and heating value of hydrogen is 39.3 kWh, the required amount of energy to produce such an amount of hydrogen is about 8600 MW of electric power. For comparison, in 2005 the electric utilities generated 4055 MW of electricity. Adding together additional energy requirements for AC–DC conversion, gas compression or liquefaction, transportation, delivery, and energy losses in the form of heat dissipation leads us to the conclusion that hydrogen production by electrolysis cannot be supported with the existing energy infrastructure.

The electrical power required for generating enough hydrogen to replace a significant portion of our country's gasoline consumption has to come from conventional power-generating facilities. Although the use of electrolytic hydrogen does not generate greenhouse gas, the HC, NO_x, and CO_2 emissions produced by conventionally generated electricity make hydrogen produced by electrolysis as "dirty" as a fossil fuel. Only hydrogen production from renewable technologies can deliver comprehensive, cradle-to-grave, zero-emissions results.

There are other sources of emission that are not related to greenhouse gas creation. One is the possible increase of hydrogen gas content in the stratosphere due to hydrogen gas leakage. Such a possibility may influence the ozone depletion process and a result of this phenomenon is unknown.

Replacement of only 50% of current gasoline consumption by electrolysis would consume 1,575,000 tons/day of water, the availability of which in many geographical areas may be problematic. Replacing the same amount of gasoline with electrolytic hydrogen would be coupled with by-product oxygen in amounts of 1,400,000 tons/day. Commercial use of such quantity of coproduced oxygen is unlikely, but oxygen releasing into the atmosphere may potentially create an impact to neighboring flora and fauna in addition to related safety consideration and energy loss.

Thermochemical reforming is likely to be a choice process for hydrogen production due to its emissions and efficiency improvements. Thermochemical processes (reforming) extracts the hydrogen from hydrocarbons such as natural gas, methane, or methanol. Reforming represents a technically viable replacement for a significant fraction of worldwide gasoline consumption, capable of generating large quantities of relatively low-cost hydrogen. Up to now, the primary focus has been on the conversion of natural gas into hydrogen.

Limited North American supplies, driven by rising consumption in multiple use areas, do not justify using natural gas reserves for this process. According to the US Energy Information Agency [46], US rates of natural gas consumption in 2005 were 22.24 Tcf (trillion cubic feet) and verified reserves were 193 Tcf. Simple calculation shows that our country has less than 9 years of natural gas reserves.

Development of biomass and other methane-producing processes is taking place to supply the hydrocarbons needed to feed the processes developed around the use of natural gas. While methane can be produced from plants and biomass, it is important to critically evaluate the complete energy cycle of this feedstock. Despite potential improvements to photoelectrolysis and biomass, neither is likely to be cost-competitive in the foreseeable future. Hydrogen production can be achieved through a number of additional methods that have not yet achieved pilot scale operation, such as exposing water to high-intensity UV rays or other (including possibly nuclear) radiation. While "breakthrough" technologies are possible, technical feasibility, scalability, and modest environmental impacts must be analyzed in assessing the proposed hydrogen economy. Development work with reforming is extensive and continues to focus on hydrogen purification, process emissions reduction, and using alternate heat sources.

5.4.4 Current Issues

5.4.4.1 Fuel Cells Current fuel cell technologies have proven their ability to generate electricity from hydrogen. Utilization of fuel cells as the source of micropower shortly will become a reality. However, the mass production of vehicles based around hydrogen fuel cells has been delayed by the number of problems that require further development. Aside from inadequate hydrogen production, undeveloped delivery infrastructure, and limited options for gaseous storage, and the current technologies employed in the construction of the fuel stack are suffering from extensive cost. This is due in part to the cost of the precious metals used as the catalyst to break the hydrogen atom from the fuel and partly due to the need to make the cell operational in the environmental conditions of those processes they are being developed to replace. There is not a sufficient quantity of platinum on earth to support the number of fuel cell systems that would be needed to fully transition to a hydrogen economy. Extensive research and development needs to continue to find ways of reducing costs and increasing reliability and durability while making the stack less sensitive to environmental conditions such as temperatures below freezing point. The current consumer market gives us many examples that large-scale production is driving cost down with concurrent performance improvement.

5.4.4.2 Hydrogen Safety As mentioned under Section 5.3, the storage, transportation, and utilization of hydrogen requires special precautions and subsequent regulations. In addition, the general public has demonstrated some psychological resistance to hydrogen due to safety fears. Widespread use of hydrogen, especially when it is stored in compressed form, requires effective safety measures.

5.4.4.3 Transportation, Storage, and Distribution Hydrogen is stored and transported in a compressed state as a gas or a condensed state as liquid. It is known that each step in transportation, storage, and distribution to the end users results in additional energy consumption and therefore in cost penalties.

5.4.4.4 Transitional Fuel Utilization of other fuels for transportation or as a transitional step to hydrogen economy can be achieved by exploring other hydrocarbon-containing sources such methanol, ethanol, methane, butanol, coal, and natural gas. For instance, methanol can be used directly as fuel in the vehicles or in a dissolved methanol fuel cell and methanol-fueled solid oxide fuel cell. Gasoline blended with ethanol already is available at many gasoline stations. Compared with the fuels mentioned above, hydrogen is easy to store, transport, and does not require expensive infrastructure development.

The extensive coal reserves in North America suggest that a 400-year (or more) supply of feedstocks exist in this form at present consumption rates. The prospects for developing new methods of coal conversion with near-zero atmospheric emissions should be a valuable step in creating an economical electricity source for hydrogen production.

5.4.5 Summary

In parallel to creation of a hydrogen economy, an evaluation of alternatives such methane or extensive coal utilization should be considered. In the transportation segment, progress in battery capacity can make the cost of electric vehicles a more attractive option in comparison to electrochemical engines because they do not require extensive infrastructure investments. Since hydrogen is considered a viable source of energy to displace the volumes of nonrenewable fuels that sustain civilization, the following issues need to be addressed:

1. Creation of the hydrogen infrastructure is very costly.
2. With current hydrogen production technologies, emission is switched from the point of use to the point of production.
3. Psychological resistance to hydrogen utilization for safety reasons.
4. Hydrogen production by electrolysis requires excessive electricity sources.
5. Manufacturing hydrogen by reforming natural gas is compromised by limited sources and rising consumption.
6. Ineffective and energy consuming current hydrogen storage technologies.
7. With current technologies, availability and cost of catalytic materials can be problematic.
8. Fuel cell stack performance at low temperature.
9. Materials (ceramic) temperature cycling.
10. Limited durability of the fuel cell stack.

The future of the hydrogen economy will be decided on economical factors related to the cost of energy required for hydrogen production, storage, and transportation. All the negative factors mentioned above may delay an introduction of sustainable hydrogen economy. Electricity generation based on zero-emission technologies from alternative energy sources, including nuclear, direct hydrogen production based on splitting water using photosynthesis processes reactions, and successful research in biological materials conversion to hydrogen, can make the hydrogen society a reality.

MULTIPLE CHOICE QUESTIONS

1. Which characteristics correctly describe electrolysis of water?
   (a) This is the process that transforms water into its elemental parts through the use of an electric current
   (b) During this process, hydronium ions capture electrons from the cathode, producing hydrogen gas
   (c) Water molecules lose electrons to the anode, producing oxygen gas
   (d) All of the above
2. What method is used for industrial production of ammonia (NH_3)?
   (a) Heat nitrogen and mix with hydrogen gas
   (b) Expose both gases to a very high pressure for extended period of time
   (c) Catalytically reduce nitrogen by hydrogen
3. What method(s) is (are) used in hydrogen production commercially?
   (a) Partial oxidation of hydrocarbons
   (b) Gasification of coal and steam reforming of natural gas
   (c) Thermal decomposition of methane or thermal cracking of methane
   (d) Any of the above

4. What is the main benefit of the thermal decomposition of methane or thermal cracking of methane?
   (a) Drastic reduction of CO_2 emission
   (b) Production of black carbon as a result of reaction
   (c) The reaction executed at relatively low temperature (<1000°C)
   (d) Easy to control reaction
5. What types of electrolyzers are used to produce hydrogen from water?
   (a) Alkaline
   (b) Polymer electrolyte membrane
   (c) Solid oxide
   (d) All of the above
6. What type of energy is mostly used for hydrogen production from renewable energy?
   (a) Electric produced from wind, solar power, and hydro
   (b) Heat directly
   (c) Applying high dose of radiation
   (d) Intensive solar exposure
7. Why the renewable energy is so attractive for hydrogen production?
   (a) Simple and direct conversion from one form into another
   (b) Abundance of renewable energy in nature
   (c) Allows to store renewable energy in the form of hydrogen for farther use
   (d) Renewable is the cleanest form of energy
8. Can the sunlight energy to be used directly for hydrogen production without conversion to electric?
   (a) Yes, using the nanotech-based metal oxide photoelectrochemical materials
   (b) Yes, highly concentrated sunlight producing high temperatures to split methane
   (c) Yes, using concentrated solar energy to generate temperatures over 2000°C at which thermochemical reaction cycles can be used to produce hydrogen
   (d) Any of the above
9. What is the approximate heating value of landfill gas (LFG) in comparison to compressed natural gas?
   (a) 100%
   (b) 75%
   (c) 50%
   (d) 25%
10. Which fuel flame is less visible?
    (a) Methane
    (b) Ethanol
    (c) Hydrogen
    (d) Propane

11. In hydrogen distribution in liquid or gaseous form, the most critical problem is
    (a) Storage
    (b) Pipelines
    (c) Compression
    (d) Liquefaction
    (e) Permeability
12. Which fuel is more toxic?
    (a) Gasoline
    (b) Diesel
    (c) Methane
    (d) Hydrogen
13. What is the LEL (lower explosion level) for hydrogen?
    (a) 52%
    (b) 14%
    (c) 4%

BIBLIOGRAPHY

Section 5.1

1. NREL: Hydrogen and Fuel Cells Research - Wind-to-Hydrogen Project www.nrel.gov/hydrogen/proj_wind_hydrogen.html (**accessed** 14 July 2016).
2. https://www.greenoptimistic.com/hydrogen-from water/#.WIFV5FweMbw (accessed 09 June 2017).
3. NREL: Hydrogen and Fuel Cells Research - Hydrogen Production and ... www.nrel.gov/hydrogen/proj_production_delivery.html.
4. https://energy.gov/eere/office-energy-efficiency-renewable-energy.
5. A 24.4% solar to hydrogen energy conversion efficiency by combining ... https://www.researchgate.net/.../283190008_A_244_solar_to_hydrogen_energy_conver...
6. Towards efficient solar-to-hydrogen conversion: Fundamentals and ... https://www.degruyter.com/view/j/nanoph.2016.5.issue-4/.../nanoph-2016-0027.xml by Chen - 2016.
7. Solar water splitting by photovoltaic-electrolysis with a solar-to-hydrogen efficiency over 30% by 2016 Jieyang Jia, Linsey C. Seitz, Jesse D. Benck, Yijie Huo, Yusi Chen, Jia Wei Desmond Ng, Taner Bilir, James S. Harris, Thomas F. Jaramillo.
8. Solar to Chemical Energy Conversion: Theory and Application. https://books.google.com/books?isbn=3319254006 Masakazu Sugiyama, Katsushi Fujii, Shinichiro Nakamura - 2016 - Science.
9. http://www.popularmechanics.com/science/energy/a19292/were-one-step-close-to-creating-fuel-cells-from-solar-energy (accessed February 2016).
10. Nuclear power plants can produce hydrogen to fuel the "hydrogen .... https://www.acs.org/.../nuclear-power-plants-can-produce-hydrogen-to-fuel-the-hydr... (accessed 09 June 2017).
11. Nuclear Hydrogen Production Technology - International Atomic ... IAEA. https://www.iaea.org/About/Policy/GC/GC57/.../English/gc57inf-2-att1_en.pdf (accessed 09 June 2017).

12. Naterer, G.F., Dincer, I., and Zamfirescu, C. (2013) *Hydrogen Production from Nuclear Energy* Vol. 8, SpringerISBN: 978-1-4471-4937-8.
13. Cheap Hydrogen from Sunlight and Water - MIT Technology Review. https://www.technologyreview.com/s/.../cheap-hydrogen-from-sunlight-and-water/Nov.2013 (accessed 09 June 2017).
14. https://en.wikipedia.org/wiki/Water_splitting (accessed 09 June 2017).
15. Lede, J., Lapique, F., Villermaux, J. *et al.* (1982) Production of hydrogen by direct thermal decomposition of water, preliminary investigation. *International Journal of Hydrogen Energy*, **7** (12), 939–950.
16. Production of hydrogen by direct thermal decomposition of water http://www.eolss.net/sample-chapters/c08/e3-13-03.pdf (accessed 09 June 2017).
17. NREL: Hydrogen and Fuel Cells Research - Wind-to-Hydrogen Project. www.nrel.gov/hydrogen/proj_wind_hydrogen.html (accessed 14 July 2016).
18. "Landfill Gas Utilization at RIT's Campus Feasibility Study" PON No. 732–02 Rochester Institute of Technology, March 2005.
19. Hydrogen from Biomass http://www.nrel.gov/docs/legosti/old/36262.pdf (accessed 09 June 2017).
20. https://en.wikipedia.org/wiki/Biohydrogen (accessed 09 June 2017).
21. http://www.ibtimes.co.uk/bacteria-used-yield-pure-water-produce-hydrogen-fuel-future-1486180 (accessed 02 February 2015).
22. New process harvests hydrogen from plant waste at record speeds. https://www.extremetech.com (accessed 09 June 2017).

Section 5.2

23. https://en.wikipedia.org/wiki/Hydrogen_infrastructure.
24. Rifkin, J. (2002) *The Hydrogen Economy*, Tarcher/Penguin, New York, NY.
25. Hydrogen Fueling Infrastructure Analysis - NREL.
26. http://www.nrel.gov/hydrogen/proj_infrastructure_analysis.html (accessed 8 July 2016).
27. http://www.fuelcelltoday.com/analysis/analyst-views/2013/13-11-20-hydrogen-infrastructure-developments-in-the-usa (accessed 20 November 2013).
28. http://www.hit-tent.eu/ (accessed 09 June 2017).
29. https://en.wikipedia.org/wiki/Hydrogen_station (accessed 09 June 2017).

Section 5.3

30. http://www.nrel.gov/hydrogen/proj_safety_codes_standards.html (accessed 09 June 2017).
31. ehs.berkeley.edu/sites/default/files/lines-of-services/...safety/80hydrogen.pdf (accessed 09 June 2017).
32. https://en.wikipedia.org/wiki/Hydrogen_safety (accessed 09 June 2017).
33. http://www.arhab.org/pdfs/h2_safety_fsheet.pdf (accessed 09 June 2017).
34. https://www.hydrogen.energy.gov/codes_standards.html (accessed 09 June 2017).
35. www.nfpa.org/codes-and-standards/...codes-and-standards/list-of-codes-and-standards (accessed 09 June 2017).

Section 5.4

36. Wilson, J.R., Burgh, G., The Hydrogen Report. *Windsor Canada*: TMG/Management Group. www.tmgtech.com (accessed October 2006).
37. Bossel, U. *Energy and the Hydrogen Economy:* and Eliasson, B., Accessed January 2005 from http://www.methanol.org/pdfFrame.cfm?pdf=HydrogenEconomyReport 2003.pdf (accessed 09 June 2017).
38. Bossel, U., Does a Hydrogen Economy Make Sense? *Proceedings of the IEEE.* Volume 94, No. 10, October 2006.
39. Wise, J., The truth about hydrogen. *Popular Mechanics*, November 2006.
40. Accessed July 2007 from http://www.eia.doe.gov/oil gas/natural gas/info glance/natural gas.html (accessed 09 June 2017)
41. http://maps.unomaha.edu/peterson/funda/sidebar/oilconsumption.html (accessed 09 June 2017).
42. https://www.eia.gov/tools/faqs/faq.cfm?id=50&t=8 (accessed 09 June 2017).
43. http://www.efn-uk.org/l-street/renewables-lib/ren-reports/index_files/H2-economics.pdf (accessed 09 June 2017).
44. Bossel, U., "Does a Hydrogen Economy Make Sense?" Proceedings of the IEEE. Volume 94, No. 10, October 2006.
45. www.iea.org/publications/freepublications/publication/technology-roadmap-hydrogen-and-fuel-cells.html (accessed 09 June 2017).
46. http://www.hubbertpeak.com/midpoint.htmh (accessed 09 June 2017).
47. https://www.eia.gov/dnav/ng/hist/n9140us2A.htm (accessed 09 June 2017).
48. http://www.renewableenergyworld.com/hydrogen/tech.html (accessed 09 June 2017).
49. www.eia.gov/electricity (accessed 09 June 2017)
50. https://en.wikipedia.org/wiki/Natural_gas (accessed 09 June 2017).

CHAPTER 6

Batteries

6.1 INTRODUCTION

Since today's technology is based on using energy stored (batteries) and energy-generating devices (fuel cells) with both of them driven by electrochemical principles, it is appropriate to consider the mechanisms involved in each of them. In essence, both are sources of electrical energy, batteries giving for a limited time and fuel cells for an extended period as long as the fuel is supplied to it. The energy and power delivered are expressed in the same units in both categories (batteries and fuel cells). Hence, an introduction to batteries at this stage would be appropriate for appreciating the fuel cell technology discussed in Chapter 7.

The materials contained in Chapter 2 on the fundamentals of electrochemistry laid the foundation for the basics in electrochemistry that is useful for developing battery technology. Electrochemistry deals with redox reactions occurring at the electrodes that can be initiated by either

a) passage of externally imposed currents (electrolytic cells)

or

b) by spontaneous chemical reactions occurring at the electrodes (galvanic or voltaic cells).

Based on these definitions, there are two types of electrochemical cells we confront today. One type is called electrolytic cells that are used for large number of fundamental and applied studies. In this category, mechanistic analytical studies are carried out using a wide variety of perturbations given to the electrode. The second type is called galvanic cells where free energy change in a chemical redox reaction is transformed to electrical energy. Thus, the two types of electrochemical cells may be rationalized as one where passage of current forces the conversion of electrical energy to chemical energy and in the other type where chemical energy is converted into electrical energy. For this latter type of galvanic cells, a general term of battery is used. In general, larger the free energy change in a chemical reaction, greater will be the energy capability of a battery than the smaller free energy one, and the energy conversion process can be evaluated from the standard potentials of the redox couples involved in the chemical reaction.

Introduction to Hydrogen Technology, Second Edition.
K.S.V. Santhanam, Roman J. Press, Massoud J. Miri, Alla V. Bailey, and Gerald A. Takacs

Book Companion site : www.wiley.com/go/santhanam/hydrogentech_2e

6.1.1 Example of a Battery

Let us consider a zinc metal placed in a solution of zinc salt (1.0 M $ZnSO_4$) in one half cell and copper metal placed in a solution of copper salt (1.0 M $CuSO_4$); this is shown in Figure 6.1. If a voltage measuring source such as a voltmeter is connected through a wire as shown in Figure 6.1, it will show a voltage of 1.10 V. You may notice in the figure that there is a bridge connecting the two solutions. The bridge allows ions from one compartment to move to the other based on the charges. As discussed in Chapter 2, all electrochemical cells consist of two half cells; in one half cell oxidation (electrons are released in a chemical reaction) occurs and in the other reduction reaction. Thus, battery oxidation occurs in one half cell and reduction occurs in the other half cell. In order to determine the half cell reactions, the standard potentials given in Table 6.1 are used.

The redox couple (oxidant and reductant in the table) having a more negative value is undergoing oxidation in one half cell and is termed as the anode compartment. Conversely, the redox couple having a less negative value will undergo reduction in the other half cell called the cathode compartment.

Figure 6.1 contains two redox couples

$$Zn^{2+} + 2e^- = Zn(s) \qquad E^\circ = -0.7618\,V \tag{6.1}$$

$$Cu^{2+} + 2e^- = Cu(s) \qquad E^\circ = +0.3370\,V \tag{6.2}$$

Of the two redox couples, zinc couple is more negative than copper couple. Hence, zinc metal will undergo oxidation and release electrons. The half cell in which the electrons are released is defined as the anode compartment. The electrons travel in the external circuit to the copper electrode. The other couple involving copper is designated as the cathode compartment and forms the other half cell. The abovementioned couples are represented

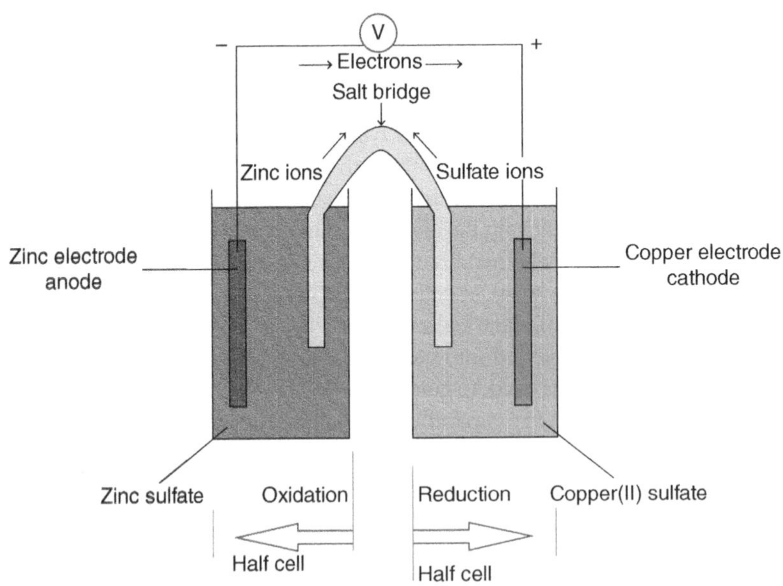

Figure 6.1 Battery made with Zn and Cu metal with electrolytes.

TABLE 6.1 Standard potentials of redox reactions

Half-reaction			
Oxidant	⇌	Reductant	$E°$ (V)
$Sr^{+} + e^{-}$	⇌	Sr	−4.101
$Ca^{+} + e^{-}$	⇌	Ca	−3.8
$Pr^{3+} + e^{-}$	⇌	Pr^{2+}	−3.1
$3\,N_2(g) + 2\,H^{+} + 2\,e^{-}$	⇌	$2HN_3(aq)$	−3.09
$Li^{+} + e^{-}$	⇌	Li(s)	−3.0401
$N_2(g) + 4\,H_2O + 2\,e^{-}$	⇌	$2\,NH_2OH(aq) + 2\,OH^{-}$	−3.04
$Cs^{+} + e^{-}$	⇌	Cs(s)	−3.026
$Ca(OH)_2 + 2\,e^{-}$	⇌	$Ca + 2OH^{-}$	−3.02
$Er^{3+} + e^{-}$	⇌	Er^{2+}	−3.0
$Ba(OH)_2 + 2\,e^{-}$	⇌	$Ba + 2OH^{-}$	−2.99
$Rb^{+} + e^{-}$	⇌	Rb(s)	−2.98
$K^{+} + e^{-}$	⇌	K(s)	−2.931
$Ba^{2+} + 2\,e^{-}$	⇌	Ba(s)	−2.912
$La(OH)_3(s) + 3\,e^{-}$	⇌	$La(s) + 3\,OH^{-}$	−2.90
$Fr^{+} + e^{-}$	⇌	Fr	−2.9
$Sr^{2+} + 2\,e^{-}$	⇌	Sr(s)	−2.899
$Sr(OH)_2 + 2\,e^{-}$	⇌	$Sr + 2OH^{-}$	−2.88
$Ca^{2+} + 2\,e^{-}$	⇌	Ca(s)	−2.868
$Li^{+} + e^{-}$	⇌	$LiC_6(s)$	−2.84
$Eu^{2+} + 2\,e^{-}$	⇌	Eu(s)	−2.812
$Ra^{2+} + 2\,e^{-}$	⇌	Ra(s)	−2.8
$Ho^{3+} + e^{-}$	⇌	Ho^{2+}	−2.8
$Bk^{3+} + e^{-}$	⇌	Bk^{2+}	−2.8
$Yb^{2+} + 2\,e^{-}$	⇌	Yb	−2.76
$Na^{+} + e^{-}$	⇌	Na(s)	−2.71
$Mg^{+} + e^{-}$	⇌	Mg	−2.70
$Nd^{3+} + e^{-}$	⇌	Nd^{2+}	−2.7
$Mg(OH)_2 + 2\,e^{-}$	⇌	$Mg + 2OH^{-}$	−2.690
$Sm^{2+} + 2\,e^{-}$	⇌	Sm	−2.68
$Be_2O_3^{2-} + 3H_2O + 4\,e^{-}$	⇌	$2Be + 6OH^{-}$	−2.63
$Pm^{3+} + e^{-}$	⇌	Pm^{2+}	−2.6
$Dy^{3+} + e^{-}$	⇌	Dy^{2+}	−2.6
$No^{2+} + 2\,e^{-}$	⇌	No	−2.50
$HfO(OH)_2 + H_2O + 4\,e^{-}$	⇌	$Hf + 4OH^{-}$	−2.50
$Th(OH)_4 + 4\,e^{-}$	⇌	$Th + 4OH^{-}$	−2.48
$Md^{2+} + 2\,e^{-}$	⇌	Md	−2.40
$Tm^{2+} + 2\,e^{-}$	⇌	Tm	−2.4
$La^{3+} + 3\,e^{-}$	⇌	La(s)	−2.379
$Y^{3+} + 3\,e^{-}$	⇌	Y(s)	−2.372
$Mg^{2+} + 2\,e^{-}$	⇌	Mg(s)	−2.372
$ZrO(OH)_2(s) + H_2O + 4\,e^{-}$	⇌	$Zr(s) + 4\,OH^{-}$	−2.36
$Pr^{3+} + 3\,e^{-}$	⇌	Pr	−2.353
$Ce^{3+} + 3\,e^{-}$	⇌	Ce	−2.336
$Er^{3+} + 3\,e^{-}$	⇌	Er	−2.331
$Ho^{3+} + 3\,e^{-}$	⇌	Ho	−2.33

(continued overleaf)

TABLE 6.1 (*Continued*)

Half-reaction			
Oxidant	⇌	Reductant	$E°$ (V)
$H_2AlO_3^- + H_2O + 3e^-$	⇌	$Al + 4OH^-$	−2.33
$Nd^{3+} + 3e^-$	⇌	Nd	−2.323
$Tm^{3+} + 3e^-$	⇌	Tm	−2.319
$Al(OH)_3(s) + 3e^-$	⇌	$Al(s) + 3OH^-$	−2.31
$Sm^{3+} + 3e^-$	⇌	Sm	−2.304
$Fm^{2+} + 2e^-$	⇌	Fm	−2.30
$Am^{3+} + e^-$	⇌	Am^{2+}	−2.3
$Dy^{3+} + 3e^-$	⇌	Dy	−2.295
$Lu^{3+} + 3e^-$	⇌	Lu	−2.28
$Tb^{3+} + 3e^-$	⇌	Tb	−2.28
$Gd^{3+} + 3e^-$	⇌	Gd	−2.279
$H_2 + 2e^-$	⇌	$2H^-$	−2.23
$Es^{2+} + 2e^-$	⇌	Es	−2.23
$Pm^{2+} + 2e^-$	⇌	Pm	−2.2
$Tm^{3+} + e^-$	⇌	Tm^{2+}	−2.2
$Dy^{2+} + 2e^-$	⇌	Dy	−2.2
$Ac^{3+} + 3e^-$	⇌	Ac	−2.20
$Yb^{3+} + 3e^-$	⇌	Yb	−2.19
$Cf^{2+} + 2e^-$	⇌	Cf	−2.12
$Nd^{2+} + 2e^-$	⇌	Nd	−2.1
$Ho^{2+} + 2e^-$	⇌	Ho	−2.1
$Sc^{3+} + 3e^-$	⇌	Sc(s)	−2.077
$AlF_6^{3-} + 3e^-$	⇌	$Al + 6F^-$	−2.069
$Am^{3+} + 3e^-$	⇌	Am	−2.048
$Cm^{3+} + 3e^-$	⇌	Cm	−2.04
$Pu^{3+} + 3e^-$	⇌	Pu	−2.031
$Pr^{2+} + 2e^-$	⇌	Pr	−2.0
$Er^{2+} + 2e^-$	⇌	Er	−2.0
$Eu^{3+} + 3e^-$	⇌	Eu	−1.991
$Lr^{3+} + 3e^-$	⇌	Lr	−1.96
$Cf^{3+} + 3e^-$	⇌	Cf	−1.94
$Es^{3+} + 3e^-$	⇌	Es	−1.91
$Pa^{4+} + e^-$	⇌	Pa^{3+}	−1.9
$Am^{2+} + 2e^-$	⇌	Am	−1.9
$Th^{4+} + 4e^-$	⇌	Th	−1.899
$Fm^{3+} + 3e^-$	⇌	Fm	−1.89
$Np^{3+} + 3e^-$	⇌	Np	−1.856
$Be^{2+} + 2e^-$	⇌	Be	−1.847
$H_2PO_2^- + e^-$	⇌	$P + 2OH^-$	−1.82
$U^{3+} + 3e^-$	⇌	U	−1.798
$Sr^{2+} + 2e^-$	⇌	Sr/Hg	−1.793
$H_2BO_3^- + H_2O + 3e^-$	⇌	$B + 4OH^-$	−1.79
$ThO_2 + 4H^+ + 4e^-$	⇌	$Th + 2H_2O$	−1.789
$HfO^{2+} + 2H^+ + 4e^-$	⇌	$Hf + H_2O$	−1.724
$HPO_3^{2-} + 2H_2O + 3e^-$	⇌	$P + 5OH^-$	−1.71

TABLE 6.1 (*Continued*)

Half-reaction			
Oxidant	⇌	Reductant	$E°$ (V)
$SiO_3^{2-} + H_2O + 4\,e^-$	⇌	$Si + 6OH^-$	−1.697
$Al^{3+} + 3\,e^-$	⇌	Al(s)	−1.662
$Ti^{2+} + 2\,e^-$	⇌	Ti(s)	−1.63
$ZrO_2(s) + 4\,H^+ + 4\,e^-$	⇌	$Zr(s) + 2\,H_2O$	−1.553
$Zr^{4+} + 4\,e^-$	⇌	Zr(s)	−1.45
$Ti^{3+} + 3\,e^-$	⇌	Ti(s)	−1.37
$TiO(s) + 2\,H^+ + 2\,e^-$	⇌	$Ti(s) + H_2O$	−1.31
$Ti_2O_3(s) + 2\,H^+ + 2\,e^-$	⇌	$2\,TiO(s) + H_2O$	−1.23
$Zn(OH)_4^{2-} + 2\,e^-$	⇌	$Zn(s) + 4\,OH^-$	−1.199
$Mn^{2+} + 2\,e^-$	⇌	Mn(s)	−1.185
$Fe(CN)_6^{4-} + 6\,H^+ + 2\,e^-$	⇌	Fe(s) + 6HCN(aq)	−1.16
$Te(s) + 2\,e^-$	⇌	Te^{2-}	−1.143
$V^{2+} + 2\,e^-$	⇌	V(s)	−1.13
$Nb^{3+} + 3\,e^-$	⇌	Nb(s)	−1.099
$Sn(s) + 4\,H^+ + 4\,e^-$	⇌	$SnH_4(g)$	−1.07
$SiO_2(s) + 4\,H^+ + 4\,e^-$	⇌	$Si(s) + 2\,H_2O$	−0.91
$B(OH)_3(aq) + 3\,H^+ + 3\,e^-$	⇌	$B(s) + 3\,H_2O$	−0.89
$Fe(OH)_2(s) + 2\,e^-$	⇌	$Fe(s) + 2\,OH^-$	−0.89
$Fe_2O_3(s) + 3\,H_2O + 2\,e^-$	⇌	$2Fe(OH)_2(s) + 2\,OH^-$	−0.86
$TiO^{2+} + 2\,H^+ + 4\,e^-$	⇌	$Ti(s) + H_2O$	−0.86
$\mathbf{2H_2O} + 2\,e^-$	⇌	$H_2(g) + 2\,OH^-$	−0.8277
$Bi(s) + 3\,H^+ + 3\,e^-$	⇌	BiH_3	−0.8
$Zn^{2+} + 2\,e^-$	⇌	Zn(Hg)	−0.7628
$Zn^{2+} + 2\,e^-$	⇌	Zn(s)	−0.7618
$Ta_2O_5(s) + 10\,H^+ + 10\,e^-$	⇌	$2\,Ta(s) + 5\,H_2O$	−0.75
$Cr^{3+} + 3\,e^-$	⇌	Cr(s)	−0.74
$Ag_2S(s) + 2\,e^-$	⇌	$2Ag(s) + S^{2-}(aq)$	−0.69
$[Au(CN)_2]^- + e^-$	⇌	$Au(s) + 2\,CN^-$	−0.60
$Ta^{3+} + 3\,e^-$	⇌	Ta(s)	−0.6
$PbO(s) + H_2O + 2\,e^-$	⇌	$Pb(s) + 2\,OH^-$	−0.58
$2\,TiO_2(s) + 2\,H^+ + 2\,e^-$	⇌	$Ti_2O_3(s) + H_2O$	−0.56
$Ga^{3+} + 3\,e^-$	⇌	Ga(s)	−0.53
$U^{4+} + e^-$	⇌	U^{3+}	−0.52
$H_3PO_2(aq) + H^+ + e^-$	⇌	$P(white) + 2\,H_2O$	−0.508
$H_3PO_3(aq) + 2\,H^+ + 2\,e^-$	⇌	$H_3PO_2(aq) + H_2O$	−0.499
$H_3PO_3(aq) + 3\,H^+ + 3\,e^-$	⇌	$P(red) + 3\,H_2O$	−0.454
$Fe^{2+} + 2\,e^-$	⇌	Fe(s)	−0.44
$2\,CO_2(g) + 2\,H^+ + 2\,e^-$	⇌	HOOCCOOH(aq)	−0.43
$Cr^{3+} + e^-$	⇌	Cr^{2+}	−0.42
$Cd^{2+} + 2\,e^-$	⇌	Cd(s)	−0.40
$GeO_2(s) + 2\,H^+ + 2\,e^-$	⇌	$GeO(s) + H_2O$	−0.37
$Cu_2O(s) + H_2O + 2\,e^-$	⇌	$2\,Cu(s) + 2\,OH^-$	−0.360
$PbSO_4(s) + 2\,e^-$	⇌	$Pb(s) + SO_4^{2-}$	−0.3588
$PbSO_4(s) + 2\,e^-$	⇌	$Pb(Hg) + SO_4^{2-}$	−0.3505
$Eu^{3+} + e^-$	⇌	Eu^{2+}	−0.35

(*continued overleaf*)

TABLE 6.1 ***(Continued)***

Half-reaction			
Oxidant	⇌	Reductant	$E°$ (V)
$In^{3+} + 3\,e^-$	⇌	In(s)	−0.34
$Tl^+ + e^-$	⇌	Tl(s)	−0.34
$Ge(s) + 4\,H^+ + 4\,e^-$	⇌	$GeH_4(g)$	−0.29
$Co^{2+} + 2\,e^-$	⇌	Co(s)	−0.28
$H_3PO_4(aq) + 2\,H^+ + 2\,e^-$	⇌	$H_3PO_3(aq) + H_2O$	−0.276
$V^{3+} + e^-$	⇌	V^{2+}	−0.26
$Ni^{2+} + 2\,e^-$	⇌	Ni(s)	−0.25
$As(s) + 3\,H^+ + 3\,e^-$	⇌	$AsH_3(g)$	−0.23
$AgI(s) + e^-$	⇌	$Ag(s) + I^-$	−0.15224
$MoO_2(s) + 4\,H^+ + 4\,e^-$	⇌	$Mo(s) + 2\,H_2O$	−0.15
$Si(s) + 4\,H^+ + 4\,e^-$	⇌	$SiH_4(g)$	−0.14
$Sn^{2+} + 2\,e^-$	⇌	Sn(s)	−0.13
$O_2(g) + H^+ + e^-$	⇌	$HO_2•(aq)$	−0.13
$Pb^{2+} + 2\,e^-$	⇌	Pb(s)	−0.126
$WO_2(s) + 4\,H^+ + 4\,e^-$	⇌	$W(s) + 2\,H_2O$	−0.12
$P(red) + 3\,H^+ + 3\,e^-$	⇌	$PH_3(g)$	−0.111
$CO_2(g) + 2\,H^+ + 2\,e^-$	⇌	HCOOH(aq)	−0.11
$Se(s) + 2\,H^+ + 2\,e^-$	⇌	$H_2Se(g)$	−0.11
$CO_2(g) + 2\,H^+ + 2\,e^-$	⇌	$CO(g) + H_2O$	−0.11
$SnO(s) + 2\,H^+ + 2\,e^-$	⇌	$Sn(s) + H_2O$	−0.10
$SnO_2(s) + 4\,H^+ + 4\,e^-$	⇌	$SnO(s) + 2\,H_2O$	−0.09
$WO_3(aq) + 6\,H^+ + 6\,e^-$	⇌	$W(s) + 3\,H_2O$	−0.09
$P(white) + 3\,H^+ + 3\,e^-$	⇌	$PH_3(g)$	−0.063
$Fe^{3+} + 3\,e^-$	⇌	Fe(s)	−0.04
$HCOOH(aq) + 2\,H^+ + 2\,e^-$	⇌	$HCHO(aq) + H_2O$	−0.03
$2\,H^+ + 2\,e^-$	⇌	**$H_2(g)$**	**0.0000**
$AgBr(s) + e^-$	⇌	$Ag(s) + Br^-$	+0.07133
$S_4O_6^{2-} + 2\,e^-$	⇌	$2\,S_2O_3^{2-}$	+0.08
$Fe_3O_4(s) + 8\,H^+ + 8\,e^-$	⇌	$3\,Fe(s) + 4\,H_2O$	+0.085
$N_2(g) + 2\,H_2O + 6\,H^+ + 6\,e^-$	⇌	$2\,NH_4OH(aq)$	+0.092
$HgO(s) + H_2O + 2\,e^-$	⇌	$Hg(l) + 2\,OH^-$	+0.0977
$Cu(NH_3)_4^{2+} + e^-$	⇌	$Cu(NH_3)_2^{+} + 2\,NH_3$	+0.10
$Ru(NH_3)_6^{3+} + e^-$	⇌	$Ru(NH_3)_6^{2+}$	+0.10
$N_2H_4(aq) + 4\,H_2O + 2\,e^-$	⇌	$2\,NH_4^{+} + 4\,OH^-$	+0.11
$H_2MoO_4(aq) + 6\,H^+ + 6\,e^-$	⇌	$Mo(s) + 4\,H_2O$	+0.11
$Ge^{4+} + 4\,e^-$	⇌	Ge(s)	+0.12
$C(s) + 4\,H^+ + 4\,e^-$	⇌	$CH_4(g)$	+0.13
$HCHO(aq) + 2\,H^+ + 2\,e^-$	⇌	$CH_3OH(aq)$	+0.13
$S(s) + 2\,H^+ + 2\,e^-$	⇌	$H_2S(g)$	+0.14
$Sn^{4+} + 2\,e^-$	⇌	Sn^{2+}	+0.15
$Cu^{2+} + e^-$	⇌	Cu^+	+0.159
$HSO_4^{-} + 3\,H^+ + 2\,e^-$	⇌	$SO_2(aq) + 2\,H_2O$	+0.16
$UO_2^{2+} + e^-$	⇌	UO_2^{+}	+0.163
$SO_4^{2-} + 4\,H^+ + 2\,e^-$	⇌	$SO_2(aq) + 2\,H_2O$	+0.17
$TiO^{2+} + 2\,H^+ + e^-$	⇌	$Ti^{3+} + H_2O$	+0.19

TABLE 6.1 (*Continued*)

Half-reaction			
Oxidant	⇌	Reductant	$E°$ (V)
$SbO^+ + 2H^+ + 3e^-$	⇌	$Sb(s) + H_2O$	+0.20
$AgCl(s) + e^-$	⇌	$Ag(s) + Cl^-$	+0.22233
$H_3AsO_3(aq) + 3H^+ + 3e^-$	⇌	$As(s) + 3H_2O$	+0.24
$GeO(s) + 2H^+ + 2e^-$	⇌	$Ge(s) + H_2O$	+0.26
$UO_2{}^+ + 4H^+ + e^-$	⇌	$U^{4+} + 2H_2O$	+0.273
$Re^{3+} + 3e^-$	⇌	$Re(s)$	+0.300
$Bi^{3+} + 3e^-$	⇌	$Bi(s)$	+0.308
$Cu^{2+} + 2e^-$	⇌	$Cu(s)$	+0.337
$VO^{2+} + 2H^+ + e^-$	⇌	$V^{3+} + H_2O$	+0.34
$[Fe(CN)_6]^{3-} + e^-$	⇌	$[Fe(CN)_6]^{4-}$	+0.36
$Fc^+ + e^-$	⇌	$Fc(s)$	+0.4
$O_2(g) + 2H_2O + 4e^-$	⇌	$4OH^-(aq)$	+0.401
$H_2MoO_4 + 6H^+ + 3e^-$	⇌	$Mo^{3+} + 2H_2O$	+0.43
$CH_3OH(aq) + 2H^+ + 2e^-$	⇌	$CH_4(g) + H_2O$	+0.50
$SO_2(aq) + 4H^+ + 4e^-$	⇌	$S(s) + 2H_2O$	+0.50
$Cu^+ + e^-$	⇌	$Cu(s)$	+0.520
$CO(g) + 2H^+ + 2e^-$	⇌	$C(s) + H_2O$	+0.52
$I_3{}^- + 2e^-$	⇌	$3I^-$	+0.53
$I_2(s) + 2e^-$	⇌	$2I^-$	+0.54
$[AuI_4]^- + 3e^-$	⇌	$Au(s) + 4I^-$	+0.56
$H_3AsO_4(aq) + 2H^+ + 2e^-$	⇌	$H_3AsO_3(aq) + H_2O$	+0.56
$[AuI_2]^- + e^-$	⇌	$Au(s) + 2I$	+0.58
$MnO_4{}^- + 2H_2O + 3e^-$	⇌	$MnO_2(s) + 4OH^-$	1.692
$S_2O_3{}^{2-} + 6H^+ + 4e^-$	⇌	$2S(s) + 3H_2O$	+0.60
$H_2MoO_4(aq) + 2H^+ + 2e^-$	⇌	$MoO_2(s) + 2H_2O$	+0.65
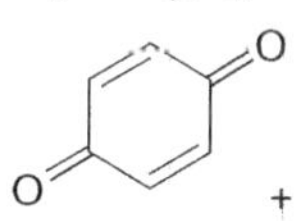 $+ 2H^+ + 2e^-$	⇌	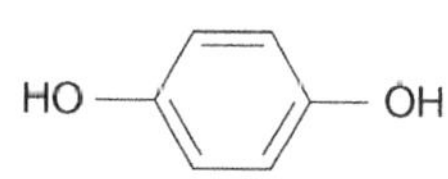	+0.6992
$O_2(g) + 2H^+ + 2e^-$	⇌	$H_2O_2(aq)$	+0.70
$Tl^{3+} + 3e^-$	⇌	$Tl(s)$	+0.72
$PtCl_6{}^{2-} + 2e^-$	⇌	$PtCl_4{}^{2-} + 2Cl^-$	+0.726
$H_2SeO_3(aq) + 4H^+ + 4e^-$	⇌	$Se(s) + 3H_2O$	+0.74
$PtCl_4{}^{2-} + 2e^-$	⇌	$Pt(s) + 4Cl^-$	+0.758
$Fe^{3+} + e^-$	⇌	Fe^{2+}	+0.77
$Ag^+ + e^-$	⇌	$Ag(s)$	+0.7996
$Hg_2{}^{2+} + 2e^-$	⇌	$2Hg(l)$	+0.80
$NO_3{}^-(aq) + 2H^+ + e^-$	⇌	$NO_2(g) + H_2O$	+0.80
$2FeO_4{}^{2-} + 5H_2O + 6e^-$	⇌	$Fe_2O_3(s) + 10OH^-$	+0.81
$[AuBr_4]^- + 3e^-$	⇌	$Au(s) + 4Br^-$	+0.85
$Hg^{2+} + 2e^-$	⇌	$Hg(l)$	+0.85
$[IrCl_6]^{2-} + e^-$	⇌	$[IrCl_6]^{3-}$	+0.87
$MnO_4{}^- + H^+ + e^-$	⇌	$HMnO_4{}^-$	+0.90
$2Hg^{2+} + 2e^-$	⇌	$Hg_2{}^{2+}$	+0.91
$Pd^{2+} + 2e^-$	⇌	$Pd(s)$	+0.915
$[AuCl_4]^- + 3e^-$	⇌	$Au(s) + 4Cl^-$	+0.93

(*continued overleaf*)

TABLE 6.1 ***(Continued)***

Half-reaction			
Oxidant	⇌	Reductant	$E°$ (V)
$MnO_2(s) + 4\,H^+ + e^-$	⇌	$Mn^{3+} + 2\,H_2O$	+0.95
$NO_3^-(aq) + 4\,H^+ + 3\,e^-$	⇌	$NO(g) + 2\,H_2O(l)$	+0.958
$[AuBr_2]^- + e^-$	⇌	$Au(s) + 2\,Br^-$	+0.96
$[HXeO_6]^{3-} + 2\,H_2O + 2\,e^-$	⇌	$[HXeO_4]^- + 4\,OH^-$	+0.99
$[VO_2]^+(aq) + 2\,H^+ + e^-$	⇌	$[VO]^{2+}(aq) + H_2O$	+1.0
$H_6TeO_6(aq) + 2\,H^+ + 2\,e^-$	⇌	$TeO_2(s) + 4\,H_2O$	+1.02
$Br_2(l) + 2\,e^-$	⇌	$2\,Br^-$	+1.066
$Br_2(aq) + 2\,e^-$	⇌	$2\,Br^-$	+1.0873
$IO_3^- + 5\,H^+ + 4\,e^-$	⇌	$HIO(aq) + 2\,H_2O$	+1.13
$[AuCl_2]^- + e^-$	⇌	$Au(s) + 2\,Cl^-$	+1.15
$HSeO_4^- + 3\,H^+ + 2\,e^-$	⇌	$H_2SeO_3(aq) + H_2O$	+1.15
$Ag_2O(s) + 2\,H^+ + 2\,e^-$	⇌	$2\,Ag(s) + H_2O$	+1.17
$ClO_3^- + 2\,H^+ + e^-$	⇌	$ClO_2(g) + H_2O$	+1.18
$[HXeO_6]^{3-} + 5\,H_2O + 8\,e^-$	⇌	$Xe(g) + 11\,OH^-$	+1.18
$Pt^{2+} + 2\,e^-$	⇌	$Pt(s)$	+1.188
$ClO_2(g) + H^+ + e^-$	⇌	$HClO_2(aq)$	+1.19
$2\,IO_3^- + 12\,H^+ + 10\,e^-$	⇌	$I_2(s) + 6\,H_2O$	+1.20
$ClO_4^- + 2\,H^+ + 2\,e^-$	⇌	$ClO_3^- + H_2O$	+1.20
$\mathbf{O_2}(g) + \mathbf{4\,H^+} + \mathbf{4}\,e^-$	⇌	$2\,H_2O$	+1.229
$MnO_2(s) + 4\,H^+ + 2\,e^-$	⇌	$Mn^{2+} + 2\,H_2O$	+1.23
$[HXeO_4]^- + 3\,H_2O + 6\,e^-$	⇌	$Xe(g) + 7\,OH^-$	+1.24
$Tl^{3+} + 2\,e^-$	⇌	Tl^+	+1.25
$Cr_2O_7^{2-} + 14\,H^+ + 6\,e^-$	⇌	$2\,Cr^{3+} + 7\,H_2O$	+1.33
$Cl_2(g) + 2\,e^-$	⇌	$2\,Cl^-$	+1.36
$CoO_2(s) + 4\,H^+ + e^-$	⇌	$Co^{3+} + 2\,H_2O$	+1.42
$2\,NH_3OH^+ + H^+ + 2\,e^-$	⇌	$N_2H_5^+ + 2\,H_2O$	+1.42
$2\,HIO(aq) + 2\,H^+ + 2\,e^-$	⇌	$I_2(s) + 2\,H_2O$	+1.44
$BrO_3^- + 5\,H^+ + 4\,e^-$	⇌	$HBrO(aq) + 2\,H_2O$	+1.45
β-$PbO_2(s) + 4\,H^+ + 2\,e^-$	⇌	$Pb^{2+} + 2\,H_2O$	+1.460
α-$PbO_2(s) + 4\,H^+ + 2\,e^-$	⇌	$Pb^{2+} + 2\,H_2O$	+1.468
$2\,BrO_3^- + 12\,H^+ + 10\,e^-$	⇌	$Br_2(l) + 6\,H_2O$	+1.48
$2ClO_3^- + 12\,H^+ + 10\,e^-$	⇌	$Cl_2(g) + 6\,H_2O$	+1.49
$HClO(aq) + H^+ + 2\,e^-$	⇌	$Cl^-(aq) + H_2O$	+1.49
$MnO_4^- + 8\,H^+ + 5\,e^-$	⇌	$Mn^{2+} + 4\,H_2O$	+1.51
$HO_2^{\bullet} + H^+ + e^-$	⇌	$H_2O_2(aq)$	+1.51
$Au^{3+} + 3\,e^-$	⇌	$Au(s)$	+1.52
$NiO_2(s) + 4\,H^+ + 2\,e^-$	⇌	$Ni^{2+} + 2\,OH^-$	+1.59
$Ce^{4+} + e^-$	⇌	Ce^{3+}	+1.61
$2\,HClO(aq) + 2\,H^+ + 2\,e^-$	⇌	$Cl_2(g) + 2\,H_2O$	+1.63
$Ag_2O_3(s) + 6\,H^+ + 4\,e^-$	⇌	$2\,Ag^+ + 3\,H_2O$	+1.67
$HClO_2(aq) + 2\,H^+ + 2\,e^-$	⇌	$HClO(aq) + H_2O$	+1.67
$Pb^{4+} + 2\,e^-$	⇌	Pb^{2+}	+1.69
$MnO_4^- + 4\,H^+ + 3\,e^-$	⇌	$MnO_2(s) + 2\,H_2O$	+1.70
$AgO(s) + 2\,H^+ + e^-$	⇌	$Ag^+ + H_2O$	+1.77
$H_2O_2(aq) + 2\,H^+ + 2\,e^-$	⇌	$2\,H_2O$	+1.78
$Co^{3+} + e^-$	⇌	Co^{2+}	+1.82

TABLE 6.1 (*Continued*)

Half-reaction			
Oxidant	⇌	Reductant	$E°$ (V)
$Au^+ + e^-$	⇌	$Au(s)$	+1.83
$BrO_4^- + 2\,H^+ + 2\,e^-$	⇌	$BrO_3^- + H_2O$	+1.85
$Ag^{2+} + e^-$	⇌	Ag^+	+1.98
$S_2O_8^{2-} + 2\,e^-$	⇌	$2\,SO_4^{2-}$	+2.010
$O_3(g) + 2\,H^+ + 2\,e^-$	⇌	$O_2(g) + H_2O$	+2.075
$HMnO_4^- + 3\,H^+ + 2\,e^-$	⇌	$MnO_2(s) + 2\,H_2O$	+2.09
$XeO_3(aq) + 6\,H^+ + 6\,e^-$	⇌	$Xe(g) + 3\,H_2O$	+2.12
$H_4XeO_6(aq) + 8\,H^+ + 8\,e^-$	⇌	$Xe(g) + 6\,H_2O$	+2.18
$FeO_4^{2-} + 3\,e^- + 8\,H^+$	⇌	$Fe^{3+} + 4\,H_2O$	+2.20
$XeF_2(aq) + 2\,H^+ + 2\,e^-$	⇌	$Xe(g) + 2HF(aq)$	+2.32
$H_4XeO_6(aq) + 2\,H^+ + 2\,e^-$	⇌	$XeO_3(aq) + 3\,H_2O$	+2.42
$F_2(g) + 2\,e^-$	⇌	$2\,F^-$	+2.87
$F_2(g) + 2\,H^+ + 2\,e^-$	⇌	$2\,HF(aq)$	+3.05
$KrF_2(aq) + 2\,e^-$	⇌	$Kr(g) + 2\,F^-(aq)$	+3.27

Source: https://en.wikipedia.org/wiki/Standard electrode potential_(data_page). Used under CC BY SA 3.0 https://creativecommons.org/licenses/by-sa/3.0/.

hereafter as Zn/Zn^{2+} and Cu/Cu^{2+} to indicate that zinc metal is in contact with zinc ion in solution; the line "/" represents a solid-solution boundary. Based on the above discussion, the above galvanic cell

$$Zn \rightarrow Zn^{2+} + \cancel{2e^-} \tag{6.3}$$

$$Cu^{2+} + \cancel{2e^-} \rightarrow Cu \tag{6.4}$$

$$\text{Adding the two equations,}\quad Zn + Cu^{2+} \rightarrow Zn^{2+} + Cu \tag{6.5}$$

This galvanic cell is an example of a battery and is often called Daniell cell.

Under the given conditions, the difference in the cathode and anode potentials established in the cell will be manifested as the battery voltage. In other words, equation (6.6) gives the battery voltage,

$$E(\text{V}) = E_c - E_a \tag{6.6}$$

where E_c is the cathode potential and E_a is the anode potential. At 1 M concentrations of the two solutions in Figure 6.1, $E_c = 0.3370$ V and $E_a = -0.7618$ V,

$$E(\text{V}) = 0.3370 - (-0.7618) = 1.0988 \ \text{V} = 1.10 \ \text{V} \ \text{(approximated)}$$

If the concentration of the solutions in the two compartments of the half cells changes or is different from 1 M, then the battery voltage will also change. If the changing concentrations are known, then it is possible to calculate the expected E_c and E_a. This is discussed in the following section.

Consider that the process shown in equation 6.3 is occurring. When a current is drawn from the battery, the zinc metal electrode will continuously dissolve and ultimately, there will be no zinc metal left in the anode compartment. At this stage, the battery is "dead" and can no longer be useful. This is an example of a **primary battery**. However, if we replace with a **new** zinc plate in the anode compartment and rewire it, we can bring it back to working condition. In a **primary battery**, this operation is unrealistic as it would require breaking open the battery and replacing the anode material. If the zinc metal is not fully oxidized (only a portion of the metal is dissolved) and the processes in equations 6.1 and 6.2 are reversed, this would result in the **dissolved** zinc metal (Zn^{2+} ions) accept the electrons and redeposit on the remaining zinc metal-regeneration process. With a battery, if there is discharging and then recharging process is involved that would be a **secondary** battery.

6.2 DEFINITIONS

6.2.1 Battery

A primary battery is a galvanic cell that provides energy and power output until the redox reactions are complete and cannot be reversed. Sometimes, the primary batteries are called voltaic cells giving emphasis on the voltage only. The primary batteries constitute 90% of the $50 billion dollar market. About 15 billion batteries are discarded after usage in the world.

A secondary battery is a rechargeable battery that can be charged and discharged; in the charged mode, electrical energy is stored and is available for usage. In the discharged mode, the electrical energy is drawn for driving devices.

6.2.2 Battery Voltage

A battery voltage can be theoretically estimated from the standard potentials as described in the previous section (Section 6.1). It is given by

$$E(\mathrm{V}) = E_{\mathrm{c}} - E_{\mathrm{a}}$$

assuming the two half cells of the battery are having 1 M concentrations of the redox species, the standard potentials shown in Table 6.1 can be used for calculating E(V). When the concentrations are other than 1 M, E_c and E_a values have to be modified using the following equation derived by Nernst (it is often called Nernst equation). Let us consider the cathodic reaction in Section 6.1 where

$$\mathrm{Cu}^{2+} + 2\mathrm{e}^- \rightarrow \mathrm{Cu}$$

The potential of this solution in other than 1 M concentration is given by equation 6.7

$$E_{\mathrm{c}} = E^{\circ} - [RT/2F]\,\ln[a_{\mathrm{Cu}}/a_{\mathrm{Cu}}{}^{2+}] \tag{6.7}$$

where E° is the standard potential of Cu/Cu^{2+} (1 M concentration of copper ion in solution), a_{Cu} and $a_{\mathrm{Cu}}{}^{2+}$ are the activities of copper metal (solid activity is 1.0) and cupric ion in solution, respectively, R is the gas constant (8.321 J/mol K), T is the absolute temperature, and F is Faraday constant. With dilute solutions, the activity coefficient tends toward one.

Hence, the activity of cupric ion and the concentration of cupric ion are equal. Equation 6.7 then can be rewritten as

$$E_c = 0.3370 - [RT/2F] \ln[1/C_{Cu}{}^{2+}] \tag{6.8}$$

In a simplified notation for temperature of 25°C and converting ln from base e to base 10 will yield

$$E_c = 0.3370 - [0.059/2] \log[1/C_{Cu}{}^{2+}] \tag{6.9}$$

Similarly, E_a can be modified for any selected concentration of zinc ion as

$$E_a = -0.7618 - [0.059/2] \log[1/C_{Zn}{}^{2+}] \tag{6.10}$$

Hence, the battery voltage as given by $[E_c - E_a]$ can be obtained by substituting E_c and E_a in equations 6.9 and 6.10,

$$E(\mathrm{V}) = \{0.3370 - [0.059/2] \log[1/C_{Cu}{}^{2+}]\} - \{-0.7618 - [0.059/2] \log[1/Zn^{2+}]\} \tag{6.11}$$

$$E(\mathrm{V}) = \{0.3370 + 0.7618\} - [0.059/2] \log[C_{Cu}{}^{2+}/C_{Zn}{}^{2+}] \tag{6.12}$$

Equation (6.12) shows that the battery voltage will change with concentrations of Cu^{2+} and Zn^{2+}. The above discussion is valid for any battery in its performance.

CRITICAL THINKING QUESTION

A galvanic cell operates at 298 K, where the following reaction takes place:

$$Zn(s) + Ni^{2+}(aq) \rightarrow Zn^{2+} + Ni(s)$$

Note: $[Ni^{2+}] = C_{Ni}{}^{2+}$
$[Zn^{2+}] = C_{Zn}{}^{2+}$

A What is the voltage of this cell under standard conditions?

B When $[Ni^{2+}] = 3.0$ M and $[Zn^{2+}] = 0.100$ M, what is the voltage?

C What is the voltage when $[Ni^{2+}] = 0.20$ M and $[Zn^{2+}] = 0.900$ M?

A

The first question is: under standard conditions which utilizes 1 M concentrations of solutions in the above reaction. We can use the standard potentials in Table 3.1 for the two redox couples in the above reaction, Zn/Zn^{2+} and Ni/Ni^{2+}.

$$Zn/Zn^{2+} \quad E^\circ = -0.7618\,\mathrm{V}$$

$$Ni/Ni^{2+} \quad E^\circ = -0.250\,\mathrm{V}$$

The galvanic cell is made of the abovementioned two redox couples, and since Zn/Zn^{2+} is having more negative potential than Ni/Ni^{2+}, it will function as anode. The other redox couple Ni/Ni^{2+} will act as the cathode. The battery voltage is given by

$$E = E_c - E_a$$

Battery voltage, $E = -0.250 - (-0.7618) = 0.512$ V

B

For nonideal condition, the potentials of E_c and E_a are to be calculated using equation 6.7 with the given concentration in the problem,

$$E_c = E^\circ - [0.059/2] \log\{1/[Ni^{2+}]\}$$

$$E_c = -0.250 - [0.0592/2] \log\{1/3.0\} = -0.250 + 0.014 = -0.236\,V$$

Similarly $E_a = -0.7618 - [0.059/2] \log\{1/0.10\} = -0.791$ V

Battery voltage $E = E_c - E_a = -0.236 - (-0.791) = \mathbf{0.555}\,V$

C

$$E_c = -0.250 - [0.059/2] \log\{1/0.20\} = -0.229\,V$$

$$E_a = -0.7618 - [0.059/2] \log\{1/0.90\} = -0.791\,V$$

Battery voltage $E = E_c - E_a = -0.229 - (-0.791) = 0.562\,V$

Write the two half cell reactions and arrive at the overall reaction given in the problem.

6.3 WORKING UNITS

6.3.1 Specific Energy

The amount of energy available from a battery can be obtained using Faraday's law of electrolysis, which states that 1 F of electricity will bring about 1 g equivalent of substance dissolution. Taking the example of Daniell cell,

$$1\,F = 96{,}494\,C = 1\,g \text{ equivalent of Zn}$$

In the case of Zn, the following reaction

$$Zn \rightarrow Zn^{2+} + 2e^-$$

occurs involving $2e^-$. Since atomic mass of Zn is 65.39 g/mol, the equivalent mass is

$$\text{Equivalent mass} = [\text{Atomic Mass} / \text{number of electrons involved}]$$
$$= [65.39/2] = 32.69\,g$$

So passage of 1 F = 32.69 g will dissolve.

$$96{,}494\,C = 96{,}494\,A\,s = 32.69\,g; \text{ note that } [1\,C = 1\,A \times 1\,s].$$

$$\text{Capacity} = [96{,}494\,A\,s / 32.69\,g] = [96{,}494\,A.s/3600\,s] \times [1\,h/32.69\,g]$$
$$= [26.8\,A\,h/32.69\,g] = 0.819\,A\,h/g$$
$$= 819\,A\,h/kg$$

Based on the above calculation, 1 kg of Zn as anode in the Daniell cell is expected to give 819 A for 1 h duration. So the general formula is to be used,

$$\text{Capacity} = [26.8\ \text{A h/Equivalent mass active material in grams}] \tag{6.13}$$

$$\begin{aligned}\text{Specific Energy} &= \text{Capacity} \times \text{Voltage}\\ &= [0.819\,\text{A h/g}] \times 1.10\ \text{V}\\ &= 0.909\,\text{V A h/g}\end{aligned} \tag{6.14}$$

Since $1\ \text{W} = 1\ \text{V} \times 1\ \text{A}$

$$\text{Specific energy} = 0.909\,\text{Wh/g for the Daniel cell}$$

6.3.2 Energy Efficiency

$$\text{Practical available energy } \varepsilon_\text{P} = \int_0^t (EI)\ \text{d}t \tag{6.15}$$

where E is the battery voltage and I is the current. The integration time is t.

$$\eta_\text{wh} = [\varepsilon_\text{P}/\varepsilon_\text{T}] \tag{6.16}$$

ε_T is the theoretical efficiency and η_wh is the ratio of total energy during discharging/total energy during charging.

6.3.3 Power Density

$$P = [E^2/R_\text{i}W] \tag{6.17}$$

where W is the weight of the battery, including electrodes, electrolyte, and ancillary components; R_i is the internal resistance of the battery; and E is the battery voltage.

6.3.4 Discharge Rate

The discharge rate is defined as the rate at which charge is drawn from the battery. t-h rate is given as $[Q_\text{theoretical}/t]$, where $Q_\text{theoretical}$ is the theoretical quantity of electricity available and t is the duration in seconds.

The battery capacity is related to discharge rate by the following relationship:

$$C_\text{P} = I^k t \tag{6.18}$$

where C_P is the ampere hour capacity at 1 A discharge rate, I is the discharge current in amperes, t is the discharge time in hours, and k is Peukert's coefficient.

6.3.5 Coulombic Efficiency

$$Q_\text{theoretical} = X\,nF \tag{6.19}$$

where X is the moles of reaction with complete discharge of the battery, n is the number of electrons involved in the battery reaction, and F is the Faraday constant.

Practical capacity (Q_P) is given by

$$Q_P = \int_0^t I(t)\mathrm{d}t = (1/R_e) \int_0^t E(t)\mathrm{d}t \tag{6.20}$$

R_e is the external load and t is the time at which $E = 0.6\ E_{CO}$, E_{CO} is the cutoff voltage.

$$\eta = [Q_P / Q_{theoretical}] \tag{6.21}$$

6.3.6 Energy Efficiency (EF)

Energy efficiency (EF) is defined as a product of voltage efficiency and coulombic efficiency

$$\mathrm{EF} = [E_D/E_c][I_D \cdot t_D/I_c \cdot t_c] \tag{6.22}$$

E_D = voltage at discharging
E_c = voltage at charging
I_D = discharge current
I_c = charging current
t_D = time of discharging
t_c = time of charging

6.3.7 Volumetric Density (VD)

Volumetric density (VD) is defined by taking the volume of the battery into account,

$$\mathrm{VD} = [\mathrm{Wh}/L] \tag{6.23}$$

Wh refers to watt-hour delivered by the battery and L is the volume of the battery.

6.3.8 *C* and *E* Rates in Discharging a Battery

C-Rate measures the rate at which the battery is discharged in relation to its maximum capacity. Let us assume that a battery has the capability of delivering a maximum of 1 A current.

Then, 1*C* rate means that the battery will be discharged in 1 h if 1 A current is drawn.

If a battery has a capacity of 500 A h, then 1*C* refers to drawing 500 A current for 1 h. A 2*C* rate for this battery would be 1000 A and a *C*/2 rate would be 250 A. Similarly, *E*-rate describes the discharge power. A 1*E* rate represents discharging the entire battery in 1 h.

6.3.9 Self-discharge and Shelf Life

A battery tends to lose its capacity with time very slowly due to internal chemical reactions. The loss of capacity with time is a function of the type of chemical reactions, temperature of the environment in which it is present, state of charge, and extent of charging.

Shelf life is the time for which the battery can be stored without becoming unfit. In the quality control of a battery, shelf-life test is carried out by drawing a small current for a period of 3 months at 37°C and monitoring the voltage of the battery during this period. If it is constant, then it is accepted for marketing.

6.3.10 Cycle Life

A secondary battery is subjected to a process of charging and discharging a number of times. The ability of the battery to survive by giving the energy and power densities in this process of cyclic charge is defined as the cycle life. Different manufacturers follow different criteria for labeling the battery life. For example, Apple Inc. uses charging the battery to half of full charged battery and then discharging the same amount of charge as one cycle. Some others use full charging and discharging to a limiting level.

6.4 EXAMPLES OF SELECTED BATTERIES

In the earlier section, we discussed a simple Daniell cell. In this section, we consider a few selected batteries for their performance. This section illustrates that the active material (may be called the fuel) at the anode can differ in performance in different chemical environments. Daniell cell uses zinc anode in an aqueous solution under slightly acidic conditions. In the following example, a Zn–Ni battery that operates under alkaline conditions gives a different performance.

6.4.1 Zn–Ni battery

At anode $$Zn + 2OH^- \rightarrow Zn(OH)_2 + 2e^-$$

At cathode $$2\ NiOOH + 2H_2O + 2e^- \rightarrow 2Ni(OH)_2 + 2OH^-$$

$E_c = 0.52$ V and $E_a = -1.12$ V, and hence the expected battery voltage under standard conditions would be $E = 0.52 - (-1.12) = 1.64$ V. Using the procedures described earlier, the following quantities have evolved:

Specific energy: 100 Wh/kg
Energy density: 280 Wh/L
Cell voltage: 1.64 V

6.4.2 Ni–Metal Hydride Battery

Based on the previous example of Zn–Ni battery, Ni–metal hydride battery has been constructed, which is considered for automobiles,

At the anode, $$MH + OH^- \rightarrow M + H_2O + e^-$$

At the cathode, $$NiOOH + H_2O + e^- \rightarrow Ni(OH)_2 + OH^-$$

MH in the above example is metal hydride; metal alloys V, Ti, Zr, Ni, Cr, Co, and Fe are used. The $E_c = 0.83$ V and $E_a = 0.52$ V would give the battery voltage of 1.35 V. This battery is expected to give specific energy in the range 60–120 Wh/kg.

6.4.3 Zinc–Air Battery

Zinc is present in alkaline environment in this battery with the following reactions occurring:

At the anode, $$Zn + 2OH^- \rightarrow Zn(OH)_2 + 2e^-$$

At the cathode, $$1/2O_2 + 2e^- + H_2O \rightarrow 2OH^-$$

With $E_c = 0.40$ V and $E_a = -1.25$ V, the battery is expected to give a voltage of 1.65 V. The specific energy of this battery is 1370 Wh/kg, and the energy density is 1480 Wh/L. This battery is used in radios, toys, and photoflash.

TABLE 6.2 Batteries and their parameters

System	Pb/acid	Ni/Fe	Na/S	Zn/Cl_2
Anode	Pb	Fe	Na	Zn
Cathode	PbO_2	NiOOH	S	Cl_2
Electrolyte	H_2SO_4	KOH	β-Alumina	$ZnCl_2$
Cell voltage (V)				
Open circuit (V)	2.1	1.4	2.078	2.12
Nominal (V)	2.0	1.2	1.7	1.9
Energy density (theoretical) (Wh/kg)	25–40 (252)	50 (340)	120–150 (700)	100 (830)
Wh/L	60	100	280	130
Operating temperature (°C)	−40 to 50	−10 to 50	210–400	20–60
Cycle life	500–1000	2000–4000	1000–4000	
Applications	Automotive Marine aircraft Diesel engine	Electric vehicle	Electric vehicle Energy storage	Electric vehicle Energy storage
Advantage	Low cost Ready availability High and low temperature operation	Very rugged Long life	Good power capability Low cost materials	Good energy density Long life
Disadvantage	Low energy density Low cycle life	Low power Energy density	Safety problems High corrosion	Difficult to manufacture

Table 6.2 shows several batteries and their expected voltage, specific energy, cycle life, and durability. The theoretically estimated specific energy is never reached in any battery.

Generally, the practically available specific energy is less by several factors, which are listed in the table.

6.4.4 Alkaline Cells

A number of alkaline cells have been reported in the literature, and the most popular one is $Zn–MnO_2$, which uses Zn as the anode and MnO_2 as the cathode. The following reactions occur:

$$\text{At the anode,} \qquad Zn + 2OH^- \rightarrow ZnO + H_2O + 2e^-$$

$$\text{At the cathode,} \qquad 2\,MnO_2 + H_2O + 2e^- \rightarrow Mn_2O_3 + 2OH^-$$

The over all reaction: $Zn + H_2O + 2MnO_2 + 2e^- \rightarrow Mn_2O_3 + 2OH^-$. Under standard conditions, $E_c = 0.25$ V and $E_a = 1.25$ V. The expected battery voltage is 1.50 V.

6.4.5 Pb–PbO_2 Acid Battery

Pb–PbO_2 acid battery is a unique battery that has provided the energy for starting automobiles for decades. Table 6.2 gives the characteristics of this battery. The battery reactions are as follows:

At the anode, $Pb + HSO_4^- \rightarrow PbSO_4 + H^+ + 2e^-$

At the cathode, $PbO_2 + HSO_4^- + 3H^+ + 2e^- \rightarrow PbSO_4 + 2H_2O$

The overall reaction: $Pb + PbO_2 + 2H_2SO_4 \rightarrow 2PbSO_4 + 2H_2O$ which is expected to give 2.1 V and an energy density of 252 Wh/kg and 60 Wh/L with a cycle life in the range 500–1000. It can operate in the temperature range −40 to 50°C. Although it has a high theoretical energy density, in practice only 25–40 Wh/kg is realized.

In Tables 6.2–6.4, some of the well-known batteries' operating temperatures are listed with their characteristics, advantages, and disadvantages. The values of energy densities in brackets refer to the theoretically estimated values and are identical to specific energies discussed in the earlier section.

Lithium batteries are considered as a possible alternative to fuel cells in automobiles. Hence, in the following section we consider the lithium batteries.

6.4.6 Lithium Batteries

Lithium metal is used as the anode, and a wide variety of cathodes are used in the construction of lithium batteries. Examples are Li–MnO_2, Li–$(CF)_x$, Li–FeS_2, Li–$SOCl_2$, Li–SO_2Cl_2, Li–Ag_2CrO_4, Li–silver oxide + Vanadium pentoxide, Li–CuO, Li–$CuCl_2$, Li–(Al–MnO_2), Li–(Al–V_2O_5), and Li–air. Generally, they are primary batteries. The battery reactions are as follows:

At the anode, $Li \rightarrow Li^+ + e^-$

At the cathode, $4Li^+ + 4e^- + 2SOCl_2 \rightarrow 4LiCl + SO_2 + S$

The overall reaction: $4Li + 2SOCl_2 \rightarrow 4LiCl + SO_2 + S$

TABLE 6.3 Secondary batteries

System	Fe/air	Zn/air	Li/air	Al/air
Anode	Fe	Zn	Li	Al
Cathode	O_2	O_2	O_2	O_2
Electrolyte	KOH	KOH	KOH	KOH
Cell voltage (V)				
Open circuit (V)	1.3	1.6	3.4	2.78
Nominal (V)	1.0	1.2	2.4	1.6
Energy density (theoretical) (Wh/kg)	95 (1250)	110– (1320)	250 (13,124)	150–200 (8135)
Wh/L	110	100	80	
Operating temperature (°C)	50	50–60	Ambient	50–80
Cycle life	1000	300–400		
Applications	Electric vehicles Energy storage	Electric vehicle Energy storage	Electric vehicle Energy storage	Energy storage Standby power
Advantage	High specific energy density	Good energy density	Good cell voltage	High energy density
Disadvantage	Low efficiency	Poor life	High corrosion of Li electrode	Low life

TABLE 6.4 Secondary batteries

System	Ni/Cd	Ni/Zn	Zn/Br$_2$	Zn/Ago
Anode	Cd	Zn	Zn	Zn
Cathode	NIOOH	NIOOH	Br$_2$	AgO
Electrolyte	KOH	KOH	ZnBr$_2$	KOH
Cell voltage (V)				
Open circuit (V)	1.29	1.73	1.82	1.85
Nominal (V)	1.2	1.6	1.7	1.5
Energy density (the theoretical) (Wh/kg)	35 (210)	70 (370)	70 (430)	100 (440)
Wh/L	110	100	80	
Operating temperature (°C)	−30 to 50	−30 to 60	20–50	−25 to 70
Cycle life	500–2000	200–300		150–300
Applications	Emergency lighting engine starting instruments	Electric vehicles military applications	Electric vehicle energy storage	Communications mines satellites space probes
Advantage	Very rugged	High energy density	Good energy density	High energy density High discharge rate
Disadvantage	Low energy density high cost	Poor cycle life	Problems of bulkiness safety	High cost low cycle life

For $E_c = 0.50$ V and $E_a = -3.04$ V in the above reactions, a battery voltage of 3.54 V is expected.

The specific energy may be calculated following the steps given earlier by equations 6.12 and 6.13:

$$\text{Capacity} = [26.8\,\text{A h}/6.94\,\text{g}] = 3.86\,\text{A h/g}$$

$$\text{Specific Energy} = \text{Capacity} \times \text{Voltage}$$

$$= 3.86 \times 3.54 = 13.67\,\text{Wh/g}$$

$$= 13{,}670\,\text{Wh/kg}$$

These batteries have high charge density and high cost per unit. Lithium ion batteries use lithium ions moving between the electrodes during charging and discharging. It uses intercalated lithium as the electrode. These batteries are rechargeable and hence finding replacement for lead acid battery discussed in the previous section. Let us consider a battery made of $LiCoO_2$ as cathode and graphite anode. This battery can give a voltage of 3.6 V, specific energy of 240 Wh/kg, and last for 500–1000 cycles. Another example of Li

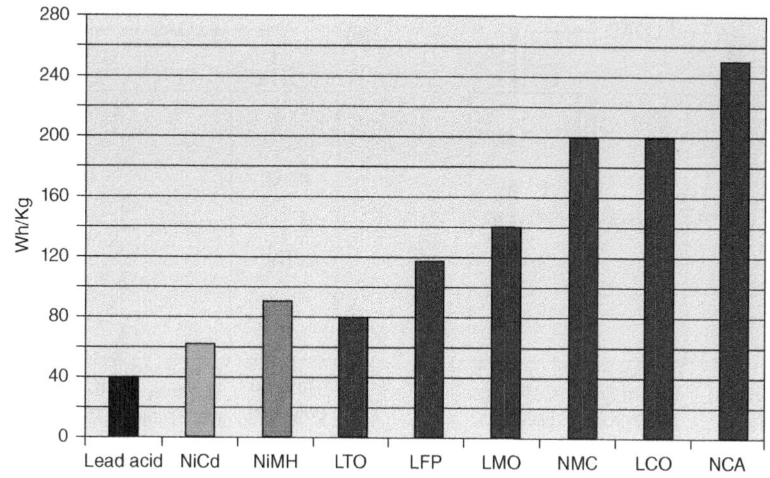

Figure 6.2 Specific energy available with different secondary batteries. NiCd = Nickel–Cadmium, NiMH = Nickel–metal hydride, LTO = Lithium titanate, LFP = Lithium iron phosphate, LMO = Lithium manganese dioxide, NMC = Lithium Nickel manganese cobalt oxide, LCO = Lithium cobalt oxide, and NCA = Lithium nickel cobalt aluminum oxide. [Courtesy Cadex.]

ion battery is Li_2MnO_4 as the cathode and graphite as anode, which can give a voltage of 3.7 V, specific energy of 100–150 Wh/kg, and last for 300–700 cycles. Recently, cycle life of 1000–2000 has been reached using $LiNiMnCoO_2$ as the cathode and graphite as the anode, with a specific energy of 150–220 Wh/kg. Figure 6.2 shows the specific energy available with different secondary batteries. Lithium ion battery has the capability of giving six times the energy density of lead acid battery and finding replacement in automobiles, electric trains, power tools, and so on.

6.5 CONDUCTING POLYMER BATTERIES (ORGANIC BATTERIES)

With the discovery of conducting polymers by MacDiarmid, Heeger, Shirakawa and coworkers [7], a new arena of activities in polymer science emerged. These polymers are also called the synthetic metals as they possess some of the important properties of naturally occurring metals, particularly electrical conductivity. The polymer state of charge opens up a variable redox potential, defined as "floating redox potential." The batteries we considered in the previous sections all carry a distinct redox potential that is independent of their state of charge or discharge. Figure 6.3 shows an interesting class of conducting polymer batteries that have been proposed and tested. In Figure 6.3a, a redox couple Li/Li^+ is combined with conducting polymer redox couple P/P^+A^-. A simple example is that lithium metal in a solution [tetrahydrofuran (THF)] containing lithium salt ($LiClO_4$) is coupled with polyaniline (P)-coated graphite electrode. The battery reaction is given as follows:

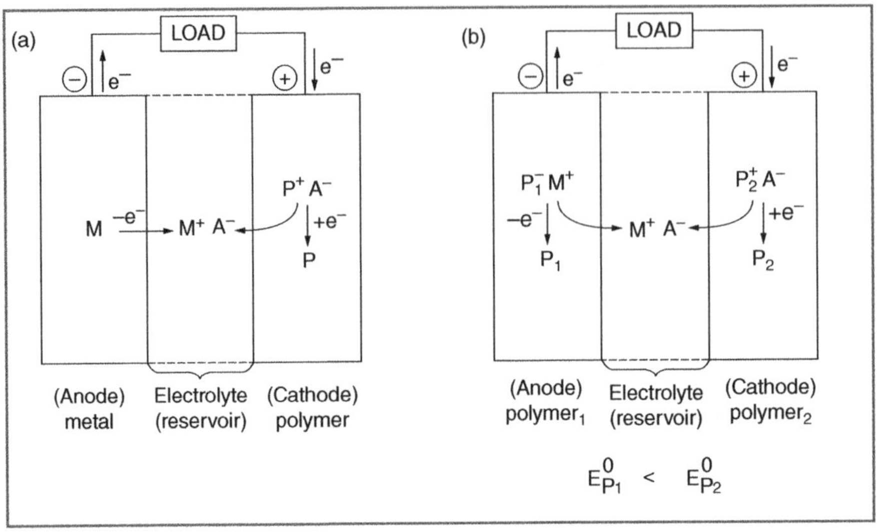

Figure 6.3 Conducting polymer batteries. (a) M is anode metal (typically Li or Zn), M^+A^- is the electrolyte filling the anode and cathode, P^+A^- is conducting polymer is in the oxidized state, and P is conducting polymer in the reduced state. (b) $P_1{}^-M^+$ is the reduced state of conducting polymer and $P_2{}^+A^-$ is the oxidized state of a conducting polymer. P_1 is oxidized state of the polymer and P_2 is the reduced state of the polymer. [*Source:* Novák *et al.* (1997) [5]. Reproduced with permission of American Chemical Society.]

During discharging,

$$\text{At the anode,} \quad Li \rightarrow Li^+ + e^-$$

$$\text{At the cathode,} \quad P^+A^- + e^- \rightarrow P + A^-$$

$$\text{The overall reaction:} \quad Li + P^+A^- \rightarrow Li^+ + P + A^-$$

Li^+ enters the solution at the anode. P^+A^- is polyaniline (PANI) in the oxidized state, ion paired with A^- ($ClO_4{}^-$). The battery voltage with polyaniline electrode is about 4 V with a specific energy in the range 87–539 Wh/kg. A number of other conducting polymers can substitute PANI (see Table 6.5).

During charging, the above processes are reversed.

In Figure 6.3b, two conducting polymers are used in constructing a battery.

$$\text{At the anode,} \quad P_1{}^-M^+ \rightarrow P_1 + M^+ + e^-$$

$$\text{At the cathode,} \quad P_2{}^+A^- + e^- \rightarrow P_2 + A^-$$

$$\text{The overall reaction:} \quad P_1{}^-M^+ + P_2{}^+A^- \rightarrow P_1 + P_2 + M^+ + A^-$$

For this battery to operate, the redox potential of $P_1{}^-M^+$ should be less than $P_2{}^+A^-$. Table 6.5 gives the conducting polymers used in organic batteries.

The colorful demonstration of the organic battery charging and discharging is depicted in Figure 6.4.

TABLE 6.5 List of conducting polymers for battery applications

Abbreviation	Name	Formula monomer unit	MW monomer	MW monomer unit in polymer	Abbreviation	Name	Formula monomer unit	MW monomer	MW monomer unit in polymer
Pln	Polyindole	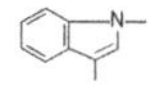	117	115	TTP	Tetrathiaporphyrin	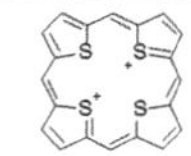	110	108
PT	Polythiophene Polythienylene		84	82					
PBT	Poly (bithiophene) Polythiophene	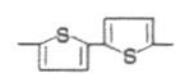	166	164 82	PPP	Poly (p-phenylene) Poly (1,4-phenylene) Polybenzene Polyphenyl	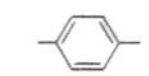	78	76
PTT	Poly (thioenothiophene)	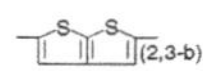	140	138	PTP	Poly (triphenylene)	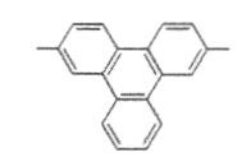	228	226
		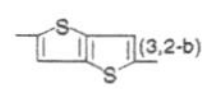	140	138					
PDTT	Poly (dithienothiophene)	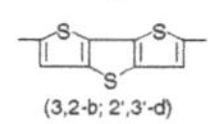	172	170	PAz	Polyazulene Poly (bicyclo(5,3,0) decapentaene)	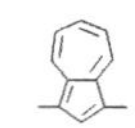	128	126
PMT	Poly (3-methylthiophene)	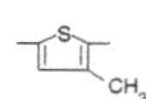	98	96	PFI	Polyfluorene	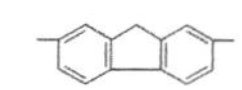	166	164
PAT	Poly (3-alkylthiophene)	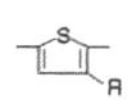	$83+MW_R$	$81+MW_R$					
POT	Poly (3-n-octylthiophene)	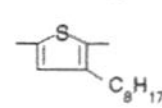	196	194	PFu	Polyfuran		68	66
PMOT	Poly (4,4′-dimethoxy- bithiophene)	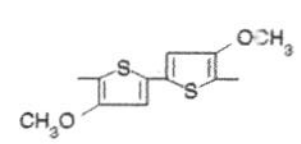	226	224	PCZ	Polycarbazole	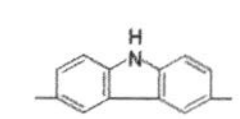	167	165
PEDOT	Poly(3,4-ethlenedioxy- thiophene)	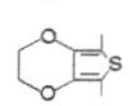	142	140	PECz	Poly (N-ethylcarbazole)	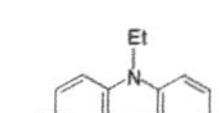	195	193

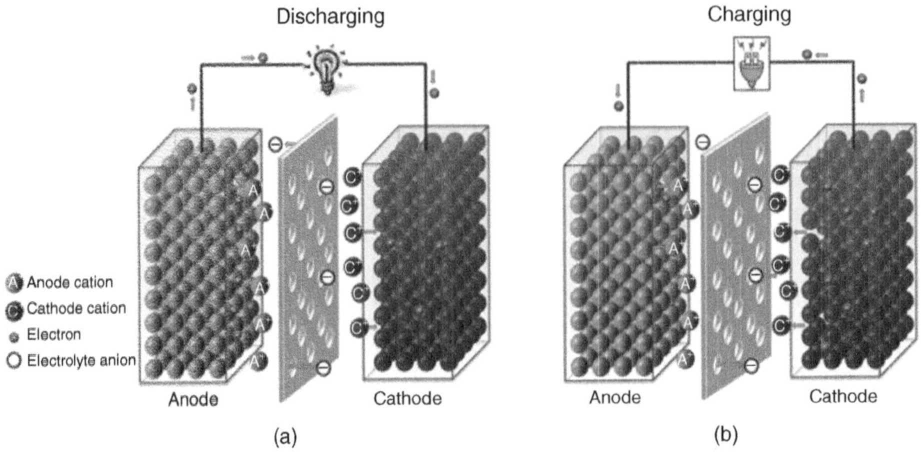

Figure 6.4 Organic battery while charging and discharging, the electron and ion movement paths are shown. The anode compartment and cathode compartment are separated by the center plate (membrane). [*Source:* Muench, S. *et al.*, (2016) [3]. Reproduced with permission of American Chemical Society]

6.6 PRACTICAL CONSIDERATIONS

6.6.1 Features of Selected Batteries

Although there are a large number of batteries available in the market, still there are several challenges faced in the implementation. We take representative examples of Ni–metal hydride battery, zinc–air battery, and lithium–polymer battery for our discussion as follows.

6.6.1.1 Ni–Metal Hydride Battery

Background: It is a room temperature, aqueous medium-based battery system that is at present considered for replacement of Ni–Cd battery as Cd is not desirable because of its toxicity. It involves low-cost components as an added attraction for marketing.

Status: This product is commercially available in small size and scalable. Argonne National Laboratory has shown that it can give a specific energy of 54 Wh/kg at the 3-h rate and a peak power of 174 W/kg at 50% discharge and is suitable for electric vehicles (EVs).

Critical issues: Self-discharge may limit its usage in EVs and thermal stability of the hydride.

These issues have been solved for use in EVs.

6.6.1.2 Lithium–Polymer Battery

Background: Solid-state polymer electrolyte system with Li or Li/C anodes. For inorganic insertion, thin film manufacturing techniques are applicable.

Status: Current technology can provide a battery at 100–200 Wh/kg with a pulse power 250–400 W/kg. Several hundred cycles are possible.

Critical issue: Thermal management problem due to high resistivity. Low temperature performance is yet to be established.

6.6.1.3 Zinc–Air Battery

Background: Aqueous system using air as the active cathode material. Theoretical specific energy of 560 Wh/kg.

Status: Low specific power due to slow kinetics. Foam metal zinc may provide some solution.

Critical issue: The lifetime of zinc and air electrodes limits the usage as they provide obstacles.

Tables 6.6 and 6.7 give the performance characteristics of primary and secondary batteries in detail. Note that the specific energy is given in MJ/kg. 1 MJ corresponds to 277.7 kW.

6.6.1.4 Applications Modern living is entirely based on new technologies as they provide speed and comfort. Batteries are used in laptops, remote controls, iphones, satellite communications, watches, and so on. For usage in mobility applications, batteries are divided into two types based on the energy density required for selected purpose and rechargeability. Besides several other considerations such as power deliverability, speed of deliverability and sustainability would be decisive factors.

1. Off-road vehicle applications
2. On-road vehicle applications

In the first category

Industrial trucks—generally, they are used for 6–8 h. Required battery energy density is 20–24 Wh/kg. Require battery to go through 1000 cycles of charge/discharge.

Industrial sweepers

Golf cars

Garden tractors. Usually cover 2–3 acres. Require battery to provide energy density of 24–32 Wh/kg.

In the second category

Family car is expected to cover 200 miles at 60 mph. Battery with energy density of 160 Wh/kg.

Commuter car is expected to cover 100 miles at 60 mph. A battery with energy density of 56 Wh/kg would be desirable.

City taxi: Covers 150 miles at 40 mph. A battery with the energy density of 120 Wh/kg is desirable.

City bus: Covers 120 miles at 30 mph. A battery with an energy density of 110 Wh/kg would be desirable.

TABLE 6.6 Battery characteristics: primary batteries and their characteristics

Battery	Anode (−)	Cathode (+)	Max. voltage, theoretical (V)	Nominal voltage, practical (V)	Specific energy (MJ/kg)	Elaboration	Shelf life at 25°C, 80% capacity (months)
Zinc–carbon	Zn	MnO_2	1.6	1.2	0.13	Inexpensive	18
Zinc–chloride			1.5			Also known as "heavy-duty," inexpensive	
Alkaline (zinc–manganese dioxide)	Zn	MnO_2	1.5	1.15	0.4–0.59	Moderate energy density Good for high- and low-drain uses	30
Nickel oxyhydroxide (zinc–manganese dioxide/nickel oxyhydroxide)			1.7			Moderate energy density Good for high-drain uses	
Lithium (lithium–copper oxide) Li–CuO	Li	CuO	1.7			No longer manufactured Replaced by silver oxide (IEC-type "SR") batteries	
Lithium (lithium–iron disulfide) $LiFeS_2$	Li	FeS_2	1.8	1.5	1.07	Expensive Used in "plus" or "extra" batteries	337
Lithium (lithium–manganese dioxide) $LiMnO_2$	Li	MnO_2	3.0		0.83–1.01	Expensive Used only in high-drain devices or for long shelf life due to the very low rate of self-discharge "Lithium" alone usually refers to this type of chemistry	
Lithium (lithium–carbon fluoride) Li–$(CF)_n$	Li	$(CF)_n$	3.6	3.0			120
Mercury oxide	Zn	HgO	1.34	1.2		High-drain and constant voltage Banned in most countries because of health concerns	36
Zinc–air	Zn	O_2	1.6	1.1	1.59	Used mostly in hearing aids	
Zamboni pile	Zn	Ag or Au		0.8		Very long life Very low (nanoamp, nA) current	>2000
Silver-oxide (silver–zinc)	Zn	Ag_2O	1.85	1.5	0.47	Very expensive Used only commercially in "button" cells	30
Magnesium	Mg	MnO_2	2.0	1.5			40

1 MJ = 277.7 kW.

TABLE 6.7 Battery characteristics: secondary (rechargeable or traction) batteries and their characteristics

Battery	Cell voltage	Specific energy (MJ/kg)	Comments
NiCd	1.2	0.14	Inexpensive
			High-/low-drain, moderate energy density
			Can withstand very high discharge rates with virtually no loss of capacity
			Moderate rate of self-discharge
			Environmental hazard due to cadmium—use now virtually prohibited in Europe
Lead-acid	2.1	0.14	Moderately expensive
			Moderate energy density
			Moderate rate of self-discharge
			Higher discharge rates result in considerable loss of capacity
			Environmental hazard due to lead
			Common use—automobile batteries
NiMH	1.2	0.36	Inexpensive
			Performs better than alkaline batteries in higher drain devices
			Traditional chemistry has high energy density, but also a high rate of self-discharge
			Newer chemistry has a low self-discharge rate, but also a ~25% lower energy density
			Used in some cars
NiZn	1.6	0.36	Moderately inexpensive
			High drain device is suitable
			Low self-discharge rate
			Voltage closer to alkaline primary cells than other secondary cells
			No toxic components
			Newly introduced to the market (2009), has not yet established a track record
			Limited size availability
AgZn	1.86	0.46	Smaller volume than equivalent Li-ion.
	1.5		Extremely expensive due to silver
			Very high energy density
			The very high drain is capable
			For many years considered obsolete due to high silver prices
			Cell suffers from oxidation if unused
			Reactions are not fully understood
			Terminal voltage very stable but suddenly drops to 1.5 V at 70–80% charge (believed to be due to the presence of both argentous and argentic oxide in positive plate—one is consumed first)
			Has been used instead of a primary battery
			Is being developed once again as a replacement for Li-ion
Lithium ion	3.6	0.46	Very expensive
			Very high energy density
			Not usually available in "common" battery sizes
			Very common in laptop computers, moderate to high-end digital cameras

Source: https://en.wikipedia.org/wiki/Battery electricity. Used under CC BY SA 3.0 https://creativecommons.org/licenses/by-sa/3.0/.

6.7 ELECTRIC TRANSPORTATION

6.7.1 Background Information

In this section, a historical development of batteries and electric transportation is discussed. As early as 200 BC, batteries were used in Iraq and Egypt for electroplating and precious metal gilding. In 1800, Alessandro Volta observed that an electrical current could be generated using two dissimilar metals in electrolytic solutions. After this invention, several other batteries were developed using a number of different metals in the next 60 years. Georges Leclanche of France developed a carbon–zinc wet cell, and Raymond Gaston Plant invented the lead-acid battery. Both batteries had short shelf life. In 1881, Émile Alphonse Faure proposed batteries using a mixture of lead oxides for the positive plate in electrolytes that would drive chemical reactions faster and provide a higher efficiency. In 1900, the nickel storage battery was developed by Thomas Alva Edison, which led to the development of nickel–iron battery in 1905 [8]. An alkaline battery was introduced in 1949, which was modified a year later for a sustainable alkaline–manganese battery. The utilization of solar energy for practical purposes had dawned by this time. In 1954, the first solar battery or solar cell was developed. The invention of a gel-type electrolyte for lead-acid battery was introduced in 1960 [9, 10] that led to the development of several new batteries until 1990. Commercial lithium ion batteries for powering portable computers, cell phones, and space probes opened up a new arena of activities. The need of a battery for running electronic chips and sensors led to a search for better performance batteries. The area of current research indicates that employing carbon nanotubes in batteries may double the life of traditional batteries.

The age-old cars built in the beginning used a bell instead of a horn to alert the people on the road and for other purposes. The head lights were gas powered. For starting the engine, mechanical cranking of the engine was requiring human energy. In 1920, invention of electric starters had promoted extensive batteries usage for transportation. Till 1971, the secondary batteries were open type in the sense that addition of electrolytes or cleaning of the electrodes, and so on could maintain the battery in a rechargeable condition. In 1971, a sealed battery was introduced that increased the battery safety and cycle life. Increased requirements for the internal combustion engines (ICEs) for speedy powering resulted in developing batteries for providing voltages of 12–13.6 V. This battery output voltage is used for starting, lighting, and ignition for ICE, and these are known as SLI batteries. Contemporary trend has been to increase SLI batteries voltage from 13.6 to 42 V to fulfill the needs of more efficient engines, electronic ignition, increased complexity in wiring harnesses, digital controls, and accumulation of different sensors and actuators. The trend in developing batteries providing several different voltages for different purposes has not been realized yet due to technical limitations in electric system components durability. The rapid progress in the development for electric propulsion systems may force the development of batteries for higher voltages.

Figure 6.5 pictorially shows the primary and secondary battery utilization schemes.

6.7.2 Primary Batteries

A cell consists of an electrolyte, typically in gel form, positive and negative electrodes, carrying external terminals that connected to the electric load. The electrolyte was an ionic conductor for the transfer of the electrons between the electrodes. The voltage or electromotive force depends on the chemical properties of the used material, without any relation to the electrodes size or the amount of electrolyte. By design, the battery is a dielectric container having internal electrolytic cells with the electrodes attached to the terminals.

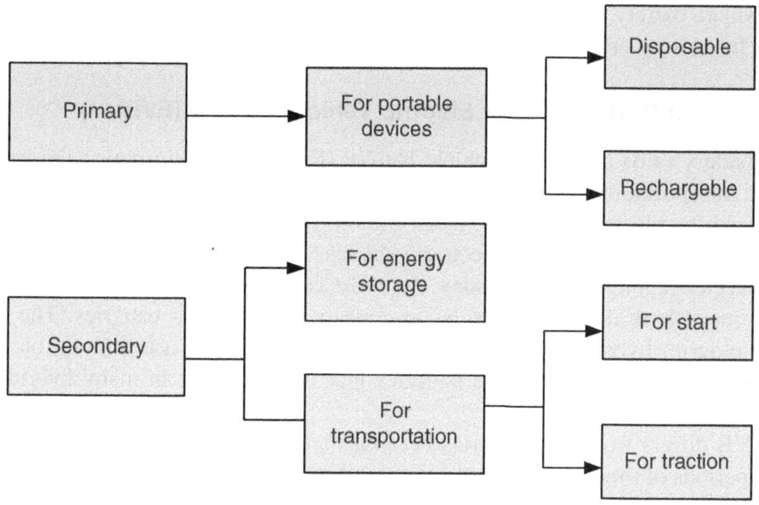

Figure 6.5 Applications of batteries.

All batteries can be separated into two main types: disposable or rechargeable. The primary cells are a disposable family of batteries that can be operational as long as the chemical reaction generating the electrical current ceases; the battery is considered dead when no electrical energy is available and should be discarded. Rechargeable primary batteries are commonly used in smaller, portable devices for telecommunications, laptop computers, and medical instrumentations for limited time use, without access to alternative power sources.

Batteries are classified as either dry cell or slow disappearing wet cell. In a dry cell, the electrolyte is restrained from flowing by absorption in a porous material. In a wet cell, the liquid electrolyte is free to move inside the enclosure. Dry cell batteries such as carbon–zinc dry cells are used to power small devices as pocket-sized radios and CD players or flashlights and mobile phones. Alkaline batteries, utilizing the hydrous alkaline solution as an electrolyte, are less expensive in production and more reliable. They are not as much corrosive than zinc-based batteries and therefore last longer.

The disposable primary cells produce current immediately after assembly but cannot reliably recharge since the chemical reactions are not easily reversible, and battery active materials may not restore to their original conditions.

6.7.3 Batteries for Portable Devices

The dissemination and miniaturization of portable consumer electronics forced the design advancement in lithium-ion batteries. The Li-ion popularity is based on their high power-to-weight ratio, high energy efficiency, good high-temperature performance, and low self-discharge and fully rechargeable. They physically consist of a group of electric (Li-ion) cells that are embedded in one enclosure. All rechargeable batteries have peripheral devices that supply the appropriate current for reversing the chemical reaction. Most of the contents of lithium-ion batteries can be recycled [28]. The current trend in the development is reducing the cost and increasing a number of charging cycles and power density. Until today, the Li-ion batteries are the most accepted in consumer's rechargeable market. Other portable rechargeable batteries include several sealed dry cell types that are useful in the portable devices. Nickel–cadmium (NiCd) batteries are widely employed for power tools, two-way radios, and medical equipment; in comparison, nickel–zinc (NiZn) and nickel metal hydride (NiMH) batteries provide a higher capacity. Efforts are underway to make smart battery packs with embedded electronics that monitors charging with a provision for battery protection.

6.7.4 Traction Battery (TB) or Electric Vehicle Battery (EVB)

Traction battery (TB) or electric vehicle battery (EVB) is used to power the propulsion of electric transportation devices. The traction batteries are used in off-road transportation such as forklifts, electric golf carts, electric motorcycles, electric cars, trucks, and vans. In the initial development for electric vehicles, traction engineers used automotive lead-acid rechargeable batteries. Today, the most favorable device for EV is lithium-ion or nickel–metal hydride battery similar to consumer electronics batteries. The majority of a new plug-in, hybrids, and all-electric vehicles (EVs) will use lithium-ion batteries instead of the nickel–metal-hydride batteries that were popular in many hybrid electric vehicles (HEVs).

The EVB differs from SLI batteries because their purpose is in providing power over extended periods of time. Therefore, the traction batteries have to provide a high amount of electric charge, which is measured in ampere-hours or coulombs, high power-to-weight ratio, energy to weight ratio, and energy density. Driving EV range is directly proportional to the electric battery specific energy. The likely success of EV in comparison to ICE depends on the improvement of battery technologies providing specific energy comparable to liquid petroleum–based fuels.

6.7.4.1 Lead-Acid Batteries Lead-acid batteries that are installed in any ICE transportation units possess high power output. They are inexpensive, safe, and reliable and are not constructed for real deep discharge as a full discharge may result in minimization of batteries' life. The gelled battery contains a semisolid electrolyte to simplify maintenance and improve the durability. The portable rechargeable batteries based on a number of fully sealed dry-cell devices are widely used in mobile phones, laptops, medical, and industrial instrumentation.

There are two main types of lead-acid batteries: automobile engine starter batteries and deep cycle batteries [28]. Automobile alternators are designed to provide SLI batteries high charge rates for fast charges, while deep cycle batteries are used for off-road electric vehicles such as forklifts or golf carts and as the auxiliary energy storing devices for airport's transportation and RVs. The lead acid battery should not be discharged below 50% of its capacity as it will tend to shorten the battery's life. In addition, old unsealed flooded batteries needed an inspection of electrolyte level and occasional replacement of water during the normal charging cycle.

The disadvantage of using the lead-acid batteries in EV applications is a significant weight that can reach up to 40% of the total vehicle mass. Similar to all batteries, they have significantly lower energy density than petroleum fuels, about 30–40 Wh/kg. Even the best performing batteries tend to lead to higher masses when applied to vehicles with a normal range. The efficiency (70–75%) and storage capacity of the current generation of common deep cycle lead-acid batteries decreases at lower temperatures as part of the power is used to run a heating coil that reduces the battery power and consequently reduces the travel range by up to 40%. It is important to mention that during charging operation of old lead-acid batteries, emissions of small amount of hydrogen, oxygen, and sulfur are naturally occurring and can be harmless if vented out. In the currently produced sealed lead-acid batteries, this problem is solved.

Each cell of a lead storage battery consists of alternate plates of lead for the cathode and lead coated with lead dioxide anode immersed in an electrolyte of sulfuric acid solution.

A 12-V lead-acid storage battery is constructed by combining six cells, each producing approximately 2 V. The chemical reactions occurring in the battery release electrons that flow through the conductors as electric current. When the battery is discharged, the acid in the electrolyte reacts with the materials of the plates producing on their surface lead sulfate. When the battery is recharged, the chemical reaction is reversed with lead sulfate converted to lead dioxide. With the plates restored to their original condition, the process will be repeated for many times [27].

The lead-acid batteries possessing a low production cost and high surge current levels have been widely used in SLI battery. In the contemporary cars, it typically delivers a peak current of 450 A. It has a capacity to reach 800 A. The limiting factors with existing lead-acid battery technology for using in EV are sensitivity to cold temperature, packaging limitation, excessive weight, and short cycle life. However, the recent progress in battery efficiency, capacity, materials, safety, toxicity, and durability can make them attractive for EV applications. The current lead-acid battery technology application for EV requires improvement in the following:

- *Limited capacity in minimizing the driving range.* They have a limited capacity; a typical lead-acid battery pack might hold 12–15 kWh of electricity, giving a car a range of only 50 miles or so, although a few models on a full charge can go 200–300 miles.
- *Recharge time.* Fully recharging the battery pack can take 4–8 h. Even a "fast charge" to 80% capacity can take 30 min. They are slow to charge (typical recharge times for a lead-acid pack range between 4 and 10 h for full charge) depending on the battery technology and the charger [30].
- *Battery cost.* The large battery packs are expensive and have to be replaced few times during vehicle's life.
- Limited number of recharge
- Short life (3–5 years, perhaps 200 full charge/discharge cycles).
- *Bulk and weight.* Figure 6.6 shows that battery packs are heavy and take up considerable vehicle space.

6.7.4.2 *Lithium-Ion Batteries* Lithium-ion (Li-I) battery is the preferable technology for currently produced EV. The conventional lithium-ion chemistry employed a lithium cobalt oxide cathode and a graphite anode resulted in the electrodes ability to hold a large number of lithium ions (and accordingly electrons) per unit surface area than in other batteries. The most attractive parameters are high energy density, excellent power density, and high up to 90% charge/discharge efficiency. The limiting factors of previously utilized lithium-ion batteries are an insufficient number of charging cycles, the degradation during lifespan, a fire safety risk if damaged or improperly charged, and sensitivity to the high-temperature environment. It is important to mention that heat is the primary cause of fire safety problem, battery failure also happened due to accelerated corrosion inside the battery. Figure 6.7 shows that the corrosion of battery terminals can lead to increase in electrical resistance, including a buildup of sulfate concentration around terminals (acid attack of the metal terminal resulting in sulfate formation) that occurs when the battery is not fully charged for a long time.

The work on lithium-ion chemistry is concentrated on trade-off optimization between safety, energy and power density, environmental requirements, quick charges, and extended shelf and duration of the batteries life. For instance, the portable A123 battery used for tools charging can stand over 1000 charging cycles and exceptional charge/discharge

Figure 6.6 Portable 12 V 50 A/10-h deep cycle sealed lead-acid battery.

Figure 6.7 Lithium-ion battery.

cycles. The lithium iron phosphate batteries can last for 10 years with the ability to stand 7000 charge cycles.

A significant density improvement has been demonstrated using lithium vanadium oxide batteries in the Subaru prototype G4e (short for "green for the earth") with doubling the energy density and increasing the driving range up to 40%. All other developments have been on replacing the traditional graphite anode for the battery performance improvement. Figure 6.8 shows the use of nanowires to increase the surface area on one or both electrodes,

Figure 6.8 2012 Chevrolet Volt T-shaped lithium-ion battery nickel–metal hydride.

silicon nanoparticles for anodes, and composite materials for cathodes promise significant improvement in the energy density.

The well-known nickel–metal hydride batteries, used in computers and medical equipment, demonstrated rare performance in specific energy and specific power capabilities. The nickel–metal hydride batteries stand a substantially longer life cycle than lead-acid batteries and are safe and abuse tolerant. They are relatively costly, having high self-discharge, heat generation at high temperatures, and the need to control hydrogen loss [28]. Nickel–metal hydride batteries (NiMH) are now considering as well-established technology. While they are less efficient (60–70%) in charging and discharging than even lead-acid battery, it has a high energy density of 40–80 Wh/kg. When properly used, nickel–metal hydride battery can have exceptionally long life, as has been demonstrated by their use in hybrid cars. In Toyota RAV4 Hybrid, NiMH batteries can operate up to 100,000 miles (160,000 km) and last for a decade. The downsides of it are the poor efficiency, high self-discharge, poor charge cycles, and poor performance in cold weather [27].

6.7.4.3 Molten Salt Battery The sodium or "zebra" battery uses molten chloroaluminate sodium ($NaAlCl_4$) as the electrolyte. This chemical process is also occasionally referred to as "hot salt." A relatively mature technology, the zebra battery can provide an energy density of 120 Wh/kg and reasonable series resistance. They have been used in several EVs. Zebras can last for a few thousand charge cycles and are nontoxic. The downsides to the zebra battery includes poor power density (<300 W/kg) and the requirement of having to heat the electrolyte to about 270°C (520 °F), which wastes some energy and presents difficulties in long-term storage of charge [27].

6.7.5 Recharging

The batteries in EV must be periodically recharged. BEV most commonly charge from the power grid (at home or using a street or shop recharging point), which receives a variety of renewable and nonrenewable resources, such as coal, hydroelectricity, nuclear, and others. Another option is to build several recharging spots, also called the charging stations, for the general public usage, or specific fleet segment as taxi stands.

The personal sources installed in a home or at public places using the photovoltaic solar cell panels, micro-hydro or wind also may be very attractive because of concerns regarding global warming. With appropriate power supplies, good battery's lifespan is usually achieved at rates not exceeding "0.5*C*," taking 2–3 h for a full charge, but faster charging is possible.

The charging time is limited by the capacity of the supply connection and maximum charge rate of the battery. A normal household outlet delivers 1.5 kW (in the US, Canada, Japan, and other countries with 110 V supply) and 3 kW (in countries with 240 V supply). Even if a higher supply of power is available, batteries charge is limit by *C*-rate, permitted for some batteries a recharge time of 20–30 min to 80%. For consumer products, usage of the short charge time is critical, and currently research is concentrated on charge time minimization up to 60 s. For transportation, the quick recharging is desirable, but not critical because it can be done during overnight or during the working time. The current trend is the construction of special parking bays for electric vehicles charging in the metropolitan area. Tesla is working on equipment manufacturing for home and office battery charging, and initiate network installation for high-powered superchargers across North America, Europe, and Asia. The charging power can be connected to the car in two ways. The first is a direct electrical connection known as conductive coupling. It can be a special high capacity cables with weatherproof socket and driver protection from high voltage. The special standard has been developed for electric car charging. An example is SAE 1772 conductive connector for plug-in vehicle.

The second approach is known as inductive charging. A special connector is inserted into a slot on the car. The paddle is one winding of a transformer, while the other is built into the car. When the paddle is inserted, it completes a magnetic circuit that provides power to the battery pack. In one inductive charging system, one winding is attached to the underside of the car, and the other stays on the floor of the garage. Figure 6.9 shows the advantage of the inductive approach is that there is no possibility of electrocution as there are no exposed conductors, although interlocks, special connectors, and ground fault detectors can make conductive coupling nearly as safe.

6.7.6 Travel Range Before Recharging

The General Motors EV1 had a range of 75–150 miles (240 km) with NiMN batteries in 1999. Contemporary BEV is having a range of 350 miles *depending on the number and type of batteries used.*

6.7.6.1 *Battery Specifications* Extensive progress in battery's technology development makes it difficult to compare different batteries' parameters [18], but some data are presented as follows.

The Chevy PHEV Volt recharges with a 120-V portable cord in approximately 13 h or with 240-V charging system in about 4.5 h. The driving range for Volt 2017 model with a fully charged Li-I battery is capable of driving up to 53 miles without using gas and up to 420 miles with a full charge and a full tank of gas.

An example of a typical specification for rechargeable EV TB used in Nissan Leaf [23] is presented in Table 6.8.

Figure 6.9 General motors EV1.

TABLE 6.8 Nissan Leaflet

Type	Laminated lithium-ion battery
Voltage	403.2 V
Nominal voltage	360 V
Total capacity	24 kWh 16 kWh available
Power output	Over 90 kW
Energy density	140 Wh/kg
Power density	2.5 kW/kg
Dimensions	1570.5 × 1188 × 264.9 mm
Weight	648 lb
Number of modules	48, each with four cells (total 192 cells)
Battery pack contents:	Positive electrodes: lithium manganate
	Negative electrodes: carbon
	Cells
	Modules
	Assembly parts
Charging times:	Quick charger DC 50 kW (0–80%): approx. 30 min
	Home-use AC 240 V charging dock (0–100%): 8 h
	Regular 110/120 V 15-A outlet: 22 h
Battery layout	Under seat and floor

6.7.6.2 Environmental Consequence The batteries are subjected to environmental pollution, and hence its disposal requires recycling. Therefore, all used batteries are considered as hazardous waste. In accordance with the EPA, the 12-V car lead acid battery is the most recycled product in the world. The recycling rates are 95% in the United States and about 97% through the lead recovered.

6.7.6.3 *Recycling Batteries* As EV will become increasingly common, the battery-recycling market will expand. A widespread battery recycling infrastructure will control hazardous materials from entering the waste stream. There are two processes available for battery recycling:

- *Smelting*. It is kind of extractive metallurgical process operated at high temperature. The valuable metals are recovered. The electrolyte and other organic components support combustion of fuel. These processes are in action now on a large scale. The recycling process is commercialized for lithium-ion and nickel–metal hydride batteries.
- *Direct recovery*. It is a low-temperature process consuming less minimal energy. Components are separated by a variety of physical and chemical processes, and all active materials and metals can be recovered [21].

6.7.7 Batteries and Transportation

Using electricity for transportation needs is not a new approach (Figure 6.10). Off-road equipment such as forklift, scooters, electric bicycles, and others were part of life long ago. Due to depletion of natural resources, search for "green" transportation resulted in the substitution of petroleum by electricity.

Figure 6.11 shows a simplified block diagram of an electric vehicle.

The EV propulsion today uses DC motors with appropriate motor controller. It was found that DC commencement tends to be simpler and less expensive than AC motors. AC motor allows the use of almost any industrial three-phase AC motor. An important feature of AC motors and controllers is an option for energy regeneration, also known as "regen," which permits the energy of a decelerating EV to be utilized either using the drive motor as a

Figure 6.10 The first electric vehicle built in 1897.

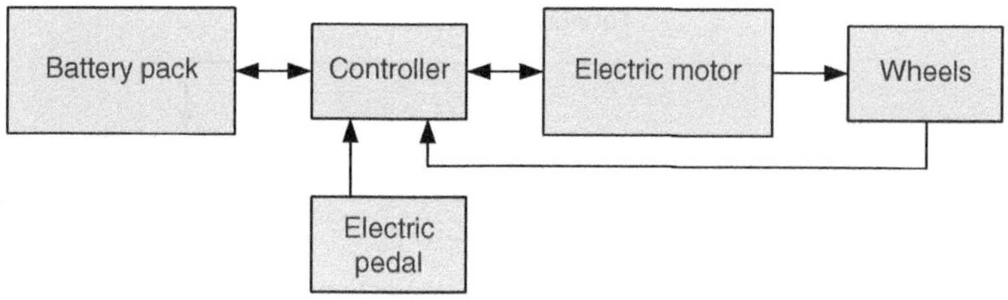

Figure 6.11 Simplified block diagram for electric vehicle.

Figure 6.12 Ford focus electric car.

generator or by driving a separate generator. During braking, the braking energy is returned to the battery pack, which provides a small improvement in the driving range. Figure 6.12 shows that another positive feature of DC motors is the ability to provide the short energy bursts for acceleration.

The EV transportation uses electricity as their primary propulsion source, or for improving the efficiency of conventional ICE vehicles. This will result in a substantial saving of natural petroleum resources and cleaner environment.

As shown in Figure 6.13, all electric drive vehicles can be divided into three categories: hybrid electric vehicles (HEVs), plug-in hybrid electric vehicles (PHEVs), and all-electric vehicles (EVs). Figure 6.14 shows that they use electricity either as their primary energy source fuel or to improve the efficiency of conventional vehicle designs.

6.7.7.1 Hybrid Electric Vehicles (HEVs) HEV named hybrid because it possesses the two sources of propulsion: an ICE that runs on petroleum based or alternative fuel and an electric motor working on energy stored in a battery. The battery is charged through regenerative braking and by the ICE [32]. The extra power provided by the electric motor allows the battery to feed auxiliary vehicle loads (audio systems and headlights) and can

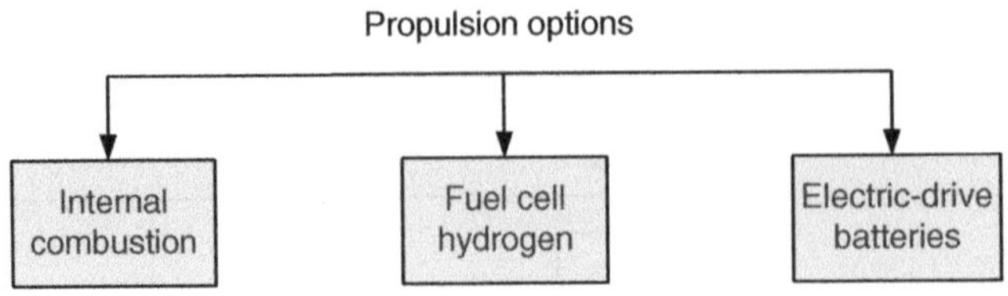

Figure 6.13 Propulsion options.

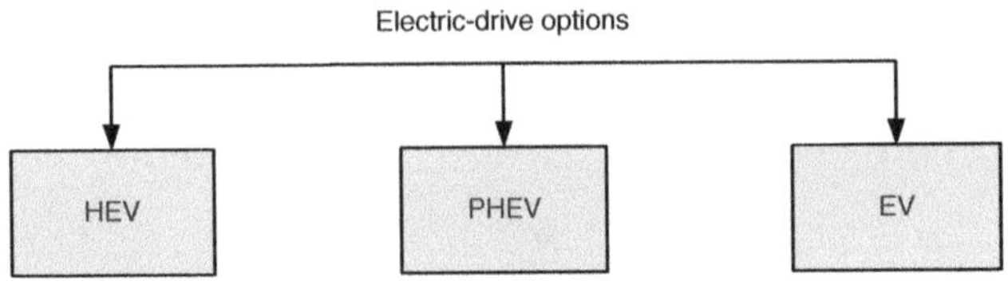

Figure 6.14 Electric-drive options.

reduce engine idling when the vehicle is stopped [36]. Some HEVs can drive short distances (40–60 miles) at low speeds on electrical power alone. All these capabilities lead to better fuel economy and lower emissions than comparable conventional vehicles.

HEV cannot be plugged in to charge the battery. HEV is working on external electric power source. Regenerative braking allows HEV to detain energy that dissipated during braking by feeding the additional electric motor as a generator and storing that captured energy in the battery [36].

6.7.7.2 Plug-In Hybrid Electric Vehicles (PHEVs) PHEV is similar to HEV because it is powered by an ICE that can run on conventional or alternative fuel and an electric motor that utilizes energy stored in a battery [36]. The difference is that an electric external power source is connected to charge the battery. It can travel up to 70 miles on electricity on full battery charge, and also can operate exclusively on gasoline (similar to a conventional hybrid). Some types of PHEVs are also called extended range electric vehicles or EREVs.

PHEV uses batteries to power an electric motor. In addition it depends on a fuel such as gasoline to power ICE. PHEV can be plugged into the grid to charge the battery and also can be charged both by the ICE and through regenerative braking system. PHEV uses a large battery packs than HEV, providing an all-electric driving range of 40-plus miles for light-duty models. As long as the battery becomes fully charged, a PHEV can draw most of its power from electricity stored in the battery during typical urban driving. When the battery is mostly depleted, during rapid acceleration, at high speeds, or when intensive heating or air conditioning has required, the vehicle will be powered by ICE. Obviously, when running on battery power alone, PHEVs produce no tailpipe emissions.

Even when the ICE is operating, PHEV consumes less gasoline and typically produce lower emissions than similar conventional vehicles do. A PHEV gasoline consumption depends on the distance traveled between charges. If the vehicle is never plugged in, its gasoline-only fuel economy will be about the same as that of a similarly sized HEV. If the vehicle is plugged into a charging mode, it may be possible to use only electric power.

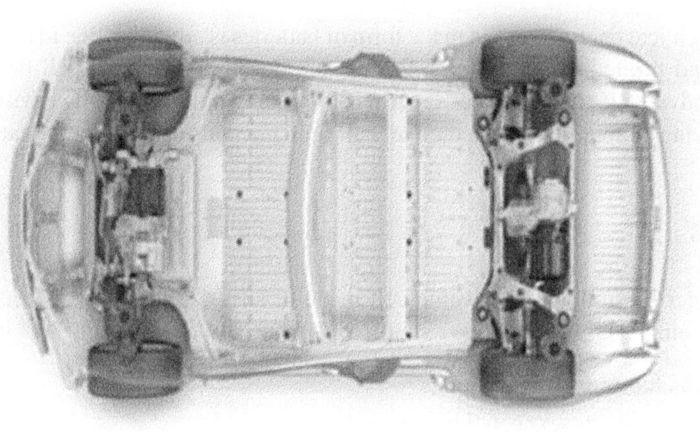

Figure 6.15 Tesla EV electric motors location.

6.7.7.3 All-Electric Vehicles (EVs) EVs use only one internal source of propulsion; the battery is driving the electric motor. EV batteries are charged by plugging the vehicle into an electric power source. EVs are sometimes referred to as battery electric vehicles (BEVs). Figure 6.15 gives Tesla EV showing the electric motor location.

6.7.8 Electric Vehicles and Internal Combustion Engine Comparison

EVs have several advantages over vehicles with ICEs. They are as follows:

- *Energy efficient.* Electric vehicles convert about 60% of the electrical energy from the grid to power at the wheels—conventional gasoline vehicles only convert about 21% of the energy stored in gasoline for the rotation of the wheel.
- *Environmentally friendly.* EVs emit no tailpipe pollutants, although the power plant utilizing petroleum sources for producing electricity may emit the small amount. Electricity from clean sources as hydro, solar, wind, and very limited greenhouse discharge from nuclear-powered plants originated no air pollutants.

They demonstrate the following:

- *Improved performance.* Electric motors provide quiet, smooth operation and quick acceleration and require lower maintenance cost than ICE.
- *Better durability.* Today there is no data on durability available for comparison. As there is limited friction in EV components, the assumption is that it should have a longer life span.
- *Ability to use an available infrastructure.* EV recharging does not require excessive resources for infrastructure development in comparison to hydrogen production and distribution network creation.
- *Reduce energy dependence.* Electricity is a domestic energy source.

6.7.9 Fuel Cells and Electric Batteries Comparison

The electrochemical energy storage in the form of batteries is attractive due to the simplicity of logistics and limited expenses for infrastructure development. Fuel cell-powered cars have greater advantage than batteries in terms of reducing pollution in the environment.

The following chapter discusses the fuel cell fundamentals and their applications.

PROBLEMS

1. Consider the following galvanic cell made of silver–zinc, which is used in space applications. It is made of two redox couples: $Zn^{2+} + 2e^- \rightarrow Zn$ and $Ag^+ + e^- \rightarrow Ag$.
   (a) Which electrode is cathode?
   (b) Determine the expected battery voltage?
   (c) What would be the battery voltage if the concentration of cathode half cell is increased by 10?
   (d) What is the battery voltage if the ion concentration in the anode half cell is increased by 10?
2. With silver–zinc battery in Problem 1, what are the theoretically expected specific energy and power density?
3. The overall reaction in the alkaline zinc battery is given as follows:

$$Zn(s) + 2MnO_2(s) \rightarrow ZnO(s) + Mn_2O_3$$

   Identify the two half cells of the battery and indicate what is the expected voltage?
   The standard potentials of anodic and cathodic reactions are $E^{\circ} = -1.28\text{ V}$ and $E^{\circ} = +0.15\text{ V}$.
4. Nickel–cadmium battery consists of an anode composed of solid cadmium and a cathode composed of NiO(OH)(s). The electrolyte is usually KOH(aq). Given the half cell potentials as follows:

$$Cd(s) + 2OH^-(aq) \rightarrow Cd(OH)_2 + 2e^- \qquad E^{\circ} = -0.824\text{ V}$$

$$2NiO(OH)(s) + 2H_2O(l) + 2e^- \rightarrow 2Ni(OH)_2(s) + 2OH^-(aq) \qquad E^{\circ} = 0.56\text{ V}$$

   Calculate
   (a) The battery voltage under standard conditions
   (b) Calculate the capacity of the battery
   (c) Calculate the specific energy of the battery.
5. A Daniell cell was used to drive a toy fan for several hours. If 100 g of Zn is consumed, calculate the maximum electrical work in joules.
6. The AA-size and D-size alkaline batteries are both made with Zn and MnO_2. What are the battery reactions that give rise to the difference between the two batteries?
7. An alkaline battery made of Zn and MnO_2 has been used in flash light. After several days of operation, 5.0 g of zinc dissolved at anode of the battery. (a) Calculate the mass of MnO_2 reduced at the cathode during this usage. (b) Since electrons

flow from zinc to MnO_2 in the battery, calculate the coulombs of electrical charge transferred.

8. Lithium–silver chromate coin battery is used in pacemakers for proper functioning of heart.

   The two half cells of the battery are made of redox couples Li/Li^+ and Ag/Ag_2CrO_4. The standard potentials of the two redox couples are −3.040 V (Li/Li^+) and 0.08 V (Ag/Ag_2CrO_4).

   (a) Write the battery reaction (b) Which metal is the anode in the battery? (c) What is the expected battery voltage? (d) Since the battery is inside the human body, calculate the battery voltage at the body temperature.

9. In cars, generally lead-acid battery is used for starting the engine. If after an operation 2.0 g of lead is oxidized, find the mass of $PbSO_4$ formed in the battery.
10. In the light weight battery, the two redox couples used are Li/Li^+ and graphite in contact with polyaniline. Write the two half cell reactions of the light weight battery. (a) Estimate the battery voltage under standard conditions. (b) Calculate the specific energy of this battery.

MULTIPLE CHOICE QUESTIONS

1. Which car is more environment friendly?
   (a) PNEV
   (b) HEV
   (c) EV
   (d) Natural gas powered
   (e) Hydrogen powered
2. For charging a car battery, the car must have what type of converter?
   (a) AC-to-DC converter
   (b) DC-to-DC converter
3. What factors limit the driving range?
   (a) Battery capacity
   (b) Electric and mechanical losses in energy conversion
   (c) Car aerodynamic configuration
   (d) Number of passengers

BIBLIOGRAPHY

Section 6.6

1. https://en.wikipedia.org/wiki/Standard electrode potential.
2. Zhang, X., Xue, Q., Li, L. *et al.* (2016) Sustainable recycling and regeneration of cathode scraps from industrial production of lithium-ion batteries. *ACS Sustainable Chemistry & Engineering*, **4**, 7041.
3. Muench, S., Wild, A., Friebe, C. *et al.* (2016) Polymer-based organic batteries. *Chemistry Review*, **116**, 9438–9484.

4. Hayner, C.M., Zhao, X., and Kung, H.H. (2012) Materials for rechargeable lithium-ion batteries. *Annual Review of Chemical and Biomolecular Engineering*, **3**, 445–471.
5. Novak, P., Muller, K., Santhanam, K.S.V., and Haas, O. (1997) Electrochemically active polymers for rechargeable batteries. *Chemical Reviews*, **97**, 209.
6. Xu, K. (2014) Electrolytes and interphases in Li-ion batteries and beyond. *Chemistry Review*, **114** (23), 11503.
7. Shirakawa, H., Louis, E.J., MacDiarmid, A.G. *et al.* (1977) Synthesis of electrically conducting organic polymers: halogen derivatives of polyacetylene, $(CH)_x$. *Journal of the Chemical Society, Chemical Communications*, 578.

Section 6.7.6

8. http://www.infoplease.com/encyclopedia/science/battery-electric.html (accessed 10 June 2017).
9. *Columbia Electronic Encyclopedia*, 6th ed. Copyright © 2016, Columbia University Press.
10. http://batteryuniversity.com/learn/article/When was the battery invented (accessed 10 June 2017).
11. http://aboutbatteries.batterycouncil.org/What is a lead battery? (accessed 10 June 2017).
12. A Short Course on Charging Systems www.carparts.com/classroom/charging.htm (accessed 10 June 2017).
13. Q & A: Car Batteries|Department of Physics|University of Illinois at… https://van.physics.illinois.edu/qa/listing.php?id=567 (accessed 10 June 2017).
14. https://en.wikipedia.org/wiki/42 volt electrical system (accessed 10 June 2017).
15. Whatever Happened to the 42-V Car? www.popularmechanics.com/cars/a2198/4226979 (accessed 10 June 2017).
16. Best Car Battery Reviews – Consumer Reports www.consumerreports.org/cro/car-batteries (accessed 10 June 2017).
17. https://www.researchgate.net/publication/236466652 Technical and Economic Feasibility of Applying Used EV Batteries in Stationary Applications.
18. http://www.electricvehiclewiki.com/Battery specs (accessed 10 June 2017).
19. www.chargingchargers.com/tutorials/charging.html (accessed 10 June 2017).
20. https://www.cargurus.com/Cars/2017-Toyota-RAV4-Hybrid-Overview-c26005 (accessed 10 June 2017).
21. 2011 leaf First Responder's Guide www.nissanusa.com/.../leaf/2011/2011 (accessed 10 June 2017).
22. www.electricvehiclewiki.com/images/f/f7/2011-leaf-service-maintenance-guide.pdf (accessed 10 June 2017).
23. https://www.nissanusa.com/ev/media/pdf/specs/FeaturesAndSpecs.pdf (accessed 10 June 2017).
24. http://www.chevrolet.com/volt-electric-car.html (accessed 10 June 2017).
25. https://en.wikipedia.org/wiki/Energy density (accessed 10 June 2017).
26. https://www.extremetech.com/extreme/184236 how does a lithium-ion-battery work (accessed 10 June 2017).
27. https://en.wikipedia.org/wiki/Electric_vehicle_battery (accessed 10 June 2017).
28. http://hypertextbook.com/facts/2001/ThiMeaganLe.shtml (accessed 10 June 2017).
29. http://wattev2buy.com/electric-vehicle-technical-terms-explained/ (accessed 10 June 2017).
30. http://auto.howstuffworks.com/electric-car3.htm (accessed 10 June 2017).

Section 6.7.9

31. http://klingbergautoshow.org/the-nations-auto-industry-started-in-hartford/ (accessed 10 June 2017).
32. http://www.afdc.energy.gov/vehicles/electric.html (accessed 10 June 2017).
33. Benefits and Considerations of Electricity as a Vehicle Fuel http://www.afdc.energy.gov/fuels/electricity_benefits.html (accessed 10 June 2017).
34. Maintenance and Safety of Hybrid and Plug-In Electric Vehicles.
35. http://www.afdc.energy.gov/vehicles/electric_maintenance.html (accessed 10 June 2017).
36. http://www.afdc.energy.gov/uploads/publication/hybrid_plugin_ev.pdf (accessed 10 June 2017).
37. http://www.electricvehiclewiki.com/Battery_specs (accessed 10 June 2017).
38. http://www.afdc.energy.gov/vehicles/electric_batteries.html (accessed 10 June 2017).
39. Plug-In Electric Vehicle Handbook for Consumers, https://www.afdc.energy.gov/uploads/publication/pev_consumer_handbook.pdf (accessed 10 June 2017).
40. Plug-In Electric Vehicle Handbook for Fleet Managers, https://www.afdc.energy.gov/uploads/publication/pev_consumer_handbook.pdf (accessed 10 June 2017).

CHAPTER 7

Fuel Cell Essentials

7.1 INTRODUCTION

A fuel cell generates electric power by oxidation of the fuel supplied to it. It is an electrochemical cell made up of two electrodes, where oxidation occurs at one electrode and reduction at another. The electrode at which the oxidation occurs is labeled as the anode. Here the electrons are released as a result of the reaction. The electrode where reduction reaction occurs is labeled as the cathode. The electrons released in the anodic reaction are consumed in the cathode reaction. With a fuel cell, the oxidation of the fuel occurs at the anode and is the source for the electrons. The electrons are consumed in the cathode reaction. These reactions will occur at the electrodes as long as the fuel is fed into the electrochemical cell. A typical fuel cell is shown in Figure 7.1. It gives a detailed illustration of the fuel cell. In a practical version, it is compacted as shown in Figure 7.2. A voltage is produced and current flow occurs as a result of the fuel oxidation at anode and a counterreaction at the cathode. In Figure 7.1, a voltmeter, *V*, is shown as the load. In practice, it could be any heavy-duty machine such as a fan, motor, or automobile. In this way, a fuel cell is similar to the batteries that we use in our daily lives now in TV remote controls, laptops, or electric shavers, except that the battery works only for a limited time after which we have to discard it (primary batteries) or recharge it (secondary batteries) for it to provide the power. This limitation is overcome in a fuel cell as it can operate continuously as long as the fuel supply is provided.

The fuel cell was invented in 1839 by W. Grove, using two platinum electrodes submerged half way in aqueous sulfuric acid with tubes filled with hydrogen over one electrode and oxygen over the other, which generated electric power. Figure 7.3 shows an early version of a fuel cell. This was subsequently modified in 1842 to be an electrolytic cell as shown on the right side of Figure 7.3. Here a series of electrochemical cells were coupled to each other. After about 47 years, the first prototype fuel cell was constructed using a diaphragm to separate the two parts of the electrochemical cell, anode and cathode, instead of the tubes as shown in Figure 7.3. The operation of the fuel cell with a supply of hydrogen and oxygen gases raises the question as to which of the two gases here is acting as a fuel in the fuel cell? Obviously from today's general knowledge, we can confidently say hydrogen is the fuel; however, a rational definition and classification of fuels are necessary.

Introduction to Hydrogen Technology, Second Edition.
K.S.V. Santhanam, Roman J. Press, Massoud J. Miri, Alla V. Bailey, and Gerald A. Takacs

Book Companion site : www.wiley.com/go/santhanam/hydrogentech_2e

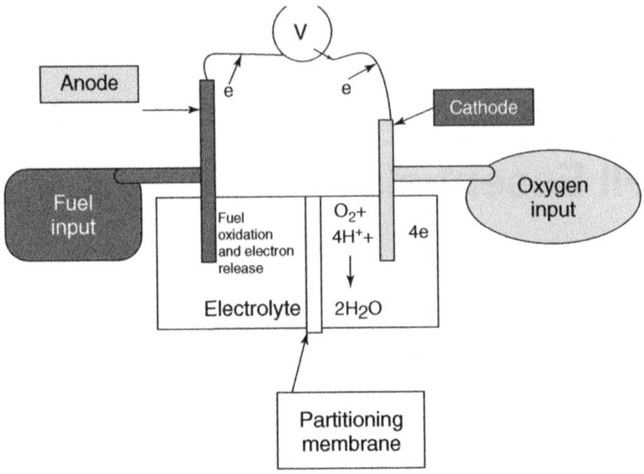

Figure 7.1 Simplified illustrative picture of a fuel cell.

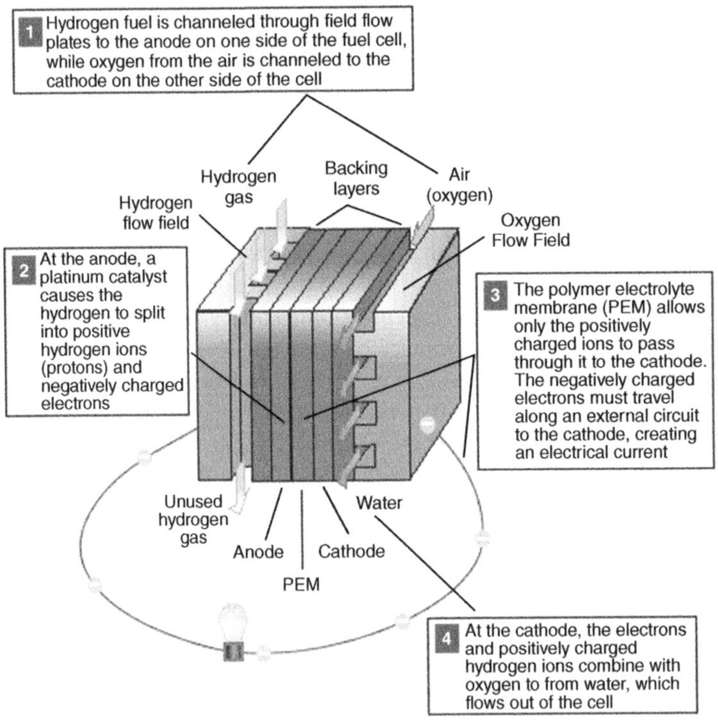

Figure 7.2 Basic sandwich configuration of compact fuel cell. [*Source*: http://en.wikipedia.org/wiki/Fuel_cell. Used under CC BY SA 3.0 https://creativecommons.org/licenses/by-sa/3.0/.]

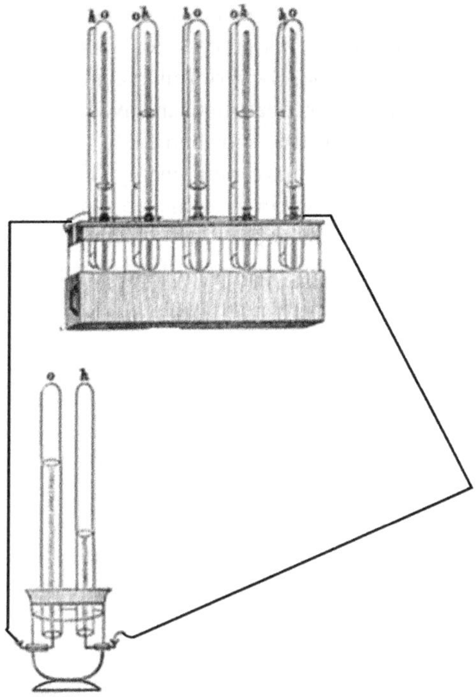

Figure 7.3 A sketch of Grove's gas battery (1839), which produced a voltage of about 1 V.

7.2 DEFINITION OF FUEL

A fuel is defined as a chemical that produces energy when it is converted from one form to the other. From ancient times, we have considered wood as a fuel. When it is burned, it produces thermal energy that can, for example, be used for cooking food. Our ancestors used this readily available fuel for centuries. Coal is another chemical that was used as a fuel. Gasoline, which is made up of a number of organic chemicals (hydrocarbons), came later as an automobile fuel. Upon burning gasoline, we produce energy to drive an automobile. There are many other fuels available in our universe that could be exploited for liberating energy. Among the several different types of fuels available in the market, it is necessary to consider which one is the most suitable for running a fuel cell. For the present considerations, we examine (a) the amount of energy available by burning the chosen fuel and (b) its mass. *Hence, what we have to calculate is the minimum amount of chemical that can liberate maximum amount of energy. Using this data with different fuels*, we can classify the fuels based on the amount of energy released that would permit us to select the most suitable fuel for a fuel cell.

7.3 WHAT IS A FUEL VALUE?

In order to determine the amount of energy liberated by a chemical upon combustion, we start with the same quantity of each chemical (fuel) and determine how much energy it produces. In other words, it is easier to standardize all fuels by determining how much energy is liberated upon combustion. This process converts chemical energy to thermal energy. This is defined as the fuel value. By performing combustion experiments with the commonly used fuels such as wood, coal, gasoline, and hydrogen, fuel values, as shown in Table 7.1, can be determined. From the data shown in Table 7.1, it is obvious that hydrogen can be ranked high, as it produces the maximum energy. The type of chemical bonding that exists in different chemicals is shown in the same table. While H–H bonding produces the highest fuel value, the next highest is gasoline, which has C–H bond. Gasoline fuel value falls very close to the hydrocarbons such as octane or butane. In all the chemicals, a common bond is C–H. The others in the list have C–O bonds in addition to the C–H bond that reduces the fuel value.

While bomb calorimeter is used for measuring the fuel values, thermogravimetric method of determining the heat of vaporization has been used for a variety of alcohols and biodiesels. Tables 7.2 and 7.3 gives the fuel values of alcohols and biodiesels as kJ/mol as determined by thermogravimetric method. Table 7.2 shows a correlation of the fuel value with the number of carbon atoms in the alcohols.

The above alcohols containing carbon atoms ranging from 7 to 16 are available in the jet-diesel fuels. The biodiesels made from canola, coconut, jatropha, palm, soybean, and tallow have fuel values of the same order as those of the alcohols in Table 7.3.

TABLE 7.1 Fuel values of different chemicals

Chemical	Types of bonds	Fuel value (kJ/g)
Hydrogen	H–H	141.8
Gasoline	C–H	48
Coal	C–H, C–O	15–27
Wood	C–H, C–O	15
Ethanol	C–H, O–H	29.7
Butanol	C–H, O–H	36.0
Octane	C–H	48.0
Butane	C–H	49.5
Methane	C–H	55.5
Methanol	C–H, C–OH	22.7

TABLE 7.2 Fuel values of series of alcohols

Alcohol	Number of carbon atoms	Fuel value (kJ/mol)
n-heptane $CH_3(CH_2)_5CH_3$	7	36.6
Isooctane $CH_3C(CH_3)_2CH_2CH(CH_3)_2$	8	38.4
Nonanol $CH_3(CH_2)_8OH$	11	55.6
n-Dodecane $CH_3(CH_2)_{10}\ CH_3$	12	50.9
1-Hexadecanol $CH_3(CH_2)_{15}\ OH$	16	59.5
n-Hexadecane $CH_3(CH_2)_{14}\ CH_3$	16	51.4

Source: Zhou *et al.* (2014) [44]. Reproduced with permission of American Chemical Society.

TABLE 7.3 Fuel values of biodiesels

Biodiesel (B100)	Fuel value (kJ/mol)
Canola biodiesel	65.1
Coconut biodiesel	39.8
Jatropha biodiesel	64.6
Palm biodiesel	83.7
Soybean biodiesel	65.7
Tallow biodiesel	78.4

Source: Zhou *et al.* (2014) [44]. Reproduced with permission of American Chemical Society.

TABLE 7.4 Fuel value of diesels

Fuel	Fuel value (MJ/kg)
Diesel fuel	45.84
B10	44.59
B15	44.11
B20	43.55
B25	43.17
B30	42.72
Butanol	35.90
P10	45.06
P15	44.69
P20	44.37
P25	43.96
Pentanol	37.97

Source: Fernández *et al.* (2012) [45]. Reproduced with permission of Elsevier.

Table 7.4 gives the fuel values obtained by thermogravimetric method. Although biofuels are increasingly being used in transportation, it certainly does not help in solving the greenhouse gas effect unless precautions are taken to remove the greenhouse gases produced by chemical methods as the combustion of biofuels produce carbon dioxide and other gases.

The chemical formula of diesel fuel is established as $C_{14.342}H_{24.75}$ [45]. The fuel values (taken as the high heat values) of the diesel blends made with diesel fuel (D100), 10% butanol/90% diesel fuel blend (v/v) (B10), 15% butanol/85% diesel fuel blend (v/v) (B15), 20% butanol/80% diesel fuel blend (v/v) (B20), 25% butanol/75% diesel fuel blend (v/v) (B25), 30% butanol/70% diesel fuel blend (v/v) (B30), 10% pentanol/90% diesel fuel blend (v/v) (P10), 15% pentanol/85% diesel fuel blend (v/v) (P15), 20% pentanol/80% diesel fuel blend (v/v) (P20) and 25% pentanol/75% diesel fuel blend (v/v) (P25) are given in Table 7.4. Based on these results, it was concluded that the diesel blends of the following compositions, 30% 1-butanol/70% diesel fuel or 25% 1-pentanol/75% diesel fuel can be successfully used in diesel engines. However, no long-term tests have been performed to prove their sustainability.

CRITICAL THINKING QUESTION

1. Explain how thermogravimetric analysis enables the determination of a fuel value. Convert the fuel value data in Table 7.2 to kJ/g.
2. Why are biodiesels not environmentally friendly?

7.4 WHY DO WE WANT TO USE HYDROGEN AS FUEL?

Hydrogen is ranked highest in the fuel value among all the fuels considered here (see Table 7.1). It liberates about three times more energy than the gasoline that we are burning today in automobiles. Its combustion produces water as by-product, as shown below:

$$2H_2(g) + O_2(g) \rightarrow 2H_2O(aq) \tag{7.1}$$

Compared to the above reaction, gasoline that is made up of several hydrocarbons produces carbon dioxide in the combustion reaction (5.2):

$$\text{Gasoline} \rightarrow CO_2(g) + H_2O(l) \tag{7.2}$$

In the case of octane combustion:

$$2C_8H_{18}(l) + 25O_2(g) \rightarrow 16CO_2(g) + 18H_2O(l) \tag{7.3}$$

Carbon dioxide must be catalytically converted to safe product(s) as it is toxic beyond a ppm. Comparing reactions (7.1) and (7.3), water is inert as compared to CO_2 and hence hydrogen technology is preferable for a pollution-free atmosphere.

7.5 CLASSIFICATION OF FUEL CELLS

The operation of a fuel cell requires the supply of a fuel (see Figure 7.2) and, hence, one way of classifying the fuel cells would be based on the type of fuel input. The fuel cells developed so far utilize mostly hydrogen, or methanol or methane. Utilizing these fuels, a large number of fuel cells have been studied in detail, with some of them reaching commercial-scale production. Figure 7.4 gives a list of fuel cells that are developed for practical applications. With hydrogen gas as a fuel, a number of fuel cells have been developed based on the operating conditions. When a polymeric membrane with an electrolyte is used to separate the anode and cathode, the fuel cell is designated as PEMFC (polymer electrolyte membrane fuel cell). The polymer electrolyte membrane (PEM) is an important part of this fuel cell and, consequently, a number of different membranes have been developed. It also continues to be an area of active research. A detailed description of this fuel cell is given later.

An alkaline fuel cell (AFC) operates in alkaline medium. It has been used in space missions. An alkaline medium separates the anode and cathode. The operation of this fuel cell is described in Section 7.19.

Phosphoric acid (PA) is also used as an electrolyte in another type of fuel cell, known as a phosphoric acid fuel cell (PAFC). In this fuel cell, phosphoric acid is filled, by capillary action, into a silicon carbide matrix that is imbedded to the electrodes.

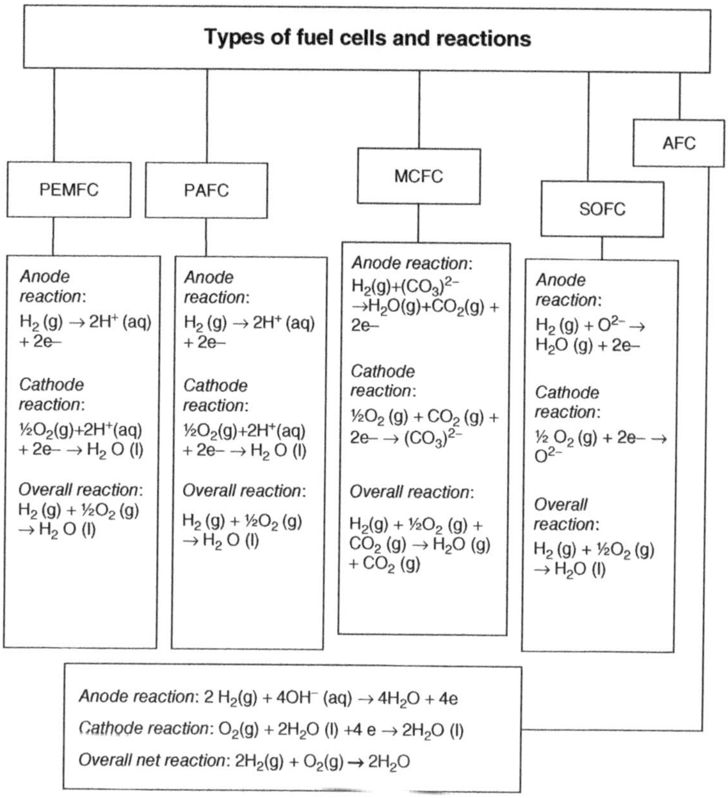

Figure 7.4 Fuel cell reactions in systems where hydrogen is the fuel.

A molten carbonate fuel cell (MCFC) utilizes carbonate along with hydrogen at the anode. Carbonate ion is generated at the cathode, where from it diffuses to the anode. With a solid oxide fuel cell (SOFC), formation of water at the anode occurs during the oxidation of hydrogen. With all of the above fuel cells that are using hydrogen as the fuel, the overall chemical reactions are given in Figure 7.4. Note that the anodic reaction with all the fuel cells involves oxidation of hydrogen (Figure 7.4).

Recently, a new fuel cell (Figure 7.5) called carbon fuel cell (CFC) has been proposed where a solid fuel is converted to electrical power. At cathode, oxygen supplied as air to the electrode is undergoing $4e^-$ conversion to produce reactive O^{2-} that migrates to the carbon anode where the oxidation of carbon occurs in the following sequence:

$$\text{At anode} \quad C + 2O^{2-} \rightarrow CO_2 + 4e^- \tag{7.3a}$$

$$\text{At cathode} \quad O_2 + 4e^- \rightarrow 2O^{2-} \tag{7.3b}$$

The overall reaction is

$$C + O_2 \rightarrow CO_2 \tag{7.3c}$$

Although this is the complete reaction, partial reaction of C (reaction 7.3a) undergoing $2e^-$ oxidation will result in the formation of CO. Other modifications of CFC are in

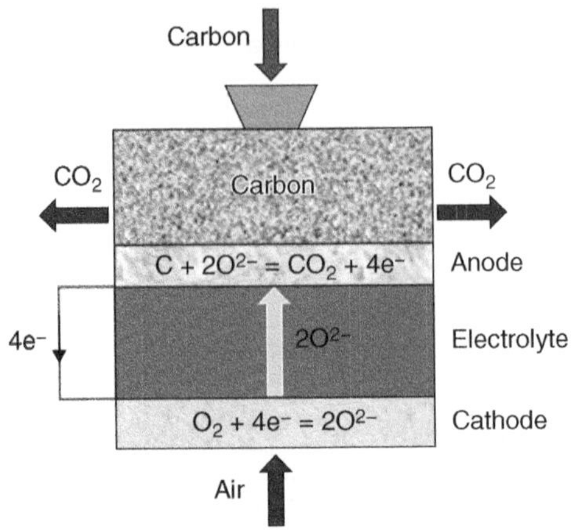

Figure 7.5 Carbon fuel cell. A solid fuel is converted to electrical power. [*Source*: Gur (2013) [46]. Reproduced with permission of American Chemical Society.]

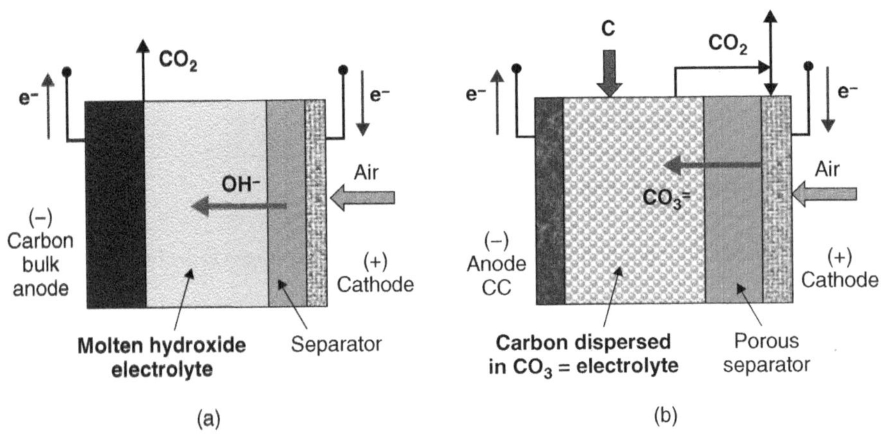

Figure 7.6 CFC with (a) molten hydroxide electrolyte and (b) molten carbonate electrolyte. [*Source*: Gur (2013) [46]. Reproduced with permission of American Chemical Society.]

using either hydroxide electrolyte or molten carbonate electrolyte. Figure 7.6 describes the hydroxide or carbonate electrolyte CFC. The transporting ions from the cathode are distinctly different in the two types of CFCs.

With molten carbonate electrolyte, the following reactions occur in CFC:

$$\text{At anode} \quad C + 2CO_3^{2-} \rightarrow 3CO_2 + 4e^- \tag{7.3d}$$

$$\text{At cathode} \quad O_2 + 2CO_2 + 4e^- \rightarrow 2CO_3^{2-} \tag{7.3e}$$

The molten carbonate electrolyte in CFC is labeled as MC–CFC. It can provide conversion efficiency of 78–80% for rates up to 1000 A/m^2 at 800°C. Hybrid CFCs are in a way similar to high temperature fuel cells such as MCFC and SOFC. Interestingly, the hybrid fuel cells give voltages greater than the thermodynamically expected voltage of 1.0 V. The reason for higher voltage produced by CFCs is under deeper investigation.

CRITICAL THINKING QUESTION

1. Consider the fuel cells discussed in this section. Compare the fuel cell reactions of MCFC and hybrid CFC. Is hybrid CFC promoting the green house gas effect for global warming?
2. What property is required in a membrane for using it in the fuel cells?
3. Chemical functional groups are used as criterion for the better performance of a membrane. How and why?

7.6 OPEN CIRCUIT VOLTAGES OF FUEL CELLS

Since a fuel cell is providing electricity through electrochemical reactions, the important electrical parameters of interest are energy and power outputs from a fuel cell. They are defined as

$$\text{energy (Wh)} = \text{voltage} \times \text{current} \times \text{time} = V \times I \times t$$

$$\text{power (W)} = \text{voltage} \times \text{current} = V \times I$$

To predict the power and energy outputs of a fuel cell, it is necessary to estimate the open circuit voltage (OCV). This can be done by two methods. In one, the redox potential data of the reactions involved are used. In the other, the thermodynamics of the fuel cell reactions are used. The two methods are somewhat related to each other, as the Gibbs free energy is related to the standard potential of the redox couple of the fuel cell.

7.6.1 Estimation of Fuel Cell Voltages Based on Redox Potentials of Fuel Cell Reactions

Any fuel cell reaction is a redox reaction, as fuel is oxidized at one electrode and oxygen is reduced at the other. The overall fuel cell reaction shown in Figure 7.4 is written in equation 7.4:

$$H_2(g) + \frac{1}{2}O_2(g) \rightarrow H_2O(aq) \tag{7.4}$$

Equation 7.4 can be viewed as a combustion reaction as it requires the supply of oxygen. One begins to wonder how the balanced equation 7.4 was arrived at. In this reaction, one species is oxidized and the other is reduced. How do we tell which species is being oxidized? This can be determined by examining the change in the oxidation number. If the oxidation number decreases in going from reactants to products, then it is a reduction process; if the oxidation number increases, then it is an oxidation process.

What is an oxidation number? It is defined as the actual charge on the atom in a substance or the charge evaluated based on hypothetical rules. What are the rules? The following rules apply:

Rule 1: For any atom in elemental form in the reaction, its oxidation number is zero.

Rule 2: For the monatomic ion, the oxidation number equals the charge on the ion; hydrogen in a compound +1. In some situations, it can have −1, as in BH_3 or CH_4 where hydrogen is electronegative as compared to when it is in H_2O.

Rule 3: Nonmetals usually have negative oxidation numbers. The oxidation number of oxygen in a compound is −2, but there are exceptions: O in H_2O_2 is −1. The oxidation number of anions Cl^-, F^-, Br^-, or I^- in a compound is −1.

Rule 4: The sum of all oxidation numbers of all atoms in a neutral compound is zero.

While these are the general rules, there are exceptions. For example, hydrogen can have an oxidation number of −1 in some compounds.

For the reactants in equation 7.4, hydrogen and oxygen gases are in elemental forms and hence the corresponding oxidation numbers are zero. When we apply the rules discussed earlier for H_2O, the oxidation number of H is +1 (as there are two hydrogen atoms in water, see Rule number 2) and the oxidation number of the oxygen atom is −2 (see Rule 3). So the net charge on water is 2 − 2 = 0. So we can write

$$H_2(g) + 1/2\ O_2(g) = H_2O(aq)$$

	$H_2(g)$	$1/2\ O_2(g)$	H_2O: H	H_2O: O	
Oxidation number	0	0 =	+2	−2	(note the value of 2H atoms is +2) (7.5)
and	↑ Rule 1	↑ Rule 1	↑ Rule 2	↑ Rule 3	

The hydrogen oxidation number increased from 0 to +1 in reaction 7.5 and the oxygen oxidation number decreased from 0 to −2. Hence, hydrogen gas is oxidized and oxygen gas is reduced in the fuel cell reaction.

In almost all combustion reactions, oxygen gas is reduced to water. Let us examine another example:

$$\begin{array}{lcccc} & CH_4(g) + & O_2(g) \rightarrow & CO_2(g) + & H_2O(aq) \\ & \uparrow & \uparrow & \uparrow & \uparrow \\ \text{Oxidation numbers:} & +4-4 & 0 & +4-4 & +2-2 \end{array} \tag{7.6}$$

Net oxidation number = Σ oxidation numbers. Take the example of CH_4. Since it does not have any charge, the net oxidation number is 0. The carbon oxidation number in a compound is +4. Hence, we can figure out the oxidation number of H in CH_4 using the above equation:

$$\begin{aligned} \text{Net oxidation number } CH_4 &= \text{oxidation number } C \text{ in } CH_4 \\ &\quad + \text{oxidation number of 4H in } CH_4 \end{aligned}$$

$$0 = +4 + \text{oxidation number of 4H in } CH_4$$

$$\text{Oxidation number of 4H in } CH_4 = -4 \text{ or oxidation number of H in } CH_4 = -1$$

In reaction 7.6, carbon oxidation state has not changed in going from CH_4 to CO_2; however, the hydrogen oxidation number has increased from −1 to +1 in going from CH_4 to

H_2O. Hence, CH_4 is oxidized. Oxygen's oxidation number decreased from 0 to −2 in going from O_2 to H_2O. Hence, oxygen is reduced. The two parts of this redox reaction are as follows:

$$CH_4(g) \rightarrow CO_2(g)$$

$$O_2(g) \rightarrow H_2O(aq)$$

7.6.2 Rules for Balancing Fuel Cell Reaction

The overall reaction shown by equation 7.4 is a balanced reaction and, in arriving at that reaction, the balancing rules have been applied. For more complex fuel cell reactions, balancing the reaction requires application of balancing rules. The simple analysis discussed as follows will elucidate the methodology applied.

A redox reaction is made up of oxidation and reduction processes. For the case of a reaction 7.4, hydrogen can be viewed as going to water, or oxygen going to water, by splitting it into an oxidation part 7 and a reduction part separately:

$$H_2(g) = H_2O(aq) \tag{7.7}$$

$$O_2(g) = H_2O(aq) \tag{7.8}$$

Rules for balancing equation 7.7 or 7.8 or any other redox reaction are given as follows:

Rule 1: Balance the atoms other than oxygen and hydrogen.

Rule 2: Balance the oxygen on either side by adding water.

Rule 3: Balance hydrogen by adding H^+ to the side that is deficient.

Rule 4: Balance charge by adding electrons to the deficient side. Each electron has −1 charge.

Rule 5: Balance the two half cells so that there are no electrons in the reaction. This is done after balancing the two half reactions first.

Taking one-half cell for balancing
Applying Rule 1 to equation 7.7:

$$H_2(g) = H_2O(aq) \tag{7.9}$$

There are no atoms other than hydrogen and oxygen and hence,

$$H_2(g) = H_2O(aq) \tag{7.10}$$

There is no oxygen on the left side and, hence, by adding H_2O to the left side oxygen balance can be reached. Applying Rule 2:

$$H_2(g) + H_2O(aq) = H_2O(aq) \tag{7.11}$$

There are more hydrogens on left side and hence, applying Rule 3:

$$H_2(g) + H_2O(aq) = H_2O(aq) + 2H^+(aq) \tag{7.12}$$

There are two positive charges on the right-hand side and none on the left side. Hence, applying Rule 4:

$$H_2(g) + H_2O(aq) = H_2O(aq) + 2H^+(aq) + 2e^- \tag{7.13}$$

This is a balanced one-half-cell reaction of the fuel cell. It can be simplified as H_2O is present on both sides of equation 7.13:

$$H_2(g) = 2H^+(aq) + 2e^- \tag{7.14}$$

Consider the other half of the fuel cell:

$$O_2(g) = H_2O(aq) \tag{7.15}$$

There are no atoms other than hydrogen or oxygen and hence, applying Rule 1 to equation 7.15:

$$O_2(g) = H_2O(aq) \tag{7.16}$$

There are two oxygen atoms on the left side and only one oxygen atom on the right side and hence, applying Rule 2:

$$O_2(g) = H_2O(aq) + H_2O(aq) \tag{7.17}$$

There are four hydrogen atoms on the right side and none on the left in equation 7.17. Applying Rule 3:

$$O_2(g) + 4H^+(aq) = H_2O(aq) + H_2O(aq) \tag{7.18}$$

There are four positive charges on the left and none on the right side of equation 7.18. Hence, applying Rule 4:

$$O_2(g) + 4H^+(aq) + 4e^- = H_2O(aq) + H_2O\ (aq) \tag{7.19}$$

This reaction is now balanced.

As we have two balanced half reactions, applying Rule 5:

$$H_2(g) = 2H^+(aq) + 2e^- \tag{7.20}$$

$$O_2(g) + 4H^+(aq) + 4e^- = H_2O(aq) + H_2O(aq) \tag{7.21}$$

The difference in the number of electrons in equations 7.20 and 7.21 can be removed by multiplying equation 7.20 by 2 and adding the two equations:

$$2H_2(g) = 4H^+(aq) + 4e^- \tag{7.22}$$

$$\underline{O_2(g) + 4H^+(aq) + 4e^- = H_2O(aq) + H_2O(aq)} \tag{7.23}$$

$$2H_2(g) + O_2(g) = 2H_2O(aq) \tag{7.24}$$

Or 1 mol of H_2 reacts with 1/2 mol of O_2 (g), as shown in equation 7.25:

$$H_2(g) + 1/2O_2(g) = H_2O(aq) \tag{7.25}$$

Direct methanol fuel cell (DMFC) uses methanol as the fuel and, in this situation, the net reaction that occurs can be arrived at using the above rules. In Appendix 7, DMFC redox reactions are explained.

7.6.3 Half-Cell Potentials

In the preceding section, the two half-cell reactions of a fuel cell have been identified as given by equations 7.20 and 7.21. Each of the half-cells develops a potential that needs to be measured. For this purpose, it needs to be combined with another half-cell with known potential such that, from the measured cell voltage, it is possible to calculate the half-cell potential of the unknown. The half-cell with known potential is chosen as the hydrogen electrode and, by convention, the half-cell potential of it is arbitrarily given a value of zero. The hydrogen half-cell is made up of a platinum black electrode in contact with 1 M HCl solution. Hydrogen gas is passed on to the platinum black electrode at 1 atmospheric pressure. The temperature of the half-cell is maintained at 25°C. A symbolic representation of this half-cell is given as follows:

$$Pt/H_2(1\text{ atm.}),\ 1\text{ M HCl} \tag{7.26}$$

The potential of this half-cell [sometimes called a standard hydrogen electrode (SHE) or a normal hydrogen electrode (NHE)] is used in the measurement of standard potentials of different redox couples. Table 7.5 gives the redox potentials of different half-cell reactions.

The voltage of the fuel cell (OCV, V_{oc}) is given as the difference in the two half-cell potentials. Thus,

$$V_{oc} = E^{\circ}_{O_2/H_2O} - E^{\circ}_{H^+/H_2} \tag{7.27}$$

$E^{\circ}_{O_2/H_2O}$ refers to the half-cell reaction 7.23 and $E^{\circ}_{H^+/H_2}$ corresponds to half-cell reaction 7.22. From Table 7.5, the values for $E^{\circ}_{O_2/H_2O}$ and $E^{\circ}_{H^+/H_2}$ are 1.229 and 0.00 V, respectively. Substitution of these values into equation 7.27 gives

$$V_{oc} = 1.229 - 0.00 = 1.229\text{ V} \tag{7.28}$$

Thus, the expected voltage for PEMFC is 1.229 V. This procedure of estimating the fuel cell voltage can be extended to other types of fuel cells.

TABLE 7.5 Standard electrode potentials at 25°C[a]

Half cell reaction	E° (V)
$O_2 + 4H^+ + 4e^- \rightarrow 2H_2O$	1.229
$Ag^+ + e^- \rightarrow Ag$	0.799
$Fe^{2+} + 2e^- \rightarrow Fe$	0.771
$Cu^{2+} + 2e^- \rightarrow Cu$	0.337
$AgCl(s) + e^- \rightarrow Ag(s) + Cl^-$	0.222
$2H^+ + 2e^- \rightarrow H_2(g)$	0.000
$Pb^{2+} + 2e^- \rightarrow Pb(s)$	−0.126

[a]In water.

Source: Bard *et al.* (2001) [1]. Reproduced with permission of John Wiley & Sons.

7.7 THERMODYNAMIC ESTIMATE OF FUEL CELL VOLTAGE

The fuel cell voltage that is expected from any selected fuel cell can be obtained from the thermodynamic Gibbs free energy data. For the basic reaction 7.4 of the fuel cell, the amount of chemical energy released can be calculated from the basic thermodynamics (see Chapter 2) using Gibbs free energy ($G°$) data of the reactants and products:

$$\Delta G_{\mathrm{f}}^{\circ} = G_{\text{f-Products}}^{\circ} - G_{\text{f-reactants}}^{\circ} \tag{7.29}$$

where G_{f}° represents Gibbs free energy of formation of reactants or products, and $\Delta G_{\mathrm{f}}^{\circ}$ is the change in Gibbs free energy in going from reactants to products. Table 7.6 gives the G_{f}° values of selected chemicals.

As discussed in Chapter 2, the free energy change in a redox reaction is related to the standard potential by

$$\Delta G^{\circ} = -n\mathrm{FE}_{\text{Fuel cell}}^{\circ} \tag{7.30}$$

where n is the number of electrons transferred in the reaction, F is Faraday constant having a value of 96,496 C/equiv., and $E_{\text{Fuel cell}}^{\circ}$ (or V_{oc}) is the voltage developed by the fuel cell under standard conditions. (Note that ΔG° refers to Gibbs free energy change under standard temperature and pressure and with pure reactants; see the definition of standard conditions

TABLE 7.6 Standard free energy of formation of selected fuel cell substances at 25°C

Substance	G_{f}° (kJ/mol)
H^+(aq)	0.00
H_2(g)	0.00
H_2(aq)	17.57
H_2O(l)	−237.2
H_2O(g)	−228.6
OH^-(aq)	−157.2
O_2(g)	0.00
O_2(aq)	16.32
CO(g)	−137.3
CO_2(g)	−394.4
CO_2(aq)	−386.23
C(s)	0.00
H_2CO_3(aq)	−623
CO_3^{2-}(aq)	−527.28
Methanol(aq) or CH_3OH(aq)	−166.2
Ethanol(aq) or C_2H_5OH(aq)	−174.8
Methane(g) or CH_4(g)	−50.79
Methane(aq) or CH_4(aq)	−34.74
Propane(g) or C_3H_8(g)	−21.60
Propane(aq) or C_3H_8(aq)	−7.32

Source: Brady *et al.* (2000) [80]. Reproduced with permission of John Wiley & Sons.

in Section 2.3, p. 60). The heat liberated in the chemical reaction is given by the change in enthalpy ($-\Delta H^\circ$), and this can be calculated from the thermodynamic quantities as shown below:

$$\Delta G^\circ = \Delta H^\circ - T\Delta S^\circ \tag{7.31}$$

where ΔH° is the standard enthalpy of the reaction and ΔG° is the free energy change in the redox reaction ΔS° is the entropy change. We can now estimate the Gibbs free energy change for the fuel cell reaction 7.4:

$$\begin{aligned} \Delta G^\circ{}_{H_2O} &= G^\circ{}_{H_2O(1)} - (G^\circ{}_{H_2} + G^\circ{}_{1/2O_2}) \\ &= -237.2 - (0 + 0) \\ &= -237.2\ \text{kJ/mol} \end{aligned} \tag{7.32}$$

Substituting in equation 7.30

$$-237.18\ \text{kJ/mol} = -[2 \times 96{,}496 \times E^\circ_{\text{fuel cell}}]$$

$$E^\circ_{\text{fuel cell}} = 237.18/2 \times 96{,}496$$

$$= 0.001228\ \text{kJ/C}$$

or

$$V_{oc} = 0.001228\ \text{kJ/C}$$

Since 1 C = J/V

$$\begin{aligned} &- 0.001\,228\ \text{kJ/J}\ [1/\text{V}] \\ &= 1.228\ \text{V} \end{aligned} \tag{7.33}$$

This gives the expected voltage of the hydrogen–oxygen fuel cell. Note that the ΔG° is calculated using the data for the free energy of formation of reactants and products at 25°C and, hence, the voltage of the fuel cell refers to this temperature. At higher temperatures, the relevant values need to be used. Thus at 80°C, $\Delta G_{H_2O} = -226$ kJ/mol, and, consequently, $E = 226 \times 103$ J/2 mol $\times$ 96,485 C/mol = 1.17 J/C = 1.17 V is to be expected. Note that, as temperature increases, the expected voltage from the fuel cell decreases.

CRITICAL THINKING QUESTION

Drunken driving is an offense. When caught on a highway, the cop gives a breath test. It uses an ethanol fuel cell. How would fuel cell work for ethanol detection?

7.7.1 Voltage Losses

The predicted voltage is very seldom realized in an actual experimental situation due to a variety of losses. These losses are categorized as (a) activation polarization losses, (b) ohmic losses, and (c) mass transport losses:

Activation polarization. This is caused by the barrier for the electron transfer kinetics at the electrode. For the electron transfer reaction, the loss due to this factor is minimal

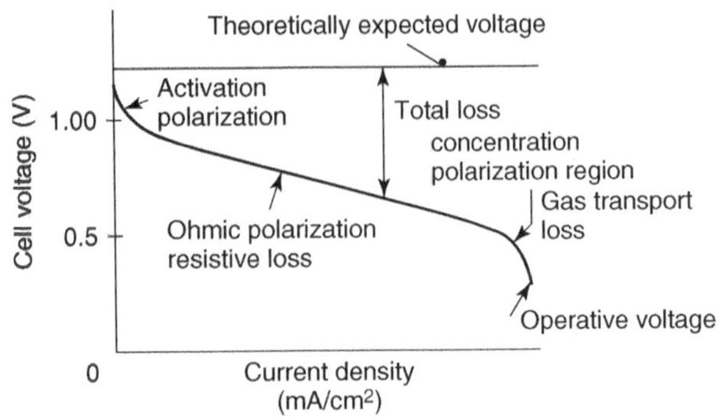

Figure 7.7 Voltage losses in the operation of a fuel cell.

at small current densities and grows with increasing current densities. It is similar to chemical reactions that have an activation barrier.

Ohmic losses. This loss is due to the resistance to both the electrolyte ion movement and the electron movement in the electrode and bipolar plates. It obeys Ohm's law.

Mass transport losses. This is caused by the supply of the reacting species to the electrode by concentration polarization. It is also called concentration polarization. The different losses that one faces in fuel cell development are shown in Figure 7.7.

7.8 EFFICIENCY OF A FUEL CELL

In almost all energy-based machines there is always a component of energy that appears as heat. Hydroelectric power supply or thermoelectric power supply engines convert mechanical energy or thermal energy into electrical energy. The conversion efficiencies are in the range of 30–60% with these machines, which suffer from large heat losses. Fuel cells operate on the principle of conversion of chemical energy to electrical energy. This conversion is never 100%, as in this process a small quantity of heat is produced as given by equation 7.31. If heat is released in an exothermic chemical reaction, then $\Delta H°$ is given a negative sign. If the chemical reaction absorbs heat, then such a reaction is defined as endothermic and $\Delta H°$ is positive. Fuel cell efficiency is defined as

$$\eta = \frac{\text{Electrical energy produced per mole of fuel}}{\text{Change in enthalpy of formation}} \times 100\% = \frac{\Delta G_f^\circ}{\Delta H_f^\circ} \times 100\% \quad (7.34)$$

This efficiency is sometimes called the calorific efficiency. This can reach 100% if entropy change in the reaction is zero. At 25°C, the electrical energy produced per mole of fuel in the case of a hydrogen–oxygen fuel cell is −237 kJ/mol and the change in enthalpy of formation is −285.84 kJ/mol. Hence, the expected efficiency is

$$\begin{aligned}\eta &= (-237 \text{ to} - 285.84 \text{ kJ/mol} \times 100\% \\ &= 82.9\% \end{aligned} \quad (7.35)$$

This means that about 82% of energy content of hydrogen is converted to electrical energy. In comparison, the overall automotive gasoline engine has a conversion of about

20%. In other words, 20% of the thermal energy of gasoline is converted to mechanical work.

7.9 EFFICIENCY AND TEMPERATURE

In equation 7.34, the values of ΔG_f° and ΔH_f° are dependent on temperature. Table 7.7 shows the values at selected temperatures. This data demonstrates that the fuel cell efficiency changes with temperature. It also shows that the theoretically expected voltage decreases with increasing temperature. Thus, PEMFC operating at 25°C is expected to give a theoretical voltage of 1.23 and 1.17 V at 100°C.

7.10 INFLUENCE OF ELECTRODE MATERIAL ON CURRENT OUTPUT

Fuel oxidation at the anode and the corresponding oxygen reduction at the cathode (see Figure 7.4) are electrochemical processes and are governed by electrode kinetics [1]. The kinetic factors will not only control the flow of current in the fuel cell but also the efficiency in equation 7.34. In other words, the electron transfer from the fuel to the electrode in the oxidation of hydrogen at the anode and corresponding electron transfer reaction from the cathode to oxygen are the important factors. The question of how the kinetic factor influences the current flow was addressed in a general way by Tafel and later by Butler–Volmer [1]. Their contributions to the understanding of the kinetics led to the concept of exchange current (i) that can be written for the equilibrium condition as

$$I_o = [i/A] = nFk^\circ C \tag{7.36}$$

where F is Faraday, A is the area of the electrode, k° is the standard heterogeneous rate constant, n is number of electrons in the electrode reaction, and C is the concentration of the fuel or oxygen I_o is the exchange current density.

When a system is shifted from its equilibrium potential by passing a current, Tafel empirically showed that the plot of over voltage (η) [defined as $\eta = (E - E_{eq})$] and current density (I) follows equation 7.37:

$$\eta = a \pm b \log I \tag{7.37}$$

TABLE 7.7 Free energies and enthalpies of fuel cell reaction at different temperatures

Reaction	ΔG_f° (kJ/mol)	ΔH_f° (kJ/mol)
$H_2(g) + 1/2\,O_2(g) \rightarrow H_2O(l)$	−237.2 (25°C)	
$H_2(g) + 1/2\,O_2(g) \rightarrow H_2O(g)$	−228.2 (80°C)	
$H_2(g) + 1/2\,O_2(g) \rightarrow H_2O(g)$	−225.3 (100°C)	−242.6 (100°C)
$H_2(g) + 1/2\,O_2(g) \rightarrow H_2O(g)$	−220.4 (200°C)	−244.5 (300°C)
$H_2(g) + 1/2\,O_2(g) \rightarrow H_2O\,(g)$	−199.6 (600°C)	−246.2 (500°C)
$H_2(g) + 1/2\,O_2(g) \rightarrow H_2O(g)$	−188.6 (800°C)	−247.6 (700°C)
$H_2(g) + 1/2\,O_2(g) \rightarrow H_2O(g)$	−177.4 (1000°C)	−248.8 (900°C)
$CO(g) + 1/2\,O_2(g) \rightarrow CO_2(g)$	−230.7 (100°C)	−283.4 (100°C)
$CO(g) + 1/2\,O_2(g) \rightarrow CO_2(g)$	−283.7 (300°C)	−232.7 (300°C)
$CO(g) + 1/2\,O_2(g) \rightarrow CO_2(g)$	−282.0 (900°C)	−178.5 (900°C)

where a and b are constants. E is the potential after polarization and E_{eq} is the equilibrium potential. The application of current in the positive direction is defined as anodically polarized and in the negative direction is defined as cathodically polarized. In equation 7.37, $\pm$ refers to the type of polarization used in the experiment. The constants were later theoretically shown by Butler–Volmer as

$$a = \{2.3\ RT/\alpha nF\} \log I_o \tag{7.38}$$

$$b = -2.3\ RT/\alpha nF \tag{7.39}$$

R is a gas constant, T is the absolute temperature, α is transfer coefficient (favoring the extent of either cathode or anode reaction), n is number of electrons transferred at the electrode, and F is Faraday. I_o is the exchange current density. For a deeper understanding of these equations and assumptions involved in the derivation of Butler–Volmer equation, the reader should refer to more advanced electrochemical books [1]. One can visualize it in the following way. Consider the fuel cell reaction

$$H_2 \rightleftarrows 2H^+(aq) + 2e^- \tag{7.40}$$

which is occurring at the anode, depicts an electron transfer reaction that results in the current flow. At equilibrium, the rate of oxidation of hydrogen and the rate of reduction of hydrogen ion are equal; in other words, the rate of forward reaction is equal to the rate of backward reaction. The current that flows at equilibrium is known as the exchange current and, when it is normalized to the area, it is the exchange current density. In a similar way, the cathode reaction

$$O_2 + 4H^+ + 4e^- \rightleftarrows 2H_2O \tag{7.41}$$

will have its exchange current density. These reactions proceed at different rates for different electrode materials and hence it is necessary to establish the electrode that is most suitable for the operation of the fuel cell.

7.10.1 Measurement of Exchange Current Densities

The Tafel experiment is the most convenient method for determining the exchange current density. The oxidation of hydrogen or reduction of oxygen in an electrolyte is carried out at the chosen electrode material by polarizing the electrode in the positive or negative direction and measuring the current flow that results in the electrochemical cell. A plot of overpotential (η) versus log current density results in a straight line, as shown in Figure 7.8, whose intercept gives the exchange current density value [see equation 7.38].

Tables 7.8–7.10 provide the measured exchange current densities for hydrogen and oxygen at different electrodes.

7.11 PRESSURE DEPENDENCE OF FUEL CELL VOLTAGE

Nernst was the first person to relate electrode potential to concentration. As explained in Chapter 2, for a reaction

$$O + n \rightleftarrows R \tag{7.42}$$

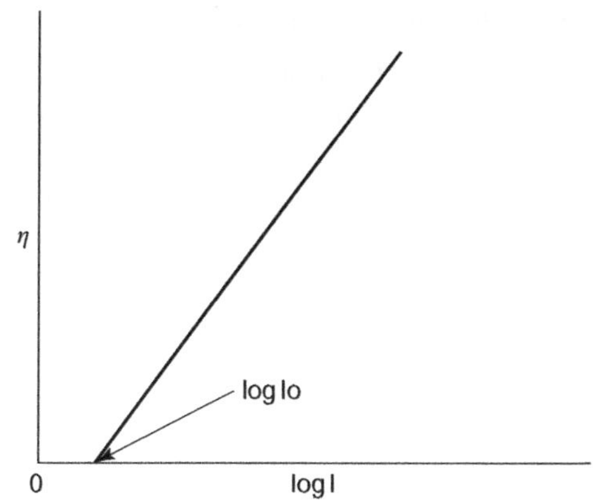

Figure 7.8 Tafel curve.

TABLE 7.8 Exchange current densities for hydrogen in acid electrolyte

Electrode	I_o (A/cm^2)
Platinum	5×10^{-4}
	7.9×10^{-4}
Silver	4×10^{-7}
	1.41×10^{-8}
Zinc	3×10^{-11}
Lead	2.5×10^{-13}
Graphite	1.2×10^{-4}
Au	3.98×10^{-6}
Rh	2.51×10^{-4}
Ir	1.99×10^{-4}
Ni	6.1×10^{-6}
W	1.25×10^{-6}
Co	4.78×10^{-6}
Cu	4.26×10^{-6}
Mo	6.51×10^{-8}
Re	1.34×10^{-3}
Nb	1.58×10^{-7}

TABLE 7.9 Exchange current densities for oxygen in acid and basic electrolytes

Electrode	I_o (A/cm^2)
Pt/C	3.16×10^{-7}
Pt–W_2C/C	5.01×10^{-5}
Pt	9.0×10^{-8}
Pt	4.0×10^{-9}
Au	5.0×10^{-9}
Graphite	1.2×10^{-4}
Carbon nanotubes	6.3×10^{-4}
Carbon nanotubes/Pd (0.08 mg/cm^2)	36.6

TABLE 7.10 Exchange current densities at silicon membranes

Electrode material	Temperature, °C H_2/cell/O_2	Pressure, atm. anode/cathode	I_o (mA/cm^2)
Control Nafion®115	90/80/88	1/1	1.7×10^{-4}
	90/80/88	3/3	3.3×10^{-3}
	130/130/130	3/3	2.4×10^{-3}
Control recast Nafion®	90/80/80	3/3	1.7×10^{-3}
	130/130/130	3/3	3.2×10^{-8}
2% Silicon oxide/Nafion®115	130/115/120	3.75/3	1.99×10^{-4}
6% Silicon oxide/Nafion®115	130/115/120	3.75/3	1.93×10^{-4}
3% Silicon oxide/Nafion®	130/130/130	3/3	5.03×10^{-5}
6% Silicon oxide/recast Nafion®	130/130/130	3/3	7.9×10^{-5}
10% Silicon oxide/recast Nafion®	90/80/88	3/3	1.0×10^{-5}
	130/130/130	3/3	9.2×10^{-5}

Source: Chenshung (2003) [81]. Reproduced with permission of The Electrochemical Society.

where O is a species that is being reduced (taking up electrons) and R is the product of reduction. An inert metal (metal not reacting with O or R) in the solution of O and R will develop an electrode potential that is given by equation 7.43.

$$E = E^\circ - \mathrm{RT}/nF \ln[a_\mathrm{R}/a_\mathrm{o}] \tag{7.43}$$

where E° is the standard potential of the redox couple O/R, R is a gas constant, T is absolute temperature, and F is the Faraday. a_R and a_o are the activities of species R and O in solution.

In other words, the potential that is established by an inert electrode (not participating in the reaction) in a solution containing species O and R is given by equation 7.43. With gaseous oxidant and reductant, the concentrations are replaced by pressures. For the fuel cell anode reaction,

$$H_2(g) \rightarrow 2H^+(l) + 2e^- \tag{7.44}$$

the potential established by the electrode is given by

$$E_{H_2} = E^\circ_{H_2} - RT/2F \ln\left[p_{H_2}/a^2_{H^+}\right] \tag{7.45}$$

where p_{H_2} is the pressure of hydrogen gas. Similarly, for the fuel cell cathode reaction,

$$1/2O_2(g) + 2H^+(aq) + 2e^- \rightarrow H_2O(l) \tag{7.46}$$

the Nernst potential for this reaction is given by equation 7.47

$$E_{O_2} = E^\circ_{O_2} - RT/2\mathrm{F} \ln\left[p_{H_2O}/p^{1/2}_{O_2} \cdot a^2_{H^+}\right] \tag{7.47}$$

where $E^\circ_{O_2}$ is the standard potential for the reaction (7.46). All the other terms have their usual significance.

The OCV of the fuel cell is written as

$$V_{oc} = E_{O_2} - E_{H_2}$$
$$= E^{\circ}_{O_2} - RT/2F \ln\left[p_{H_2O}/p^{1/2}_{O_2} a^2_{H+}\right] - E^{o}_{H_2} + RT/2F \ln\left[p_{H_2}/a^2_{H^+}\right]$$
$$V_{oc} = [E^{o}_{O_2} - E^{o}_{H_2}] + RT/2F \ln\left[p^{1/2}_{O_2} p_{H_2}/p_{H_2O}\right] \quad (7.48)$$

Equation 5.48 describes the temperature and pressure dependence of OCV. Under standard conditions, the $E^{\circ}_{H_2}$ is zero. Taking the values of $E^{\circ}_{O_2}$ and $E^{\circ}_{H_2}$ from Table 7.5,

$$V_{oc} = [1.229 - 0.0] + RT/2F \ln\left[p^{1/2}_{O_2} \cdot p_{H_2}/p_{H_2O}\right] \quad (7.49)$$

Another approach for obtaining the V_{oc} is from the Gibbs free energy by writing the overall reaction of the fuel cell:

$$H_2 + 1/2O_2 = H_2O \quad (7.50)$$

The equilibrium constant for this reaction can be written as

$$K = \left[a_{H_2O}/a_{H_2}.a^{1/2}_{O_2}\right] \quad (7.51)$$

Here, activities can be replaced by partial pressures or pressures by assuming ideal behavior.

$$a_{H_2O} = [p_{H_2O}/P];\ a_{O_2} = [p_{O_2}/P] \text{ and } a_{H_2} = [p_{H_2}/P]$$

where P = standard pressure. Assuming all pressures are in bar, we can substitute in equation 7.51.

$$K = \left[p_{H_2O}/p_{H_2}p^{1/2}_{O_2}\right] \quad (7.52)$$

Chapter 2 relates the Gibbs free energy to the equilibrium constant as

$$\Delta G = \Delta G^{\circ} + RT \ln K \quad (7.53)$$

where ΔG° is the standard free energy change and others are the standard constants. Substituting (7.52) into (7.53),

$$\Delta G^{\circ} = \Delta G^{\circ} + RT \ln\left[p_{H_2O}/p_{H_2}p^{1/2}_{O_2}\right]$$

Since $\Delta G^{\circ} = -nFE^{\circ}$ where $n = 2$, for the reaction 5.50

$$E = E^{\circ} + (RT/2F) \ln\left[p_{H_2}p^{1/2}_{O_2}/p_{H_2O}\right] \quad (7.54)$$

E° refers to the difference in the standard two half-cell potentials. Note the relation of equation 7.54 to equation 7.48. The voltage calculated by equation 7.48 or 7.54 is Nernst's

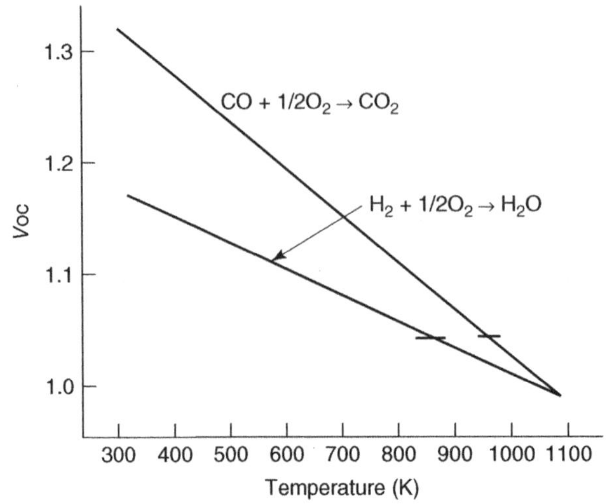

Figure 7.9 Expected open circuit voltage of fuel cells at different temperatures.

reversible voltage for the fuel cell and is a function of both temperature and pressure. It is obvious that if a fuel cell is operated at a higher temperature, the voltage will decrease, as shown by Figure 7.9.

7.12 THERMODYNAMIC PREDICTION OF HEAT GENERATED IN A FUEL CELL

The OCV of a fuel cell calculated from the free energy is given by

$$V_{oc} = [-\Delta G°/2F] \tag{7.55}$$

From Chapter 2, the enthalpy, $\Delta H°$, is related to the free energy

$$\Delta H° = \Delta G° + T\Delta S° \tag{7.56}$$

and the OCV based on $\Delta H°$ can be calculated as

$$V_{oc} = [-\Delta H°/2F] \tag{7.57}$$

For the fuel cell reaction, $\Delta H° = 285.84$ kJ/mol (product is $H_2O(l)$) and hence

$$\begin{aligned} V_{oc} &= [+285.84\ \text{kJ/mol}/2 \times 96496\ \text{C/mol}] \\ &= 1.48\ \text{V} \end{aligned}$$

This is considered as the higher heating value; in other words, if all the fuel (hydrogen) is converted to electrical energy, the expected voltage will be 1.48 V. Based on the free energy data, use of equation 7.55 gives $V_{oc} = 1.25$ V and is called the lower heating value.

7.12.1 Factors that Contribute to Fuel Cell Heat

In any process, there are inefficiencies and/or losses. In a fuel cell, the useful work is electricity; however, not all of the energy contained in the hydrogen and oxygen can be turned into electricity. Inefficiencies in the fuel cell turn some of the available energy into heat. In a fuel cell, the inefficiencies are associated with four distinct processes: (a) dissociation of hydrogen, defined as activation losses; (b) crossover of fuel from anode to cathode; (c) ohmic resistance; and (d) mass transport losses.

7.13 FUEL CELL MANAGEMENT

Three important factors for efficient fuel cell operation are (a) amount of hydrogen supply, (b) amount of oxygen supply, and (c) amount of water produced. As fuel cell powering unit is made up of a stack of fuel cells (either 21 cells or multiples thereof), management requires enough supply of hydrogen and oxygen into each of the fuel cell and removal of water that is formed in the reaction. The following discussion focuses on the methodology adopted in the evaluation of all the three quantities.

7.13.1 Hydrogen Supply

In this section, we discuss how much hydrogen gas supply is required for any selected output of a fuel cell.

By applying Faraday's law (the law states passage of 1 Faraday (F) of electricity will liberate or consume one chemical equivalent of the reactant at the electrode) for the electrochemical reaction:

$$H_2(g) \rightarrow 2H^+(aq) + 2e^- \quad (7.58)$$

1 F = 1 gram equivalent of hydrogen. The definition of equivalent is

$$1 \text{ equiv.} = \text{molecular mass } (M)/\text{number of electrons } (n) \quad (7.59)$$

and

$$1 \text{ F} = 96{,}494 \text{ Coulombs (C)} \quad (7.60)$$

or

$$\text{Coulomb (C)} = I \text{ (A)} \times t \text{ (s)} \quad (7.61)$$

Let us calculate, for generating Q coulombs of electricity, what mass of hydrogen gas should be supplied to the fuel cell. Combining the definitions (7.59)–(7.61), for any selected coulombs (Q), the mass of hydrogen required can be arrived at as

$$\text{Mass of hydrogen} = \frac{Q \cdot M_{H_2}}{n \cdot F} g \quad (7.62)$$

or substituting $Q = I\text{ (A)} \times t\text{ (s)}$ [see equation 7.61] in equation 7.62

$$\text{Mass of hydrogen} = \left(\frac{I(\text{A}) \cdot t(\text{s}) \cdot M_{H_2}(\text{g/mol})}{n \cdot F(\text{C/mol})} \right) \text{g} \quad (7.63)$$

g stands for unit grams.

In equation 7.63, n does not have a unit, and the unit of "Amperes x time" is a coulomb. This is the mass of hydrogen required for generating the required current from each fuel cell. Equations 7.62 and 7.63 are very useful in estimating the number of cylinders of hydrogen gas required in running the fuel cells. If the fuel cells are stacked, then

$$\text{Number of stacked cells} = Y$$

Equation 7.63 will have to be multiplied by Y to get the total supply of hydrogen gas:

$$\text{Mass of hydrogen gas required for } Y \text{ cells} = Y\left(\frac{I(\text{A}) \cdot t(\text{s}) \cdot M_{\text{H}_2}(\text{g/mol})}{n \cdot F(\text{C/mol})}\text{g}\right) \quad (7.64)$$

Let us now take an example for estimating the number of gas cylinders required for running one single fuel cell (no stacking). Here, we have to decide on the current we want to get from the fuel cell and the time for which it is needed. Assuming we need 100 A current for 1 h,

$$Q = 100\,\text{A} \times 1\,\text{h} = 100\,\text{A} \times 3600\,\text{s} = 3.6 \times 10^5\,\text{C}$$

$$M_{\text{H}_2} = 2.016\,\text{g/mol}$$

$$n = 2$$

$$F = 96{,}494\,\text{C/mol}$$

Substituting these values into equation 7.63,

$$\text{Mass of hydrogen gas required} = \{3.6 \times 10^5 \times 2.016\}/\{2 \times 96{,}494) = 3.76\text{ g}$$

The mass of hydrogen gas can be converted into volume using the relationship [mass/volume] = density. Using the density of hydrogen as 0.084 kg/m^3 or 0.084×10^{-3} g/cm^3, the above mass can be converted into volume of hydrogen:

$$\text{Volume of hydrogen} = \{3.76\text{ g}/0.084 \times 10^{-3}\text{ g/cm}^3\} = 44.7 \times 10^3\text{ cm}^3$$

$$= 44.76\text{ L}$$

Table 7.11 gives the specifications of hydrogen gas cylinders used in the laboratories and commercial establishments. Each cylinder carries approximately about 330 L. So, for getting 100 A current from the fuel cell for 1 h, we will utilize (44.76/330), 0.135, or 1/7th of the gas in the cylinder.

Table 7.12 gives the hydrogen gas consumption for different current values and durations. It also gives the theoretically expected consumption of hydrogen, assuming that all hydrogen gas that enters the fuel cell is oxidized at anode with 100% current efficiency. Equation 7.63 expresses the relationship between current and consumption of hydrogen. This equation can be modified to give the relationship between power output of a fuel cell and hydrogen consumption by straightforward substitution of current. Since

$$P_{\text{fuel cell}} = V \cdot I \quad (7.65)$$

Equation 7.63 is written as

$$\text{Mass of hydrogen} = \frac{P_{\text{fuel cell}} \cdot t \cdot M_{\text{H}_2}}{V \cdot nF} \quad (7.66)$$

TABLE 7.11 High-pressure cylinder technical specifications

Gas storage	Approx 330 std. (L)
Length (with valve) × diameter	21 × 4 in.
Mass	12 lbs
Stage regulator technical specifications	
Max inlet pressure	4000 psig
Operating temperature	0 to +140 F
Body inlet connection	1/8″ NPT female
Body outlet connection	1/4″ NPT female
Outlet valve connection	1/4″ NPT male

Source: Reproduced with permission of FuelCellStore.com.

TABLE 7.12 Estimated consumption of hydrogen in the fuel cell for selected currents and durations

Current (A)	Duration (s)	Expected Q (C)	Mass of hydrogen (g)	Volume of hydrogen (L)
0.5	3,600	1,800	0.0186	221
1.0	3,600	3,600	0.0373	444
2.0	7,200	14,400	0.1492	1,776
5.0	1,800	90,000	0.9327	11,100
10.0	3,600	36,000	0.3730	4,440
20.0	1,800	36,000	0.3730	4,440
100.0	3,600	360,000	3.7300	44,400

In this equation, V refers to the average voltage of the fuel cell. For a hydrogen–oxygen fuel cell, it can be taken as about 0.60 V. We can use a model calculation to see how much hydrogen will be consumed for generation of 100 W for 1 h.

$$\text{Mass of hydrogen} = \frac{100 \times 3600 \times 2.016}{0.60 \times 2 \times 96{,}494}\ \text{g}$$

$$= 6.26\ \text{g or}$$

$$\text{Volume of hydrogen} = 74.5\ \text{L}$$

Thus, in this example, hydrogen gas level in the cylinder will be reduced by about a fourth. A power table has been constructed using equation 7.66, giving the hydrogen gas consumption for a selected power output of a fuel cell. Table 7.13 depicts the details.

For a fuel cell stack, equation 7.66 should be modified to give a power output in the same manner described earlier by substituting $P_{\text{fuel cell}} = V \cdot I \cdot Y$. The power output expression for a stacked fuel cell will be

$$\text{Mass of hydrogen for stacked fuel cell} = \times[P_{\text{fuel cell}} \times t \times M_{H_2} = V \times nF] \qquad (7.67)$$

where × is the number of stacks. For a kilowatt of power generation using a stack of 200 cells in series at an average fuel cell voltage of 0.60 V, hydrogen consumption expected is

TABLE 7.13 Power output and hydrogen consumption in PEMFC

Power output (kW)	Duration (s)	Average voltage of a cell (V)	Mass of hydrogen (g)	Volume of hydrogen (L)
0.10	3600	0.60	6.26	74.52
0.50	3600	0.60	31.30	372.62
1.00	1800	0.60	31.30	372.62
10.00	900	0.60	156.69	1863.10

given as follows:

$$\text{Mass of hydrogen for stacked fuel cell} = \{\{1000 \times 3600 \times 2.016\}/0.6 \times 2 \times 96{,}494]\ \text{g}$$
$$= 62.68\ \text{g}$$

For generating 94 kW power, hydrogen required will be 5892 g.

This supply is provided by three cylindrical tanks weighing 75 kg containing 6 kg of hydrogen at about 350 bars (http://auto.howstuffworks.com/hy-wire2.htm).

7.13.2 Oxygen Supply

The oxygen requirement for a fuel cell can be calculated as follows:

$$\text{Mass of oxygen} = \frac{I(\text{A}) \cdot t(\text{s}) \cdot M_{O_2}\,(\text{g/mol})}{n \cdot F\,(\text{C/mol})} \qquad (7.68)$$

In PEMFC, the reaction at the cathode is 4e reduction of oxygen. By applying Faraday's law, M is molecular mass of oxygen, I is expected fuel cell current, n is number of electrons involved in the reaction, F is Faraday constant, and t is duration for which the current is demanded (or drawn from the fuel cell). For example, for generating 100 A current for 1 h, how much of oxygen is required? The following calculation shows the result:

$$\text{Mass of oxygen} = \frac{100 \times 3600 \times 32}{4 \times 96{,}494}$$
$$= 29.86\ \text{g}$$

Assuming the density of oxygen gas as 1.429 kg/m^3,

$$\text{Volume of oxygen} = 20.89\ \text{L}$$

Oxygen requirement for any power output can be obtained in the same way as discussed in the earlier section (hydrogen requirement). Substituting $P_{\text{fuel cell}} = V \cdot I$ or $I = P_{\text{fuel cell}}/V$:

$$\text{Mass of oxygen} = \frac{P_{\text{fuel cell}} \cdot t \cdot M_{O_2}}{V \cdot n \cdot F}\ \text{g} \qquad (7.69)$$

For 100 W power output, assuming an average fuel cell voltage of 0.60 V, the oxygen requirement is given as follows:

$$\text{Mass of oxygen} = (100 \times 3600 \times 32)/(0.6 \times 4 \times 96{,}494)$$
$$= 49.74\ \text{g}$$

$$\text{Volume of oxygen} = 49.74\ \text{g}/1.429 \times 10^{-3}\ \text{g/cm}^3$$
$$= 34.80 \times 10^3\ \text{cm}^3 \text{ or } 34.80\ \text{L}$$

For a stacked fuel cell, assuming number of individual cells as Y, equation 7.68 is modified as

$$\text{Mass of oxygen} = Y \frac{I(\text{A}) \cdot t(\text{s}) \cdot M_{O_2}\,(\text{g/mol})}{n \cdot F\,(\text{C/mol})}\text{g} \tag{7.70}$$

7.13.3 Air Usage

Since fuel cells can operate with air supply instead of oxygen, the following discussion pertains to the air usage for a fuel cell. Air contains 21% oxygen and calculations should be based on this amount. Molar mass of air is 28.97 g/mol (this is arrived at by taking the components of air molar masses). Hence, M value in equations 7.69 and 7.70 should be 28.97 g/mol instead of 32 g/mol. Also, since only 21% of air supply is taking part in the fuel cell reaction, equations 7.69 and 8.70 are rewritten as

$$\text{Mass of air} = \frac{I(\text{A}) \cdot t(\text{s}) \cdot 28.97\,(\text{g/mol})}{0.21 \cdot n \cdot F\,(\text{C/mol})}\text{g}$$

for the current flow in the fuel cell. The air flow needed for generating a selected power output is given as follows:

$$\text{Mass of air} = \frac{P_{\text{fuel cell}}(\text{W}) \cdot t(\text{s})28.97\,(\text{g/mol})}{0.21 \cdot V(\text{V}) \cdot n \cdot F\,(\text{C/mol})}$$

A sample calculation similar to the one discussed earlier on oxygen usage can be made using the above two equations. For 1 kW power output, assuming an average fuel cell voltage of 0.60 V, the oxygen requirement is given as follows:

$$\text{Mass of air} = \frac{1000 \times 3600 \times 28.97}{0.21 \times 0.60 \times 4 \times 96{,}494}$$
$$= 2144.47\ \text{g or } 2.144\ \text{kg}$$

7.13.4 Water Output

The stoichiometry of the PEMFC chemical reaction

$$H_2 + 1/2O_2 \rightarrow H_2O$$

states that every mole of hydrogen in the fuel cell reaction will produce 1 mol of water. Alternatively, every half a mole of oxygen usage will result in 1 mol of water. As water is produced as output during the fuel cell reaction, necessary precautions should be taken so as not to flood the fuel cell. In order to calculate the amount of water produced during the running of the fuel cell, it is the generally accepted practice to calculate the amount of water produced by the fuel cell for any selected power output. Let us calculate the amount of water produced by a 1 kW power generating fuel cell and the moles of hydrogen required:

$$\text{Moles of hydrogen} = \frac{P_{\text{fuel cell}} \cdot t}{V \cdot n \cdot F}\text{moles} \tag{7.71}$$

1 mol H_2 = 1 mol H_2O (see equation 5.25).

$$\text{Moles of water} = \frac{P_{\text{fuel cell}} \cdot t}{V \cdot n \cdot F} \text{moles} \tag{7.72}$$

Using $P_{\text{fuel cell}} = 1$ kW, $t = 3600$ s, average fuel cell voltage = 0.60 V, $n = 2$, and $F = 96{,}494$,

$$\text{Moles of water produced in the fuel cell} = \frac{1000 \times 3600}{0.60 \times 2 \times 96{,}494} = 31.09 \text{ mol}$$

$$\text{Mass of water} = 31.09 \text{ mol} \times 18.02 \text{ g/mol} = 559.62 \text{ g}$$

Using density of water as 0.998 g/cm^3 at 20°C,

$$\text{Volume of water} = 559.62 \text{ g}/0.998 \text{ g/cm}^3 = 560.74 \text{ cm}^3$$

Thus, 1 kW of power generation will produce an output of about 561 cm^3 of water.

It is possible to calculate water production from the oxygen requirement of the fuel cell.

$$1/2 \text{ mol } O_2 = 1 \text{ mol water or } 1 \text{ mol } O_2 = 2 \text{ mol water}$$

Using equation 7.69

$$\text{Moles of oxygen} = \left(\frac{P_{\text{fuel cell}} \cdot t}{V \cdot n \cdot F}\right) \tag{7.73}$$

$$\text{Moles of water} = 2\frac{P_{\text{fuel cell}} \cdot t}{V \cdot n \cdot F} \tag{7.74}$$

$$\text{Moles of water produced} = 2 \times \frac{1000 \times 3600}{0.60 \times 4 \times 96{,}494} = 31.09 \text{ mol}$$

$$\text{Volume of water} = 31.09 \text{ mol} \times 18.02 \text{ g/mol}/0.998 \text{ g/cm}^3$$
$$= 560.74 \text{ cm}^3$$

Thus, both calculations, one using hydrogen and another using oxygen, produce the same quantity of water.

7.14 RATE OF CONSUMPTION OF HYDROGEN AND OXYGEN

In the previous section, equations 7.71 and 7.73 describe the amounts of hydrogen or oxygen needed to generate current or power. They also describe the quantity of water produced at a given power output. These equations can be modified to the rate equations by considering how much quantities are involved per second. Taking the amount of hydrogen required for PEMFC operation, equation 7.63 gives the mass of hydrogen consumed in *t* seconds. If this equation is transformed to 1 s, it gives the rate of consumption of hydrogen as shown by equation 7.75. Note that current units, A (amperes) = C/s:

$$\text{Rate of hydrogen consumption} = \frac{I(\text{A}) \cdot M_{H_2}(\text{g/mol})}{n \cdot F(\text{C/mol})} \text{g/s} \tag{7.75}$$

Similarly,

$$\text{Rate of hydrogen for stacked fuel cell} = Y \cdot \frac{I(\text{A}) \cdot M_{H_2}(\text{g/mol})}{n \cdot F(\text{C/mol})} \text{g/s} \tag{7.76}$$

$$\text{Rate of hydrogen consumption} = \left(\frac{P_{\text{fuel cell}} \cdot M_{H_2}}{V \cdot n \cdot F}\right) \text{g/s} \tag{7.77}$$

$$\text{Rate of oxygen consumption} = \left(\frac{I(\text{A}) \cdot M_{O_2}}{n \cdot F}\right) \text{g/s} \tag{7.78}$$

$$\text{Rate of oxygen consumption} = \frac{P_{\text{fuel cell}} \cdot M_{O_2}}{V \cdot n \cdot F} \text{g/s} \tag{7.79}$$

The consumption of gases and moles of water produced are given by the above equations; in the real operation of a fuel cell, nearly two times the theoretically calculated values are used because of a variety of experimental factors such as leaks and nonideal behavior.

7.15 RATE OF PRODUCTION OF WATER

In the previous section, the total moles of water produced at any power output in a fuel cell were given by equations 7.72 and 7.74; equation 7.72 is based on hydrogen used ($n = 2$) and equation 7.74 is based on oxygen used ($n = 4$) by the fuel cell. Both the equations reduce to a common equation 7.80 when the n values are substituted into equations 7.72 and 7.74:

$$\text{Rate of production of water} = \left(\frac{P_{\text{fuel cell}}}{2 \cdot V \cdot F}\right) \text{g/s} \tag{7.80}$$

The amount of water formed is based on the stoichiometry of the reaction.

CRITICAL THINKING QUESTION

1. Water management problem in a fuel cell was overlooked by several automobile companies.

   Toyota Motor Company (TMC) has formally introduced a new, advanced version of its Highlander-based fuel cell hybrid vehicle (FCHV) equipped with a newly designed higher-performance Toyota fuel cell stack. The FCHV-adv provides a 25% improvement in fuel efficiency through improved fuel cell unit performance, enhancements to the regenerative braking system, and a reduction in energy consumed by the auxiliary systems. Equipped with TMC-developed 70 MPa (10,000 psi) high-pressure hydrogen tanks, the FCHV-adv has a range of approximately 830 km (516 miles) based on the Japanese 10–15 cycle on a single fueling—more than double that of the older FCHV. Under the more rigorous JC08 test cycle, approximate range is 760 km (472 miles). In testing last year, an earlier version of the FCHV-adv successfully completed a real-world 560-km (348-mile) long-distance road test by traveling from Osaka to Tokyo on a single fueling of hydrogen.

The building block of the Toyota FC Stack is the membrane electrode assembly (MEA), where engineers focused on the basic problem of *internally produced water interfering with electrical generation within the MEA at low temperatures.*

Fundamental research, such as internal visualization tests, was carried out to understand the behavior and amount of water generated in the fuel cell, allowing engineers to optimize the MEA design to improve low-temperature startup.

As a result, the TOYOTA FCHV-adv can start and operate in cold regions at temperatures as low as −30°C, meaning the vehicle can be used in a wider variety of conditions and climates.

During development, TMC analyzed results and data from various utilization studies by the Japan Hydrogen & Fuel Cell Demonstration Project organized by Japan's Ministry of Economy, Trade and Industry, tests conducted by the California Fuel Cell Partnership in the United States and cold-weather tests in Timmins, Canada.

PROBLEM

The following are the fuel cell hybrid specification.

Toyota fuel cell hybrid specification

		FCHV-adv	FCHV
Length/width/height (mm)		4735/1815/1685	4735/1815/1685
Weight (kg)		1880	1880
Seating capacity		5	5
Max. range (km) 10–15 test cycle/JC08 test cycle		Approx. 830/approx. 760	Approx. 330
Max. speed (kph)		155	155
Fuel cell	Name	Toyota FC Stack	Toyota FC Stack
	Type	PEM	PEM
	Output (kW)	90	90
Motor	Type	Perm. magnet	Perm. magnet
	Max. output (kW)	90	90
	Max. torque (Nm)	260	260
Max. storage pressure (MPa)		70	35
Tank capacity (L)		156	148
Battery type		NiMH	Ni

CRITICAL THINKING QUESTION

1. Based on the fuel cell hybrid specification given earlier, how much water would be accumulated in the fuel cell if there were no mechanisms of water evaporation?
2. How is the water accumulation in the fuel cell overcome in the fuel cell hybrid cars?

7.16 FUEL CROSSOVER PROBLEM

In the operation of the fuel cells, a general problem is fuel crossing over from the anode side to the cathode side. This results in short circuiting the fuel cell, where the fuel directly combines with oxygen to produce water. This crossover reduces the voltage as the concentration of hydrogen in the Nernst equation is reduced (equation 7.45); as the number of hydrogen molecules oxidized at the anode decreases and the current flow that is associated with it is also reduced. Hence, the efficiency of the fuel cell decreases. The membrane that is used in the fuel cell is expected to reduce this fuel crossover. In one method, void spaces are provided in the membrane such that the fuel is sequestered into this space, thereby preventing it from reaching the cathode [40]. The fuel could subsequently be removed by a physical process.

7.17 POLYMER MEMBRANES FOR PEMFC

7.17.1 General Aspects

As discussed in the earlier section, electrodes play key roles in the electron transfer processes in all fuel cells wherein oxidation and reduction reactions occur. This being the case, what is the purpose of the partitioning membrane? Its purpose is (a) to separate the two electrodes from physical contact and (b) to prevent the fuel crossover from one compartment to the other. For example, hydrogen fuel going from the anode compartment to the cathode compartment (fuel crossover) or oxygen moving from the cathode compartment to the anode will result in the loss of fuel cell efficiency. This free flow of molecules or ions is controlled or prevented by having a thin polymeric film called a membrane between the two electrodes. The membrane acts as a separator in the electrolytic cell. As in any electrolytic process, when the electrons are transported through the external circuit, the ions migrate between the electrodes. The membrane serves a useful purpose in allowing ions to be transported from one compartment to the other. Here, the nature of the membrane is important in maintaining the ionic conductivity.

In the initial phase of the development of PEMFC, sulfonated polystyrene–divinylbenzene (SPSDVB) copolymers were used. These polymers were costly but, nevertheless, a few space ships carried PEMFC made with SPSDVB, which led to greater confidence about the practicality of fuel cells. The first commercial synthesis of a membrane was reported by DuPont in 1960, and this opened up an innumerable number of applications in electrochemistry and polymers. DuPont's registered trademark for perfluorinated ion exchange membrane is Nafion®. It is a tetrafluoroethylene polymer containing sulfonic acid groups. It has hydrophobic and hydrophilic groups present in it; the hydrophobic entity is $–CF_2–CF_2–$ and hydrophilic part is $–SO_3H$ attached permanently. The polymer is very similar in chemical structure to Teflon, which is perfluorinated polymer having the structure $(–CF_2–CF_2–)_n$. It is substituted with a sulfonic acid group for producing the ion transport. Figure 7.10 shows the $–SO_3H$ entity.

$$
\begin{array}{l}
(CF_2–CF_2)_n–CF_2–CF_2– \\
| \\
OR \\
R = SO_3^- H^+
\end{array}
$$

Figure 7.10 Nafion® structure.

The sulfonic acid group makes the membrane electrolyte suitable for fuel cell applications. The sulfonic acid group is ionically bonded and it dissociates into $-SO_3^-$ and H^+ with the $-SO_3^-$ still attached to the C–F network. In other words,

$$
\begin{array}{l} SO_3^- - CF_2 - CF_2 - O - \underset{\displaystyle CF_2}{\underset{|}{CF}} - CF_3 \end{array}
$$

is called the ionomer (polyelectrolyte containing not more than 15% ionic units). The ionomers have a property of attracting oppositely charged ions. Thus, in the above structure, the negative charge on $-SO_3^-$ attracts the positively charged hydrogen ion. Hence it is defined as having hydrophilic character. Ionomers are key substances in the development of sensors using membranes. Membrane-based sensors have an innumerable number of applications. The most common one is a glass membrane that is in the glass electrode used in pH measurements. Among others, bicarbonate sensors, calcium ion selective electrodes, fluoride ion selective electrodes, and oxygen sensors find usage in biomedical areas.

DuPont developed a large number of Nafion® membranes with differing molecular mass and classified them by a series. Thus, Nafion®-115 or Nafion®-117 membranes, which were initially developed for fuel cell applications, continue to find large number of applications. Dupont extended the 100 series (115, 117, etc.) to 900 by developing new membranes having both sulfonic acid and carboxylic acid groups in fluoropolymers. One such Nafion® membrane, 90209 in 900 series, finds extensive usage in chloralkali industries wherein hydrogen is produced at cathode and chlorine at the anode in the electrolysis of brine (sodium chloride solution). This is one of the very versatile methods of producing hydrogen as a side product in the industrial-scale production of caustic soda (sodium hydroxide).

Although Nafion® membranes find extensive applications in fuel cells, they do have limitations. They do not have good conductivity at low moisture content, which results in the failure of the fuel cell. With a view to overcome this problem, several new membranes have been synthesized. A good developmental effort in this direction is reviewed by Hickner *et al.* [10] and Wang *et al.* [11]. These efforts could be rationalized as efforts to synthesize membranes: (a) modification of existing polymers, (b) polyimide-based membranes, (c) improved polymeric backbones, (d) polyphosphazenes, (e) membranes without sulfonic acid group, and (f) polymer composites. In the following sections, these different types are discussed.

A large number of membranes have been prepared that have silicon incorporated into them for fuel cell applications [8]. The preparation details are summarized as follows.

A control Nafion® 115 membrane was prepared by first refluxing it for 6–8 h with 50:50 water:concentrated nitric acid to remove metal and organic impurities. It was later refluxed with 50:50 water:concentrated sulfuric acid to remove trace metal impurities. The membrane was refluxed in water until the pH of water was about 6.5. It was dried for 24 h and later immersed in a mixture of methanol:water for 1 min followed by immersion in a mixture of tetramethylsilane:water (3:2). The membrane was dried at 100°C for 24 h. The membrane was later refluxed with 3% hydrogen peroxide for 1 h followed by distilled water wash, sulfuric acid wash, and distilled water wash. Silicon dioxide/recast membranes were prepared by mixing × mL of 5% Nafion® solution, 2× ml of isopropyl alcohol and tetraethyl ortho silicate in hydrochloric acid. This solution mixture was kept at 90°C when the membrane was formed. The platinum/carbon (Pt/C) electrodes supplied by E-Tek Inc. with a platinum loading of 0.4 mg/cm^2 is generally recommended for fuel cell work. The electrodes were sandwiched with Nafion® (0.6 mg/cm^2). The electrode–membrane–electrode assembly was heated to 90°C in carver press at 2 MPa for 1 min. The details of the preparation and experimental conditions are available in the literature [6].

7.17.2 Modification of Existing Polymers

A large number of polymers used as membranes have been modified by persulfonation to examine the ability to store water and for improved proton conductivity. The aromatic polymers were generally sulfonated to produce polymers desirable for PEMFC. A few examples of this class are sulfonated Victrex poly(ether ether ketone). Poly (ether sulfone) and bisphenol A-based polymers are other examples in this class. Poly(ether ether ketone) has been modified by adding sulfonic acid group to the backbone. This modification produced increased solubility but decreased the crystallinity of the polymer. Figure 7.11 gives the structures of some of the sulfonic acid polymers that are disulfonated. The sulfonated monomers have been copolymerized to produce random (statistical) copolymers. Bisphenol sulfone H^+ form (BPSH series of copolymers) shows conductivity and water uptake (Figures 7.12 and 7.13). This figure demonstrates that the conductivity and water uptake of this series of copolymers increases with disulfonation. At 60 mol%, a semicontinuous hydrophobic phase develops and the membrane swells dramatically, forming a hydrogel that is not useful for fuel cell applications.

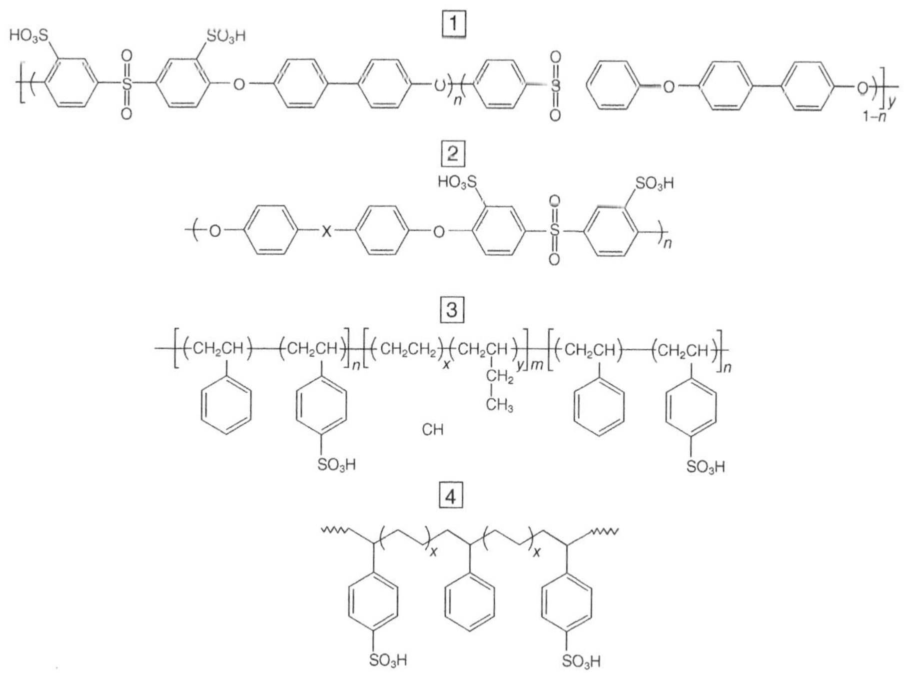

Figure 7.11 Disulfonated polymers. (1) sulfonated poly(arylene ether sulfone); (2) sulfophenylated polysulfone; (3) sulfonated styrene–ethylene–butylene–styrene (SEBS) block copolymer; (4) sulfonated styrene–ethylene interpolymer. [*Source*: Hickner *et al.* (2004) [10]. Reproduced with permission of American Chemical Society.]

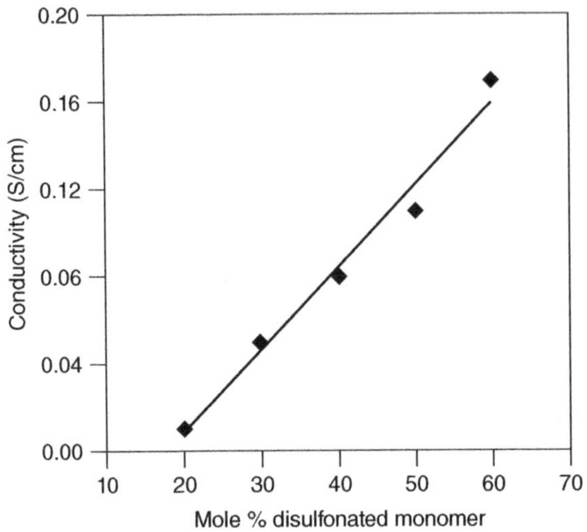

Figure 7.12 Conductivity of disulfonated monomer. [*Source*: Hickner *et al.* (2004) [10]. Reproduced with permission of American Chemical Society.]

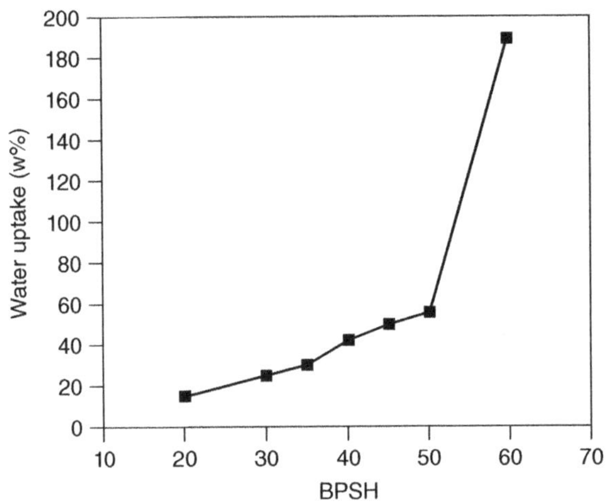

Figure 7.13 Water uptake by BPSH at 30°C. [*Source*: Hickner *et al.* (2004) [10]. Reproduced with permission of American Chemical Society.]

Nafion has been modified by several physical and chemical methods as shown in Table 7.14. Grafting the polymer has been successful, and the preparative scheme is given in the following paragraph.

Here, the fluoropolymer film is irradiated with various energy sources. This process is assumed to generate radicals on the surface. The next stage involves grafting that results in grafted fluoropolymer, which is sulfonated to produce fluoropolymer-based PEM. While the schematic shown in Figure 7.14 appears simple, it involves several chemical steps such as refluxation and extraction. The composite membrane made of perfluorinated polymer

TABLE 7.14 Modifications of polymer membranes

Base membrane	Modified methods	Main conclusions and comments
PFSI membranes treated by physical or chemical methods		
Nafion	Carboxylate/sulfonate bifunctional treatment	Dramatically reduced water permeability and modestly decreased proton conductivity
Nafion	Depositing polydopamine nanolayer	About 50% decrease of methanol permeability and slightly decreased proton conductivity
PFSI membrane	Perfluoro heptane and argon plasma deposition	Greatly improved mechanical strength and greatly decreased methanol permeability
Nafion	Depositing ultrathin palladium or its alloy layer	Improvements in methanol permeability are often weakened by the cracks deriving from the different swelling ratios between the ultrathin metal layer and Nafion membrane
Nafion solution	Applying an electric field during casting	Better PEMFC performance than Nafion117
Aciplex	Ion irradiation	Conductivity increased three orders of magnitude in dry conditions
Nafion	Thermal annealing treatment	Dependent on annealing temperature and not always resulting in improvement
PFSI membrane	Supercritical carbon dioxide treatment	Decreases methanol permeability 41.8% without sacrificing proton conductivity
Nafion	Dissolving in dimethylsulfoxide (DMSO) at 190°C and recasting	Properties similar to Nafion115
Nafion	Adopting modified Pt electrocatalyst supports	Increases power density of a H_2/air PEMFC from 345 to 456 mW/cm^2 at 100°C and 30% RH
Reinforced PFSI membranes by porous supporting materials		
Nafion/PTFE	Conventional impregnation process	Better PEMFC performance than that of pure Nafion membrane
Nafion/ePTFE	Impregnation process employing surfactant under low inner gas pressure	Rather better durability and stability with an open circuit voltage (OCV) degradation rate of 3.3 mV/h
Nafion/ultrahigh molecular weight polyethylene (UHMWPE)	Conventional impregnation process	Significantly lowered methanol crossover and thin membrane thickness

(continued overleaf)

TABLE 7.14 *(Continued)*

Base membrane	Modified methods	Main conclusions and comments
Reinforced PFSI membranes by porous supporting materials		
Nafion/poly(tetrafluoroethylene) (PTFE)/hygroscopic particles	*In situ* incorporation of a third composition (nanosilicate or nano-ZrP particles) combined with conventional impregnation process	Possesses a self-humidified functionality suitable for low relative humidity (RH) and elevated temperature operation
Nafion/poly(vinylidene fluoride) (PVDF)	Conventional impregnation process	Good improvements in DMFC performance over plain Nafion
Nafion/poly(tetrafluoroethylene) (PTFE)/hygroscopic particles	*In situ* incorporation of a third composition (nanosilicate or nano-ZrP particles) combined with conventional impregnation process	Possesses a self-humidified functionality suitable for low relative humidity (RH) and elevated temperature operation
PFSI composite membranes incorporated with organic or inorganic compounds		
Nafion/organic methanol-blocking fillers	Impregnating Nafion membranes into the solution of organic fillers or their precursors (e.g., PANI, PVI, PPV, chitosan (CS), PEDOT, SSt, and sulfonated poly(tetrafluoroethylene-*co*-hexafluoropropylene) (FEP))	Improvements in methanol permeation are mostly offset by reduced proton conductivity
Nafion/polypyrrole/Pt	*In situ* polymerization of pyrrole monomers with Pt precursors in Nafion membrane	Significantly enhanced PEMFC performance; large reduction in methanol permeability with moderate sacrifice of proton conductivity
Nafion/PVI/Pd	Complexation followed by impregnation	Better performance of DMFC fed with 5 M methanol and oxygen only at 30°C
Nafion/organic polymer constituents	Direct blending of solution between Nafion and sulfonated poly(aryl ether ketone) (SPAEK), sulfonated poly(ether ether ketone) (SPEEK), nitrated SPEEK, or PBI	Reduction in proton conductivity and improved methanol crossover; readily undergoes phase separation due to the differences in specific gravity and solvent concentration profile
Nafion/PAH-PSS layer-by-layer (LbL) film	LbL technique	Very small methanol permeation improvements with large reduction in proton conductivity

Nafion/inert hygroscopic inorganic fillers	Blending method or sol–gel process using Nafion and inert hygroscopic inorganic particles or their precursors (e.g., fullerene, WO_3, SiO_2, TiO_2, $Sr(OH)_2$, zeolite, MMT, ORMOSIL, PVOS, zirconia, GPTMS, and TEOS)	Enhancing the water retention property, improving thermal stability, and reducing methanol crossover; however, often decreasing proton conductivity due to negative effects of these inert hygroscopic fillers such as insulation for proton and poor compatibility between fillers and Nafion
Nafion/inorganic proton-conductive fillers	Blending method or impregnation process using Nafion and proton-conductive particles or their precursors (e.g., α-ZrP, heteropoly acid (HPA), BPO_4, and ErTfO)	Proton conductivity similar to Nafion; improved MEA and methanol crossover, but HPA leaching often occurs
Nafion/inorganic binary component fillers	Incorporating two kinds of inorganic particles such as zeolite–Pt, SiC_2–$Cs_{2.5}H_{0.5}PWO_{40}$, SiO_2–SZ, ZrO_2–SiO_2, TiO_2–WO_3	Better water management and reasonable proton conductivity at high temperatures and low humidity
Nafion/bifunctional fillers	Incorporating inorganic particles with both hydrophilic and proton-conductive functions (e.g., acid-functionalized multiwalled carbon nanotubes (MWCNTs), acid- or CS-functionalized methylcyclopentadienyl manganese tricarbonyl (MMT), sulfonic acid-functionalized zeolite beta, sulfonic acid-functionalized silica, and polysiloxanes)	Great improvements in methanol permeability without a large conductivity sacrifice, some of them providing reasonable proton conductivity above 100°C
PFSI composite membrane	NA	Operating >6000 h at 120°C and 50% RH fed with H_2/air and only 1% to 1‰ degradation rate of conventional Nafion MEA

Partially fluorinated acid ionomer membranes by radiation-grafting polymerization

Monomers	Grafting methods	Main conclusions and comments
PTFE base film		
Styrene	γ-Ray preirradiation grafting	Undergoes substantial structural changes; increased ionic sites; enhanced hydrophilicity; reduced hydrophobicity; decreased crystallinity, and without mechanical properties examined

(continued overleaf)

TABLE 7.14 (*Continued*)

Monomers	Grafting methods	Main conclusions and comments
Partially fluorinated acid ionomer membranes by radiation-grafting polymerization		
Styrene	UV-induced photografting	Avoids significant degradation of the PTFE main chains; proton conductivity in the thickness direction similar to that of Nafion even if the degree of grafting is <10%
cPTFE base film		
Styrene	γ-Ray preirradiation grafting	Improved radiation resistance of PTFE; 50% lower self-diffusion coefficient of water than that of Nafion®-112; excellent thermal stability up to 200°C
Cross-linked PTFE/PFA polymer-alloy film		
Styrene	Soft-electron beam (EB)-induced grafting	Proton conductivity of 0.17 S/cm at 60°C and 95% RH; delamination of the three-phase interface of MEA occurs when operated above 80°C
FEP base film		
Styrene or acrylic acid	Preirradiation grafting	Cross-linking of the FEP matrix leads to remarkable improvements in the thermal durability and mechanical properties, whereas the cross-linking of the polystyrene phase profoundly influences the surface composition of PEM
Styrene	Simultaneous irradiation grafting	Has a considerably higher value of proton conductivity and less methanol crossover than for the corresponding properties of Nafion; the proton conductivity ranges up to 0.25 S/cm, but it has a greater susceptibility to oxidative degradation
Styrene/DVB	EB-induced grafting	PEMFC single-cell durability test over 7900 h at a cell temperature of 80–85°C at 500 mA/cm^2 of current density shows negligible degradation during the first 4000 h
AMS and MAN	Preirradiation grafting	Proton conductivity between 50 and 100 mS/cm; PEMFC durability ~10 times as long as that of a FEP-*g*-PSSA membrane (both are un-cross-linked); different chemical degradation than FEP-*g*-PSSA membranes

ETFE base film		
Styrene/DVB	γ-Ray preirradiation grafting	Proton conductivity in thickness direction and the chemical stability of PEM are strongly affected by the choice of grafting solvent; an optimum proton conductivity up to 0.11 S/cm at 25°C
Styrene	Swift heavy ion-beams of Xe^{23+} creating latent tracks through ETFE films – γ-Ray preirradiation grafting	Has highly anisotropic proton conductivity: with protons the conductivity in the thickness direction is three times higher than that in the surface direction
Styrene/DVB	Selective exposures of ETFE with hard X-rays to create patterns of free radicals – Grafting	Slightly lower performance than that of conventional FEP-*g*-PSSA membrane but significantly longer lifetime
Styrene/DVB	Preirradiation grafting	Significantly influenced graft level, conductivity, and water uptake by DVB concentration, while slightly influenced crystallinity and thermal stability
Styrene	Preirradiation grafting	Melting temperature (T_m) stays almost constant with increasing irradiation dose, whereas crystallinity increases slightly with irradiation dose; weight loss and crystallinity strongly affected by graft level, while degradation temperatures are independent of graft level
MeSt/^{t}BuSt/DVB	Preirradiation grafting	Significantly improved chemical stability with a durability time of ~200 h in H_2O_2 solution at 60°C
Styrene and MAN	Preirradiation grafting	Proton conductivity is >50 mS/cm with lower hydration numbers and considerably improved chemical stability with a single fuel cell lifetime exceeding 1000 h
StSi	– γ-Ray preirradiation grafting – Hydrolysis and condensation – Sulfonation	Enhanced chemical and thermal stabilities superior to DVB-cross-linked FEP-*g*-PSSA membrane; an optimum membrane shows an appropriate proton conductivity of 0.06 S/cm; a durability time of ~200 h in H_2O_2 solution at 60°C; a methanol permeability almost six times as low as Nafion
4VP or *N*-vinylformamide (NVF) or 1VIm	Electron beam irradiation grafting – Doping with phosphoric acid	Exceptional flexibility and mechanical stability and a thermal resistance up to at least 160°C in air; proton conductivity up to 0.10 S/cm at 120°C without humidification; power density of a H_2/O_2 fuel cell up to 150 mW/cm^2 at 120°C without humidification

(*continued overleaf*)

TABLE 7.14 (*Continued*)

Cross-linked PTFE/PFA polymer-alloy film		
Monomers	Grafting methods	Main conclusions and comments
PVDF base film		
Sodium styrene sulfonate (SSS)/*N*,*N*-dimethyl formamide (DMF)/H_2SO_4	EB-induced grafting	Grafting level up to 65%; proton conductivity of 0.114 S/cm; a very good combination of basic physicochemical properties for PEMFC
Styrene	Simultaneous EB-induced grafting	Grafting level up to 46%; proton conductivity of 5.59×10^{-2} S/cm at room temperature
Styrene	ATRP	Properties of PEMs are significantly affected by their nanoarchitectured morphology; interconnected network of small ionic clusters of 2–3 nm in size; ion-exchange capacity (IEC) up to 2.22 mmol/g without excessive swelling and dissolution

ATRP, atom transfer radical polymerization; PEM, polymer electrolyte membrane; EB, electron beam; PEMFC, polymer electrolyte membrane fuel cell; PVDF, poly(vinylidene fluoride); DVB, divinylbenzene; FEP, poly(tetrafluoroethylene-*co*-hexafluoropropylene); *g*-PSSA *graft*-polystyrene sulfonic acid; MAN, methacrylonitrile; MeSt, methylstyrene; ^{t}BuSt, *t*-butylstyrene, StSi, *p*-styryltrimethoxysilane; DVB, divinylbenzene; ETFE, poly(ethylene-*co*-tetrafluoroethylene); MEA, membrane electrode assembly; AMS, α-methylstyrene; MAN, methacrylonitrile; cPTFE, cross-linked poly(tetrafluoroethylene); PFSI, perfluorosulfonic acid ionomer; α-ZrP, zirconium hydrogen phosphate; BPO_4, benzoyl peroxide phosphate; ErTfO, erbium triflate salts; PAH, poly(allylamine hydrochloride); PSS, polystyrene sulfonic acid; PVI, poly(1-vinylimidazole); DMFC, direct methanol fuel cell; PANI, polyaniline; PPV, poly(*p*-phenylene vinylene); PEDOT, poly(3,4-ethylenedioxythiophene); SSt, sulfonated styrene; FEP, poly(tetrafluoroethylene-*co*-hexafluoropropylene); and PFSI, perfluorosulfonic acid ionomer.

Source: Zhang (2012). Taken with permission from: Chem. Rev., 112, 2780 (2011).

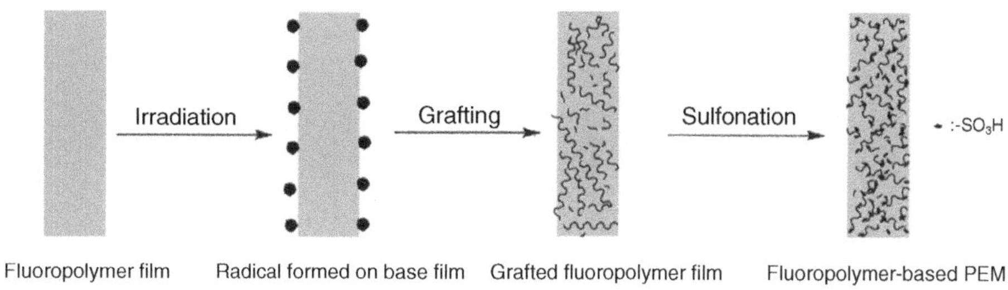

Figure 7.14 Process for the preparation of PEMs by radiation. [*Source*: Zhang, H. and Shen, P.K. [48] Reproduced with permission of American Chemical Society.]

composite has been operated for >6000 h at 120°C and 50% RH (relative humidity). The degradation rate is reduced to (1/1000)th of conventional MEA. Among other membranes examined, sulfonated polyimides are showing promise. The following membrane shows water uptake of 80% (w/w) and proton conductivity of 0.168 S/cm.

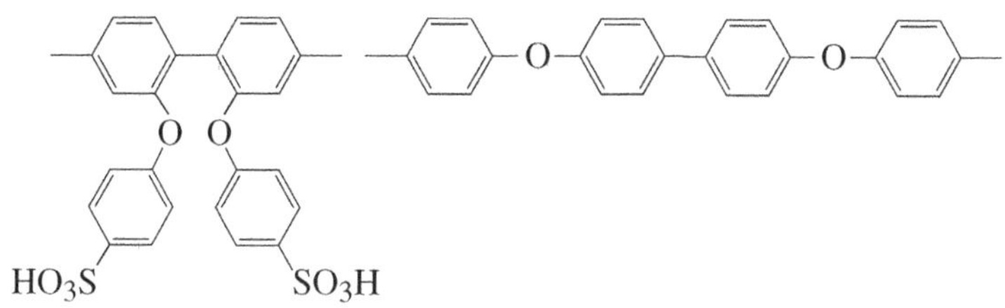

7.17.3 Ion Transport

Ion transport is an important aspect of any fuel cell operation. Let us consider the reactions that occur in PEMFC. At anode, hydrogen ion is produced by oxidation of the fuel (hydrogen) and, as a result, the anode compartment will contain hydrogen ions, molecular hydrogen, and water molecules. A corresponding reaction at the cathode will bring oxygen reduction to water and, hence, this compartment will contain oxygen and water. The Nafion® membrane that separates the anode and cathode permits the transport of hydrogen ion from the anode compartment to the cathode compartment. As discussed earlier, the membrane is a cationic type and hence it restricts the movement of any anionic species from one compartment to the other. This transport scheme is shown in Figure 7.15, where the membrane does not permit the hydroxyl ion (OH^-) to pass through it. It does, however, allow the hydrogen ion to pass through from one side of the compartment to another.

7.18 PARTS OF PEMFC AND FABRICATION

The main parts of the fuel cell are the anode, the cathode, and the membrane. Figure 7.16 shows the three important parts in the fuel cell. All the three units of the PEMFC are specially prepared as described below. The fuel input to the anode (hydrogen gas), oxygen or air input to the cathode, and gas humidifiers all form auxiliary parts of the fuel cell. When the fuel cells are stacked in order to get a higher output, bipolar plates are important (Figure 7.17) and the stacked cells are shown in Figure 7.18. In the following sections, we discuss the details of the individual parts of the PEMFC.

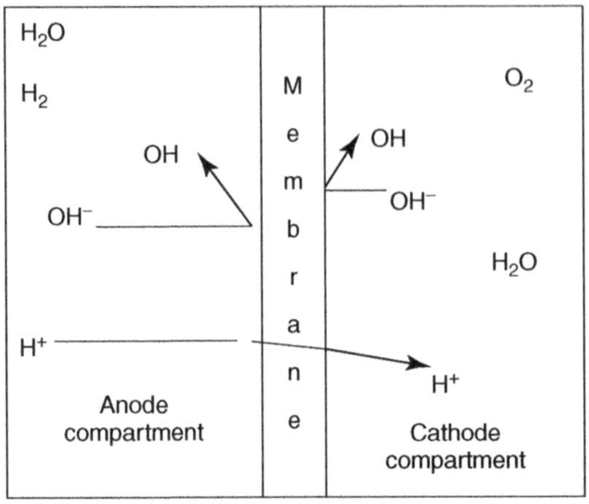

Figure 7.15 Hydrogen and hydroxyl ion movement in the membrane. Although Nafion membrane has been successfully used in PEMFC, there are several new membranes that can stand high temperatures have been developed. Phosphoric acid-doped polybenzimidazole (PBI) can be used up to temperatures of 200°C. A membrane that requires new humidification has been developed by Celtec using PBI and phosphoric acid. It has higher tolerance against carbon monoxide gas. Gyner Electrochemical Systems developed perflurosulfonic acid that has higher conductivity than Nafion. Thus, the activity in membranes has been focused on high temperature operations with lower relative humidity, preferably at less than 10%.

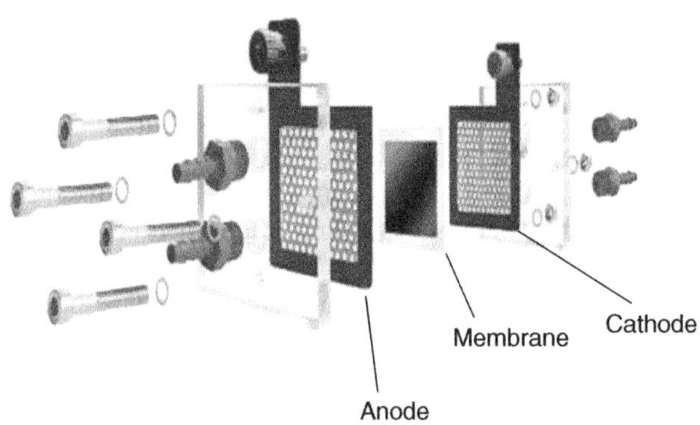

Figure 7.16 A view of a single PEMFC. [*Source*: Reproduced with permission of FuelCellStore .com.]

7.18.1 Anode

Graphite powder (generally Cabot) in a finely divided form with a catalyst (generally platinum) is used as the anode material. The incorporation of the catalyst is highly technical and many patented methods are available. Carbon fiber cloth with incorporated catalyst is used as the anode in several PEMFC. The purpose of the catalyst is to drive the electron transfer reaction faster than on the normal graphite. In other words, the electron transfer reaction is about three orders of magnitude higher on a platinum incorporated graphite electrode. In the initial stages of the development of PEMFC, 30 mg/cm^2 of platinum was used.

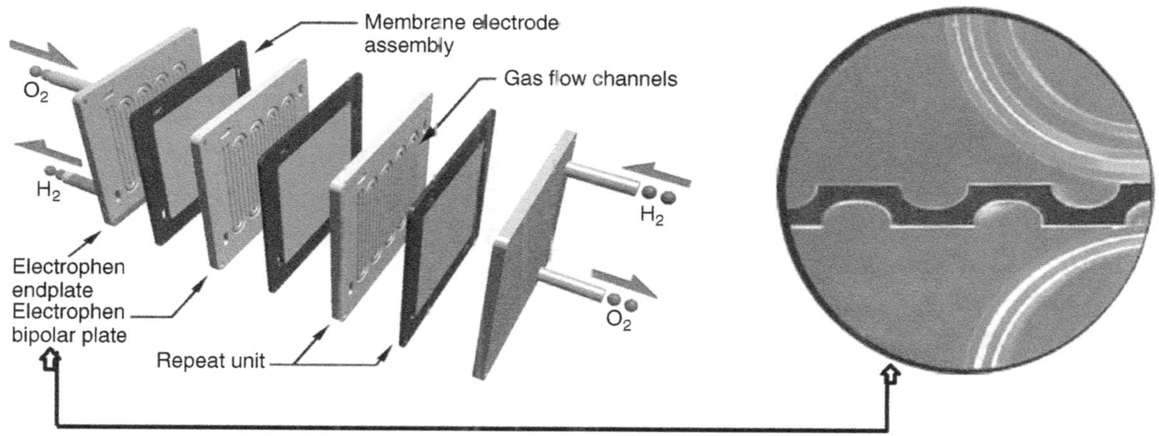

Figure 7.17 Bipolar plate. [*Source*: Reproduced with permission from http://www.tech-etch.com/photoetch/fuelcell.html.]

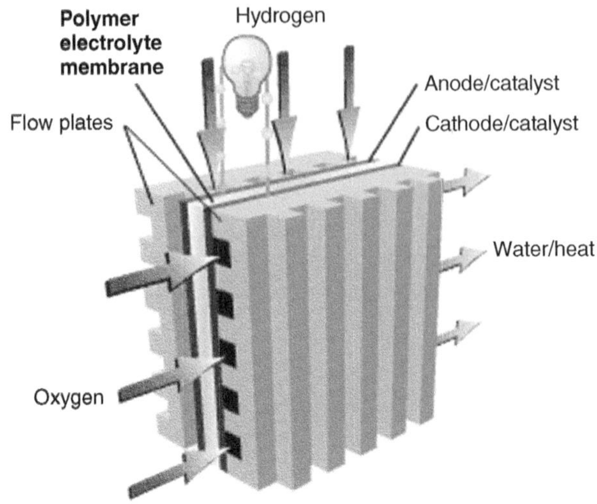

Figure 7.18 Stacked PEMFC. [*Source*: Reproduced from https://energy.gov/eere/fuelcells/types-fuel-cells.]

However, with the discovery of nanosized particles of graphite powder and better incorporation methods, platinum content has been reduced to about 10 mg/cm^2. This is a great advancement, as platinum cost is high and it would be difficult to imagine that fuel cells with such expensive material would ever become practical for universal applications. It is estimated that the basic platinum material cost in 1 kW PEMFC would be about $12. The amount of platinum loading has been further reduced [9] with the addition of additional catalyst. Figure 7.19 shows the current–voltage curve for various amounts of the catalyst on

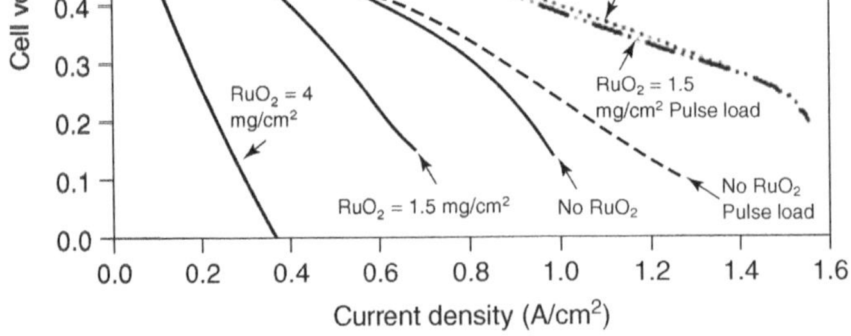

Figure 7.19 Current–voltage curves for PEMFC with different amounts of RuO_2. [*Source*: Chunsheng and Appleby (2003) [8]. Reproduced with permission of The Electrochemical Society.]

the electrode. PEMFC is constructed with hydrogen/air with platinum anode and cathode electrocatalyst loading of 0.17 mg/cm^2 with Nafion®-112 membrane. The current–voltage curve is drastically different at low and high loadings of RuO_2 catalyst [13] and increased the pulse power output as shown in Figure 7.19.

Research is being carried out in finding efficient metal catalysts in PEMFC. Typically, catalysts such as Pt, Pt/Ru, and Pt/Mo alloys were studied extensively. In order to reduce the Pt requirement in the anode, graft polymerization of electrolyte polymer onto carbon was attempted with very little gain in the performance of PEMFC. In a recent development, a monolayered Pt nanoparticle structure on the membrane interface has been achieved and shows promise for reducing low Pt content ion the fuel cells [14].

Two methods are generally used in making the anode. After incorporating the catalyst (Pt) into carbon, it is made into an electrode by hot pressing as discussed below. Alternatively, the catalysts incorporated carbon particles are impregnated into the polymer electrolyte and used as an electrode. The process involves mixing the carbon-containing catalyst with a hydrophobic binder polytetrafluoroethylene (PTFE) and transferring the mixture into polymer electrolyte with one of the following three methods: rolling [14], spraying [15, 16], or printing [17]. Following the fixing of the catalyst to the polymer electrolyte, a carbon cloth or carbon paper is fixed on to the surface to allow the gas (hydrogen or oxygen) entry on to the catalyst surface. Sometimes, this is called applying a "gas diffusion layer" [17]. It also acts as contact between the catalyst and the current collector. In addition, water permeability is provided that will reduce the flooding of the membrane.

7.18.2 Cathode

Oxygen reduction that occurs on a graphite surface is very slow and, hence, incorporation of a catalyst is needed to enhance the rate of reduction. Here again, as with anode preparation, incorporation of Pt-based material such as platinum itself or Pt alloys or metal carbides increases the performance of PEMFC. Increasing the oxygen reduction rate is still an active area of research, and combinatorial chemistry methods are currently used in finding suitable catalysts. Pt–Ir alloy has recently been found to increase the catalytic activity by a factor of 1.5 [18]. The durability of the catalyst such as Pt–Co alloy has also been investigated [19, 20] and was found to show an improved performance over Pt.

7.18.3 Membrane

The anode and cathode are separated by a membrane, and, in practice, several different membranes are used for this separation. The Nafion® 115 membrane has been used successfully in the performance of PEMFC.

The preferential choice for PEMFC is perfluoro sulfonic acid-based ionomer. This imposes a humidification system incorporated into the fuel cell system. The challenge is to discover a high-temperature membrane with low humidity [48].

The commercially available membranes are given in Table 7.15. The membrane N-1135 has remarkable properties such as high conductivity and low water content.

Table 7.16 shows the performance of PEMFC with three different membranes.

The Nafion-coated sulfonated FEP membrane provides the maximum power density that is not available with Nafion®-112.

TABLE 7.15 Commercially available PEMs and their properties

Membranes	Manufacturer	Equivalent weight (g/mol/–SO_3H)	Thickness/ day (μm)	Conductivity (mS/cm)	Water content (wt%)
Nafion	DuPont				
N-112		1100	50	165	~21
N-1135		1100	89	1100	~21
N-115		1100	125	90	~22
N-117		1100	175	80	~23
Aciplex	Asahi Chemicals	1000	120	110	43
Flemion-T	Asahi Glass	1000	120	N/A	N/A
Flemion-S		1000	80	N/A	N/A
Flemion-R		1000	50	140	38
Dow	Dow Chemicals	800	125	114	54
Gore Select	Gore & Associate	900–1100	5–20	28–96	32–43
Fumapem F-930	Fumatech GmbH	900	30	>100	~35
Fumapem F-1050		1000	50	>90	~31
Aquivion PFSA	Solvay	850–890	90–110	N/A	<40
R1010	IonClad	835	40	70	N/A
R4010	IonClad	835	20	80	N/A
61AZL386-389	Ionics Inc., Watertown	400–450	50	67–81	N/A
MC 3470-3142	Ionac Chemical Co.	650–950	60–80	75–114	N/A
OEM	Prochema	N/A	50	100	50

Source: Nasef (2014) [49]. Reproduced with permission of American Chemical Society.

TABLE 7.16 PEMFC performance with modified membranes

Membrane	OCV (mV)	Power density (mW/cm^2)	
		At 500 mA/cm^2	Maximum
FN	759	325	861
s-FEP	930	310	491
Nafion®112	970	320	467

FN, Nafion-coated sulfonated FEP-based membrane; s-FEP, sulfonated FEP membrane; FEP, poly(tetrafluoroethylene-*co*-hexafluoropropylene).
Source: Nasef (2014) [49]. Reproduced with permission of American Chemical Society.

7.18.4 Bipolar Plates

Each PEMFC is capable of giving an output of about 0.7 V. This amount of voltage output is not sufficient for many practical applications. When we need to increase the output voltage, it is desirable to connect several cells in series (see Figures 7.17 and 7.18). Let us visualize a possible scheme shown as follows. The cathode of each cell is connected to the anode of the next cell in series such that the total voltage will be about 1.4 V (two times 0.70 V) under ideal conditions. However, since there will be ohmic and other resistances (see Section 7.7.2) in the circuit, the output voltage will be slightly less.

7.18.5 The Need for a Bipolar Plate

Connecting each cell to the other is cumbersome and unwieldy when hundreds of cells have to be connected for obtaining a high-output voltage. Hence, bipolar plates with a compacted arrangement are used in practice. In this arrangement, a plate made of a conducting material is used. This plate has dual polarity, as on one side it will be in contact with the cathode side and on another side it will be in contact with the anode side. The use of bipolar plate makes stacking of cells easy. These bipolar plates carry grooves for the passage of gases such as hydrogen and oxygen. The side that is in contact with cathode will carry grooves for air or oxygen supply and the side in contact with anode will carry grooves for hydrogen supply. The bipolar arrangement of PEMFC is shown in Figure 7.17.

What are the Functions of Bipolar Plates?

a) It is a support for MEA.
b) It serves the purpose of humidification of the membrane.
c) It distributes the heat–heat management.
d) It distributes the reactants in the cell package.
e) It channels the fuel cell reaction product (water) and residual gases from the cell package.
f) It conducts electrons to support the electrochemical reaction.
g) It transfers heat from the cell package.

Bipolar Plate Should Fulfill the Following Requirements

a) It should be corrosion resistant in an oxidizing and reducing atmosphere under varying potentials and electrolytes.
b) There should be no leaching of metal/metal oxidants.
c) It should be stable under operating pressures, temperatures, and compressive loads.
d) It should be in sizes and shapes that would be amenable for use in the fuel cells.
e) It should withstand operating temperature of about 80°C.
f) It should tolerate operating pressure (ambient to 100 psig).
g) It should stand the operating loads.

The material thickness is dependent on the design of the fuel cell; thinner the better. It is generally made of 0.100″ thick with length/width as system defined. The life of the bipolar plate depends on the usage of the fuel cell. When it is used in automobiles, it is expected to last for 5000 h (many thermal cycles). If it is used in stationary applications, it is expected to last for anywhere from 40,000 to 80,000 h (limited thermal cycles). There are six designs commonly used in bipolar plates. They are (a) pin-type flow fields, (b) straight or parallel design, (c) serpentine flow fields, (d) integrated flow fields, (e) flow fields created in metal sheets, and (f) interdigitated flow field. A brief description of the above six bipolar plates is given in the following paragraph.

Figure 7.20 shows the flow fields design. In pin-type flow field, the cathode and anode plates have an array of circular or rectangular pins protruding from the plates. The reactant gases flow through the grooves formed by the pins, which is actually a network of series and parallel flow paths. This design results in low reactant pressure drop. The straight or parallel design uses a number of separate parallel flow channels connected to inlet and

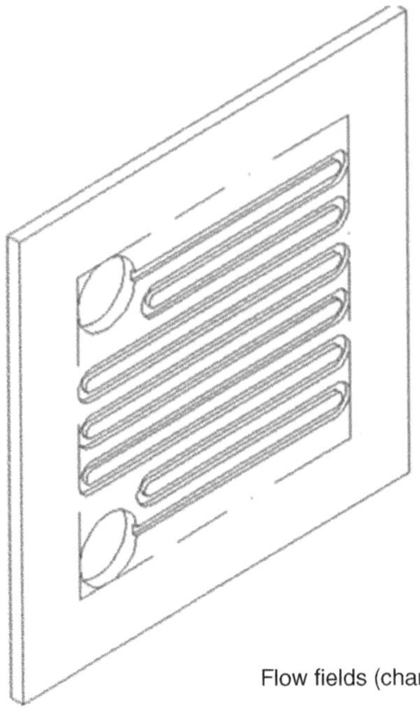

Figure 7.20 Flow fields design.

exhaust headers. This design produces a low and unstable voltage after long periods of operation. The size and number of droplets formed are different and the reactant gases follow the least resistant path. With serpentine flow design, there is a continuous fluid flow channel for avoiding the flooding of water. This design forces the reactant to flow over the entire active area, which eliminates the stagnant flow problem areas. In this design, substantial pressure drop occurs due to long flow path and as a consequence, concentration gradients develop from inlet to outlet. In integrated flow field, both reactant and cooling flow fields are on the same plate surfaces. The gas flow field faces an active area of MEA and as a result the cooling flow field surrounds the gas field. It eliminates a separate cooling layer. The drawback is that it cannot maintain a uniform temperature distribution over the entire fuel cell surface. With metal sheets created flow design, thin metal sheets of metals such as Cr, Ti, and stainless steel are pressed into required form and brazed in order to obtain thin lightweight plates. The cooling fluid flows are inside while the reactants are on the outside surface. With the interdigitated flow field, the reactants flow to the catalyst layer for initiating the electrochemical reaction. The power generation occurs by molecular diffusion through backing layer and it is a slow process. Thus, this flow field differs from the conventional flow field designs where the reactant flow is running parallel to the electrode surface.

7.18.6 Materials for Bipolar Plates

Bipolar plates of several designs have been evolved and they are proprietary of their manufacturers. A boiler plate is made of a conducting material and flow channels for hydrogen and oxygen. The corrugated flow channels, as shown in Figure 7.17, are recommended by the manufacturer. Typical materials used for bipolar plates are carbon, stainless steel, or titanium. Carbon, while convenient for designing flow channels, is costly and, hence, many fuel cell companies prefer stainless steel material.

7.19 ALKALINE FUEL CELLS (AFCS)

One of the most practical fuel cells that has found usage in space missions is the AFC. It was successfully used in the Apollo mission. In principle, it is very similar to PEMFC but differs in the medium employed and the product of oxygen reduction. The reactions at the electrodes are given in equations 7.81–7.83.

$$\text{At anode: } H_2(g) \rightarrow 2H^+(aq) + 2e^- \tag{7.81}$$

$$\text{At cathode: } O_2(g) + 4e^- + 2H_2O(aq) \rightarrow 4OH^- \tag{7.82}$$

$$\text{Overall reaction: } H_2(g) + 1/2O_2(g) \rightarrow H_2O(aq) \tag{7.83}$$

The technology of AFC is well advanced for practical usage. Figure 7.21 shows an AFC operating principle. Oxygen supply in the cathode compartment could be replaced by air. However, since air contains small quantities of carbon dioxide, the fuel cell tends to be poisoned by it. Hence, precaution should be taken to minimize or eliminate carbon dioxide in the fuel system. The cathode and anode are separated by aqueous alkaline potassium hydroxide. This is done by soaking an anion exchange membrane with potassium hydroxide. The electrolyte region in the figure is a membrane region.

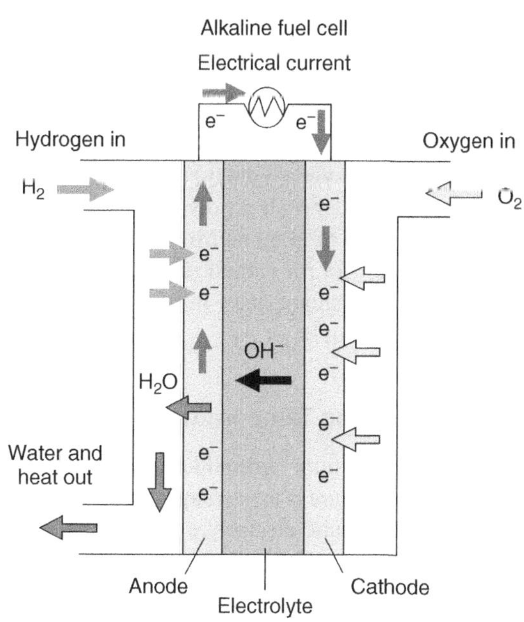

Figure 7.21 Alkaline fuel cell. [*Source*: Reproduced from https://energy.gov/eere/fuelcells/types-fuel-cells.]

7.19.1 Electrodes

The electrodes for AFCs are specially fabricated to withstand the alkaline medium.

7.19.2 Anode

As described earlier, nickel was used as the anode in the early stages of development. With the development of metal hydride alloys such as $LaNi_3$ (AB_3 type), these materials are preferred as anodes. The material is coated with an electrocatalyst layer such as lanthanide–mischmetal-based lanthanum (Lm) alloy [$LmNi_{4.1}Co_{0.4}Mn_{0.4}Al_{0.3}$]. The gas diffusion layer is formed on both the cathode and the anode using PTFE-acetylene black on nickel mesh. The anode material catalysts are Ni, Pt, and Pd. Recently, a modified method was proposed in which a mixture of [$Ni(NO_3)_2{\cdot}6H_2O$, $Co(NO_3)_2{\cdot}6H_2O$, $ZrO(NO_3)_2{\cdot}6H_2O$, and $Y(NO_3)_3{\cdot}6H_2O$] were heated to form a homogeneous sample of $Ni_{1-x}Co_xO_y$ and YSZ (Yttria-stabilized zirconia). The combustion is carried out for 1–2 min to produce solid solutions of YSZ and $Ni_{1-x}Co_xO_y$. By this method, typical crystallites of 30–40 nm are produced. The powders are agglomerates of porous green compacts that form ceramics at high temperatures of 1450°C.

7.19.3 Cathode

In the early stages of the development of AFCs, metal catalysts such as platinum were avoided. During that period (1940–1960), nickel was known to be a catalytically active electrode and, hence, nickel electrodes were fabricated. They were prepared from powdered nickel in two layers using two different sizes of nickel powders for creating a material with fine pores for gas diffusion. This material was suitable for use as a cathode as well as an anode. This electrode material was later replaced with lithiated nickel (the nickel plate was lithiated with lithium metal) for use as the cathode. Apollo mission fuel cells used cathode material of this type.

As the cathode material is required to reduce oxygen in alkaline medium, the problem of oxygen reduction is similar to that faced with alkaline batteries. As a result, MnO_2 was found to be a very good catalyst for oxygen reduction. Graphite mixed with MnO_2 (1:1) is mechanically milled. A polymeric binder such as poly(tetrafluoroethylene) (PTFE) is added to the mixture. Typically, a mass ratio of graphite:MnO_2:PTFE of 45:45:10 is suitable.

Siemens AFC uses silver catalyst for the cathode and Raney nickel for the anode. Such fuel cells find useful applications for submarines. In general, for AFC, the recommended cathode catalyst materials are NiO, Au, Pt, or Ag.

7.19.4 Electrolyte and Operating Temperature

The electrolyte used in AFC is potassium hydroxide. Its concentration is in the range of 35–85 wt% and the operating temperature is in the range of 100–250°C. AFC operation with continuously flowing electrolyte or static electrolyte has also been developed. Table 7.17 gives the conditions of AFC used in space and other missions. A typical AFC that has found extensive applications in space missions is shown in Figure 7.22. The AFCs operate in the temperature range of 80–260°C, with potassium hydroxide concentration ranging from 35 to 85 wt%. In general, the low-temperature AFCs use lower concentrations as compared to the high-temperature ones.

TABLE 7.17 AFC conditions of operation

Usage	Temperature (°C)	Pressure (bar)	KOH electrolyte (%)	Anode catalyst	Cathode catalyst
Bacon	200	45	30	Ni	NiO
Apollo	230	3.4	75	Ni	NiO
Orbiter	93	4.1	35	Pt/Pd	Au/Pt
Siemens	80	2.2		Ni	Ag

Source: Fuel Cell Systems Explained: J. Larminie and A. Dicks, W Larmine (2000). Reproduced with permission of John Wiley & Sons.

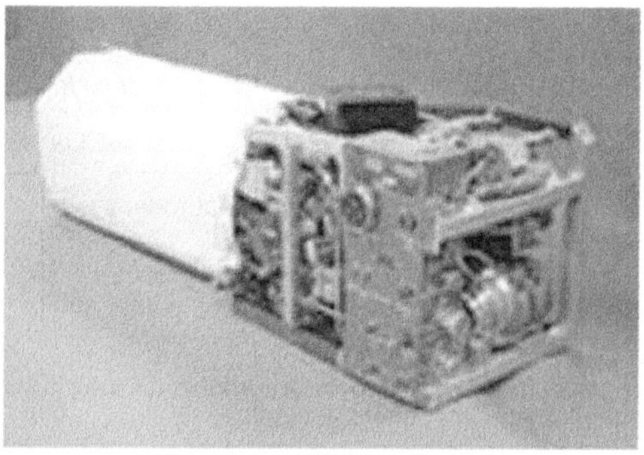

Courtesy of UTC fuel cells

Figure 7.22 Alkaline fuel cell used in space missions. [*Source*: Reproduced with permission from UTC.]

7.19.5 Membraneless AFC

A recent development in the AFC is to operate it without a membrane. A membrane is a vital part of all the fuel systems that have been discussed earlier. The membraneless fuel cell is based on the principle of "laminar flow." It requires that liquids do not mix in ultra-narrow channels. Hence, the solution in the cathode compartment does not mix with the solution in the anode compartment, and a clear boundary is maintained during the flow of the electrolytes in the two compartments. This was developed at the University of Illinois and would cost 40% less than the conventional fuel cell. The practicality of membraneless AFC is currently being examined.

7.20 MOLTEN CARBONATE FUEL CELL (MCFC)

The MCFC is another high-temperature fuel cell that is finding practical usage. It operates in the temperature range of 700–900°C. Carbonate ion is generated at the cathode and migrates to the anode. The following reactions typically occur in MCFC:

$$\text{At anode: } H_2(g) + CO_3^{2-}(l) \rightarrow H_2O(g) + CO_2(g) + 2e^- + \text{heat} \tag{7.84}$$

$$\text{At cathode: } 1/2O_2(g) + CO_2(g) + 2e^- \rightarrow CO_3^{2-}(l) \tag{7.85}$$

$$\text{Overall reaction: } H_2(g) + 1/2\ O_2(g) \rightarrow H_2O(g) \tag{7.86}$$

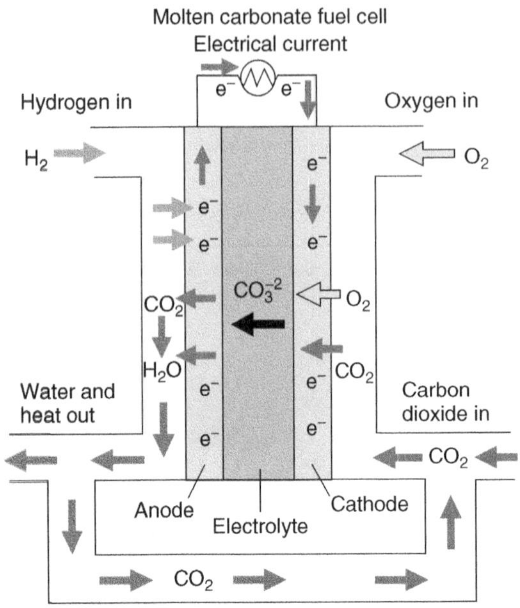

Figure 7.23 Molten carbonate fuel cell. [*Source*: Reproduced from http://en.wikipedia.org/wiki/Image:Fcell_diagram_molten_carbonate.gif.]

The operation of MCFC is shown in Figure 7.23. At 600–700°C, the alkali carbonates are in a molten state and carbonate ion (CO_3^{2-}) is transported into the anode compartment through the matrix separating the anode and cathode compartments.

7.20.1 Electrodes

In the early stages of development for MCFCs, noble metals and metal oxides were used as the electrodes. With the development of low-cost new materials, they have been replaced with several new materials. Current technology uses the following electrode materials.

7.20.2 Anode

Ni–Cr/Ni–Al anode of 3–6 μm pore size having a porosity of 45–70% is used. It has a thickness of 0.20–1.5 mm.

7.20.3 Cathode

Lithiated NiO of 7–15 μm pore size is a good electrode material. It has a porosity of 60–65% and a thickness of 0.5–1 mm.

7.20.4 Electrolyte and Electrolyte Support

In 1965, MgO was used as the matrix in the construction of MCFCs. It was replaced with $LiAlO_2$ in 1975. Now the technology has improved for greater durability using a matrix

40 wt% γ-LiOAlO$_2$ having a diameter of less than 1 μm. It is used in the form of fibers. This technology is proprietary and hence details are not available. The electrolyte employed in the matrix is ^{62}Li–^{38}K or ^{50}Li–^{50}Na. A tape cast method is employed in 50 wt%.

MCFCs provide higher fuel to electricity efficiency than the other fuel cells that operate at lower temperatures. The efficiency is near 60% and has the advantage of converting the waste heat for electricity generation. The thermal efficiency for this system reaches nearly 85%.

7.20.5 Electrode Poisoning

Fuel cells are made up of electrodes that are sensitive to carbon monoxide (CO) poisoning. A PEMFC is generally built with electrodes made with platinum alloy catalyst as anode that is highly sensitive to CO poisoning. This results in the failure of the fuel cell performance. This is explained as due to the catalyst having an active site in the anode that is blocked by CO adsorption. Hence, the cell voltage and current output of the fuel cell drop during CO poisoning. Figure 7.24 shows the current–voltage curves recorded with various levels of CO poisoning [27]. In this work, the pressure and temperature dependence of cell voltage and current have been examined. The fuel cell output at platinum anode relative to platinum–ruthenium anode with mixtures of gases such as N_2/H_2, and CO_2/H_2 has been investigated [77]. The poisoning of the anode made of platinum and platinum/ruthenium was studied by impedance spectroscopy [29]. It showed a strong dependence of pseudoinductive capacitance due to CO adsorption. The water–gas reaction produces CO and it has been estimated that CO concentration can range from 10 to 170 ppm in these experiments. The CO coverage on the anode was estimated at 5×10^{-7} mol/cm^2 for a 0.4 mg/cm^2 platinum anode. The tolerance of CO by a platinum/ruthenium catalyst was examined by nuclear magnetic resonance (NMR) [78]. NMR evidence showed that there is a rapid exchange of CO between two sites, one site with CO molecules on the

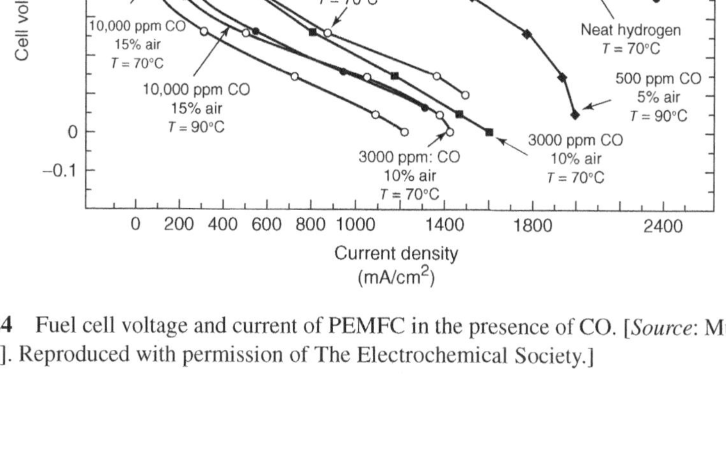

Figure 7.24 Fuel cell voltage and current of PEMFC in the presence of CO. [*Source*: Murthy *et al.* (2003) [27]. Reproduced with permission of The Electrochemical Society.]

platinum domain and another with O_2 molecules on the platinum/ruthenium domain. CO on the latter domain showed highly shifted ^{13}C NMR resonance. The combining of the techniques such as electrochemical NMR, cyclic voltammetry, and potentiostatic current generation has been suggested to optimize the catalyst performance.

Several attempts have been made to eliminate or reduce CO poisoning of PEMFC. Platinum black catalyst has been prepared by exposing it to fluorine gas with the idea that CO poisoning is prevented by fluorinated platinum catalyst [31]. The results showed that the catalyst treated in 1% fluorine showed a higher CO tolerance as compared with 10% fluorine gas. Platinum deposited on carbon nanotubes showed less resistance than platinum/ruthenium on carbon nanotubes [32]. Ruthenium appears to oxidize adsorbed CO to CO_2 on the surface of platinum and hence the platinum/ruthenium anode in PEMFC is protected from CO poisoning. Thermogravimetric analysis has been carried out for understanding CO poisoning [33]. Platinum catalyst decomposed in air completely at 400°C as platinum catalyzed the carbon oxidation.

Nanometal catalysts have been examined for fuel cell applications. Nanosized platinum, platinum/ruthenium colloids, and gold/platinum nanoparticles on carbon support showed the existence of platinum(0) and ruthenium(0) with traces of +4 state of the metals [34, 35]. Nanoscopic colloidal precursors of platinum and ruthenium were deposited on conducting carbon (Vulcan XC72) and annealed at 250–300°C [38]. Two types of reactive annealing were carried out; one in the presence of tetra-alkyl ammonium and the other with triorganoaluminium. Platinum/ruthenium catalyst is formed in a zero valent state of the metals with tetra-alkyl ammonium annealing. With triorganoaluminium annealing, platinum metal exists in the oxidized state.

CO poisoning on porous platinum in 100% phosphoric acid has been found to be linearly dependent on the ratio of concentrations of CO and H_2 [36]. The thermodynamic quantities such as free energy and entropy of CO adsorption were evaluated as −14.5 and −39 cal/mol K. Carbon dioxide poisoning has been examined in AFCs [37]. This study showed that poisoning could come from the use of impure hydrogen or oxygen contaminated with carbon dioxide. Polarization tests revealed that the fuel cell performance deterioration is dependent on potassium hydroxide concentration. The conductivity decrease has been observed due to conversion of hydroxide to carbonate by reaction with carbon dioxide.

While CO and CO_2 effects on the performance of fuel cells have been discussed in the literature, the notable influence of these gases on global warming (see Chapter 1) causes concern about the use of the low-temperature fuel cells. In this respect a high-temperature fuel cell such as SOFC offers an advantage as it uses CO as a fuel [38]. With this fuel cell, Cr poisoning has been examined in SOFC constructed with a mixture of $LaNi_{0.6}Fe_{0.4}O_3$ (LNF) and $La_{0.8}Sr_{0.2}MnO_3$ (LSM) as cathode and yttria-stabilized zirconia (YSZ) or alumina-doped, scandia-stabilized zirconia (ScSZ) as the electrolyte has been studied. While LNF showed the absence of Cr poisoning, the LSM cathode suffered from chromium poisoning [40]. These studies were carried out in the presence of the chromium-containing alloy Inconel 600.

In summary, the electrode poisoning by CO or any other species in a fuel cell is a general concern but has been tackled in a number of ways to prevent the failure of the fuel cell. For details, see Refs [36–40]. As a result, we have now fuel cells being used in space crafts, submarines, and vehicle applications.

7.21 SOLID OXIDE FUEL CELL (SOFC)

This fuel cell operates with the supply of hydrogen and oxygen to the electrodes but the mechanism and temperature of operation are different from the PEMFC. In this fuel cell, the following reactions occur:

$$\text{At anode: } H_2(g) \rightarrow 2H^+ + 2e^- \quad (7.87)$$

$$2H^+ + O^{2-} \rightarrow H_2O(g) \quad (7.88)$$

$$\text{At cathode: } 1/2O_2(g) + 2e^- \rightarrow O^{2-} \quad (7.89)$$

Since the fuel cell is operated at 800–1000°C, the water formed at the anode evaporates into the atmosphere. The hydrogen ion produced at the anode is scavenged by reaction 7.82. However, for reaction 7.88 to occur, O^{2-} species are transported from the cathode. A typical SOFC that is driving a small fan is shown in Figure 7.25. The fuel flow (H_2)

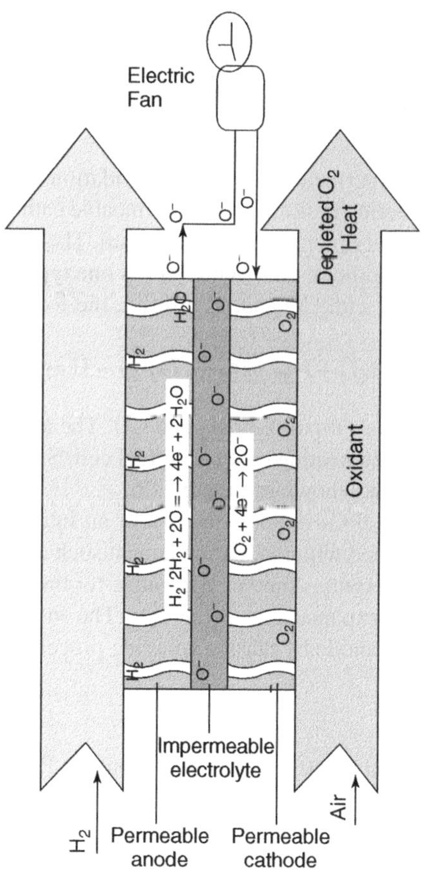

Figure 7.25 Solid oxide fuel cell. [*Source*: Reproduced from https://energy.gov/eere/fuelcells/types-fuel-cells.]

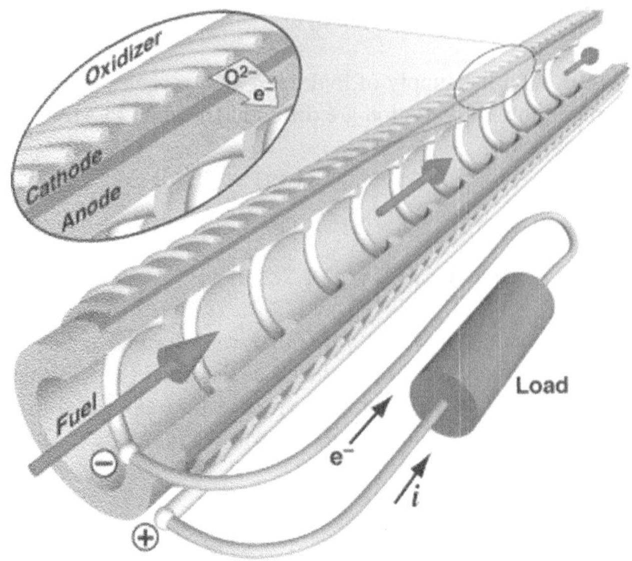

Figure 7.26 Tubular SOFC. [*Source*: Zhu and Kee (2007) [79]. Reproduced with permission of Elsevier.]

is shown through the porous permeable anode; the oxidation of the fuel occurs first. The corresponding reduction reaction is shown on the permeable cathode. This process requires the supply of air or oxygen and is depicted in the diagram. The transport of charged oxygen (O^-) is shown through the middle region. While this is one type of SOFC, there are others where the fuel (H_2) is mixed with CO and, as a result, the following exothermic reaction occurs:

$$CO(g) + H_2O(g) \rightarrow CO_2(g) + H_2(g) \tag{7.90}$$

This type uses a fuel that is a mixture of H_2 and CO. The advantage of using this composition is that it cuts down the heating cost of the fuel cell. SOFC has also been designed in a cylindrical configuration as shown in Figure 7.26.

The stacking of several SOFCs requires the use of an interconnect (see Figure 7.27), which is made of $LaCrO_3$ doped with rare earth element such as Sr, Ca, or Mg. The doping is done to improve the conductivity. Another alternative for this is yttrium chromite–doped Ca. That has better thermal expansion than $LaCrO_3$. The interconnect, that is shown in Figure 7.27, is applied to the anode by plasma spraying process.

7.21.1 Materials

The materials used in SOFC are distinctly different from the PEMFC as they have to withstand the high temperature operation of the fuel cell (Table 7.18).

7.21.2 Anode

In the early stages of the development of SOFC, platinum was used as the anode. Due to the high cost of platinum, it was replaced by less expensive materials. Now it is made up of a metallurgical powder of nickel and yttria-doped zirconia having permeability to the hydrogen gas. It has a high porosity and good thermal and electrical conductivities.

INEEL (Idaho National Laboratory) regenerative solid oxide fuel cells stack

Figure 7.27 SOFC stacking arrangement. [*Source*: Reproduced with permission from http://www.blackwell-synergy.com/doi/pdf/10.1111/j.17447402.2005.02031.x?cookieSet=1.]

It is generally made from a composite of nickel–yttria-doped zirconia rather than using the pure form of nickel and yttria-doped zirconia as the thermal expansion of nickel ($13.3 \times 10^{-6}\,C^{-1}$) is different from the yttria-doped zirconia ($10 \times 10^{-6}\,C^{-1}$). This difference in expansion results in sintering of the product, which seals the porosity that is required for gas diffusion in the operation of the fuel cell. Use of composites prevents this problem and also improves the adherence to the electrolyte. Thus, the main requirement of the anode material is thermal expansion compatibility, electrical conductivity, and permeability of the fuel (hydrogen). In addition, it is necessary that the materials stand high temperature without degradation, especially in the hydrogen atmosphere (note hydrogen is a reducing agent or reducer). Thus, in the above examples of cermet of nickel and yttria-doped zirconia, the starting material is a composite that holds the porosity.

In the practical preparation of the anode, a slurry of nickel is applied first, followed by electrochemical vapor deposition of yttria-doped zirconia. To maintain the porosity, pore formers such as thermosetting resins or starch or carbon are used. The material is sintered before use. Alternatively, nickel oxide–yttria-doped zirconia slurries are made and fired to produce nickel particles. The use of pore formers produces a strained pore that is not desirable in the efficient operation of the fuel cell. Also this produces sometimes cracking of the material. In the modified method, the pore former is replaced by a freeze–drying process. Here, the material forms an ice-like structure and produces acceptable pores.

Several other materials are being considered for the anode. One such material is copper-cerium oxide. Copper has a high electrical conductivity but is a poor catalyst for reforming hydrocarbons. Cerium oxide is a good material for hydrocarbon oxidation.

7.21.3 Cathode

A p-type perovskite such as lanthanum manganate ($LaMnO_4$) doped with Sr, Pr, or Ce is used as the cathode. Doping enhances the conductivity of the material; Sr-doped material, $La_{1-x}\,Sr_xMnO_3$, has electronic conductivity (no ionic conductivity). The cathode is a

TABLE 7.18 Methods of making SOFC components

Component	Synthetic method/substrate used	Properties of synthesized powder/performance of SOFC component	References
Anode	Glycine nitrate process (GNP) Substrate: $La_{0.8}Sr_{0.2}CrO_3$	LSC powders became more porous and had smaller particles as the amount of glycine was increased. The reduction of the pH values of nitric acid decreased the particle size and the aggregation of the LSC powders	[65]
	Solid-state reaction Substrate: NiO–YSZ + Graphite	Cell (graphite 1 μm) achieved a maximum power density of 0.766 W/cm^2 at 850°C and showed better homogeneous microstructure	[66]
	GNP Substrate: NiCo + (Cu, V, and Fe in different amounts)	Increased specific surface area and reduced particle size	[67]
	Sol–gel Substrate: Ni–YSZ	Minimal carbon deposition	[68]
	Solid-state reaction Substrate: Ni–SDC	Synthesized powder became homogenous. Cells with anodes obtained via the ball-milling technique showed maximum power density	[69]
Cathode	Solid-state reaction Substrate: LSCF48–GDC10	The perovskite phase was found in the LSCF–GDC composite powder with surface area of 0.64 m^2/g	[70]
	Solid-state reaction Substrate: $Sr_{0.95}Nb_{0.1}Co_{0.9}O_{3\text{-}\delta}$	A stable and compatible cathode layer structure was obtained. The oxygen reduction reaction was observed to be higher in the cathode layers at operating temperatures below 600°C	[71]
	Solid-state reaction Substrate: LSCF	Particle size was smaller between 20 and 30 nm, and the mixture became more homogenous	[72]
	Solid-state reaction Substrate: LSGM	The XRD pattern showed the presence of a single phase of LSGM powders	[73]

TABLE 7.18 (*Continued*)

Component	Synthetic method/substrate used	Properties of synthesized powder/performance of SOFC component	References
	Solid-state reaction	LSGM–B36 powders (ball-milled for 24 h, precalcined for 24 h at 900°C, and calcined for 36 h at 1300°C) had smaller and finer particle sizes and less agglomeration	[74]
	Substrate: LSGM		
	Sol–gel	The XRD pattern showed that $(Li/Na)_2CO_3$ in amorphous phase was presented and coated over the SDC particles. No changes occurred to the electrolyte layer after being subjected to reduction and oxidization at atmospheric conditions	[75]
	Substrate: SDC–$(Li/Na)_2CO_3$		
Electrolyte	GNP	Sintered SDC powders were dense and closed pores	[76]
	Substrate: SDC		

Source: Mahmud *et al.* (2017) [62]. Reproduced with permission of Elsevier.

vital part of the fuel cell and contributes to 90% of its mass. Siemens Westinghouse has constructed a tubular design of SOFC (see Figure 7.26).

7.21.4 Electrolyte Bridging the Cathode and Anode

The electrolyte bridging the anode and cathode should have a high ionic conductivity with no electronic conductivity at the temperature of operation. At the cathode, O^{2-} is produced and it should migrate to the anode through the electrolyte. A high ionic conductivity will reduce the resistive losses in the cell. The material chosen should have high thermal, chemical, and structural stability to high temperature.

Yttria strontium–doped zirconia, abbreviated as YSZ, is used successfully as the electrolyte. Doped cerium and doped bismuth oxides are also suitable as the bridging electrolyte. YSZ has the advantage of having a stable zirconia cubic structure and providing oxygen vacancies. It has one oxygen vacancy per mole at 10% doping of yttria. A 40-μm-thick electrolyte layer is applied to the cathode of SOFC. Two methods are followed. In one method, spray and dip coatings followed by sintering is used. Thin layers of YSZ are applied using colloidal suspensions of YSZ. The colloidal particles are generally kept in the range of 5–10 nm. In the second method, a thin, dense electrolyte film is deposited by electrochemical vapor deposition method developed by Singhal and Kendall [21]. It is an expensive method compared to the first but produces very dense films on a porous substrate.

Tables 7.18 and 7.19 shows the electrolyte conductivity, temperature, and ceramic particle size. In selecting the electrolyte, it is necessary to have area-specific resistivity of

TABLE 7.19 Parameters used in simulation of SOFC load curves

Symbol	Meaning	Values	References
L	Fuel cell length	200 mm	
N	Segment number	10	
Δx	Segment length	20 mm	
r_a	Inside radius of the cell	3.1 mm	
r_b	Outside radius of the cell	3.9 mm	
r_c	Inside radius of the insulator	7.0 mm	
r_d	Outside radius of the insulator	9.0 mm	
M_{H_2}	Molecule mass of hydrogen	2.016 g/mol	Achenbach [25]
c_{pH_2}	Specific heat of hydrogen (constant pressure)	14,821 J/kg K	Sonntag *et al.* [23]
R_{H_2}	Hydrogen gas constant	4,124.3 J/kg K	Achenbach [25]
c_{vH_2}	Specific heat of hydrogen (constant volume)	10,697 J/kg K	Sonntag *et al.* [23]
ρ_{H_2}	Hydrogen density at 900 K	0.02723 kg/m^3	Sonntag *et al.* [23]
μ_{H_2}	Dynamic viscosity of hydrogen	18.78×10^{-6} kg/m s	Sonntag *et al.* [23]
ΔH	Low heating value of hydrogen	1.196×10^8 J	
M_{H_2O}	Molecule mass of water vapor	18.02 g/mol	Achenbach [25]
c_{pH_2O}	Specific heat of water vapor (constant pressure)	2,186 J/kg K	Sonntag *et al.* [23]
R_{H_2O}	Water vapor gas constant	188.5 J/kg K	Achenbach [25]
c_{vH_2O}	Specific heat of water vapor (constant volume)	1,997.5 J/kg K	Sonntag *et al.* [23]
ρ_{H_2O}	Water vapor density at 850 K	0.2579 kg/m^3	Sonntag *et al.* [23]
μ_{H_2O}	Dynamic viscosity of water vapor	29.69×10^{-6} kg/m s	Sonntag *et al.* [23]
M_{O_2}	Molecule mass of oxygen	32 g/mol	Achenbach [25]
c_{pO_2}	Specific heat of oxygen (constant pressure)	988.1 J/kg K	Sonntag *et al.* [23]
R_{O_2}	Oxygen gas constant	259.8 J/kg K	Achenbach [25]
c_{vO_2}	Specific heat of oxygen (constant volume)	728.3 J/kg K	Sonntag *et al.* [23]
ρ_{O_2}	Oxygen density at 550 K	0.7096 kg/m^3	Sonntag *et al.* [23]
μ_{O_2}	Dynamic viscosity of oxygen	31.97×10^{-6} kg/m s	Sonntag *et al.* [23]
h_a	Convective coefficient in anode channel	2,987 W/m^2 K	*a*
h_c	Convective coefficient in cathode channel	1,322.8 W/m^2 K	*a*
h_∞	Convective coefficient insulator/surrounding	10 W/m^2 K	Sonntag *et al.* [23]
ρ_{PEN}	Average density of PEN	6,337.3 kg/m^3	Holman [24]
c_{pPEN}	Average specific heat of PEN	594.3 J/kg K	Holman [24]
k_{PEN}	Average thermal conductivity of PEN	2.53 W m/K	Holman [24]
ρ_{Ins}	Density of insulator	480 kg/m^3	Sonntag *et al.* [23]
ε_{Ins}	Average emissivity of PEN	0.33	Sonntag *et al.* [23]

TABLE 7.19 ***(Continued)***

Symbol	Meaning	Values	References
c_{pIns}	Specific heat of insulator	1,047 J/kg K	Sonntag *et al.* [23]
k_{Ins}	Thermal conductivity of insulator	0.059 W/m K	Sonntag *et al.* [23]
ε_{Ins}	Thermal emissivity of insulator	0.09	Sonntag *et al.* [23]
σ	Stefen–Boltzmann constant	5.669×10^{-8} W/m^2 K^4	Sonntag *et al.* [23]
P_s	Fuel/gas source pressure	3.0×10^5 Pa	[b]
T_s	Fuel/gas source temperature	1,073 K	[b]
T_∞	Surrounding temperature	303 K	[b]
K_{a_in}	Inlet flow coefficient of anode channel	8.7641×10^{-10} kg/Pa s	[c]
K_{a-out}	Outlet flow coefficient of anode channel	4.3821×10^{-7} kg/Pa s	[c]
K_{c-in}	Inlet flow coefficient of cathode channel	3.1626×10^{-8} kg/Pa s	[c]
K_{c-out}	Outlet flow coefficient of cathode channel	1.5813×10^{-5} kg/Pa s	[c]
R_Ω	Fuel cell ohmic resistance	0.0257 Ω	Sonntag *et al.* [23]
E_A	Anode activation energy	110 kJ/mol	Sonntag *et al.* [23]
E_C	Cathode activation energy	160 kJ/mol	Sonntag *et al.* [23]
k_A	Anode preexponential factor	2.13×10^8 A/m^2	Sonntag *et al.* [23]
k_C	Cathode preexponential factor	1.49×10^{10} A/m^2	Sonntag *et al.* [23]

[a]The coefficients are calculated by assuming a fully developed laminar flow at constant wall temperature.
[b]Parameters are assumed.
[c]Parameters are assumed and adjustable.
Source: Achenbach (1994) [25]. Reproduced with permission of Elsevier.

the electrolyte at 0.15 W cm^2. Based on this requirement, $Ce_{0.9}Gd_{0.1}O_{1.95}$ and $(ZrO_2)_{0.9}(Y_2O_3)_{0.1}$ show the required value at 500 and 700°C. Hence for SOFC, these two electrolytes are preferred for reaching a satisfactory performance.

7.21.5 Load Curves

Xue *et al.* [22] have developed a dynamic modeling of singular SOFC using nickel and yttria–stabilized zirconia (Ni-YSZ) cermet as anode and a thin YSZ electrolyte and lanthanum strontium manganate as cathode. In this control volume method of simulation, the effects of heat transfer, electrochemical reaction, and species transportation are taken into account. The agreement between the theoretical simulation and experimental data is impressive. The cell parameters are given in Table 7.19. The load and power curves obtained for this cell at 850°C are given in Figures 7.28 and 7.29. This fuel cell is operated with pure hydrogen and oxygen drawn from air.

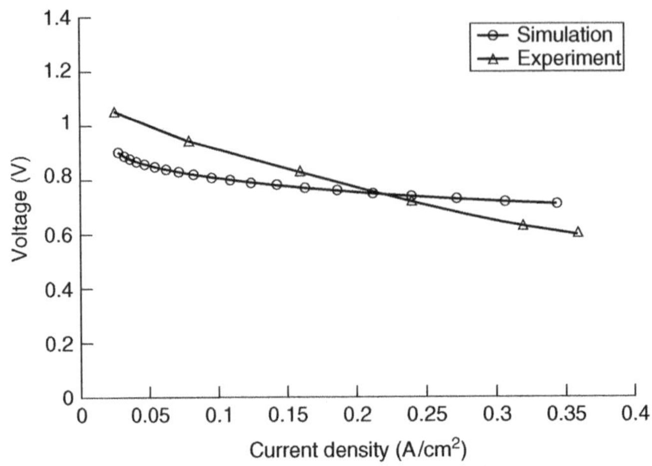

Figure 7.28 Current–voltage curve for SOFC fuel cell. [*Source*: Source: Xue *et al.* (2005) [22]. Reproduced with permission of Elsevier.]

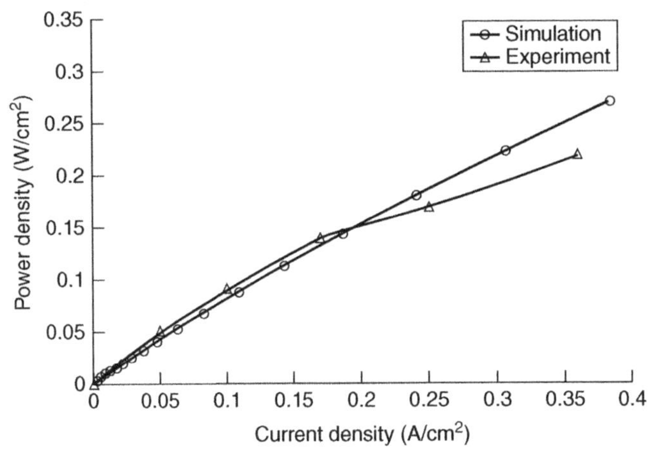

Figure 7.29 Power density curve for SOFC fuel cell. [*Source*: Source: Xue *et al.* (2005) [22]. Reproduced with permission of Elsevier.]

7.21.6 Applications of SOFC

Several SOFCs have been designed and constructed in the last several years. Pioneering work in this direction was carried out by Siemens Westinghouse. The tubular design, shown in Figure 7.26, has been tested for periods ranging from 2 to 8 years successfully, with a voltage loss of less than 0.1% in 1000 h of operation. SOFC is ideally suited for primary and auxiliary power sources. It is suitable for homes, office buildings, and military installations and in other applications. It could also be used in automobiles and trucks.

SOFC presently operates at temperatures of 800–1000°C. Several attempts are being made to reduce the temperature of the fuel cell without any sacrifice in the performance of the fuel cell. This fuel cell is free from some of the problems faced with PEMFC such as fuel purity, water management, and membrane stability.

Fuel cell size used in cars replacing the combustion engine:

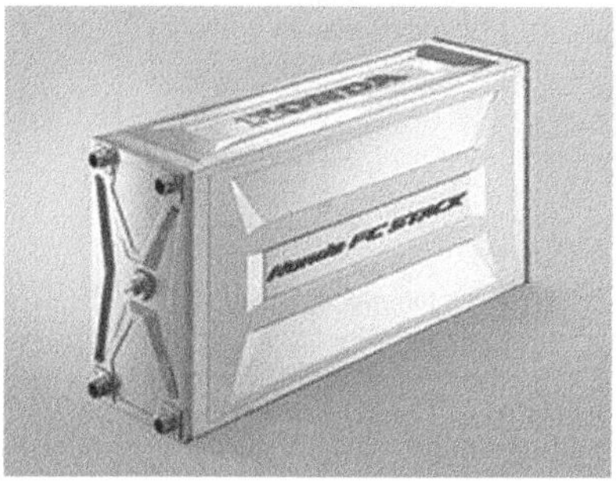

Fuel cell housed in the car

Fuel cell–powered car

7.22 FLOWCHART FOR FUEL CELL DEVELOPMENT

Fuel cell development involves a large number of chemical, electrochemical, and electrical parameters to be evaluated. The flowchart in Figure 7.30 gives the steps to consider in developing a new fuel cell. The first step is to identify a fuel that is available in large abundance (world's reserve) so that the fuel could be used for a very long time. The second step is to determine the fuel value by combustion techniques. For comparison, the value of the new fuel should be compared with the existing fuel. Determine the advantages over the other fuels in the market. At this stage, it is appropriate to determine the combustion products involved in the reaction and determine chemically the moles of the reactants and products involved. The next step is to use the oxidation number principle and balance the fuel cell reaction. From here onward, electrochemistry plays the key role in determining the success of the fuel cell. This requires identification of electrode materials and determining exchange current densities, fuel cell efficiency, fuel requirements, heat management, and membrane design. If all these parameters are satisfactory, electrical measurement and stacking of the fuel cells are the appropriate steps to be examined for the success of the fuel. The flowchart marks the various steps in the ultimate success of a fuel cell. This flowchart is constructed based on the past experience in the literature in the development of a fuel cell.

7.23 RELATIVE MERITS OF FUEL CELLS

In the previous sections, we discussed several different fuel cells that are developed to the stage of technological applications. Table 7.20 summarizes the performance characteristics of the fuel cells. PEMFC is a low-temperature fuel cell that is being planned for use in space applications. Earlier space missions used AFC in space ships (see Table 7.20). Because of its operation in alkaline media and the corrosiveness involved with this medium, NASA is planning on changing to PEMFC. This changeover will result in AFC finding more applications in submarines. PAFC is a popular choice for use in buildings and large vehicles such as buses. The two high-temperature fuel cells, SOFC and MCFC, are expected to play a major role in power stations. The technological developments of SOFC are faster today as compared to MCFC. Several attempts are being made to reduce the temperature of its operation below 800°C; this would require new electrode materials to be developed in future.

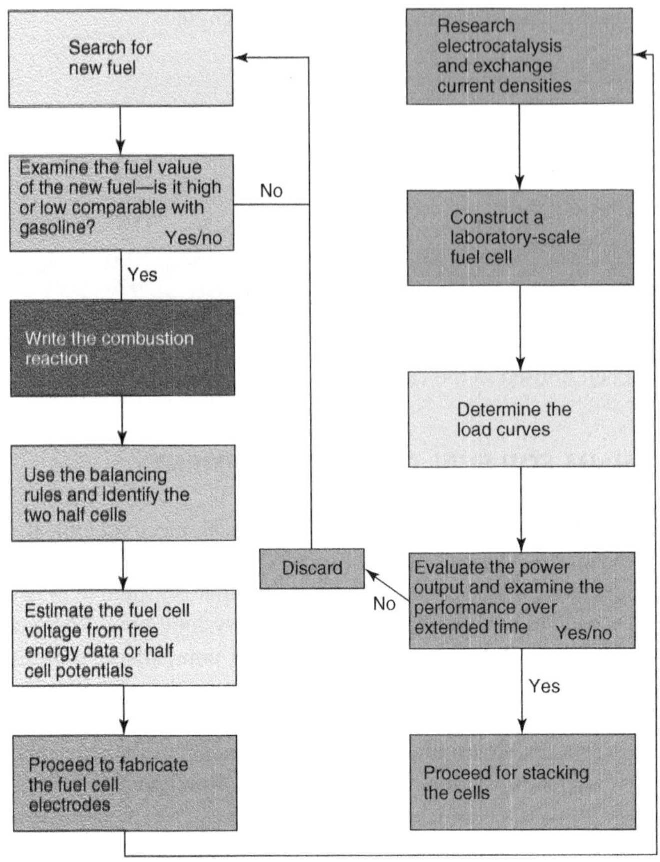

Figure 7.30 Flowchart for developing and designing a fuel cell.

TABLE 7.20 Performance characteristics and applications of fuel cells

Type of fuel cell	Operating temperature (°C)	Electrolyte	Efficiency	Power output	Possible applications
PEMFC	80	Ion-exchange membrane	40–70%	250 kW	Automobiles, small gadgets
AFC	60–90	KOH	45–60%	Up to 20 kW	Submarines, space crafts
PAFC	200	Liquid phosphoric acid immobilized	35–40%	>50 kW	Power stations
MCFC	650–800	Molten carbonate	45–60%	>1 MW	Power stations
SOFC	800–1000	Ceramic	50–65%	>200 kW	Power stations

A new class of fuel cells called "regenerative fuel cells" is being examined. Here, electrolysis of water is carried out to produce hydrogen and oxygen in the first step. This is followed by utilization of hydrogen and oxygen in the fuel cell. The first step requires the use of electricity to split water and the second step generates electrical power. The challenging problems with regenerative fuel cells are (a) finding electrode material that would be efficient for both operations, (b) storage of hydrogen and oxygen gases, and (c) technology for recycling of the gases. A few examples in the following section also fall into this class of fuel cells.

7.24 FUEL CELL TECHNOLOGY

The fuel cells have been used in a large number of consumer applications such as computers, cordless phones, cameras, and cordless tools. The power supplies for notebooks and mobile phones are based on direct methanol fuel cell (DMFC) and PEMFC in the power region of about 5 and 75 W. The demonstration units of these instruments have been developed by companies such as Toshiba, NEC, Hitachi, Panasonic, Samsung, Sanyo, and LG.

The question that one would ask is—how does one decide the suitability of any selected fuel cell for a selected application? The answer to this question is based on the following parameters:

A) The working temperature of the fuel cell and the electrolyte determines the efficiency of the fuel cell as with increasing temperature, the internal resistivity and polarization decrease, which affects the voltage.
B) Start-up time—time required to reach the optimum performance.
C) Dynamic behavior—load changes lead to temperature changes in the stack material especially with high temperature fuel cells. For mobile applications, the primary parameters are short start-up time and high load; obvious choice is PEMFC.

In essence the decision in the selection of a fuel cell for a given application depends on the essential requirements on the chosen application.

So far, fuel cells have not found applications to replace 4C batteries. Li-ion batteries (e.g., Panasonic NCR18650B, 691 Wh/L, 266 Wh/kg) has been very profitable for this use and hence fuel cells have not overtaken this market yet. However, mobile phones, tablet computers, and so on use external chargers supplied by MiniPak, PowerTrekk, and Upp Fuel Cell. PEMFC 2–5 W, 5 V USB priced at $100–230, weighing 120–235 g and are fueled by H_2-catridges based on metal hydride and water-activated NaSi. SOFC with a start-up energy density of 0.3 kW/kg is being developed using a portable liquid propane gas (LPG). It is predicted that stationary fuel cells usage will grow from nearly 40,000 in 2014 to 1.25 million in 2022 [63].

Industrial applications are growing with MCFC having a capacity of 100 MW/annum and is in operation in more than 65 sites. A 11.2 MW plant was installed in Daegu city in South Korea and 14.9 MW plant in Connecticut (USA). A largest 59 MW plant was being built in Hwasung City (South Korea) at a cost of approximately \$2500–3000/kW$_{el}$ [W$_{el}$ refers to electrical watts] and is expected to decrease to \$1500/kW$_{el}$ in future installations. Table 7.21 shows the fuel cell market in small and industrial scales for a variety of uses.

TABLE 7.21 Fuel cell market

Shipment by application						
	Fuel cell today (as published)					Forecast
1000 Units	2009	2010	2011	2012	2013	2014
Portable	5.7	6.8	6.9	18.9	13.0	21.8
Stationary	6.7	8.3	16.1	24.1	51.8	45.6
Transport	2.0	2.6	1.6	2.7	2.0	2.9
Total	14.4	17.7	24.6	45.7	66.8	70.2

Megawatts by application						
	Fuel cell today (as published)					Forecast
Megawatts	2009	2010	2011	2012	2013	2014
Portable	1.5	0.4	0.4	0.5	0.3	0.5
Stationary	35.4	35.0	81.4	124.9	186.9	147.3
Transport	49.6	55.8	27.6	41.3	28.1	28.2
Total	86.5	91.2	109.4	166.4	215.3	176.0

Market development 2009–2014 for different applications based on shipments (pieces) and power (MW)[a].
[a]Uncorrected Fuel Cell Today forecast from 2013[b].
[b]About 80% of the power was delivered by Fuel Cell Energy (FCE) and Bloom Energy for the stationary industrial market, Panasonic and Toshiba for the residential CHP market, and Plug Power for the material handling market. In 2013, worldwide fuel cell industry sales surpassed \$1 billion for the first time, reaching \$1.3 billion.
Source: Reproduced with permission of The Electrochemical Society.

The megawatt-powered fuel cells for portable applications showed a decline, whereas for stationary applications it has nearly tripled. The small-scale-powered fuel cells shown in the upper part of Table 7.21 has steadily increased in demand (Figures 7.31 and 7.32).

7.24.1 Status in Automobile Industry

Since 2014, fuel cell vehicles have started to enter into the market. Hyundai engineered innovative ix35 car along with the fuel cell electric vehicles (FCEVs) of Toyota and Honda. Toyota MIRAI FC-EV has 100 kW PEFC stack with a power density of 3 kW/L and is fueled by two 700 bar H_2 tanks. These cars have a range of 650 km. The system is having 1.6 kW h Ni-MH battery. Several transport buses use fuel cells; for example, 27 Mercedes-Benz Citaro buses use PEFC made by Ballard giving 250 kW energy with 40 kg compressed hydrogen at 350 bar—giving a range of 200 km. Among other companies making buses are HyFLEET (Europe) and 3 Gillig (USA), APTS (60 ft, 125 kW PEFC), EvoBus (40 ft, 120 kW PEFC), Van Hool (43 ft, 150 kW PEFC), Wrightbus/Bluways (40 ft, 75 kW PEFC), and New Flyer (41 ft, 150 kW PEFC).

Figure 7.33 gives a comparison of market share of internal combustion engine (ICE) automobiles and the electric vehicles. Up to 2010, the percentage of battery-powered vehicles was negligible with the result carbon dioxide enrichment and greenhouse gas effect had reached a high value. Battery-powered vehicles shared on the road started to increase from 2010 onward along with fuel cell–powered vehicles. A noticeable prediction based on the curves in Figure 7.33 is that hybrid vehicles on road will drop down by 2050. It also

Figure 7.31 Fifty-nine-megawatt Gyeonggi Green Energy fuel cell park in Hwasung City. [*Source*: Reproduced with permission of Fuel Cell Energy.]

Figure 7.32 Fuel cell energy completes 14.9 MW fuel cell park in Connecticut (USA). [*Source*: Reproduced with permission of The Electrochemical Society.]

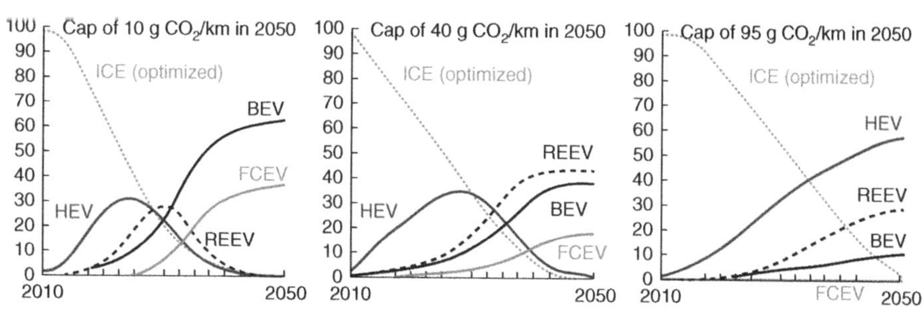

Figure 7.33 Market share of carbon dioxide generating vehicles. ICE (internal combustion engine), BEV (battery-operated vehicle), FCEV (fuel cell electric vehicle), HEV (hybrid electric vehicles) and REEV (range extender electric vehicle). [*Source*: Reproduced with permission of The Electrochemical Society.]

compares CO_2 output produced by ICE vehicles. In this comparison, one HEV vehicle generates 10 g CO_2/km while in others 40 g CO_2/km or 95 g CO_2/km. The prediction based on the above simulation is that the ICE vehicles moving on the road will be negligibly small in number by 2050.

7.25 FUEL CELLS FOR SPECIAL APPLICATIONS

Several other hydrogen-based fuel cells have been examined in the past. They are hydrogen–chlorine fuel cell (HCFC), hydrogen–bromine fuel cell (HBFC), and hydrogen–iodine fuel cell (HIFC). The development of these fuel cells has been limited, based on the toxicity of chlorine and bromine. Iodine fuel cell is of very little interest, as its output voltage is very small. Table 7.22 gives the OCV that is expected from these fuel cells. In this list, there is only one that resembles closely the hydrogen–oxygen fuel cell—hydrogen–chlorine. Instead of oxygen, chlorine is continuously supplied to the cathode. Both the gases are stored in cylinders and amenable for convenient operation of HCFC. However, chlorine is a toxic gas and will not be acceptable for regular usage.

The other fuel cells use either liquid or solid reactant at the cathode. Besides, the output voltages of these fuel cells are smaller than hydrogen–chlorine and it would be a herculean job to stack such fuel cells to reach kW levels.

In the following pages, we will focus on HCFC that would have industrial impact. Today, caustic soda is manufactured by three electrolytic processes; mercury cells, diaphragm cells, and membrane cells. It is produced by electrolysis of brine and is a successful industrial process. The by-products of this electrolysis are hydrogen at the cathode and chlorine at the anode. The demand for caustic soda production is strong and has increased steadily over the years. In 2001, the requirement was 12,250 short tons and went up to 12,460 short tons. It is projected that 13,165 short tons would be required in 2006. A large number of industries require caustic soda, including the manufacture of both organic and inorganic chemicals. The manufacturing of soaps and detergents, pulp and paper, textiles, propylene oxide, polycarbonate, ethyleneamines, epoxy resins, and sodium or calcium hypochlorite are some examples requiring caustic soda.

The fluctuations in the demand for caustic soda and chlorine cause instability in the operation of the caustic soda plants. In 1997, the demand was 13,410 short tons and, in

TABLE 7.22 Fuel cells with halogens

Fuel cell	Open circuit voltage (V)	Fuel cell reaction
Hydrogen–chlorine	1.36	Anode: $H_2(g) \rightarrow 2H^+(aq) + 2e^-$ Cathode: $Cl_2(g) + 2e^- \rightarrow 2Cl^-(aq)$ Overall: $H_2(g) + Cl_2(g) \rightarrow 2HCl(aq)$
Hydrogen–bromine	1.06	Anode: $H_2(g) \rightarrow 2H^+(aq) + 2e^-$ Cathode: $Br_2(l) + 2e^- \rightarrow 2Br^-(aq)$ Overall: $H_2(g) + Br_2(l) \rightarrow 2HBr(aq)$
Hydrogen–iodine	0.536	Anode: $H_2(g) \rightarrow 2H^+(aq) + 2e^-$ Cathode: $I_2(s) + 2e^- \rightarrow 2I^-(aq)$ Overall: $H_2(g) + I_2(s) \rightarrow 2HI(aq)$

2001, it fell to 12,250 short tons. In 2002, it is started to pick up to 12,460 tons. The average price per ton in 1997 was $127 and in 2002 it fell down to $114.

The demand for chlorine and caustic soda has been increasing year by year. However, hydrogen is generally burned off at the site, as storage and transportation problems are not economical. However, this may prove to be one of the methods of hydrogen production if we enter into "hydrogen technology."

The management of caustic soda plants requires very careful balancing of industrial needs; unfortunately, the industries that utilize chlorine do not fall in sync with the industries that utilize caustic soda. With the electricity tariff increasing, the electrolysis cost has been on the increase. These problems were severe during 1995–1998. At that time, the need was felt to utilize chlorine and hydrogen by-products to produce electricity. The fuel cell developed would then reduce the cost of electricity and also be a way to augment the industrial demands.

The construction of a HCFC requires the anode and cathode to be made up of materials that would withstand the corrosive atmosphere of chlorine gas. In addition, the separation of the anode and cathode would require a membrane that will not be attacked by chlorine gas. Both these problems have been solved using RuO_2-based electrodes on Ti substrates and Nafion 90209, with the fuel cell working successfully for a period of 28 days producing a voltage of 1.30 V [42]. The voltage obtained here is very close to the estimated OCV. In comparison, a hydrogen–oxygen fuel cell produces a voltage of 0.8–0.9 V, whereas the expected voltage is 1.23 V.

Hydrogen–bromine and hydrogen–iodine fuel cells are classified as batteries, as one of the components in these fuel cells is a solid or a liquid.

7.26 FUEL CELL REFORMERS

All the fuel cells discussed in earlier utilize hydrogen, and it is important now to discuss the sources of hydrogen and how it could be incorporated into the fuel cell structure. Chapter 1 discusses the possible pathways of producing hydrogen. Being a gas, it occupies a large volume and is inconvenient to couple with the fuel system. For example, we could use gas cylinders to store the hydrogen gas and couple them to the fuel cell system. The number of cylinders required would be enormous, as shown by calculations in Section 7.13. For a continuous supply of electric power, these cylinders need to be constantly replenished. To overcome these problems of storage and distribution, reformers are used that provide readily available fuels such as natural gas or alcohol that could be converted to hydrogen. Reformers may be considered as a chemical process by which the hydrocarbons or alcohols are decomposed to hydrogen. This conversion produces heat and other decomposition products. These products need to be removed to protect the hydrogen gas from contaminants.

By volume comparison, a liquid fuel such as gasoline provides more energy per unit volume than hydrogen, which is a gas. If we liquefy hydrogen, then it has a higher energy per unit volume. However, keeping hydrogen in a liquefied state requires a very low temperature and is a task by itself for transportation. A reformer does not have this problem. Currently, a steam reformer is used to produce pure hydrogen from methanol or natural gas. The mechanism of breakdown of methanol or natural gas is discussed in the following paragraphs.

7.26.1 Steam Reformer

Methanol can be reformed to produce pure hydrogen using steam. When a mixture of methanol and water is passed through a heated chamber containing a catalyst, methanol decomposes to hydrogen.

$$CH_3OH(l) \rightarrow CO(g) + 2H_2(g) \tag{7.91}$$

Since water vapor is present in the reaction chamber, CO splits the water to hydrogen and carbon dioxide by the following reaction:

$$CO(g) + H_2O(g) \rightarrow CO_2(g) + H_2(g) \tag{7.92}$$

Thus, the toxic CO is converted to less toxic CO_2 by this reforming process. Hydrogen gas is supplied to the fuel cell.

Natural gas or methane is also converted to hydrogen by a similar process. Here, methane is decomposed to hydrogen by the following reactions:

$$CH_4(g) + H_2O(g) \rightarrow CO(g) + 3H_2(g) \tag{7.93}$$

$$CO(g) + H_2O(g) \rightarrow CO_2(g) + H_2(g) \tag{7.94}$$

Hydrogen gas produced by this reforming process always carries trace amounts of CO and CO_2, and they need to be removed before supplying hydrogen to the fuel cell.

7.27 FUEL CELL SYSTEM ARCHITECTURE

A single fuel cell provides a low voltage and high current but is not sufficient for practical purposes. For running an automobile or a home, significantly higher power output is required. In order to achieve this goal, a combination of fuel cells and convenience of operation is required. A fuel cell power system (FCPS) has evolved for this purpose. An FCPS is a blend of subsystems integrated together to generate electrical energy. The fuel cell system's goal is to control and supply reactants for electrical energy generation by managing and processing water, gases, and excessive heat removal.

The fuel cell's system configuration is developed based on its usage. At present, there are four main industrial applications contemplated and, accordingly, the following fuel cell system' configurations have been developed:

1. Micropower source for computer and communication devices, usually limited to 100 W
2. Residential or auxiliary power unit (APU) with upper range of 10 kW
3. FC for mobile applications, typically limited to 100 kW
4. Stationary power that can cover wide range in MW.

For low-power systems such as micropower, the following system configuration (Figure 7.34) is considered. The fuel is stored as a liquid or gas in the fuel storage section.

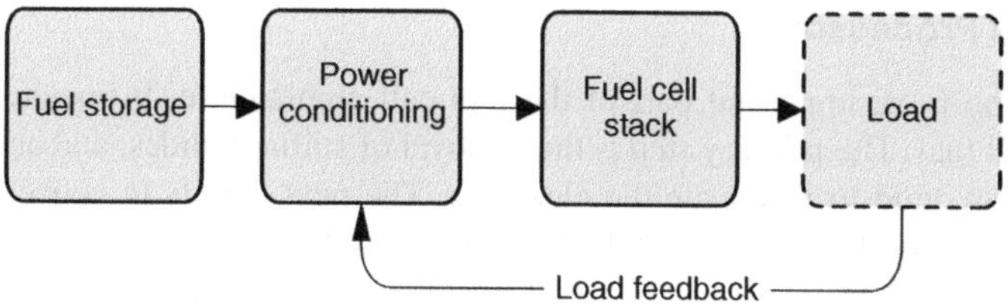

Figure 7.34 Micropower system configuration.

This fuel storage system can easily be replaced with a new one when the fuel has exhausted. The power conditioning unit that links the fuel storage and the fuel cell acts as an interface between the two systems. The micropower system may include a fuel delivery, and peak energy storage devices such as a small accumulator or an ultracapacitor that regulates the power consumption during start time or during load change.

The residential or APU is more complicated and integrated from various subsystems. The subsystems are configured in a way that the fuel cell stack operates effectively for load change, providing necessary fuel preparation, and metering.

Typically, an FCPS consists of a fuel processor, a fuel cell stack with subsystems controlled by a balance of plant section, and a power conditioner. For maintaining maximum performance, it is necessary to remove excess heat and water from the FCPS; water will be used for humidification of fuel cell stack. A block diagram of the basic system is shown in Figure 7.35. It contains two basic parts: the first part contains a fuel processor and the second part contains a fuel cell stack and power conditioner. The processor produces hydrogen using natural gas, gasoline, methanol, LPG, or gaseous hydrocarbons such as CNG, LFG, or methane, and purifies it before supplying to the fuel cell stack. It should be pointed out that the fuel cell stack is sensitive to impurities. The power conditioner converts the direct current (DC) voltage to alternating current (AC) voltage. In general, the fuel cell stack utilizes a proton exchange membrane, and such cell stacks are used in automobile applications. PEMFC operate at 45% efficiency, with low operating temperature and high volumetric (kW/L) and gravimetric (kW/kg) power density. For residential applications, the fuel cells, which operate with natural gas, are preferred and are free from environmental problems.

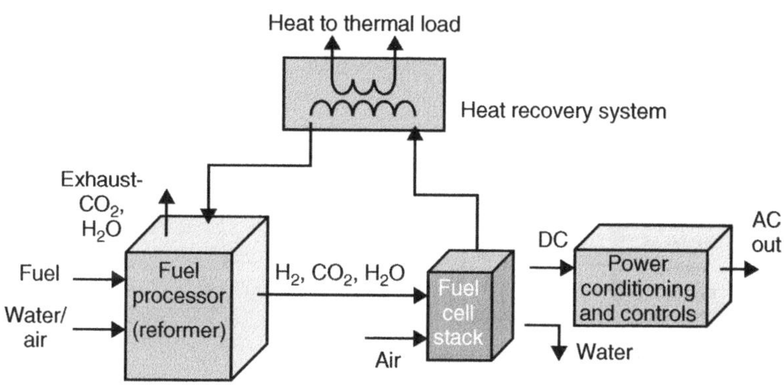

Figure 7.35 Fuel cell power system principal configuration.

7.27.1 Fuel Processor

Fuel processing is an important part of the fuel cell architecture. It operates using gas, liquid, or solid fuel. The primary step is the removal of sulfur, halides, and ammonia from the fuel, as they tend to inactivate the electrode. The next step is to convert the fuel to hydrogen and remove traces of CO and CO_2 that are produced by water–gas shift reaction. The fuel processor is a cost-effective and integral part of the system. Figure 7.36 provides an example of temperature requirements for a fuel processor unit.

7.27.2 Power Conditioning and Controls

For mobile transportation or residential use, conversion from low-voltage, high-current output to DC or AC is required. The requirements for automotive propulsion or residential use are 400 V DC and 110 V AC. In addition, conditioning of excessive heat is required. All these factors are taken care of by power conditioning and controls. The control functions based on the electric data interpretation include current, voltage, water conductivity, and material properties such as pressure, temperature, gases consumption, and humidity.

7.27.3 Balance of Plant

The critical, expensive, and power consuming part of any FC system is the balance of plant. This block is a microprocessor-controlled array of subsystems that is based on thermodynamic equations that provide the FC stable stack performance during load and environmental condition change. The balance of plant contains all the direct stack support subsystems, reformer, compressors, pumps, and the recuperating heat exchangers. The balance of plant operates on data of material and informational flows. The material flow of gases, coolant, and water at different heat contents are controlled by the balance of plants. Informational flows are "operator–machine" interfaces and are controlled by different sensors.

7.27.4 Fuel Cell Power System (FCPS) Subsystems

A typical PEM FCPS should contain the following major subsystems:

1. Fuel cell stack subsystem (FCSS)—containing fuel cell stack and all major controls to maintain optimum operating conditions, interface with other subsystems, devices, and appropriate controls and structural hardware.

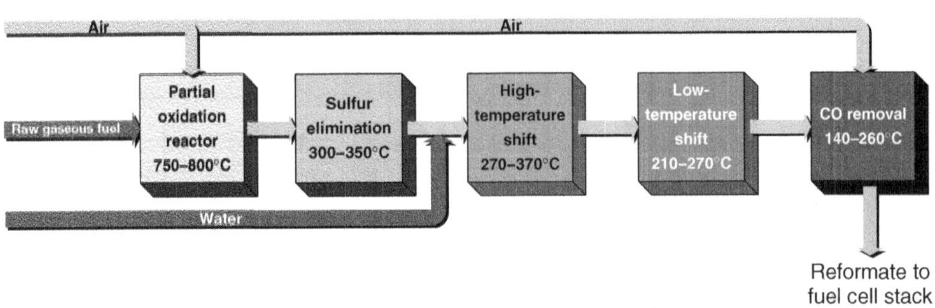

Figure 7.36 Fuel processor temperature requirements.

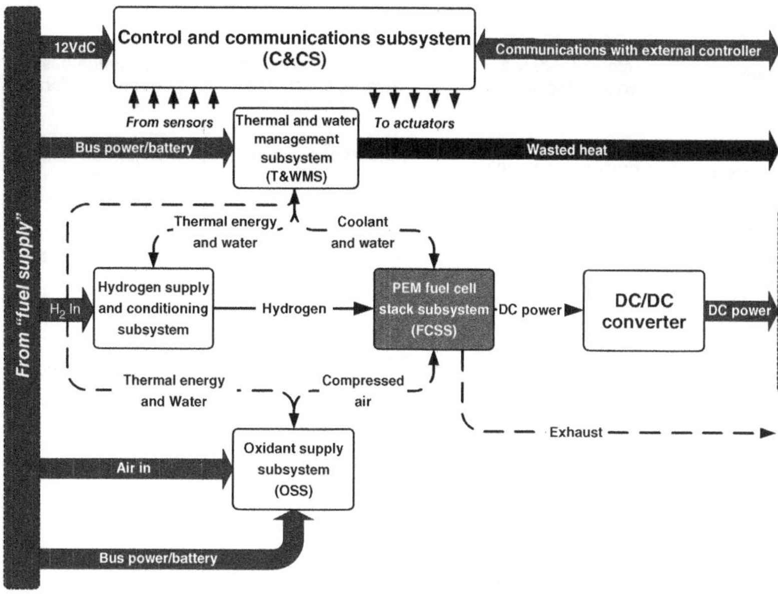

Figure 7.37 Fuel cell power system for transportation.

2. Hydrogen supply and conditioning subsystem (HS&CS)—responsible for the fuel conversion and purification from HS-based fuel into reformate (if required), flow metering, pressure, temperature, and mass flow control.
3. Thermal and water management subsystem (T&WMS)—maintaining required operating temperature throughout entire FCPS, removing excess water from the gas stream, and providing necessary gas humidification levels.
4. Oxidant supply subsystem (OSS)—usually contains air compressors, filters, and var ious devices maintaining required pressure, temperature, RH, and mass flow control.
5. Controls and communications subsystem (C&CS)—a combination of hardware and software controlling all FCPS functionality, using specific algorithm adjustable to various operating conditions and geared toward performance, durability, and efficiency optimization.
6. DC/DC or DC/digital to analog (DA) converter (power conditioner subsystem)—provides interface between FC stack output and electrical parameters that make available effective use of generated electric power.

All the above subsystems are interfaced together into FCPS as shown in Figure 7.37.

7.27.5 Fuel Cell Power System (FCPS) Functions and Features

The fuel cell system functions and features represent only the major solutions for converting one type of energy into the other, removing excess heat and water. Every function must be supported in the system by an adequate combination of hardware elements and software controls integrated into various subsystems. The design of a fuel cell system involves the optimization of the fuel cell section based on internal energy power minimization,

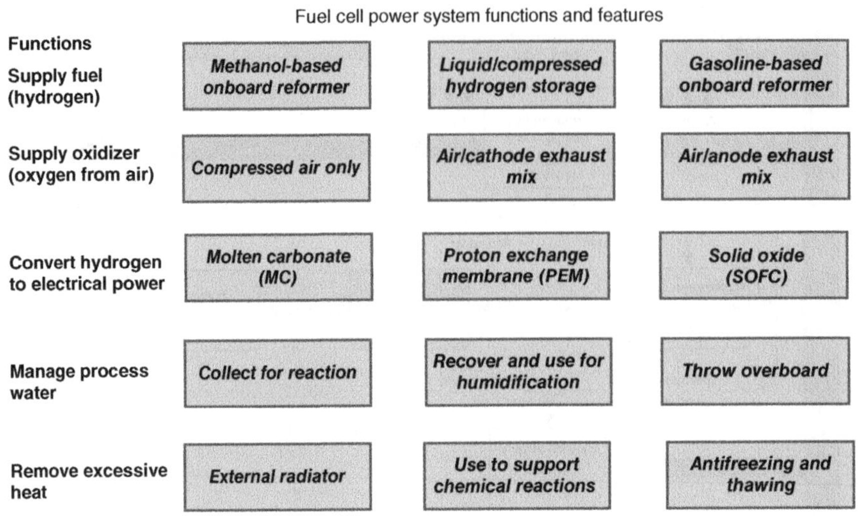

Figure 7.38 Fuel cell power system functions and features.

efficiency, and economics. In addition, any FCPS design has to cover the project-specific objectives, such as desired fuel, emission levels, potential uses of rejected heat (electricity, steam, or heat), desired output levels, and volume or mass criteria (volume/kW or mass/kW). The system functions and features are shown in Figure 7.38.

7.27.6 FCPS Performance Characteristics

FCPS gross power is equal to the maximum power in watts generated by the fuel cell stack at given operating conditions and specific load. FCPS net power is the power generated at the output of the FCPS and equal to FCPS gross power minus the power used by parasitic load in the FCPS.

FCPS efficiency is described as

$$\eta = \frac{P_{\text{electrical out}}}{P_{\text{LHV Hydrogen in}}} \times 100\% \tag{7.95}$$

where $P_{\text{electrical out}}$ is the electrical power generated by FCPS. As a result of electrochemical reaction, it is equal to the "net" electrical power at the FCPS load as shown in

$$P_{\text{electrical out}} = V_{\text{Fuel Cells Stack}} \times I_{\text{Load}} \tag{7.96}$$

$P_{\text{LHV Hydrogen in}}$ is the lower heating value of the hydrogen gas (fuel) prior to reaction and it is equal to

$$P_{\text{LHV}}H = 120 * m_{\text{Avg. Hydrogen Flow}} \tag{7.97}$$

where $m_{\text{Avg. Hydrogen Flow}}$ is an average hydrogen mass flow at the fuel cell stack inlet and hydrogen lower heating value is 120 J/g.

Fuel is defined as a hydrogen gas at specific pressure, temperature, mass flow rate, RH, and purity entering or exiting the fuel stack at any given period of time during FCPS operations. An oxidizer is defined as an oxygen gas (as a part of air stream) at specific pressure, temperature, mass flow rate, and RH entering or exiting fuel stack at any given period of time during FCPS operations. Electrical power output is defined as a product of voltage and current available at the input of FCPS load at any given period of time during FCPS operations. Different FCs require different anode gases and oxidants; can be poisoned by the presence of CO; may require expensive catalysts, working at different temperatures; and can use oxidant at atmospheric pressure or pressurized air.

7.27.7 Fuel Choice

FCPS architecture is a function of used fuel. As an example, in 1990 the development of fuel systems for automotive applications was focusing on onboard methanol fuel processing either as the fuel of choice or as a development step toward processing gasoline. The reason is that methanol was the easiest fuel to convert to hydrogen for vehicle use (Figure 7.39).

Methanol is oxidized to CO and hydrogen at temperatures below 400°C and can be catalytically steam reformed at 250°C or less. This provides a quick start-up advantage. Methanol can be converted to hydrogen with efficiencies of <90%. But methanol is produced primarily from natural gas, requiring energy, and it is less attractive than gasoline in terms of "well-to-wheel" efficiency. Another limitation is that methanol is corrosive, and its affinity for water results in difficulties in fuels distribution. Methanol is more acutely toxic than gasoline. Additives are likely to be needed for safety and health reasons. This will impact the fuel. In comparison to methanol, gasoline has a well-developed production and distribution infrastructure, is less toxic, and has higher energy density, which can lead to better vehicle range. The negative factors for gasoline reforming to hydrogen are high temperature conversion of 650°C associated with greater amounts of CO, methane, coke, and system complexity. Additional detrimental factors are ability for sulfur and trace amounts

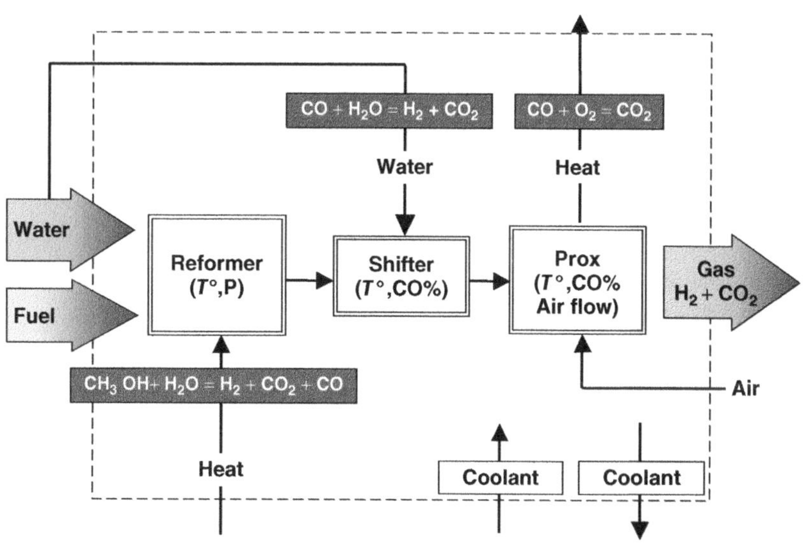

Figure 7.39 Methanol fuel processor subsystem.

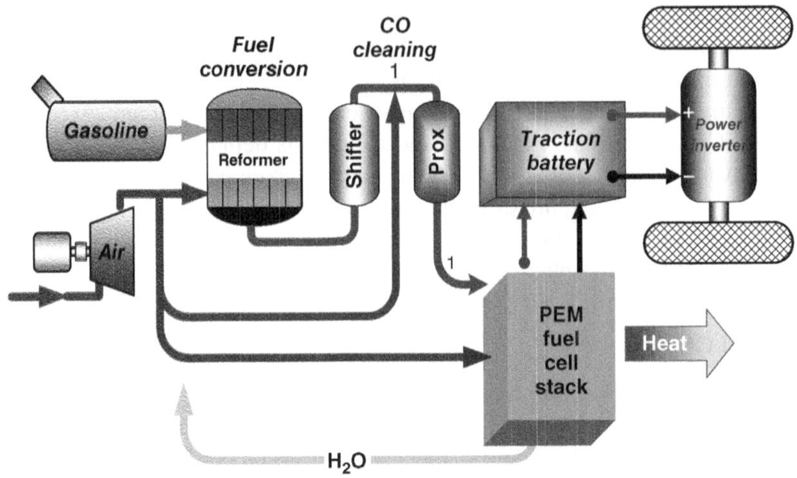

Figure 7.40 Gasoline fuel cell vehicle architecture.

of metal to poison the fuel cell stack and onboard packaging difficulties associated with component complexity. Both methanol and gasoline are less attractive due to environmental considerations, health concerns, compatibility with other materials, and system complexity. The current trend is in the utilization of hydrogen stored onboard. There are three basic options for this onboard stored hydrogen: (a) liquid H_2, (b) compressed H_2, and (c) metal hydride.

The simplified block diagram for gasoline fuel cell system architecture is presented in Figure 7.40. To provide the required functionality fuel cell system must meet various requirements as follows:

- Required power
- Available fuel
- Durability and reliability
- Cost
- Power density (volumetric (kW/L) and gravimetric (kW/kg))
- System efficiency
- Start-up time
- Capacity
- Heat utilization
- Number of on/off cycles
- Based on these requirements, the proper FC technology should be selected.

Currently, the technology development efforts are focused on the following:

1. Direct methanol FC development for portable energy sources
2. PEMFC as the main candidate for automotive propulsion and residential power production
3. SOFC technology as a source of distributed power and on-site generation

4. MCFC and PAFC for stationary power generation. Today, the PEMFC and SOFC are the two most commonly used technologies. PEMFC operates at low temperatures, providing a quick response to load changes, but requires extensive water management for humidity control and radiators for heat rejection. To reduce the problems related to the heat rejection and increase the CO tolerance, higher temperature (above 100°C) membranes are under development. The SOFC-based system does not use the high-cost precious metals that are used in PEMFC, and is less sensitive to CO presence in the fuel. It operates with simpler oxidant supply. However, the SOFC system is based on a more complicated manufacturing process for the electrode, electrolyte, and high-temperature stable metal parts. Other difficulties in SOFC systems are related to protection from heat losses and the lower power density. However, a combination of an SOFC power system with a gas turbine utilizing high-temperature off-gas may increase this hybrid system efficiency up to 70%. The design of FCPS architecture is a very complex process based on trade-offs that are affected by cost, efficiency, internal components' power consumption, complexity, and durability. In addition, every design having project-specific objectives such as choice of fuel, accepted emission, possibility to utilize the rejected heat, desired output levels, power density volume, or mass criteria (volume/kW or mass/kW) is challenging in several respects.

APPENDIX 7

Redox Reactions in DMFC

In this fuel cell, the reactants are $CH_3OH(aq)$ and $O_2(aq)$. The products are $CO_2(g)$ and water. $CH_3OH(aq)$ is oxidized and $O_2(g)$ is reduced. The two half reactions can be written as

$$CH_3OH(aq) = CO_2(g) \tag{A.1}$$

$$O_2(g) = H_2O(aq) \tag{A.2}$$

We may take equation A.1 and apply the five rules discussed in Section 5.6.2, on balancing redox reactions. Applying Rule 1, since C is the atom other than O and H in equation A.1, C is balanced on both sides:

$$CH_3OH(aq) = CO_2(aq) \tag{A.3}$$

Applying Rule 2, since there is more O on right side of the equation, add H_2O to the left side of the equation. Balancing O

$$CH_3OH(aq) + H_2O(aq) = CO_2(aq) \tag{A.4}$$

Applying Rule 3, hydrogen balance requires H^+ to be added to the right side of the equation:

$$CH_3OH(aq) + H_2O(aq) = CO_2(aq) + 6H^+(aq) \tag{A.5}$$

Applying Rule 4, charge balance is necessary and, hence, electrons are to be added to the right side of the equation:

$$CH_3OH(aq) + H_2O(aq) = CO_2(aq) + 6H^+(aq) + 6e^- \tag{A.6}$$

Considering the DMFC the other half cell reaction,

$$O_2(g) = H_2O(aq) \tag{A.7}$$

application of the four rules will result in

$$O_2(g) + 4H^+(aq) + 4e^- = H_2O(aq) + H_2O(aq) \tag{A.8}$$

(see Section 7.6.1)

Balancing the two half reactions to arrive at the net reaction requires

$$CH_3OH(aq) + H_2O(aq) = CO_2(aq) + 6H^+(aq) + 6e^- \tag{A.9}$$

$$O_2(g) + 4H^+(aq) + 4e^- = H_2O(aq) + H_2O(aq) \tag{A.10}$$

Multiplying equation A.9 by 2 and equation A.10 by 3,

$$2CH_3OH(aq) + 2H_2O(aq) = 2CO_2(aq) + 12H^+(aq) + 12e^- \tag{A.11}$$

$$3O_2(g) + 12H^+(aq) + 12e^- = 3H_2O(aq) + 3H_2O(aq) \tag{A.12}$$

$$2CH_3OH(aq) + 3O_2(g) = 2CO_2(g) + 4H_2O(aq) \tag{A.13}$$

or

$$CH_3OH(aq) + 3/2O_2(g) = CO_2(g) + 2H_2O(aq) \tag{A.14}$$

PROBLEMS

1. Consider the hydrogen technology that is based on utilization of hydrogen as fuel. The fuel cell reaction is

   $$2H_2(g) + O_2(g) = 2H_2O(l)$$

   What does this equation convey? How is it useful to predict the amount of water produced in a fuel cell?
2. What will happen if there is no membrane in a fuel cell?
3. What is the advantage of applying oxidation number rules to the fuel cell reaction?
4. Examine the scientific literature and determine the toxicity levels of carbon dioxide and carbon monoxide.
5. Survey the literature on the disposal conditions enforced in the United States and Europe for primary and secondary batteries. Are there any disposal conditions to be followed for the fuel cells?
6. Calculate the amount of oxygen required for running a 1 kW power for 1 h. If oxygen gas is replaced with air calculate how much air mass would be required?
7. In low-temperature fuel cells, water management is important. How much water is produced in a 5 kW power fuel cell in 1 h? Calculate the water using hydrogen utilized route and also by oxygen utilization.
8. Membranes in PEMFC is made of Nafion. What is the chemical formula of the repeating unit of this membrane. What is the function of the sulfonic acid group?
   The function of sulfonic acid is to absorb water.

9. What factors are considered in the selection of a membrane?
10. Currently what type of modifications is being done in membrane for utilization in the fuel cells?

MULTIPLE CHOICE QUESTIONS

1. In a fuel cell, the barrier to electron transfer is called
   (a) activity of a substance
   (b) activation polarization
   (c) electrode sensor
   (d) fuel oxidation
   Ans. ___________
2. The concentration polarization in a fuel cell causes
   (a) mass transport loss
   (b) material loss
   (c) fuel reacting with electrolytes
   (d) transport number
   Ans. ___________
3. Name three reasons for a fuel cell not practically giving the expected theoretical voltage.
   Ans. ___________
4. The three electrode materials that are under consideration for the construction of PEMFC have the following exchange current densities for hydrogen in an acid electrolyte:
   Zinc: 3×10^{-11} A/cm^2, gold: 3.98×10^{-6} A/cm^2, graphite: 1.2×10^{-4} A/cm^2, and platinum: 2.5×10^{-3} A/cm^2. Which electrode material will you select for its construction?
   (a) Zinc
   (b) Gold
   (c) Graphite
   (d) Platinum
   Ans. ___________
5. The resistance and ion movement in a fuel cell causes
   (a) heat loss
   (b) Ohmic loss
   (c) disobeying Ohm's law
   (d) evaporation of the fuel
   Ans. ___________
6. Mass of hydrogen requirement of a fuel cell in running a 6 A current gadget for half an hour is (Molar mass of hydrogen = 2.01 g/mol; $n = 2.0$ and $F = 96{,}500$ C/mol)
   (a) 0.112 g
   (b) 0.224 g
   (c) 1.12 g

(d) 2.24 g

Ans. ____________

7. An automobile used 0.5 kW of power at 0.60 V. The volume of hydrogen required for half an hour of operation is (density of hydrogen = 0.084 kg/cm^3 or 0.084×10^{-3} g/cm^3)

(a) 740 L

(b) 186 L

(c) 370 L

(d) 500 L

Ans. ____________

8. Oxygen consumption in running a fuel cell for 1 h in delivering 500 W of power at 0.70 V is

(a) 497 g

(b) 213 g

(c) 300 g

(d) 100 g

Ans. ____________

9. A fuel cell running on an average voltage of 0.60 V produced 1 L of water after 2 h of operation. Calculate the power consumed in this operation of the fuel cell (density of water: 0.99 g/cm^3).

Ans. ____________

10. Fuel crossover in a fuel cell is prevented by

(a) controlling the fuel supply

(b) making the anode stronger

(c) using a membrane

(d) reducing the size of the fuel cell

Ans. ____________

11. Nafion membrane allows

(a) OH^- ions to pass through it

(b) H^+ ion to pass through it

(c) H_2 gas to pass through it

(d) prevents all ions to pass through it

Ans. ____________

12. What is the function of a bipolar plate?

(a) Makes the stacking of fuel cells compact

(b) A plate used to separate anode and cathode

(c) Provides electrical contact

(d) A decorative piece

Ans. ____________

13. What is the preferred anode material in an alkaline fuel cell?

(a) Platinum

(b) Nickel

(c) $LaNi_3$
(d) Cu
Ans. ____________

14. The operating temperature of MCFC is
(a) 100°C
(b) 200°C
(c) 400°C
(d) 700°C
Ans. ____________

15. In the MCFC, at the cathode the following reaction occurs
(a) $O_2(g) + 4H^+(aq) + 4e = H_2O(l)$
(b) $\frac{1}{2}O_2(g) + CO_2(g) + 2e = CO_3^{2-}(l)$
(c) $CO_2(g) + H_2O(g) = H_2CO_3(g)$
(d) $H_2(g) = 2H^+(aq) + 2e$
Ans. ____________

16. Among the following gases select the one that would poison the electrode in PEMFC
(a) Oxygen
(b) Hydrogen
(c) Carbon monoxide
(d) Air
(e) Carbon dioxide
Ans. ____________

17. What material is used as an interconnect in SOFC?
(a) Platinum
(b) Nickel
(c) Doped $LaCrO_3$
(d) Carbon
Ans. ____________

18. What is the electrolyte used in bridging the anode and cathode in SOFC?
(a) Yttria-doped Zirconia
(b) Carbonate
(c) Colloidal particles of alumina
(d) Sulfate
Ans. ____________

19. In some SOFC fuel cells, hydrogen gas is mixed with carbon monoxide (CO), what is the purpose?
(a) To poison the anode
(b) CO reacts with hydrogen to produce water
(c) CO reacts with water to produce carbon dioxide and hydrogen
(d) CO cools the fuel cell
Ans. ____________

20. In the SOFC fuel cell, at the cathode the following reaction occurs:
    (a) $O_2 + 4H^+ + 4e = 2H_2O$
    (b) $2H^+ + O_2^- = H_2O$
    (c) $½O_2 + 2e = O_2^-$
    (d) $O_2 + 2H^+ + 2e = H_2O_2$
    Ans. ____________
21. The following reaction occurs in the anode compartment in SOFC:
    (a) $O_2 + 4e + 4H^+ = 2H_2O$
    (b) $2H^+ + O_2^- = H_2O$
    (c) $H^+ + e = ½H_2$
    (d) $2H_2O = 4H^+ + 4e + O_2$
    Ans. ____________
22. Write the reactions that produce energy in a hydrogen–chlorine fuel cell.
    Ans. ____________
23. The preferred material for bipolar plate in the fuel cell is
    (a) Carbon
    (b) Stainless steel
    (c) Titanium
    (d) Nickel
    Ans. ____________
24. In a membraneless fuel cell, the fuel crossover problem is reduced/avoided by
    (a) laminar flow of hydrogen
    (b) turbulent flow of hydrogen
    (c) stirring the anode
    (d) stirring the cathode
    Ans. ____________
25. What is the important characteristic of the electrolyte bridging the anode and cathode in the SOFC?
    (a) Good electronic conductivity
    (b) Good ionic conductivity
    (c) Good thermal conductivity
    (d) Good tensile strength
26. Water is produced in the fuel cell reaction at the cathode in
    (a) PEMFC
    (b) SOFC
    (c) MCFC
    (d) AFC

BIBLIOGRAPHY

1. Bard, A.J. and Faulkner, L.R. (2001) *Electrochemical Methods*, Wiley, New York.
2. Bloom, H. and Cutman, F. (eds) (1981) *Electrochemistry*, Plenum Press, New York, p. 121.

3. Trassati, S. (1972) Work function, electronegativity, and electrochemical behavior of metals: III. Electrolytic hydrogen evolution in acid solutions. *Journal of Electroanalytical Chemistry*, **39**, 163.
4. Britto, P.J., Santhanam, K.S.V., Rubio, A. *et al.* (1999) Improved charge transfer at carbon nanotube electrode. *Advanced Materials*, **11**, 154.
5. Bockris, J.O.M. and Reddy, A.K.N. (1970) *Modern Electrochemistry*, vol. **2**, Plenum Press, New York.
6. Norskov, K., Bligaard, T., Logadottir, A. *et al.* (2005) Trends in the exchange current for hydrogen evolution. *Journal of the Electrochemical Society*, **152**, J23–J26.
7. Adjemian, L.S.J., Srinivasan, S., Benzigen, J., and Bocarsly, A.B. (2002) Silicon oxide Nafion composite membranes for proton-exchange membrane fuel cell operation at 80–140°C. *Journal of the Electrochemical Society*, **149**, A256–A261.
8. Chunsheng, W. and Appleby, A.J. (2003) High-peak-power polymer electrolyte membrane fuel cells. *Journal of the Electrochemical Society*, **150**, A493–A498.
9. Meng, H. and Shen, P.K. (2005) Tungsten carbide nanocrystal promoted Pt/C electrocatalysts for oxygen reduction. *Journal of Physical Chemistry*, **109** (48), 22705–22709.
10. Hickner, M.A., Ghassemi, H., Kim, Y. *et al.* (2004) Alternative polymer systems for proton exchange membranes (PEMs). *Chemical Reviews*, **104** (10), 4587–4611.
11. Wang, F., Hickner, M., Kim, Y.S. *et al.* (2002) Direct polymerization of sulfonated poly(arylene ether sulfone) random (statistical) copolymers: candidates for new proton exchange membranes. *Journal of Membrane Science*, **197** (1–2), 231–242.
12. Uhling, H. and Revie, R.W. (1985) *Corrosion and Corrosion Control*, Wiley, New York.
13. Tang, H.L., Luo, Z.P., Pan, M. *et al.* (2005) Synthesis of platinum nanoparticles and then self-assembly on Nafion membrane to give a catalyst coated membrane. *Journal of Chemical Research*, S (7), 449–451.
14. Bevers, D., Wagner, N., and VonBradke, M. (1998) Innovative production procedure for low cost PEFC electrodes and electrode/membrane structures. *International Journal of Hydrogen Energy*, **23**, 57–63.
15. Giorgi, L., Antolini, E., Pozio, A., and Passalacqua, E. (1998) Influence of the PTFE content in the diffusion layer of low-Pt loading electrodes for polymer electrolyte fuel cells. *Electrochimica Acta*, **43** (24), 3675–3680.
16. Mosdale, R., Wakizoe, M., and Srinivasan, S. (1994) Proceedings of the Electrochemical Society. p 9423 (Electrode Materials and Processes for Energy Conversion and Storage), 179–89; New Jersey: Electrochemical Society.
17. Ralph, T.R., Hards, G.A., Keating, J.E. *et al.* (1997) Low cost electrodes for proton exchange membrane fuel cells. *Journal of the Electrochemical Society*, **144**, 3845–3857.
18. Gulla, A.F., Saha, M., Sudan, A. *et al.* (2005) Dual ion-beam-assisted deposition as a method to obtain low loading-high performance electrodes for PEMFCs. *Electrochemical and Solid-State Letters*, **8** (10), A504–A508.
19. Ioroi, T. and Yasuda, K. (2005) Platinum–iridium alloys as oxygen reduction electrocatalysts for polymer electrolyte fuel cells. *Journal of the Electrochemical Society*, **152** (10), A1917–A1924.
20. Antolini, E., Salgado, J.R.C., Giz, M.J., and Gonzalez, E.R. (2005) Effects of geometric and electronic factors on ORR activity of carbon supported Pt–Co electrocatalysts in PEM fuel cells. *International Journal of Hydrogen Energy*, **30** (11), 1213–1220.
21. Singhal, S.C. and Kendall, K. (2003) FCN/SOFC Book, http://www.fuelcellmarkets.com/article default view. fcm?articleid=3649&subsite=447 (accessed 27 May 2017)
22. Xue, X., Tang, J., Sammes, N., and Du, Y. (2005) Dynamic modeling of single tubular SOFC combining heat/mass transfer and electrochemical reaction effects. *Journal of Power Sources*, **142**, 211–222.

23. Sonntag, R.E., Borgnakke, C., and Van Wylen, G.J. (2003) *Fundamentals of Thermodynamics*, 6th edn, Wiley, New York.
24. Holman, J.P. (1997) *Heat Transfer*, 8th edn, McGraw-Hill, Boston.
25. Achenbach, E. (1994) Three-dimensional and time-dependent simulation of a planar solid fuel cell stack. *Journal of Power Sources*, **49**, 333–348.
26. Koyama, O.M., Wen, C., Yamada, K., and Takahashi, H. (2003) Object-based modeling of SOFC system: dynamic behavior of micro-tube SOFC. *Journal of Power Sources*, **118**, 430–439.
27. Murthy, M., Esayian, M., Lee, W., and Van Zee, J.W. (2003) The effect of temperature and pressure on the performance of a PEMFC exposed to transient CO concentrations. *Journal of Electrochemical Society*, **150**, A29–A34.
28. Gu, T., Lee, W.-K., Van Zee, J.W., and Murthy, M. (2004) Effect of reformate components on PEMFC performance dilution and reverse water gas shift reaction. *Journal of the Electrochemical Society*, **151**, A2100–A2105.
29. Wagner, N. and Schultze, M. (2003) Change of electrochemical impedance spectra during CO poisoning of the Pt and Pt–Ru anodes in a membrane fuel cell (PEFC). *Electrochimica Acta*, **48**, 3899–3907.
30. Tong, Y.Y., Kim, H.S., Babu, P.K. *et al.* (2002) An NMR investigation of CO tolerance in a Pt/Ru fuel cell catalyst. *Journal of the American Chemical Society*, **124**, 468.
31. Yoshida, K., Ishida, M., and Okada, T. (2005) Electrochemical performance of Na2Ti3O7/C negative electrode. *Electrochemistry*, **73**, 298–300.
32. Chong Luo, Hui Xie, Qin Wang, Geng Luo, and Chao Liu (2015) A Review of the Application and Performance of Carbon Nanotubes in Fuel Cells. *Journal of Nanomaterials*, **2015**.
33. D. Young Chung, H. Kim, Y. Chung, M. Lee, S.J. Yoo, A.D. Bokare, W. Choi, and Y. Sung (2014) Inhibition of CO poisoning on Pt catalyst coupled with the reduction of toxic hexavalent chromium in a dual-functional fuel cell. *Scientific Reports*, **4**, 7450.
34. Zhong, C., Luo, J., Maye, M.M., *et al.* Fuel cell catalysts: Gold-platinum nanoparticles. 228th American Chemical Society, 2004, Philadelphia, PA.
35. Liu, Z., Ling, X., Lee, J.Y. *et al.* (2003) Nanosized Pt and PtRu colloids as precursors for direct methanol fuel cell catalysts. *Materials Chemistry*, **13**, 3049–3052.
36. Dhar, H.P., Kush, A.K., Patel, D.N., and Christner, L.G. (1986) Performance Study of a Fuel Cell Pt-on-C Anode in Presence of CO and CO_2, and Calculation of Adsorption Parameters for CO Poisoning. *Proceedings Electrochemical Society*, **86** (12), 284–297.
37. Tewari, A. and Sambhy, V. (2006) Quantification of carbon dioxide poisoning in air breathing alkaline fuel cells. *Journal of Power Sources*, **153**, 1–10.
38. Atkinson, A., Barnett, S., Gorte, R.J. *et al.* (2004) Advanced anodes for high-temperature fuel cells. *Nature Materials*, **3**, 17–27.
39. Boennemann, H., Endruschat, U., Hormes, J. *et al.* (2004) Activation of Colloidal PtRu Fuel Cell Catalysts via a Thermal "Conditioning Process". *Fuel Cells*, **4**, 297–308.
40. Komatsu, T., Arai, H., Chiba, R. *et al.* (2006) Cr Poisoning Suppression in Solid Oxide Fuel Cells Using LaNi (Fe) O3 Electrodes. *Electrochemical and Solid State Letters*, **9**, A9–A12.
41. Salinas, C., Simpson, S.F., Murphy, O.J., *et al.*, Membrane and electrode structure for methanol fuel cell U.S. Patent 5,958,616. 1999.
42. Santhanam, K.S.V., Chemopol Industries 1996 (unpublished work). *Fuel Cell Handbook*, 5th edn. EG&G Services Parsons, Inc., Science Applications International Corporation Under Contract No. DE-AM26-9FT40575 http://www.fuelcells.org/info/library/fchandbook.pdf National Technical Information Service, U.S. Department of Commerce, 5285 Port Royal Road, Springfield, VA 22161.
43. Larminie, J. and Dicks, A. (2002) *Fuel Cell Systems Explained*, John Wiley & Sons, NJ.

44. Zhou, G., Roby, S., Wei, T., and Yee, N. (2014) Fuel Heat of Vaporization Values Measured with Vacuum Thermogravimetric Analysis Method. *Energy and Fuels*, **28**, 3138–3142.

45. Fernández, J.C., Arna, J.M., Gómez, J., and Dorado, M.P. (2012) A comparison of performance of higher alcohols/diesel fuel blends in a diesel engine. *Applied Energy*, **95**, 267.

46. Gur, T. (2013) Critical Review of Carbon Conversion in Carbon Fuel Cells. *Chemical Reviews*, **113**, 6179.

47. Gur, T. (2010) Mechanistic Modes for Solid Carbon Conversion in High Temperature Fuel Cells. *Journal of the Electrochemical Society*, **157**, B571.

48. Zhang, H. and Shen, P.K. (2011) Recent development of polymer electrolyte membranes for fuel cells. *Chemical Reviews*, **112**, 2780.

49. Nasef, M. (2014) Radiation-grafted membranes for polymer electrolyte fuel cells: current trends and future directions. *Chemical Reviews*, **114**, 12278–12329.

Section 7.21.1

50. Schafbauer, W., Menzler, N.H., and Buchkremer, H.P. (2014) Tape casting of anode supports for solid oxide fuel cells at Forschungszentrum Jülich. *International Journal of Applied Ceramic Technology*, **11**, 125.

51. He, R., Fang, D., He, F., and Wu, Y. (2014) Microstructure and cell performance of NiO–YSZ composite anode-supported solid oxide fuel cells using graphite as anode pore-former: effects of graphite particle size and HNO_3 treatment. *International Journal of Applied Ceramic Technology*, 1–7.

52. Yildiz, O., Soydan, A.M., Ata, A. *et al.* (2013) Properties of ceria based novel anode nanopowders synthesized by glycine–nitrate process. *Acta Physica Polonica A*, **123**, 432–435.

53. Macek, J., Novosel, B., and Marinsek, M. (2007) Ni-YSZ SOFC anodes – minimization of carbon deposition. *Journal of the European Ceramic Society*, **27**, 487–491.

54. Liu, Q., Dong, X., Yang, C. *et al.* (2010) Self-rising synthesis of Ni–SDC cermets as anodes for solid oxide fuel cells. *Journal of Power Sources*, **195**, 1543.

55. Sivasankaran, V., Combemale, L., Pera, M.C., and Caboche, G. (2014) Initial preparation and characterization of single step fabricated intermediate temperature solid oxide fuel cells (IT-SOFC). *Fuel Cells*, **14**, 533–536.

56. Zhu, Y., Chen, Z.G., Zhou, W. *et al.* (2013) An a-site-deficient perovskite offers high activity and stability for low-temperature solid oxide fuel cells. *ChemSusChem*, **6**, 2249–2254.

57. Zawadzki, M., Grabowska, H., and Trawczyński, J. (2010) Effect of synthesis method of LSCF perovskite on its catalytic properties for phenol methylation. *Solid State Ionics*, **181**, 1131–1139.

58. Naiqing, Z., Kening, S., Derui, Z., and Dechang, J. (2006) Study on properties of LSGM electrolyte made y tape casting method and applications in SOFC. *Journal of Rare Earths*, **24**, 90–92.

59. Ji, Y., Liu, J., He, T. *et al.* (2003) Single intermedium-temperature SOFC prepared by glycine-nitrate process. *Journal of Alloys and Compounds*, **353**, 257–262.

60. Zhu, X.D., Zhang, N.Q., Wu, L.J. *et al.* (2010) Preparation and performance of large-area $La_{0.9}Sr_{0.1}Ga_{0.8}Mg_{0.2}O_{3-\delta}$ electrolyte for intermediate temperature solid oxide fuel cell. *Journal of Power Sources*, **195**, 7583–7586.

61. Bodén, A., Di, J., Lagergren, C. *et al.* (2007) Conductivity of SDC and $(Li/Na)_2CO_3$composite electrolytes in reducing and oxidising atmospheres. *Journal of Power Sources*, **172**, 520–529.

62. Mahmud, L.S., Muchtar, A., and Somalu, M.R. (2017) Challenges in fabricating planar solid oxide fuel cells: A review. *Renewable and Sustainable Energy Reviews*, **72**, 105.

63. Garche, J. and Jörissen, L. (2015) *Applications of Fuel Cell Technology: Status and Perspectives, The Electrochemical Interface*, Wiley, NJ.

64. Ehsani, M., Gao, Y., and Emadi, A. (2010) *Modern Electric, Hybrid Electric, and Fuel Cell Vehicles: Fundamentals, Theory, and Design*, CRC Press, Boca Raton, FL, ISBN:978-1-4200-5398-2.

Anode

65. Schafbauer, W., Menzler, N.H., and Buchkremer, H.P. (2014) Tape casting of anode supports for solid oxide fuel cells at Forschungszentrum Jülich. *International Journal of Applied Ceramic Technology*, **11**, 125–135.
66. He, R., Fang, D., He, F., and Wu, Y. (2014) Microstructure and cell performance of NiO–YSZ composite anode-supported solid oxide fuel cells using graphite as anode pore-former: effects of graphite particle size and HNO_3 treatment. *International Journal of Applied Ceramic Technology*, **67**, 1–7.
67. Yildiz, O., Soydan, A.M., Ata, A. *et al.* (2013) Properties of ceria based novel anode nanopowders synthesized by glycine-nitrate process. *Acta Physica Polonica A*, **123**, 432–435.
68. Macek, J., Novosel, B., and Marinsek, M. (2007) Ni-YSZ SOFC anodes-Minimization of carbon deposition. *Journal of the European Ceramic Society*, **27**, 487–491.
69. Liu, Q., Dong, X., Yang, C. *et al.* (2010) Self-rising synthesis of Ni–SDC cermets as anodes for solid oxide fuel cells. *Journal of Power Sources*, **195**, 1543–1550.
70. Sivasankaran, V., Combemale, L., Pera, M.C., and Caboche, G. (2014) Initial preparation and characterization of single step fabricated intermediate temperature solid oxide fuel cells (IT-SOFC). *Fuel Cells*, **14**, 533–536.
71. Zhu, Y., Chen, Z.G., Zhou, W. *et al.* (2013) An a-site-deficient perovskite offers high activity and stability for low-temperature solid oxide fuel cells. *ChemSusChem*, **6**, 2249–2254.
72. Zawadzki, M., Grabowska, H., and Trawczyński, J. (2010) Effect of synthesis method of LSCF perovskite on its catalytic properties for phenol methylation. *Solid State Ionics*, **181**, 1131–1139.
73. Naiqing, Z., Kening, S., Derui, Z., and Dechang, J. (2006) Study on properties of LSGM electrolyte made y tape casting method and applications in SOFC. *Journal of Rare Earths*, **24**, 90–92.
74. Zhu, X.D., Zhang, N.Q., Wu, L.J. *et al.* (2010) Preparation and performance of large-area $La_{0.9}Sr_{0.1}Ga_{0.8}Mg_{0.2}O_{3-\delta}$ electrolyte for intermediate temperature solid oxide fuel cell. *Journal of Power Sources*, **195**, 7583–7586.
75. Bodén, A., Di, J., Lagergren, C. *et al.* (2007) Conductivity of SDC and $(Li/Na)_2CO_3$composite electrolytes in reducing and oxidising atmospheres. *Journal of Power Sources*, **172**, 520–529.
76. Ji, Y., Liu, J., He, T. *et al.* (2003) Single intermedium-temperature SOFC prepared by glycine-nitrate process. *Journal of Alloys and Compounds*, **353**, 257–262.
77. Tao, S. and Haile, S.M. (2004) A high-performance cathode for the next generation of solid-oxide fuel cells. *Nature*, **431**, 170–173.
78. Ye, X. and LeVan, M.D. (2002) Water transport properties of Nafion membranes: Part II. Multi-tube membrane module for air drying. *Journal of Membrane Science*, **221**, 163–173.
79. Kee, R.J. (2007) The influence of current collection on the performance of tubular anode-supported SOFC cells. *Journal of Power Sources*, **169**, 315–326.
80. Brady, J.E., Russell, J.W., and Holm, J.R. (2000) *Chemistry: Matter and Its Changes*, Wiley, New York.
81. Chenshung, W. and Apple, A.J. (2003) High-peak power polymer electrolyte membrane fuel cells. *Journal of the Electrochemical Society*, **150**, A493–A498.

CHAPTER 8

Fuel Cells Applications

This chapter provides an overview of the technical and economic provision of fuel cell applications. The most comprehensive examination of this topic is presented in market reports [1, 2].

8.1 STATIONARY POWER PRODUCTION

The fuel cell stationary power generation systems market includes remote sites, telecommunications, commercial and residential buildings, backup electric sources, and auxiliary power units (APUs), with individual unit size ranging from 3 to 300 kW [3]. By application, all stationary FC systems can be divided into combined heat and power (CHP), uninterruptible power systems (UPSs), and primary power units (PPUs). CHP units used either proton exchange membrane (PEM) or solid oxide fuel cell (SOFC) stack and benefited from the fact that besides electricity, fuel cells generate heat, making overall efficiencies to 80–95%. Residential CHP units have been deployed extensively in Japan and South Korea for home power and heating for residents.

- *Molten-carbonate fuel cells (MCFCs).* MCFCs have the highest efficiencies of any type fuel cell and are not subject to the high-temperature material issues that affect solid-oxide technology. Molten carbonate fuel cells are currently being developed for natural gas and coal-based power plants for electrical utility, industrial, and military applications. MCFCs are high-temperature fuel cells that use an electrolyte composed of a molten carbonate salt mixture suspended in a porous, chemically inert ceramic matrix of β-alumina solid electrolyte (BASE). Since they operate at temperatures of 650°C and above, nonprecious metals can be used as catalysts at the anode and cathode, reducing costs. MCFCs can reach efficiencies approaching 60%. This is considerably higher than the 37–42% efficiencies of a phosphoric acid fuel cell (PAFC) plant. When the waste heat is captured and used, the overall fuel efficiencies can be as high as 85%. The primary disadvantage of current MCFC technology is durability. The high temperatures at which these cells operate and the corrosive electrolyte used accelerate component breakdown and corrosion, decreasing cell life [1, 2, 4].

Introduction to Hydrogen Technology, Second Edition.
K.S.V. Santhanam, Roman J. Press, Massoud J. Miri, Alla V. Bailey, and Gerald A. Takacs

Book Companion site : www.wiley.com/go/santhanam/hydrogentech_2e

- *Phosphoric acid fuel cells (PAFCs).* PAFCs use as an electrolyte the liquid phosphoric acid and carbon paper coated with a finely dispersed platinum catalyst on the electrodes. PAFCs are not sensitive to impurities in the hydrogen flow. The phosphoric acid solidifies at 40°C, making start-up very difficult. However, at an operating range 150–200°C, the expelled water can be converted to steam for air and water heating. PAFCs have been used for stationary applications with a CHP efficiency of about 80%, and they continue to dominate the on-site stationary fuel cell market [5]. In the United States, the producer of PAFC technology is UTC Fuel Cells, having hundreds of "PureCell" 200 kW units in service globally. The negative feature of PAFCs is that the fuel-cell stack and the reformer required to be replaced every 5 years. Therefore, the recent developments concentrated on durability and cost improvement [4, 5].
- *Solid oxide fuel cells (SOFCs).* SOFCs are being developed by many companies. For example, GE Energy is developing a multi-MW system with an SOFC gas turbine hybrid that operates on coal. The current trend in development is to decrease operational temperature from 1000 to 800°C. Building a large fuel cell unit of multi-MW capacity is much more challenging due to limited test data and the soaring cost of resources required for setup.
- *Proton exchange membrane fuel cell (PEMFC).* PEMFC technology has limited popularity in the stationary market due to the necessity to utilize pure hydrogen. However, work on PEM fuel cell systems that combine fuel cell power modules with multifuel processing technology for reformed hydrogen is in progress for electric power in the range of 50 kW. Such a process is capable of converting a variety of fuels to hydrogen, including natural gas, propane, kerosene, JP-8, biodiesel, and alcohol-based fuels.

In 2005, the largest fuel cell installation was developed by Verizon Communication, which generates 1.4 MW of primary electrical power and 30,000 kWh of useable heat. Using electricity from the fuel cells and reclaiming the heat and water for building usage, Verizon eliminates some 50,000 tons of carbon dioxide that would have been emitted into the atmosphere by a similar-sized fossil fuel–based power plant during 1 year [1].

8.1.1 Fuel Cell and Gas Turbine Hybridization

Integration of modular fuel cells with gas turbine (FC–GT) cycle increases the effectiveness of large stationary stations. Recovering high-temperature wasted heat by FC–GT hybrid cycle may lead to electric efficiency above 70% and less significant cost of produced energy. As an illustration, Fuel Cell Energy is developing a technology called DFC/T, a hybrid version of the direct fuel cell (DFC) combined with a gas turbine that works at ambient pressure. This approach allows the fuel cell to produce approximately four-fifths of the power, while the turbine produces one-fifth. The turbine also recovers the waste heat and feeds back the air for the fuel cell. Large, multi-MW systems may be expected to produce energy with 80% efficiency.

Although the majority of fuel cell installations utilize natural gas, there are an increasing number of units powered by alternative fuels, such as coal gas, anaerobic digester gas, and gasified wooden waste and waste plastics [6].

8.2 FUEL CELL TRANSPORTATION

By definition, a fuel cell power plant is an electrical power source converting the energy of hydrogen gas into electrical power by the means of fuel cell technology (Figure 8.1). A fuel

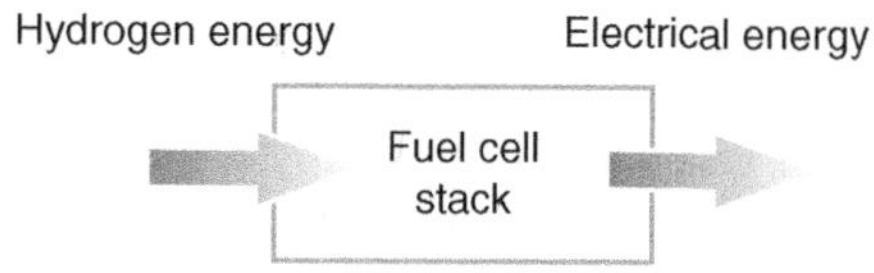

Figure 8.1 Energy conversion for transportation.

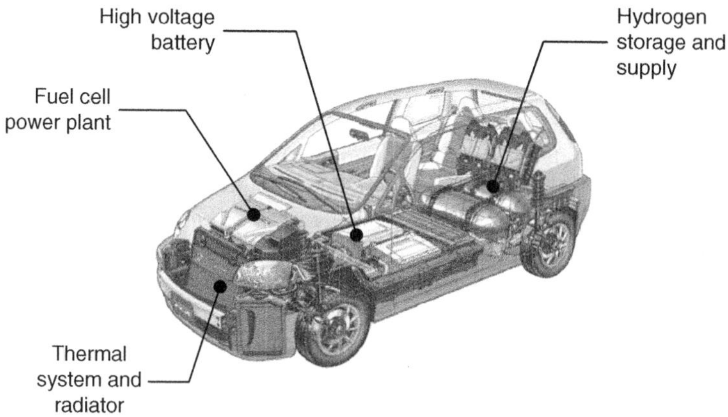

Figure 8.2 Fuel-cell vehicle design option.

cell alone is not capable of providing useful electrical power, and, therefore, implementation of a fuel cell for transportation needs requires a combination of subsystems and components with specific functions (Section 7.7).

A fuel cell vehicle (FCV) is a system that combines all required functions of a terrain vehicle, using a fuel cell power plant as a means of propulsion. The system converts electrical energy into motion and electrical power to traction conversion, vehicle motion control (i.e., steering, brakes, and suspension), passenger cabin, platform/chassis, and convenience and safety packages. In essence, the FCV is very similar to an electrical vehicle (Figure 8.2); however, the regular batteries are replaced by an electrochemical battery, which includes a fuel cell stack using the reaction between hydrogen and oxygen gases to generate electrical energy. The key FCV parameters are similar to the internal combustion engine (ICE)-based vehicles and include vehicle efficiency, fuel economy, driving range, vehicle emission, acceleration, and many others.

8.2.1 Vehicle Efficiency

FCVs use the energy generated from hydrogen gas and convert this energy into electrical power and, subsequently, into the mechanical power of the wheels. Based on this architecture, the FCV's electrical efficiency can be described as a ratio of the net electrical power (P_e) to the inlet of the electrical traction subsystem to the hydrogen latent heat value (LHV) at the inlet of the fuel cell stack subsystem.

8.2.2 Fuel Economy

Today, producing hydrogen is more expensive than gasoline on a dollar per gallon gasoline equivalent. However, FCVs exhibit outstandingly higher fuel economy compared to the

ICE vehicles due to higher efficiency of electrochemical conversion and less heat losses typical for ICE. Vehicle fuel economy is defined as the distance traveled by a vehicle on a hydrogen gas equivalent of gasoline [7]. By energy content, 1 US gallon of gasoline is equal to 1.09 kg of hydrogen gas at normal temperature. FCV fuel economy is measured as MPGe, where Ge is a gasoline gallon equivalent to hydrogen (mi/kgH_2). In the automotive industry, fuel economy is defined as a combination of city and highway driving at specific driving profiles established by the Environmental Protection Agency (EPA-USA) or the International Standard Organization (ISO-Europe).

8.2.3 FCV Range

The FCV range of driving is a critical parameter defining FCV design and performance as equivalent to an ICE vehicle [7]. It is also an important factor for calculation and planning of the future hydrogen infrastructure and refueling stations. The FCV range is limited by the amount of stored hydrogen mass and typically in the range of 300 miles. However, in June 2008, Toyota Motor Corporation reported that a new version of its FCV can travel about 515 miles on a single refueling.

8.2.4 Emission

Different from ICE-based vehicles, the tailpipe emission of an FCV is in the form of pure water as a product of the reaction of hydrogen and oxygen in the fuel cell stack. Not all hydrogen introduced into the fuel cell stack is consumed during this electrochemical reaction; very small amount of hydrogen gas is discharged from the exhaust into the atmosphere. From a safety point of view, the amount of emitted hydrogen should never exceed 50% of the lower explosion limit (LEL) by volume in all operating ranges. This value of hydrogen is usually defined in percent volume per liter of air. However, tangible emissions of other gases, such as carbon dioxide, sulfur gas, and carbon monoxide, harmful to the environment or health are generated during hydrogen production from fossil fuels. Only producing hydrogen from renewable resources can lead to true emission-free transportation [4, 5].

8.2.5 Cost

The current cost of FCVs is rather high because they are produced in limited number or as prototypes. Mass production often correlates to cost reduction and automakers worldwide believe that FCVs produced in high quantities will be competitive with conventional ICE technology. Increasing experience with advanced hybrid-electric vehicle technology and production will help lower the cost of future FCVs [8].

8.2.6 Well-to-Wheel Analysis

General Motors Corporation, with support from various organizations, conducted a "well-to-wheel" analysis (Figure 8.3), a comprehensive study for evaluating the effect of greenhouse gas emissions and energy consumption for transport-related issues [9]. The study that assessed fuel sources, processing techniques, and propulsion systems was conducted in two parts: (a) well-to-tank for fuel production and (b) tank-to-wheel for fuel utilization.

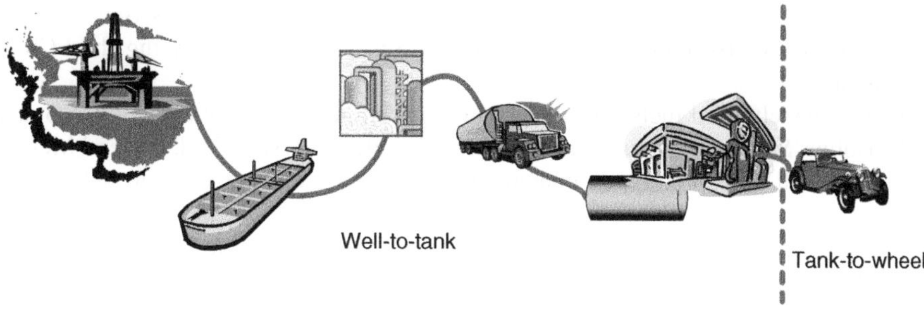

Figure 8.3 Well-to-wheel diagram.

8.2.6.1 Well-to-Tank In terms of total energy use, petroleum-based fuels and compressed natural gas (CNG) produced the lowest well-to-tank energy losses. Fuels with the highest losses included liquid hydrogen produced from central plants and hydrogen produced via electrolysis, electricity, and cellulosic ethanol. With regard to greenhouse gases, producing petroleum-based fuels and natural gas-based methanol resulted in fewer well-to-tank emissions, whereas producing hydrogen from natural gas resulted in high greenhouse gas emissions as a result of reformation of hydrocarbon fuels into hydrogen and carbon dioxide (greenhouse gas). Using electrolysis for hydrogen gas production only minimizes emission during production if nuclear or renewable energy is used in the process [9].

8.2.6.2 Tank-to-Wheel The tank-to-wheel analysis was based on EPA-defined driving cycles data collected from a Chevrolet Silverado full-sized pickup truck for the North American study and an Opel Zafira minivan for the European study. The fuel of choice included diesel, ethanol, CNG, methanol, and hydrogen in corresponding ICEs and fuel cell systems, plus conventional drives using state-of-the-art mechanical transmissions and hybrid electric drives. All systems had to meet the appropriate emissions standards—whether US or European—that will be in place in 2010 [10].

In terms of fuel consumption, diesel engines showed improvements over conventional technology, and fuel cell systems proved even better. In addition, diesel hybrids offered a significant reduction in greenhouse gas emissions compared to conventional gasoline engines. Ethanol-fueled vehicles yielded the lowest greenhouse gas emissions, followed by fuel cell hybrids.

A principal finding of the study was that FCVs' attractiveness with regard to well-to-wheel gas emission depends on the source of the hydrogen. Optimum results are achieved when renewable energies are used to produce the hydrogen [8].

8.2.7 Practical Transportation Applications

The hydrogen can be used directly in ICEs as fuel for combustion. This work continues under development in the United States and Europe [1, 3, 11]. In fuel cell transportation, segment oxygen is derived from air, but production, distribution, and storage of hydrogen gas is a more problematic and complicated issue. In the early stages of the FCV development, conversion of the hydrocarbon-based fuels onboard (both liquid and gas) was investigated, and prototypes were built. The main functions of the onboard fuel processing were (a) HC fuel reformation—conversion to hydrogen plus other gases (the so-called reformate gas) and (b) reformate cleanup from gases capable of damaging the FCS or environment.

The most widely used fuels for this type of application are liquid methanol, CNG, and liquid gasoline. Some car manufacturers, such as Daimler-Chrysler, Ford, General Motors, and others, built few prototypes containing fuel conversion and cleanup subsystems onboard.

The complexity, cost, weight, and low efficiency of this approach made the use of onboard fuel conversion prohibitive for vehicle applications. In the last few years, Shell Hydrogen, BP, Exxon-Mobil, Chevron, and many other oil companies made a commitment to create worldwide infrastructure for the production, storage, and distribution of hydrogen for vehicular applications.

Accordingly, FCV utilizing hydrogen fuels becomes a reality. In 2017, there are three companies producing FCV. The progress in developing vehicles for personal use is demonstrated in Table 8.1.

The advances in public transportation using fuel cells in buses have been demonstrated throughout the world and will most likely be the general public's first experience with fuel cell–powered vehicles, since buses use central fueling stations and maintenance facilities. Fuel cell buses are operating in California, Europe, and Japan [12].

The fuel cell buses are actively developing and tested by many companies, mostly by Daimler AG, based on Ballard Power Systems; Thorn Industries, based on UTC Power fuel cell technology; and in Japan, powered by Nino Motors and Toyota Motors. The fuel cell buses have a higher fuel economy than diesel, or natural gas fueled buses, and free from propulsion. The demonstration models on test in many cities, including Hamburg Germany, Shanghai China, London England, Sao Paulo Brazil, and Palm Springs California [13]. Today, procuring and maintenance cost are limiting factors for wide usage of hydrogen-fueled buses.

FCVs hold great promise. They offer a zero-emission mobility option that is critical for the healthy development of the world's urban centers. They also have great economic potential, as shown by the significant investments made by the private sector to develop and demonstrate FCVs.

8.2.8 Other Transport Applications

8.2.8.1 Aviation Aircraft engines have two functions: one is to provide propulsion to the craft and the other, which is secondary, is to supply comfortable conditions for passengers. At the present time, the propulsion still will be provided by liquid fuels because design of hydrogen-fueled aircraft possesses technological and societal barriers [14]. However, development work on small autonomous planes and pilot controlled planes is in progress. For instance, on September 2016 aviation engineers in Germany have successfully tested

TABLE 8.1 FCV produced in Asia

Manufacturer	Driving range (EPA estimation)	Efficiency in gasoline equivalent	Battery
Hyundai Tucson	265	49	180 V Li-ion
Toyota Mirai	312	66	245 V Ni-Mn
Honda FCX Clarity	366	67	366 V Li-ion

the four-passenger plane named HY4 that uses emission-free hybrid fuel cells/batteries to fly [15]. To provide comfort to passengers during a different part of the flight that aircraft is performing, such takeoff, cruising, landing, taxiing, etc.), the engine's load varies for heating and lighting. An electrical portion of this load can be provided by fuel cells. Calculation shows that by removing the electrical loading off the engines, the fuel consumption can be cut by three quarters when the plane is on the ground and up to 40% during flight [1, 7].

Such economy can be achieved by a fuel cell APU. Both Airbus (EU) and Boeing (USA) are working on the APU development. The scale of these units is large, 200–600 kW, and the design goal is to minimize APU weight and size to the level that onboard installation does not interfere with fee-paying passenger space. Therefore, the possibility of reforming aviation's fuel JP8/JP12 onboard is particularly attractive for reducing system complexity, fuel costs, and pollution. The first onboard prototype was demonstrated by Airbus at a 2008 Berlin air show.

8.2.8.2 Off-Road Equipment Small industrial vehicles, such as forklifts, tugs, loaders, and luggage transporters, are also attracting the attention of developers as a potential "window of opportunity" for fuel cell introduction (Figure 8.4). Airport off-road equipment and supporting vehicles with repetitive day-to-day tasks are another opportunity for hydrogen fueling. Currently, they utilize batteries that require a substantial storage capacity, are time-consuming to change, and are complex to recycle. Fuel cell–powered off-road equipment offering zero emissions, quick refueling, and better dynamic performance [1, 16] would be very attractive.

Electrification of the railway's network is a high priority for the European Union. As possible technical solution, Germany's Federal Railway Office is expected by the end of 2017 start to trial the passenger train running near Hannover. Technically, the tested railway's

Figure 8.4 Fuel cell fork lift. [*Source*: Reproduced with permission of Fuel Cell Today.]

configuration is a hybrid system integrated with a PEM stack with Li-ion battery for pick power. The hydrogen propulsion and storage equipment are mounted on the car roofs, where tanks are carrying 94 kg of hydrogen per car, providing 700 km of operation. The fuel cells were supplied by Hydrogenics (Canada). The tested trains will provide a real comparison between two propulsion systems: conventional diesel and fuel cells, utilizing the same bodies and drive equipment [17, 18]. Another example is the utilization of Ballard FCveloCity® heavy duty power modules in China with 2 tested trains and possibility to test another 10 in 2016 [19].

8.2.8.3 *Railways* Railways are another potential arena for fuel cell applications. Following are a few examples. The EU consortium initiated the Hydrogen Train Project, which hopes to demonstrate a hydrogen fuel cell–powered train by 2010. Technically, it will be a hybrid system integrating a 150–200 kW PEM stack with 50–100 kW battery for peak power. Japan has concentrated on developing reliable and environmentally friendly fuel cell commuter trains, with an on-rail demonstration by 2010, which will employ a heavy-duty PEM fuel cell/battery hybrid system, with a 300–400 km range and speed of 120 km/h [1].

In the United States, in parallel to developing the PEM power module for locomotive applications, attention has turned to mine loader applications. Nearly all mines employ diesel power, requiring expensive underground ventilation equipment to deal with harmful exhaust. Replacing diesel with hydrogen-powered vehicles would save an estimated 30–40% in ventilation costs, easily offsetting the cost of the fuel cells. For such an application, the Sandia National Lab built a metal hydride system that absorbed hydrogen onto a powdered metal alloy at the pressure of 150 psi [20].

8.2.8.4 *Miscellaneous* Bicylce, locomotives and scooters are popular vehicles in cities and urban areas all over the world but predominately in Asia (Figure 8.5). Reducing

Figure 8.5 The fuel cell/battery hybrid railway vehicle. http://w3.gorge.net/eclipse/projects/proj_wtw.html.

the emissions and energy use of these vehicles would be a great step in improving air quality and preserving the environment, including the reduction of noise. Highly efficient fuel cells running on hydrogen and feeding an electric motor form the preferred technology for the longer term, when fossil fuels will be replaced more and more by sustainable hydrogen as an energy carrier. Honda, Yamaha, and many other companies were successful in testing hydrogen fuel cell scooters based on the PEM or direct methanol fuel cell (DMFC) power plant that store hydrogen in an onboard tank or in nickel metal hydride cylinders. Applications include power-assisted bicycles, scooters, and motorcycles [1, 21].

Wheelchairs are a typical example where the technology meets societal needs. The present electric wheelchairs are heavy and have low speeds due to the weight of the battery. Adapting a wheelchair to take an early-development fuel cell onboard has not been difficult, and examples already exist in Japan and Europe. Using this technology increases the customer's mobility by providing greater torque and acceleration than ever before. Typical design configuration includes a combination of PEM stack with one time filled hydrogen storage permits to operate the unit over 12 h.

8.3 MICROPOWER SYSTEMS

In comparison to existing batteries, the adoption of portable fuel cells guaranteed high-energy density and long run time, high number of recharging cycles, and less environmental contamination during recycling. The currently adopted technologies for micropower sources are direct methanol and PEMFCs [22]. The most promising applications are power sources for consumer electronics and communication devices including digital cameras, phones, computers, radios, portable television, and night-vision goggles.

Japanese companies DoCoMo and Aquafairy [23] have proposed water-based micro-fuel cell systems. The operation of PEMFC for this purpose is shown in Figure 8.6. The right side shows a typical PEMFC that is operated with a fuel cartridge. This cartridge

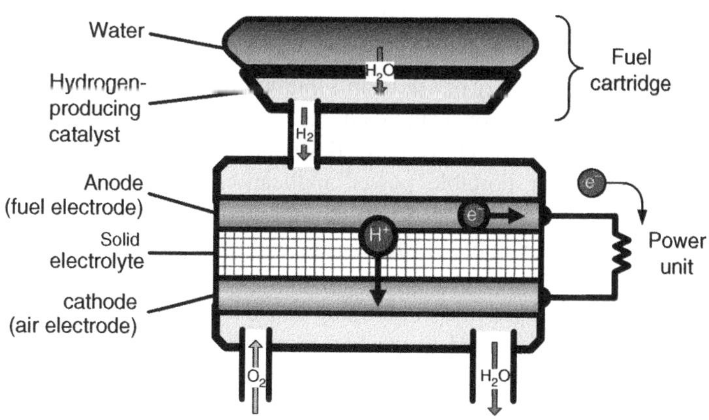

Anode	$H_2 \rightarrow 2H^+ + 2e^-$
Cathode	$\frac{1}{2}O_2 + 2H^+ + 2e^- \rightarrow H_2O$
Whole reaction	$H_2 + \frac{1}{2}O_2 \rightarrow H_2O$

Figure 8.6 The chemical reaction in micropower PEMFC.

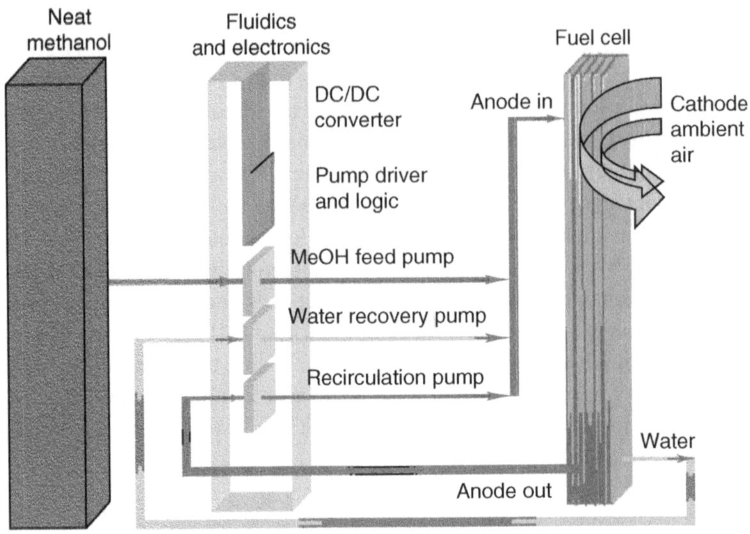

Figure 8.7 Low content direct methanol fuel cell (DMFC) active system. [Courtesy of MTI Micro-Fuel Cells.]

contains a hydrogen-producing catalyst. Hydrogen gas produced by the cartridge is supplied to the anode compartment of PEMFC. Oxygen supply comes from air and is supplied to the cathode. The power generated by this fuel cell is supplied to the mobile phone or laptop. Figure 8.6 shows the chemical reactions that occur in the fuel cell [9]. The right side of the figure shows the description of the cartridge. The power output of the mobile power generator is about 2 W. The portable recharger can charge the device several times. The recharging time is about the same as done in the normal procedure of charging a battery.

Miniaturization is another positive size of micropower FC supply. In 2005, the Guinness Book of World Records has certified Toshiba's compact DMFC as the world's smallest. It is designed for integration into devices as small as MP3 digital music players. A compact 22 × 56 × 4.5 mm DMFC is the size of a woman's thumb, weighing 8.5 g, and operates many electronic devices for about 20 h on a single charge.

One illustration of micropower systems developed in the United States includes Motorola 20 W DMFCs for portable power applications. A fuel processor based on ceramic technology has been integrated with an elevated temperature fuel cell unit and balance of plant. The other example is Mobion technology developed by MTI MicroFuel Cells, also based on DMFC that allows air breathing operation, direct supply of 100% methanol and no water collection and/or pumping and is less than 40 cc in size [24]. Medis Technologies (USA/Israel) intended to produce portable fuel cells called Medis "24/7" Power Pack for feeding diversity of electronic devices [25]. The DMFC system diagram is presented in Figure 8.7.

8.4 MOBILE AND RESIDENTIAL POWER SYSTEMS

Fuel cell technology allows for design and construction of different size power-generating units, especially in certain geographical locations where typhoons, earthquakes, and floods occur very often, causing interruption of electrical power. The other important application of FCS is as stationary backup systems where uninterrupted power supply is critical. Examples of applications are in telecommunications, utilities, hospitals, industrial, and information technology industries [26, 27].

For residential applications, industry is in the process of developing small stationary systems units of 10 kW or less, based on SOFC or PEM technologies. The majority of 1-kW units are being used to provide electricity for home appliances and hot water heating. The larger 5-kW units are being designed to provide continuous power to households with only minimum periods of peak demand being met by the grid. Most of these units are being developed to operate on natural gas, propane, or liquid petroleum gas (LPG). Current trend is to incorporate the use of biomass [1].

In the United States, GenCore® fuel cell continuous off-grid power systems from Plug Power are available for residential use and as source of backup power. The GenCore systems deliver up to 5 kW of electrical power over a wide range of operating environments [14].

8.5 FUEL CELLS FOR SPACE AND MILITARY APPLICATIONS

8.5.1 Space Exploration

The importance and performance of fuel cells in space flight have given a high level of encouragement for developing fuel cells. The Apollo mission used alkaline fuel cell technology (discussed in Chapter 7). More recently, the space shuttles have used PEMFC for powering some of their equipment [28]. NASA is developing fuel cells that would withstand the harsh space transportation environments. Fuel cell systems are currently being investigated that could provide energy and volume densities of >500 Wh/kg and >400 Wh/L, as they will enable space missions to be extended [19]. NASA is considering the fuel cells application for a manned Mars mission, a moon base, and an unmanned high-altitude satellite aircraft. A new design for the space shuttle replacement will vigorously make use of fuel cell technology.

8.5.2 The Army

Fuel cell technology provides the possibility for military to improve logistics, functionality, and specific military functions such as stealth functions or quick start. For instance, a low heat signature and low noise makes detection by the enemy more difficult. Water as a by-product of fuel cell operation can be useful for many purposes. Various experiments are conducted to examine alternative power for weapons, electronics systems, and even equipment that would allow soldiers to carry heavy loads. Examples include the night vision of global positioning systems, laser range finders, digital communication systems, intelligence-gathering sensors, and target designators. Utilization of hydrogen-based transportation enhances fuel efficiency and acceleration compared to diesel or gasoline vehicles. To accumulate the test and maintenance information data on FCV, the US Army tested a fleet of 16 General Motors FCV [29].

8.5.3 Aerospace

A promising area for fuel cell use is its application in unmanned aerial vehicles (UAVs). One example is AeroVironment [17] that developed the Helios—a solar-powered flying wing with a fuel cell system for night flight. The fuel cell generates approximately 15 kW to power the aircraft's 10 electric motors that turn the propellers. AeroVironment has also demonstrated the fuel cell–powered micro air vehicle "Hornet," which can be equipped with a video camera for military surveillance purposes. Unmanned aircraft systems (UASs), also known as a drone, developed by AeroVironment can be used for commercial and military applications, especially for collecting the tactical intelligence or deliver the tactical weapon system. It is hybrid fuel cell/battery devices such the Raven®—the most widely used UAS in the world today. The Raven B DDL system, a lightweight solution designed for rapid deployment and high mobility for low-altitude surveillance and reconnaissance intelligence during day and night.

8.5.4 Naval

The most promising field in the naval arena is fuel cell–powered submarines. Application of the fuel cells to propulsion offers high efficiency, low operating temperature, and complete silence, making submarine detection via sonar, noise, or infrared sensors a difficult task. Many other military benefits such as start-up time, endurance, and safety can be evaluated for manned and unmanned submarines and weapons systems for undersea warfare and by equipping different ships with fuel cell–powered systems. As an example, the US Navy is attempting to use General Motors automotive fuel stack, similar to those used in Equinox FCV fleet test, to power a UUV—an unmanned undersea vehicle [30]. As a source of oxygen, UUV used submerged closed-loop oxygen storage system. For hydrogen storage, UUV utilizes polymer fiber gaseous tank with 700 bars internal pressure. Another application is to power remote instruments and reconnaissance sensors requiring very low power levels [1].

8.6 CONCLUSION

In this book, several aspects related to hydrogen technology have been discussed in detail. It is attractive to visualize a hydrogen-based society to come into existence sometime in future. Such a society will have the following advantages:

- No greenhouse effect
- High energy conversion efficiency
- Localized power generation
- Integrated power sources
- An unlimited energy source of power.

Assuming that we are launching a hydrogen society soon, it is necessary to consider the development of an infrastructure. A careful consideration of hydrogen production from renewable energy sources and an effective method for hydrogen storage technology would be necessary.

Let us examine the advantages of the hydrogen technology over the existing technology. The existing technology is based on a wired network to distribute electric power through a centralized energy network and distribution systems that operate with transmission lines. The interface between the distribution system and consumer utilizes transformers and protection devices. The current cost of the system and its maintenance is high and is still in use as there are no other alternatives. However, we do observe a preference for going from massive production plants to distributed power generation systems. Switching from phone lines to wireless mobile phones and World Wide Web connecting in seconds to unlimited number of users at an unbelievable speed via then have been accepted by the society. Assuming that fuel cell transportation becomes a reality, it is not difficult to imagine a fuel cell stack installed in a vehicle can generate electricity when the vehicle is sitting in a parking lot or a garage and connected to hydrogen source. This, in future, can provide an interesting new approach to power production decentralization. As mentioned by Rifkin [31], our centralized flow of energy, now controlled by oil companies and utilities, may be converted into a new strategy in which everyone will become not only a consumer but also a producer of energy. To do this, every household having access to one or another renewable energy source can contribute to hydrogen energy webs (HEWs) and begin to share energy with other customers, similar to information exchange through the Internet. In order for the hydrogen society to materialize, government resources and the full-fledged efforts of scientists and engineers are required. Use of hydrogen can protect our environment, our energy resources, and the economic stability and prosperity of the world. We hope that this book will inspire you to become a part of hydrogen society.

MULTIPLE CHOICE QUESTIONS

1. What the FC technology used for personal transportation?
   (a) SOFC
   (b) PAFS
   (c) PEMFC
   (d) MCFC
2. What the main limitation for FC vehicles driving range?
   (a) Hydrogen storage
   (b) Ambient temperature
   (c) Electrochemical engine durability
3. What the main source of emission?
   (a) Well-to tank
   (b) Tank-to wheel
   (c) Tank charging
4. What the limitation for hydrogen use in aircraft?
   (a) Safety
   (b) Lock of effective hydrogen storage
   (c) Not developed infrastructure

BIBLIOGRAPHY

Fuel Cell Science, Engineering and Technology, International Conference on Fuel Cell Science, Engineering and Technology, Rochester, NY, Apr. 21–23, 2003.

1. http://energy.gov/eere/fuelcells/downloads/2015-fuel-cell-technologies-market-report (accessed 09 June 2017).

2. 2007 Fuel Cells Technologies Market Report NREL www.nrel.gov/docs/fy09osti/46023 (accessed 09 June 2017).
3. http://www.fchea.org/stationary/ (accessed 09 June 2017).
4. http://www.fuelcellmarkets.com/fuel_cell_markets/molten_carbonate_fuel_cells_mcfc/4,1,1,2505.html (accessed 09 June 2017).
5. http://en.wikipedia.org/wiki/Phosphoric-acid_fuel_cells (accessed 09 June 2017).
6. http://www.sciencedirect.com/science/article/pii/S0378775313019745 (accessed 09 June 2017).
7. http://standards.sae.org/j2572_201410/ (accessed 09 June 2017).
8. http://www.c2es.org/technology/factsheet/HydrogenFuelCellVehicles (accessed 09 June 2017).
9. http://www.transportation.anl.gov/transtech/v2n2/well-to-wheel.html (accessed 09 June 2017).
10. http://w3.gorge.net/eclipse/projects/proj_wtw.html (accessed 09 June 2017).
11. http://www.c2es.org/technology/factsheet/HydrogenFuelCellVehicles (accessed 09 June 2017).
12. https://en.wikipedia.org/wiki/Fuel_cell_bus (accessed 09 June 2017).
13. https://www.fueleconomy.gov/feg/fuelcell.shtml (accessed 09 June 2017).
14. AIAA 5th Aviation Technology, Integration, and Operations (ATIO) Conference … 26–28 September 2005 Hydrogen-Fueled General Aviation Airplanes, www.gbv.de/dms/tib-ub-hannover/504382314.pdf (accessed 09 June 2017).
15. http://phys.org/news/2016-09-world-seater-fuel-cell-plane-germany.html (accessed 09 June 2017).
16. http://www.plugpower.com/products/ (accessed 09 June 2017).
17. Fuel Cells for Rail Applications – International Hydrail Conference, http://hydrail.org/sites/hydrail.org/files/11_laperche-riteau.pdf (accessed 09 June 2017).
18. http://hydrail.org/sites/hydrail.org/files/11_laperche-riteau.pdf fuel cell rail & buses (accessed 09 June 2017).
19. http://www.railwaygazette.com/news/traction-rolling-stock/single-view/view/hydrogen-fuel-cell-train-order-expected-this-year.html (accessed 09 June 2017).
20. http://www.sandia.gov/LabNews/LN05-03-02/key05-03-02_stories.html(accessed 09 June 2017).
21. https://ec.europa.eu/research/energy/pdf/efchp_fuelcell18.pdf (accessed 09 June 2017).
22. http://onlinelibrary.wiley.com/doi/10.1002/er.1281/abstract (accessed 09 June 2017).
23. http://www.cellular-news.com/tags/fuel-cell/docomo/ (accessed 09 June 2017).
24. http://www.mechtech.com/company/mtimicrofuelcells.asp (accessed 09 June 2017).
25. https://www.cnet.com/products/medis-24-7-fuel-cell-power-pack/review (accessed 09 June 2017).
26. https://en.wikipedia.org/wiki/Home_fuel_cell (accessed 09 June 2017).
27. Dokupil, M., Spitta, C., Mathiak, J. *et al.* (2006) Compact propane fuel processor for auxiliary power unit application. *Journal of Power Sources*, **157** (2), 906–913.
28. https://www.nasa.gov/topics/technology/hydrogen/fc_shuttle.html (accessed 09 June 2017).
29. https://www.army.mil/article/74304/Army_unveils_world_s_first_military_fleet_of_fuel_cell_vehicles (accessed 09 June 2017).
30. http://media.gm.com/media/us/en/gm/news.detail.html/content/Pages/news/us/en/2016/jun/0623-gm-us-navy.html(accessed 09 June 2017).
31. Rifkin, J. (2002) *The Hydrogen Economy*, Tarcher/Penguin, New York, NY.
32. http://www.aerovironment.com (accessed 09 June 2017).

INDEX

Introduction to Hydrogen Technology, Second Edition.
K.S.V. Santhanam, Roman J. Press, Massoud J. Miri, Alla V. Bailey, and Gerald A. Takacs

Book Companion site : www.wiley.com/go/santhanam/hydrogentech_2e

Printed and bound by CPI Group (UK) Ltd, Croydon, CR0 4YY
17/09/2026
14964315-0004